The Biology of Lungfishes

The Biology of Lungfishes

Editors

Jørgen Mørup Jørgensen
Department of Zoophysiology
Biological Institute
University of Aarhus
Denmark

Jean Joss
Department of Biological Sciences
Macquarie University
Sydney
Australia

CRC Press
Taylor & Francis Group
Boca Raton London New York

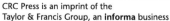

CRC Press is an imprint of the
Taylor & Francis Group, an **informa** business

Science Publishers
Enfield, New Hampshire

CRC Press
Taylor & Francis Group
6000 Broken Sound Parkway NW, Suite 300
Boca Raton, FL 33487-2742

First issued in paperback 2017

ISBN-13: 978-1-57808-431-9 (hbk)
ISBN-13: 978-1-138-11407-4 (pbk)

Cover Illustrations

Semon, R. 1893, Die äussere Entwickelung von *Ceratodus Forsteri*. Denkschriften der Medicinisches und Naturwissenschaftlisches Gesellschaft zu Jena Vol. 1, Tafel 8.

The evolutionary series (right) was reproduced by kind permission of Dr. Jennifer A. Clack.

Library of Congress Cataloging-in-Publication Data

The biology of lungfishes / editors, Jørgen Mørup Jørgensen, Jean Joss.
p. cm.
Includes bibliographical references and index.
ISBN 978-1-57808-431-9 (hardcover)
1. Lungfishes. I. Jørgensen, Jørgen Mørup. II. Joss, Jean.
QL638.3.B565 2010
597.3'9--dc22

2010013780

Visit the Taylor & Francis Web site at
http://www.taylorandfrancis.com

and the CRC Press Web site at
http://www.crcpress.com

Preface

Since the last comprehensive monograph on lungfishes appeared in 1987, edited by Bemis, Burggren and Kemp, much new information has appeared concerning this little clade (Nelson 2006). This is the main reason for the present collection of reviews on some of the most important aspects of lungfish biology. We believe that the unique position between fishes and tetrapods will make this book of interest not only to scientists but also to the general reader with an interest in evolution and biology of the vertebrates.

It is a pleasure to thank the contributors of the chapters for devoting their time to create a treatise like this to share their knowledge with everyone. We are also indebted to the reviewers who committed time and talent to ensure the excellent quality of each contribution.

Jean Joss and Jørgen Mørup Jørgensen

Contents

List of Contributors

Abdala, Virginia

Instituto de Herpetología, Fundación Miguel Lillo-CONICET, Facultad de Ciencias Naturales (UNT), Miguel Lillo, 251 4000 Tucumán, Argentina.
E-mail: virginia.abdala@gmail.com

Ahlberg, Per E.

Subdepartment of Evolutionary Organismal Biology, Department of Physiology and Developmental Biology, Evolutionary Biology Centre, Uppsala University, Norbyvägen 18A, 752 36 Uppsala, Sweden.
E-mail: Per.Ahlberg@ebc.uu.se

Alibardi, Lorenzo

Department of Evolutionary Experimental Biology, University of Bologna, Bologna, Italy.

Almeida-Val, Vera Maria Fonseca de

Laboratory of Ecophysiology and Molecular Evolution – LEEM, Department of Ecology - CPEC, Instituto Nacional de Pesquisas da Amazônia - INPA, Av. André Araújo, 2936 – Aleixo, 69.060-020, Manaus, Amazonas, Brazil.
E-mail: veraval@inpa.gov.br

Aride, Paulo Henrique Rocha

Laboratory of Gene Expression, Centro Universitário Nilton Lins – UNINILTON LINS, Manaus, Brazil.

IFAM Campus Presidente Figueiredo, Brazil.
E-mail: paride@niltonlins.br

Bailes, Helena

Faculty of Life Sciences, Michael Smith Building, Oxford Road, Manchester, M13 9PT, England.

Ballantyne, James S.

Department of Integrative Biology, University of Guelph, Guelph, Ontario N1G 2W1, Canada.
E-mail: jballant@uoguelph.ca

Brandt, Christian

Institute of Biology, University of Southern Denmark, Campusvej 55, DK-5230 Odense M, Denmark.

Christensen-Dalsgaard, Jakob

Institute of Biology, University of Southern Denmark, Campusvej 55, DK-5230 Odense M, Denmark.
E-mail: jcd@biology.sdu.dk

Clack, Jennifer A.

University Museum of Zoology, Downing Street, Cambridge CB2 3EJ, United Kingdom.
E-mail: j.a.clack@zoo.cam.ac.uk

Collin, Shaun

The University of Western Australia, School of Animal Biology & The Oceans Institute, Crawley, WA 6009, Australia.

Diogo, Rui

Center for the Advanced Study of Hominid Paleobiology, Department of Anthropology, The George Washington University, 2110 G St. NW, Washington DC 20052, USA.
E-mail: ruidiogo@gwmail.gwu.edu

Dunbrack, Robert L.

Department of Biology, Memorial University, St. John's, NL, Canada, A1B 3X9.
E-mail: dunbrack@mun.ca

Ericsson, Rolf

Department of Biological Sciences, Macquarie University, Sydney, New South Wales 2109, Australia.
E-mail: rericsson@bio.mq.edu.au

Ferreira-Nozawa, Mônica S.

Laboratory of Gene Expression, Centro Universitário Nilton Lins – UNINILTON LINS, Manaus, Brazil.
E-mail: mnozawa@niltonlins.br

Frick, Natasha T.

Department of Biochemistry, University of Toronto, 1 King's College Circle, Medical Science Building, Toronto, Ontario M5S 1A8, Canada.
E-mail: nfrick@uoguelph.ca

Glass, Mogens L.

Departamento de Fisiologia, Faculdade de Medicina de Ribeirão Preto, Universidade de São Paulo, Avenida Bandeirantes 3900, 14.049-900 Ribeirão Preto, SP, Brazil.
E-mail: mlglass@rfi.fmrp.usp.br

Green, John M.

Department of Biology, Memorial University, St. John's, NL, Canada A1B 3X9.
E-mail: jmgreen@mun.ca

Hart, Nathan

The University of Western Australia, School of Animal Biology & The Oceans Institute, Crawley, WA 6009, Australia.

Hassanpour, Masoud

Department of Biological Sciences, Macquarie University, Sydney, New South Wales 2109, Australia.
E-mail: masoudhg@gmail.com

Honda, Rubens Tomio

Laboratory of Gene Expression, Centro Universitário Nilton Lins - UNINILTON LINS, Manaus, Brazil.
E-mail: horubens@gmail.com

Jensen, Frank Bo

Institute of Biology, University of Southern Denmark, Campusvej 55, DK 5230 Odense M, Denmark.
E-mail: fbj@biology.sdu.dk

Johanson, Zerina

Department of Palaeontology, Natural History Museum, Cromwell Road, London SW7 5BD, United Kingdom.
E-mail: z.johanson@nhm.ac.uk

Joss, Jean

Department of Biological Sciences, Macquarie University, Sydney, New South Wales 2109, Australia.
E-mail: jjoss@rna.bio.mq.edu.au

Jørgensen, Jørgen Mørup

Department of Zoophysiology, Building 1131, Biological Institute, University of Aarhus, DK-8000 Århus C, Denmark.
E-mails: joergen.moerup.joergensen@biology.au.dk; biojmj@biology.au.dk

Kind, Peter K.

Principal Scientist, Freshwater, Queensland Primary Industries and Fisheries, Primary Industries Building, 80 Ann St. Brisbane, Queensland, Australia 4001.
E-mail: Peter.kind@deedi.qld.gov.au

Long, John A.

Museum Victoria, P. O. Box 666, Melbourne, Australia 3001.
E-mail: jlong@nhm.org

Lopes, Nívia Pires

Centro Universitário do Norte–UNINORTE, Manaus, Brazil.
E-mail: nivia@inpa.gov.br

Madsen, Peter Teglberg

Department of Zoophysiology, Building 1131, Biological Institute, University of Aarhus, DK-8000 Århus C, Denmark

Marshall, Justin

Sensory Neurobiology Group, School of Biomedical Sciences and Queensland Brain Institute, The University of Queensland, Brisbane, QLD 4072, Australia.
E-mail: justin.marshall@uq.edu.au

Mesquita-Saad, Lenise Socorro

Department of Food Sciences, Institute of Biochemistry, Universidade Federal do Amazonas, UFAM, Manaus, Brazil.
E-mail: lsaad@inpa.gov.br

Mlewa, Chrisestom M.

Biological Sciences, Pwani University College, P. O. Box 195-80108, Kilifi, Kenya.
E-mail: mlewa2001@yahoo.com

Northcutt, R. Glenn

Laboratory of Comparative Neurobiology, Scripps Institution of Oceanography and Department of Neurosciences, School of Medicine, University of California, San Diego, La Jolla, Ca 92093-0201, USA.
E-mail: rgnorthcutt@ucsd.edu

Nozawa, Sérgio Ricardo

Laboratory of Gene Expression, Centro Universitário Nilton Lins - UNINILTON LINS, Manaus, Brazil.
E-mail: srnozawa@gmail.com

Olsson, Lennart

Institut für Spezielle Zoologie und Evolutionsbiologie, Mit Phyletischem Museum, Friedrich-Schiller-Universität, Erbertstrasse 1, D-07743 Jena, Deutschland, Germany.
E-mail: Lennart.Olsson@uni-jena.de

Paula da Silva, Maria de Nazaré

Laboratory of Ecophysiology and Molecular Evolution – LEEM, Department of Ecology - CPEC, Instituto Nacional de Pesquisas da Amazônia - INPA, Av. André Araújo, 2936 – Aleixo, 69.060-020, Manaus, Amazonas, Brazil.
E-mail: npaula@inpa.gov.br

Popper, Arthur N.

College of Chemical and Life Sciences, Department of Biology, University of Maryland, College Park, MD 20742, USA.
E-mail: apopper@umd.edu

Sharp, Esther L.

University Museum of Zoology, Downing Street, Cambridge CB2 3EJ, United Kingdom.
E-mail: esther.sharp@cantab.net

Smith, Moya Meredith

King's College London, MRC Centre for Developmental Neurobiology, London SE1 1UL, United Kingdom.
E-mail: moya.smith@kcl.ac.uk

Toni, Mattia

Department of Evolutionary Experimental Biology, University of Bologna, Bologna, Italy.

Val, Adalberto Luis

Laboratory of Ecophysiology and Molecular Evolution - LEEM, Department of Ecology - CPEC, Instituto Nacional de Pesquisas da Amazônia - INPA, Av. André Araújo, 2936 - Aleixo, 69.060-020, Manaus, Amazonas, Brazil.
E-mail: dalval@inpa.gov.br

Wahlberg, Magnus

Fjord&Belt, Margrethes Plads 1, DK-5300 Kerteminde, Denmark.

Weber, Roy E.

Department of Zoophysiology, Building 1131, Biological Institute,University of Aarhus, DK-8000 Århus C, Denmark.
E-mail: roy.weber@biology.au.dk

Wilson, Maria

Department of Zoophysiology, Building 1131, Biological Institute, University of Aarhus, DK-8000 Århus C, Denmark.

Introduction

The first modern book devoted to the biology of lungfishes was published in 1987 as a result of a conference held in 1985. This conference was primarily inspired by the seminal publication of Donn Rosen, Peter Forey, Brian Gardner and Colin Patterson (1981), which reanalysed the morphological characters of lungfish for comparison with other sarcopterygians, both living and fossil, using cladistic anayses of rigorously selected synapomorphies to the exclusion of any pleisomorphic characters. These four eminent fish systematists/evolutionary biologists concluded that lungfish and not "rhipidistian" fish, were the direct ancestors of tetrapods. Moreover, they exposed the rhipidistians as being a paraphyletic clade, requiring the separate consideration of the groups contained within such as the porolepiformes and osteolepiformes. These conclusions were greeted with horror from most researchers in the area of the fish-tetrapod transition but they did stimulate renewed interest in the Dipnoi (lungfishes) from a phylogenetic point of view. The first section of the 1987 book was entirely given over to consideration of the phylogenetic position of lungfish.

At about the same time new molecular techniques were beginning to be applied to phylogenetic questions, including those of the fish-tetrapod transition. Of course these techniques could only consider the relationships between living species, of which there are only 7-8 – six lungfish and two coelacanths. As it became more and more apparent that the lungfish grouped with tetrapod species to the exclusion of all others, there began a considerable resurgence of interest in study of the extant lungfishes. This interest was fuelled a decade or so later by increased access to living lungfish from a Facility for breeding the Australian lungfish, which was established at Macquarie University in 1993. This species of lungfish was and still is protected by CITES, which requires all lungfish material being sent to researchers outside Australia to be justified as legitimate research by purchase of an appropriate permit from the CITES-registered Authority. This requirement has helped to filter out the more frivolous interests in lungfish from those seriously investigating how fish ancestors transformed into the first tetrapods, during the mid to Late Devonian.

As it is now more than 20 years since the publication of "The Biology and Evolution of Lungfishes" and areas of study such as physiology, development, and behaviour of lungfish have variably flourished during this time, a book bringing most of this new data together is over due. Living lungfishes comprise a small group of sarcopterygian fishes of just three genera, each genus being confined to a separate continent in the Southern Hemisphere. As had been noted in 1987, they combine many features from both fishes and tetrapods. In our invitations to authors we have put emphasis on research fields that have progressed most in the last 20 years, but also we have tried to find authors in areas, which have not been covered extensively in recent reviews. So, all authors in the present book are different from the authors in the 1987 treatise "The Biology and Evolution of Lungfishes" with the exception of Dr. R Glenn Northcutt. His expertise in neural and sensory structures and his broad knowledge of lobe finned fishes (sarcopterygian fishes) made him the most appropriate author to present recent progress in this area.

The first part of the present book is an updated account of the fossil record by Prof Jenny Clack and colleagues, followed by a chapter on the current phylogeny of lungfishes by Dr Zerina Johanson and Prof Per Ahlberg. The next three chapters concern the natural history of the three genera: the Australian *Neoceratodus* by Dr Peter Kind, the four African *Protopterus* species by Dr Chrisestom Mlewa and colleagues and the South American *Lepidosiren* by Dr Vera Almeida-Val and her colleagues. The rest of the book contains chapters that describe morphology or physiology of various organ systems. These primarily cover areas that were not covered in the first book, such as the skin by Dr Lorenzo Alibardi and colleagues, head muscles by Dr Rui Diogo and Virginia Abdala, development of the head by Dr Rolf Ericsson, Prof Jean Joss and Prof Lennart Olsson, the teeth by Prof Moya Meredith Smith and Dr Zerina Johanson. We have not included some active areas of current research such as the "evo-devo' projects, that we expect will be very informative in the next few years and should comprise the raw material for a further update in much less than 20 years!

Also, we have not included a bibliography as in the first book, which gave us access to Babs Conant's magnificent bibliography of all published works on lungfish prior to 1986. The reference lists at the end of each chapter in this volume are intended to at least partially update this bibliography for readers.

We hope that the widespread interest in these fantastic and highly significant fishes will lead to even more research investigations as well as intensified protection. Most of the lungfishes are vulnerable and if this book can contribute to more understanding of the uniqueness and importance of these fishes resulting in intensified protection of their environments, to the benefit of coming generations of human beings, our most sincere hopes will have been fulfilled.

Bemis W.E., Burggren W.W., Kemp N.E. eds. (1987) "The Biology and Evolution of Lungfishes" AR Liss.

Rosen D., Forey P., Gardner B., Patterson C. (1981) Lungfishes, tetrapods, paleontology, plesiomorphy. Bulletin of the American Museum of Natural History 167: 159-276.

Jean Joss and Jørgen Mørup Jørgensen

CHAPTER 1

The Fossil Record of Lungfishes

Jennifer A. Clack[1,*], Esther L. Sharp[1] and John A. Long[2]
[1]University Museum of Zoology, Downing St., Cambridge, CB2 3EJ UK,
(Current address for Dr. Sharp: 83 Woodlands Road, Charfield, Wotton-under-Edge, Gloucestershire, UK GL12 8LT)
[2]Museum Victoria, P.O. Box 666, Melbourne, Australia 3001,
(Current address: Natural History Museum of Los Angeles County, Los Angeles, California, 90403 USA)

ABSTRACT

The fossil record of lungfish is reviewed. Some of their unique characteristics are explained and illustrated, including some less well-known anatomical elements that are often misidentified or unidentified in museum collections of Paleozoic vertebrates. Lungfish records from the Early Devonian to the Late Mesozoic are illustrated and described chronologically. Their major diversification occurred during the Devonian Period, with more than 70 named species. Subsequently they declined in diversity during the Carboniferous and Permian to a low of three genera at the present day. Some Carboniferous and Permian forms show intermediate morphologies between the Late Devonian and Mesozoic forms. Some of the intriguing questions of lungfish biology that can be addressed from the fossil record are outlined, such as the modification of their skeletons from a more or less primitive sarcopterygian pattern to their specialised form with loss of cosmine on bones and scales, few dermal skull bones, continuous mid-line fins and reduction of the paired fins to narrow, whip-like appendages. Late Devonian and Carboniferous lungfish are frequently found associated with tetrapods, and several lungfish anatomical elements are sometimes mistaken for those of tetrapods. The reduction in ossification of lungfish skeletons creates problems for understanding their fossil record: it may be explicable in terms of retention of a juvenile state. Reduction

*Corresponding author: E-mail: j.a.clack@zoo.cam.ac.uk

in ossification takes place alongside increase in the genome size: it has been suggested that these two phenomena may be related. Their varied patterns of dentition may be explicable by the interaction of only a few developmental processes, but the present pattern had been established by the Devonian. The evolution of air-breathing, aestivation, biogeographical distribution and change in habitat from fully marine during the Early Devonian to entirely freshwater at present are reviewed.

Keywords: evolutionary history, anatomy, Devonian, Palaeozoic, Mesozoic, Cenozoic.

INTRODUCTION

Lungfish have left some of the most characteristic as well as some of the most baffling elements in the fossil record. In part because of the richness of detail found in their Devonian fossil record they have been the subjects of deep disagreements concerning their relationships and functional morphology, whereas because of the paucity of their post-Devonian record, they have been, for the most part, neglected or misidentified in studies of more recent faunas. This chapter aims to present an overview of their diversity, anatomy and evolutionary history in the hope of promoting wider interest in the story they can tell us.

Lungfish first appear in the Early Devonian and are one of only three sarcopterygian groups to survive to the present day (the others being coelacanths and tetrapods), thus they have a range of about 400 million years. From their beginnings in the Early Devonian, their diversity in terms of morphology and species richness increased to reach its acme in the Late Devonian. More than seventy species are described, exemplified in particular by faunas in Australia. Subsequent to the Devonian, though they were distributed more or less worldwide throughout the Late Paleozoic, their fossil record declined. This may to a large extent be unconnected with their actual diversity and distribution, and more to do with the fact that the animals themselves became less amenable to preservation in the fossil record. Lungfish show a gradual reduction of ossification of both the internal and external skeletons, beginning in the Late Devonian and continuing through the Carboniferous, meaning not only that skeletal material becomes correspondingly less well represented, but that which does remain is usually disarticulated, making attribution to genus or species difficult or impossible in many cases.

The most durable and characteristic elements of the post-Devonian lungfish anatomy are their unique tooth plates (Figure 1), commonly found in isolation in many fossil localities representative of freshwater environments from the Carboniferous onward. In Devonian forms, and in Carboniferous ones where they are associated with other skeletal elements, it is evident that tooth plate morphology can be diagnostic to genus or species, and the same is assumed for later examples in which little or no skeletal material is available (Kemp 1997; Cavin *et al.* 2007).

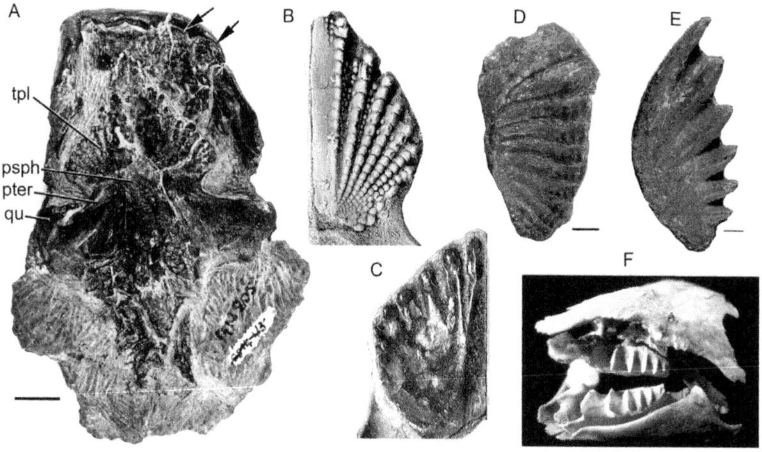

Fig. 1 Lungfish tooth plates. A. Palate of *Dipterus valenciennesi*, showing tooth plates (tpl), pterygoids (pter), parasphenoid (psph), quadrates (qu) and notches for the nares (arrows). Middle Devonian, UMZC (University Museum of Zoology, Cambridge) GN.805 (Photograph J. A. Clack). B. *Andreveyichthys epitomus,* a ridged tooth plated form. PIN (Palaeontological Institute, Moscow) 2921/1976. Late Devonian). C. Tooth plate of *Chirodipterus australis*, a dentine-plated form. AMF (Australian Museum, Sydney) 120082, Late Devonian. D. Tooth plate of *Ctenodus interruptus*, mid-Carboniferous, specimen MM (Manchester Museum) L10412 (from Sharp 2007). E. Tooth plate of *Sagenodus inaequalis*, Late Carboniferous, specimen NEWHM (Hancock Museum, Newcastle upon Tyne) G172.32. F. Skull of *Neoceratodus forsteri* showing tooth plates in lateral view, Recent (Photograph, J. A. Long). B, C from Ahlberg *et al.* (2006), scale not given; A, D, E. Scale bars 10 mm. D, E. from Sharp (2007).

Thus apart from a few instances, tooth plates alone are our guide to Mesozoic and Cenozoic lungfishes.

As well as their external morphology that is widely used in their systematics (e.g. Kemp 1993, 1994a, 1997), lungfish tooth plates have been subjected to detailed microstructural and histological analyses, to examine their growth patterns, to compare them with modern forms, and to try to assess the primitive condition (e.g. Denison 1974; Campbell and Barwick 1995, 1998, 1999, 2000; Campbell and Smith 1987; Campbell *et al.* 2002, Kemp 1995, 2002a; Reisz *et al.* 2004). To summarize this extensive literature is beyond the scope of this review.

In addition to tooth plates, a number of unique cranial features distinguish lungfish from any other osteichthyan groups. Rather than a symmetrical series of paired bones along the midline of the skull, fossil lungfish show what is best described as a 'hopscotch' pattern, in which pairs alternate with single midline bones (Figure 2). So difficult have these proved to be to equate to those in either actinopterygians or sarcopterygians that Forster-Cooper (1937) proposed a

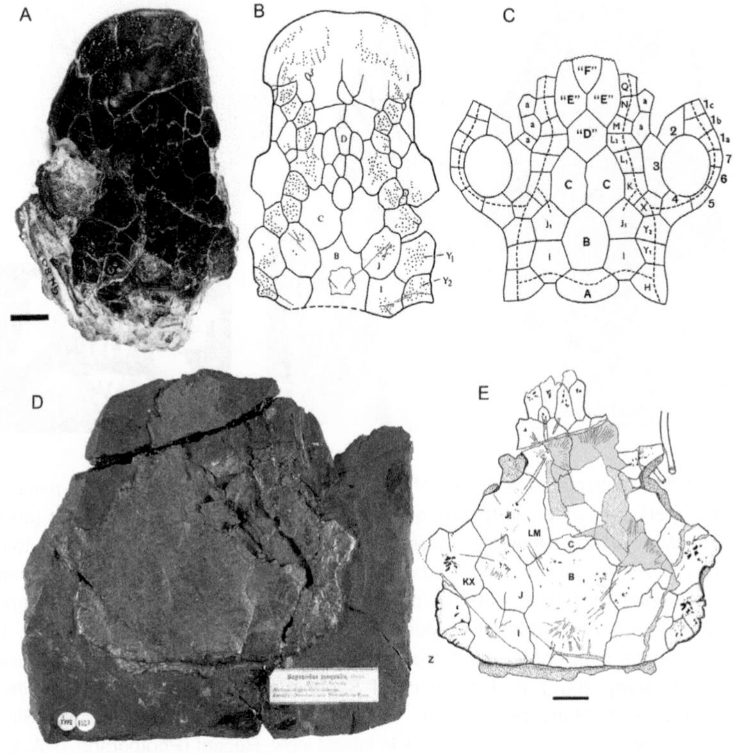

Fig. 2 Lungfish skull roofs. A. *Dipterus valenciennesi*, Middle Devonian, specimen UMZC GN.805 (Photograph J. A. Clack). Scale bar 10 mm. B. Skull roofing bones of *D. valenciennesi*, with lettering system on some bones (from Jarvik 1980). C. Forster-Cooper's alpha-numerical system of bone identification in an idealized skull roof (from Forster-Cooper 1937). D, E. *Sagenodus inaequalis* skull roof. NMS (National Museum of Scotland) 1878.45.7, Late Carboniferous (Photograph and specimen drawing respectively from Sharp 2007). Scale bar for D and E, 20 mm

system of letters and numbers to facilitiate discussion of patterns among lungfishes (Figure 2C). Though he intended this to be a temporary measure, it is still in use today. However, even this scheme is difficult to apply to more derived forms from the Mesozoic (Cavin *et al.* 2007), though several workers have proposed schemes (e.g. Schultze 1981; Kemp 1998a).

In primitive forms like *Dipterus* the skull roof is inrolled along the snout margins, where there are two embayments for the anterior and posterior nostrils on each side (Figure 1, arrows). Other unique characters include the loss of both the premaxilla and maxilla of other osteichthyans, and all but the most primitive have lost the dentary. The dentary is still present in hatchling *Neoceratodus*, but is

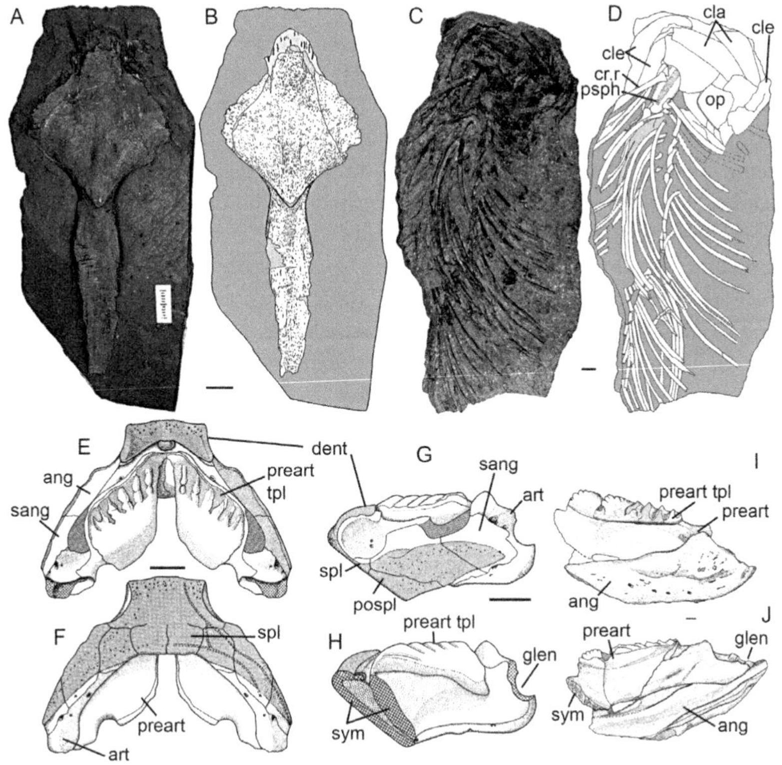

Fig. 3 Lungfish anatomy. A, B. *Sagenodus inaequalis*, isolated parasphenoid. NMS 1878.45.13, Late Carboniferous, photograph and specimen drawing respectively. C, D. Partial skeleton of *Ctenodus* sp., Early Carboniferous showing ribs and other elements. NMS 1906.108, photograph and specimen drawing respectively. E, F. Jaws of *Chirodipterus australis* in dorsal and ventral views respectively. G, H. Lower jaw of *C. australis* in external and internal views respectively. I, J. Lower jaw of *S. inaequalis* in extermal and internal views respectively. External view reversed for ease of comparison. Note that some bones are missing from the external surface. A–D and I–J from Sharp (2007). Grey fill, matrix; hatching, broken bone. E–H from Miles (1977). Close grey stipple in E–H cosmine-covered regions; coarse crosshatch, unfinished bone; fine cross-hatch, Meckelian fenestra; coarse stipple, articular surface. Scale bars, 10 mm. Abbreviations: ang, angular; art, articular; cla, clavicle; cle, cleithra; cr.r, cranial rib; dent, dentary; glen, glenoid; op, operculum; psph, parasphenoid; pospl, postsplenial, preart, prearticular; tpl, tooth plate; sang, surangular; spl, splenial.

resorbed during early development (Smith and Krupina 2001; Kemp 1995, 2002b). Dentitions are carried on the pterygoid (and sometimes the vomers) in the upper jaw, and on the prearticular and dentary (where present) in the lower jaw (Figure 3E -J). Most lungfish have a characteristically shaped parasphenoid lying between the pterygoids (Figures 1, 3A, B), though in earlier forms, such as *Diabolepis*

(Chang 1995) or *Uranolophus* (Schultze 1992a; Denison 1968), it was situated more posteriorly, separating only the posterior parts of the pterygoids. It has a lozenge-shaped anterior portion with a stem posteriorly (and is not infrequently mistaken for a tetrapod interclavicle, even by knowledgeable vertebrate paleontologists eg. Jarvik (1996, Plate 52, Fig 3)). The palatal bones are firmly attached to the braincase producing an 'autostylic' skull in which the hyomandibula is not involved with jaw suspension. In lungfish the hyomandibula is reduced as a result. The operculum is a conspicuous bone, usually almost circular in outline, and the subopercular an almost featureless elongate oval (Figure 4). Also recognizably lungfish are characters of the hyobranchial system such as the robust ceratohyal, and the shoulder girdle bones such as the anocleithrum, cleithrum and clavicle (Figure 4). These bones are often found in isolation especially in Carboniferous rocks, and are not always identifiable by other than fossil lungfish specialists (as also are isolated lower jaws that have lost their tooth plates). Lungfish skulls are highly modified from the usual sarcopterygian pattern in response to the unique mechanism of air-breathing and suction feeding employed by lungfish, as is the existence of cranial ribs attached to the occipital portion of the braincase (Figure 3C, D). The hyoid arch is anchored to the shoulder girdle and the cranial ribs are used to help the animal raise its head out of water to gulp air (Bishop and Foxon 1968; Long 1993).

Among the most commonly found postcranial remains of lungfish, especially from Carboniferous deposits, are the ribs (Figure 3C, D). These are robust, with a single head bearing a comma-shaped articular facet, and a shaft that is strongly curved, parallel-sided, and with longitudinal grooves along its entire length. The latter gives the ribs either a figure-of-eight or comma-shaped cross-section. The length and curvature of these ribs indicate that they almost completely encircled the body, which would have had an almost cylindrical cross-section as a result. These ribs are not infrequently mistaken for those of tetrapods. In addition, the well ossified centra of some Devonian lungfishes could be mistaken for those of tetrapods, and closely resemble the pleurocentra of Carboniferous embolomeres (see for example figure 5, Campbell and Barwick 2002).

From what can be seen of their early fossil record, lungfish evolution underwent a number of parallel anatomical changes, most of which ended with a gradual reduction of morphological and ecological diversity. These will be illustrated in more detail below, but briefly they are: reduction of the skull roofing bones and dermal scale cover; a loss of diversity in dental patterning and tissue modelling; restriction of their midline fin morphology; the reduction of both cranial and postcranial endoskeletons; restriction to freshwater habitats from a base of marine and marginal marine origins; and elaboration of their air-breathing adaptations.

Lungfish as a group have only recently been subjected to strict cladistic analyses (eg. Schultze and Marshall 1993; Schultze 2001; Schultze and Chorn 1997; Friedman 2007a, b; Sharp 2007), with some of the most prolific researchers on the group explicitly rejecting this methodology (eg. Campbell and Barwick 1988, 1990).

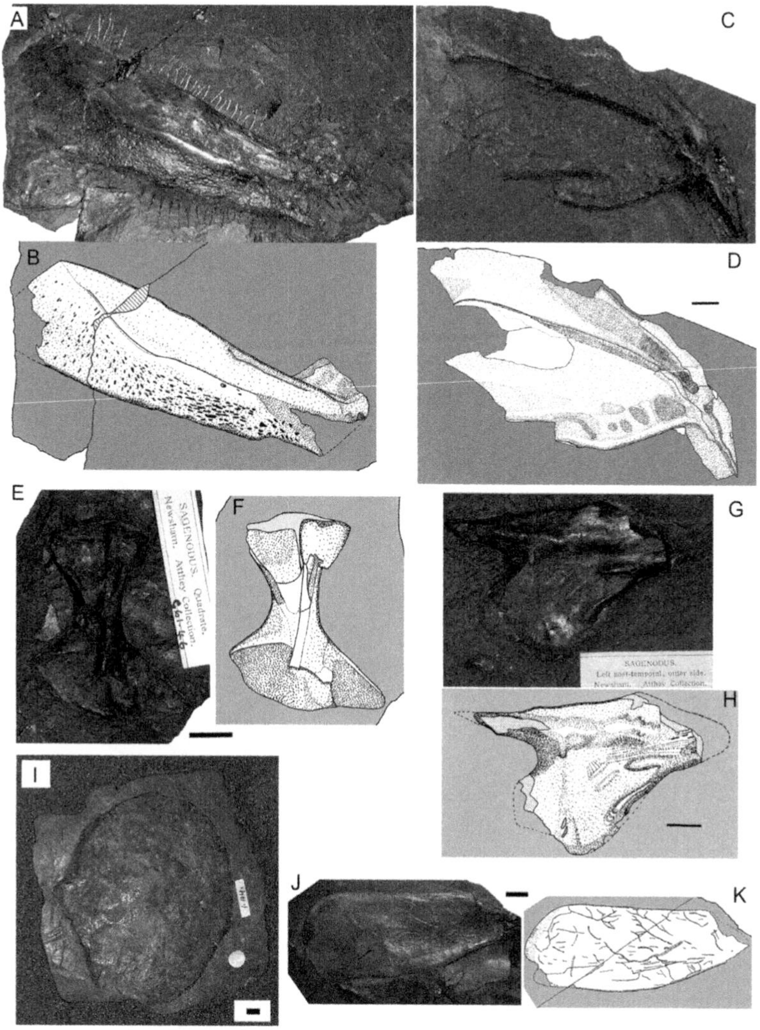

Fig. 4 Isolated elements of *Ctenodus* and *Sagenodus*. A, B. *Ctenodus* sp., right clavicle. MM unregistered, photograph and specimen drawing respectively. Size about 90 mm. C, D. *Ctenodus* sp. left cleithrum in external view, NMS 1968.17.47, photograph and specimen drawing respectively. E, F. *Sagenodus inaequalis*, left ceratohyal in lateral view. NEWHM G61.46, photograph and specimen drawing respectively. G, H. *Sagenodus inaequalis*, left anocleithrum in external view. NEWHM G61.64, photograph and specimen drawing respectively. I. *Ctenodus* sp., operculum in external view. NEWHM G40.97, photograph. J, K. *Ctenodus* sp., suboperculum. CAMSM (Cambridge, Sedgwick Museum) E4524, photograph and specimen drawing. All from Sharp (2007).

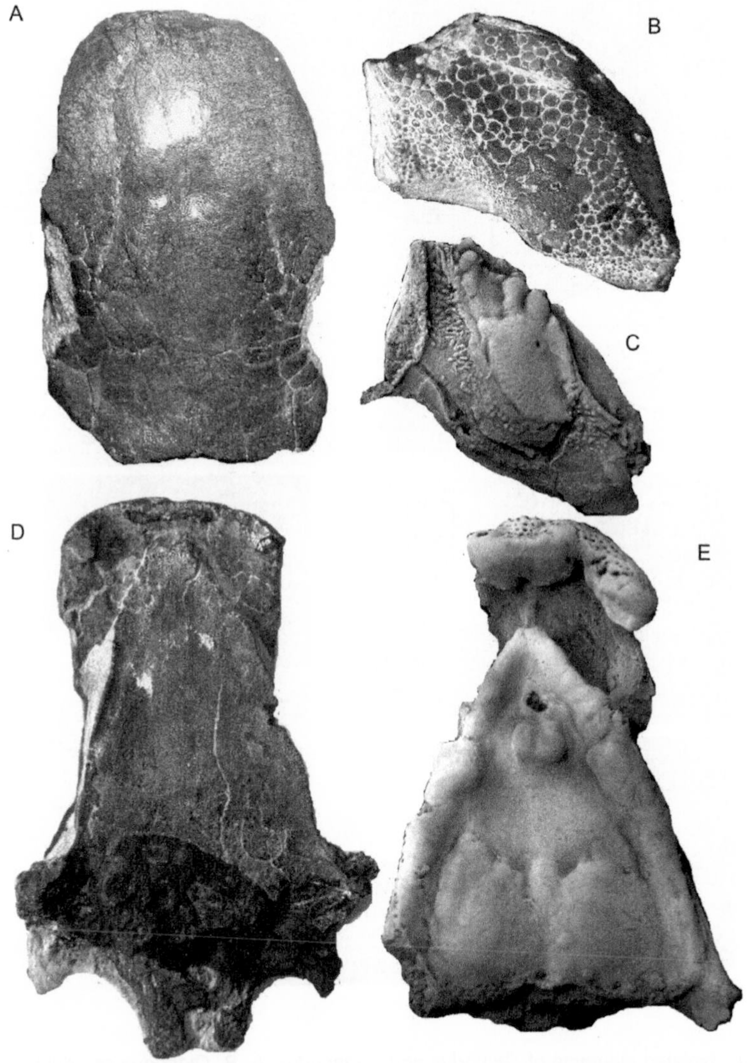

Fig. 5 Early Devonian lungfishes. A. *Diabolepis speratus* skull and B. lower jaw, Lochkovian, China. IVPP (Institute of Vertebrate Palaeontology, Bejing) V7238, length of specimen about 16 mm. C. *Ichnomylax kurnai* lower jaw, Pragian, Australia. NMV (Museum Victoria) P188479, maximum width of jaw about 30 mm. D. *Uranolophus wyomingensis* palate, Pragian, USA. Field Museum PF3792, length of specimen 195 mm. E. *Dipnorhynchus kurikae* palate, Emsian, Australia. ANU (Australian National University) 48676, length of specimen 94 mm. A–E, photographs by J.A. Long.

Thus their internal relationships have been controversial. Though not the main focus of this chapter, it is inevitable that some discussion of relationships will be necessary, for without this, discussion of topics such as their palaeobiogeography and directions of morphological evolution are very difficult to understand. For a more in depth discussion of lungfish phylogenetic relationships, readers should consult Chapter 2.

Their relationship to other sarcopterygian groups is also debated, though many researchers on the subject now accept them as belonging to a group known as 'Dipnomorpha' that also includes the Devonian Porolepiformes. Sister group to the Dipnomorpha in this scheme is the Tetrapodomorpha, making dipnoans the closest living relatives to tetrapods to the exclusion of coelacanths (Ahlberg 1991; Cloutier and Ahlberg 1996; Ahlberg and Johanson 1998). Not all paleoichthyologists accept this scheme (Zhu and Schultze 2001), and furthermore, molecular studies are equivocal, with some studies placing coelacanths and dipnoans as sister groups to the exclusion of tetrapods (Zardoya and Meyer 2001).

Marshall (1986) and Schultze (1992b) gave almost comprehensive lists of fossil and extant taxa of lungfish up to those dates. After a brief and not exhaustive chronological review of fossil lungfishes this chapter will address some issues of current interest in their biology and ecology.

EARLY DEVONIAN LUNGFISHES

Some controversy surrounds the deep relationships of lungfish, with two taxa central to the debate. *Youngolepis* and *Diabolepis* (Figure 5A, B) are two taxa from Yunnan, China that lived during the Early Devonian, and consist of disarticulated skull elements. There is a fairly widely held consensus that *Diabolepis* is basal to other dipnoans, but whether *Youngolepis* is a stem member of the Dipnoi, or is closer to the porolepiforms is not clear (eg. Cloutier and Ahlberg 1996; Smith and Chang 1990; Zhu *et al.* 1999; Zhu and Schultze 2001; Zhu and Yu 2002). *Diabolepis* shares some unique features of the structure of the lower jaw and dentition with dipnoans, including radiating rows of rounded teeth of similar structure to those of true lungfish, and reduction of the lower jaw bones. It shows the beginnings of the modern dipnoan structure of the skull roof, palate and braincase that so clearly characterize later lungfish (Chang 1992). It also shows more general characters such as a median parallel-sided and denticulate parasphenoid, pierced by a hypophyseal foramen. On the other hand, some characters of *Diabolepis* are shared with porolepiforms, corroborating the suggestion of a relationship between dipnoans and porolepiforms (Chang 1995). Campbell and Barwick (2001) contested the relationship between *Diabolepis* and Dipnoi on the grounds that the morphological and structural similarities between them were not genuine homologues, but had evolved in parallel. This view is not widely accepted.

As well as typical tooth plates, in the Devonian there were also two other forms of dentition found among lungfishes: those with more or less solid dentine plates with or without tubercules, and those with sheets of denticles only. Among the earliest of lungfish in the fossil record is *Dipnorhynchus*, a tooth plated form found in Pragian and Emsian age deposits (Middle and Late Early Devonian) of Australia (Figure 5E). Several species have been recognized, from a range of localities including Wee Jasper and Taemas (Campbell and Barwick 1982a, b, 1999, 2000). This primitive form shows the mosaic pattern of skull bones that characterizes lungfish throughout their history, though in this genus, there are more and smaller bones, particularly around the snout region. It shows that the tooth plates grew by addition of new dentine around the margins of the pterygoids. Campbell and Barwick (2000) gave a reconstruction of the skull in lateral view showing a very high, deep skull, based on a very well preserved large specimen. *D. kurikae* from Wee Jasper has a well preserved braincase, that will be helpful in elucidating the relationships of the genus, but unfortunately, no postcranial remains are known from the genus.

Speonesydrion is another tooth plated form from Wee Jasper in Australia (Emsian), that shows teeth in radial rows and a mosaic of bones on the snout. It is known from exceptionally well preserved material but which is limited to skull and jaw elements (Campbell and Barwick 1984). However, both juvenile and larger specimens are known, that shed some light on the growth of the teeth. *Ichnomylax*, from the Pragian of Australia (Long *et al.* 1994) and Emsian of Siberia (Reisz *et al.* 2004), is a form closely-related to *Speonesydrion* with well-defined radial rows of teeth coalescing onto a thick dentine heel on the lower jaw (Figure 5C). *Tarachomylax* from the Emsian of Russia is another tooth plated form with a combination of primitive and derived characters (Barwick *et al.* 1997). It was cosmine covered, with a mosaic of small bones in the skull roof. Barwick *et al.* (1977) suggested that it was more derived than *Speonesydrion*, but could not place it exactly in a phylogeny.

Both *Sorbitorhynchus* and *Erikia* from Emsian age deposits in China show cosmine on all their dermal bones. Cosmine is a composite tissue, formed by dentine, a pore-canal system, and an enameloid cover, found in most primitive lungfishes and in other primitive sarcopterygians such as osteolepidids (basal tetrapodomorphs). It has been considered to house a sensory system, and especially so in the snouts of lungfish (see below). The palatal dentition consisted of flat, heavy dental plates, and in the mandible, between these plates is a depression that has been interpreted as housing a soft tissue pad on the tongue, used in an elaborate suction feeding mechanism. Wang *et al.* (1993) also suggested several other explanations for the purpose of this depression, including that it was a pathology, a position supported by Kemp (1994a). Much of the hyobranchial system of *Sorbitorhynchus* is unknown, so whether or not it was modified in concert with that aspect of the mandible cannot be judged (Wang *et al.* 1993). *Erikia* from Yunnan Province is a dipnorhynchid related to the Australian forms (Chang and Wang 1995), whereas *Sorbitorhynchus* has thick dentine plates and is allied to the chirodipterid lineage (Wang *et al.* 1993).

Uranolophus from the Pragian of Wyoming, USA, and *Melanognathus* (Schultze 2001) from the Emsian of arctic Canada are both of the denticulated type. *Uranolophus* has some claim to being the most primitive lungfish known from well-preserved material (Campbell and Barwick 1995) (Figure 5D). Both these forms had short jaws and a short snout, and in *Uranolophus* the jaws met in a strong contact zone, suggesting the capability of a powerful bite (Long 1995).

MIDDLE DEVONIAN LUNGFISHES

Probably the best known Middle Devonian lungfish, certainly the one most widely distributed in world museum collections is *Dipterus valenciennesi* (Figures 1A, 2A, 6). This species is found most commonly in the Caithness and Moray deposits in Scotland and Orkney — the Scottish Old Red Sandstone. Many other species are known from Russia, the USA and other parts of the world, for example, *D. oervigi* from the Bergisch Gladbach region of Germany (see Marshall 1986; Schultze 1992a). *Dipterus* was first discovered in the early 19th century, though it was not until 1871 that Günther first realised that it was related to the modern lungfishes, (Günther 1871), thus it was the first fossil lungfish to be recognized as such.

Its skull is very well known from numerous specimens and it carries cosmine on all dermal bones, including scales and fin rays (lepidotrichia), though at least on the skull, this is variable in its expression. In some specimens, the snout is completely covered with cosmine, whereas in others, individual bones of the snout mosaic are obvious. Periodic resorption of cosmine, perhaps seasonally or annually, has been inferred to explain the series of concentric lines running round each skull bone, called 'Westoll lines' (Westoll 1936; Thomson 1975) in those specimens in which individual bones are visible. These contrasts originally caused confusion as to how many species actually existed. Forster-Cooper's study (1937) showed that there was only one present in Scotland, and his alphanumeric system of bone identification was formulated on the basis of this species (Figure 2C). In many ways *Dipterus* is a fairly primitive genus, and is often used to provide a picture of primitive lungfish postcranial anatomy.

The pectoral fin skeleton was long, consisting of 7-9 mesomeres, with jointed radials either side. The pelvic fin skeleton is poorly known, but was probably of a similar type to the pectoral. The anterior dorsal and anal midline fins were fairly conventional compared with those of other primitive sarcopterygians: the anterior dorsal fin consisted of a basal plate with unjointed radials and lepidotrichia; the anal fin had a basal plate and four jointed radials and lepidotrichia. The second dorsal fin is of interest because this is the one which first begins to show modifications through lungfish evolution. In *Dipterus*, it was of a primitive form and consisted of a basal plate and several branching and jointed radials. The tail was conventionally heterocercal as in other primitive sarcopterygians (Ahlberg and Trewin 1995) (Figure 6).

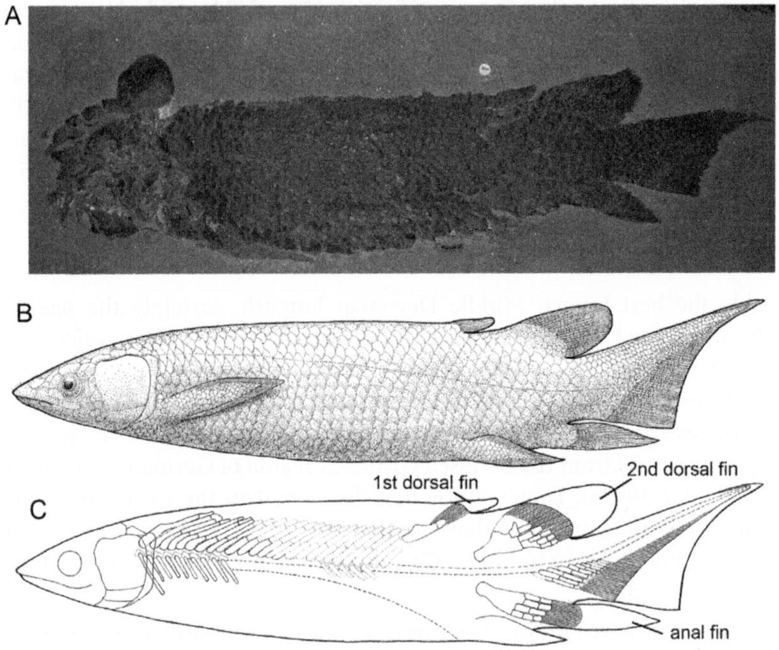

Fig. 6 *Dipterus valenciennesi*. A. Specimen photograph UMZC GN.804, 265 mm snout to tail length (Photograph, J. A. Clack). B. Life reconstruction. C. Skeletal reconstruction. B and C from Ahlberg and Trewin (1995).

Much of the postcranial skeleton was poorly ossified even in large individuals, showing the beginning of a trend through lungfish evolution towards less and less ossification. In *Dipterus*, though the anterior parts of the vertebral column including the supraneural spines that sat above each centrum, were reasonably well ossified, the posterior parts were not, and neither were the paired fin skeletons (Figure 6). The vertebral centra retained a large central hole for passage of the notochord, a primitive osteichthyan state also seen in most other Paleozoic lungfishes. There were short ribs on the vertebrae, but, contrary to some earlier assessments, it probably had no cranial ribs. Growth series for *Dipterus* from Caithness show changes to body proportions during ontogeny, showing how the head became relatively shorter compared with the length of the body (Ahlberg and Trewin 1995).

The skull had a bluntly rounded snout, exhibiting the typical embayments for the nostrils (Figure 1A). The palate and lower jaw tooth plates bore radiating rows of bluntly rounded teeth. The jaw hinge, as in other lungfish, lay far forward in the skull, in contrast to its posterior position in other osteichthyans (Figure 1A). The lower jaw was somewhat 'underslung' suggesting that *Dipterus* was a bottom-feeder. The opercular bone was large and round suggesting a substantial orobranchial chamber and that *Dipterus* was still reliant on gill breathing. Moreover, the distribution of *Dipterus* through the sediments of the Caithness Old Red Sandstone at Achanarras show that it was the first fish to become established in the locality, and was the last to disappear (Trewin 1986), corroborating the idea that it was the fish most tolerant of adverse (i.e. oxygen-depleted) conditions. Possibly it could breathe air when necessary.

There has been some suggestion that a small enigmatic fossil creature known as *Palaeospondylus gunni*, found only in deposits at Achanarras and on Orkney, can be identified as the 'larval' form of *Dipterus* (sic, Thomson *et al.* 2003). (Since lungfish do not truly metamorphose, hatchlings should strictly not be referred to as 'larvae' (A. Kemp pers. comm. 2008).) However, work on *Dipterus* itself, and on the ontogeny of the modern lungfish *Neoceratodus*, has shown that this cannot be the case (Miller 1930; Joss and Johanson 2006). Features that led to the original suggestion included the absence of teeth and the possession of a cranial rib. However, specimens of *Dipterus* as small as those of *Palaeospondylus* have been found that do not show the odd characteristics of that curious animal. These specimens, as well as fossils of very small juveniles of the Devonian lungfish *Andreyevichthys* (Krupina and Reisz 1999), and hatchling *Neoceratodus* (Kemp 1999), do carry full dentitions. Furthermore, *Dipterus* almost certainly did not possess cranial ribs and did not have fully ossified ring centra, as *Palaeospondylus* clearly does (Joss and Johanson 2006).

Another Middle Devonian lungfish from the Caithness and Orkney basin is *Pentlandia*, and though its dental anatomy is much less well known, its body morphology was more like that of the Late Devonian *Fleurantia* in having a separate but small first dorsal fin and an elongated second dorsal (pers. obs. JAC, National Museum of Scotland specimen NMS 1995.4.121).

Iowadipterus is a Middle Devonian lungfish from the USA (Schultze 1992a). This was a fairly primitive form retaining cosmine, with a relatively long head, but not a long snout, contrasting with long snouted forms commoner in the Late Devonian. In the single specimen, the end of the snout is completely cosmine-covered, but further back a bone mosaic is retained. The dentition is not visible. Schultze (1992a) reconstructed the musculature that might have operated its short lower jaw in combination with its long and deep skull roof, to provide powerful adductors. The dentition of this form is unfortunately unknown, but Schultze (1992a) suggested that in operation, it fell between the long snouted, denticle bearing forms that probably used suction feeding and rasping, and the short snouted tooth plated forms that used suction feeding and crushing mechanisms.

Stomiahykus from the Eifelian of Canada was a form with flat, dentine-covered tooth plates. *Dipnotuberculus*, from the Givetian of Morocco, had dental plates bearing a few large rounded tuberosities: a palate, a partial mandible and a dental plate are all that are preserved of this genus (Campbell *et al.* 2002).

Mount Howitt, in Victoria, is a Givetian age site preserving complete body fossils of lungfish in all stages of growth. Here are found *Howidipterus* and

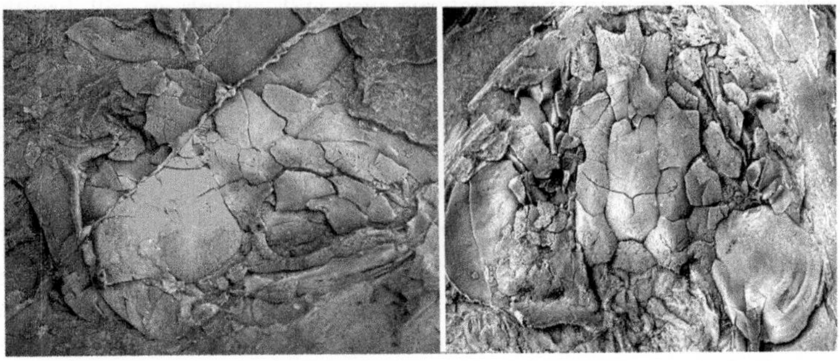

Fig. 7 Middle Devonian lungfishes. A. *Barwickia downunda* MV P181782, width of operculum, 20 mm. B. *Howidipterus donnae* NMV P118790, length of skull roof from posterior of A bone to tip of fused DE bones is 30 mm. Both casts whitened with ammonium choride. Photographs by J. A. Long.

Barwickia (Figure 7). *Howidipterus* had unusual tooth plates with well-developed teeth along the margins of each plate, coupled with smooth crushing denticulated surfaces more centrally. *Barwickia* was a denticulated form, but had an almost identical body form to *Howidipterus*. On closer comparison, it emerged that their dentitions were part of a spectrum of possibilities: in *Barwickia* there were fewer rows of teeth and more denticles compared with *Howidipterus* that had more rows of teeth and fewer denticles (Long 1992, 1993; Long and Clement 2009).

Eoctenodus from the Givetian of Victoria, Australia is known from several isolated bones and tooth plates. It had a long stalked parasphenoid, *Dipterus*-like tooth-plates and robust shoulder girdle (Long 1987). The oldest lungfish record from South America is a *Dipterus*-like toothplate and some scales from the Givetian-Frasnian of Venezuela (Young and Moody 1992).

LATE DEVONIAN LUNGFISHES

This is the time when lungfishes reached their peak of diversity. Several localities have yielded extensive material of a variety of genera, perhaps the richest being the Frasnian locality of Gogo in Australia. At least eight genera are present there, within the very narrow stratigraphic range represented by this locality. The rocks in which the lungfish are found are fully marine, and represent a community dwelling either on and around a limestone reef or in deep water between reefs. The fish were buried rapidly and limestone accumulated around the specimens, preserving them in three dimensions and often in complete states of skeletal articulation. In some instances, for example some placoderms, even soft tissue was replaced in detail allowing study of muscle structure (Trinajstic *et al.* 2007).

The two commonest genera are *Chirodipterus* and *Griphognathus*, exemplifying two ends of the spectrum of morphologies of skull shape and dentition seen in Devonian lungfish (Figures 8A, 9A). They have been described in minute detail by Miles (1977) and Campbell and Barwick (1999, 2002), including neurocrania and postcrania. *Chirodipterus* is a short snouted form, bearing massive dentine-covered plates on its palate and lower jaw, whereas *Griphognathus* is a long snouted, form bearing sheets of denticles on those surfaces (Figure 9A). *Griphognathus* had a massively ossified branchial and postcranial skeleton with complete ring centra. It may have been a bottom-dweller searching the sediment with a suction-feeding technique, or using its plier-like beak to snap off branching corals and stromatoporoids and rasp them away with the denticle-covered palate. *Griphognathus* is also found in other parts of the world during this time, including the USA and Europe, and in deposits that may not all have been fully marine. *Chirodipterus* was originally named for a form from Europe, and recent work suggests that the Australian form may in fact belong to a different genus (Long 2005; Friedman 2007b). *Holodipterus* is less common, but includes several species at first put into several subgenera by Pridmore *et al.* (1994), but later some of these were erected as separate genera by Long (in press) (Figure 8B). This genus had the deepest head profile of any Late Devonian lungfish, which combined with its long lower jaw symphysis and massive adductor muscle fossae, testifies to its ability as a powerful crusher of hard-shelled food items. Some *Holodipterus* species had tooth plates carrying bulbous crushing surfaces (*H. gogoensis*, Miles 1977), but others seem to show mainly a denticulated plate with toothlike projections along the margins (new genus A, Long in press) (Figure 8D). *Holodipterus* continually remodelled its dentition by resorption and new growth at the labial margins. *Robinsondipterus* shows the elongated snout similar to *Griphognathus*, but with tooth-like blebs of dentine along the biting margins. New genus B had powerful crushing surfaces but retained sharp teeth on separate dentaries on the lower jaws (Long in press) (Figure 8C).

Other Gogo taxa include the chirodipterids *Gogodipterus*, a form with tooth plates carrying strongly developed ridges and grooves, and *Pillararhynchus*, a long, deep headed form with narrow concave tooth plates (Figure 8F, G). *Adololophas* is another tooth plated form, in this case the material includes an almost complete body with articulated scale cover. Interestingly, this shows a diminution of scale thickness and cosmine cover from front to back of the animal (Campbell and Barwick 1998). A strong case has been made that all these genera were marine, bottom dwelling, and did not breathe air on a regular basis, based on having a full complement of functional gill-arches (Campbell and Barwick 1988; Long 1993). However, *Griphognathus* also occurs in shallow water, marginal marine environments in Europe (Schultze and Chorn 1997). Recent work has shown that the Gogo *Griphognathus* might be considerably different from the type material and should be placed in a different genus (Long 2005; Friedman, 2007a, b).

Another Frasnian locality that has yielded fossil lungfish is Miguasha in Canada. Here, two genera are described, *Fleurantia* and *Scaumenacia*, the latter named after an anglicized version of Escuminac Bay, in which Miguasha is situated (Figure 9D, E). These two genera show the beginnings of a trend seen in lungfish

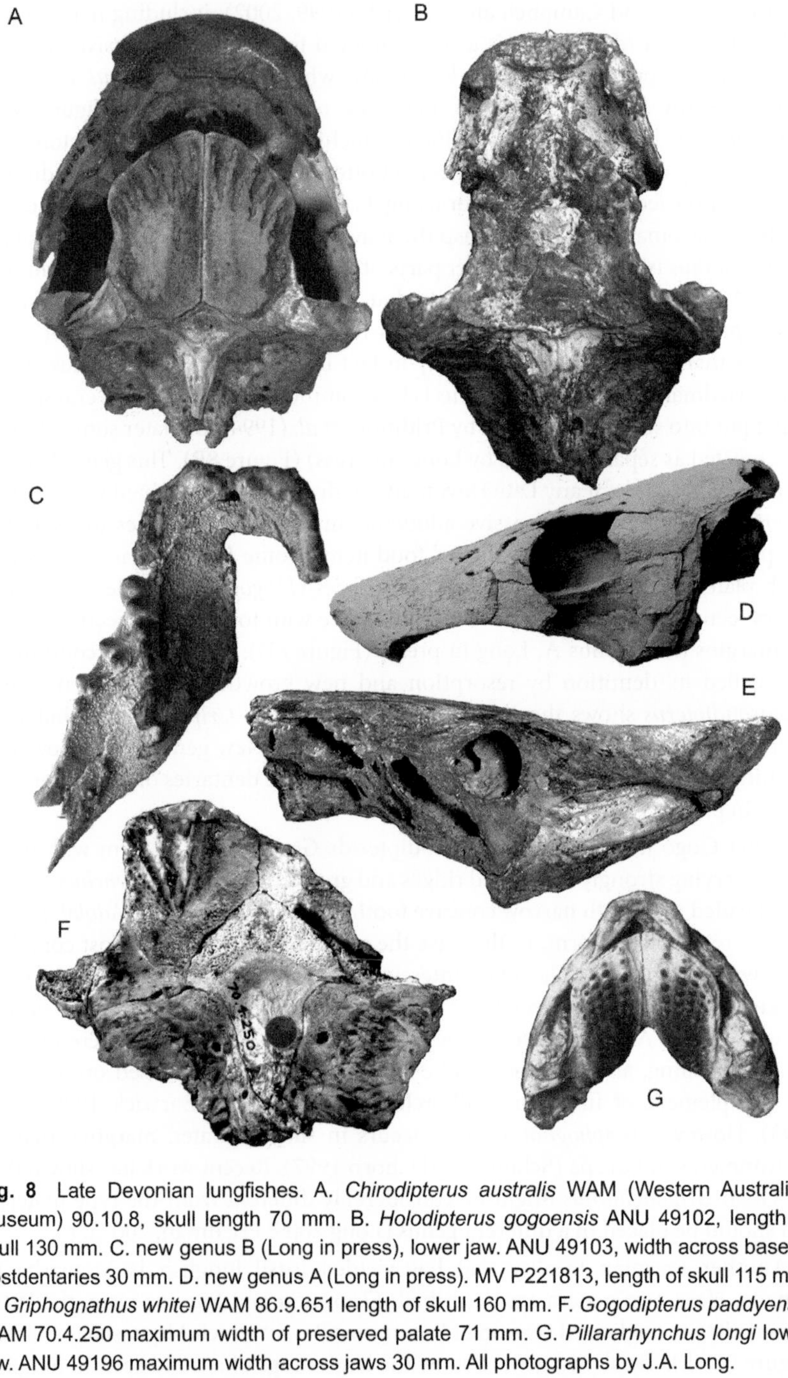

Fig. 8 Late Devonian lungfishes. A. *Chirodipterus australis* WAM (Western Australian Museum) 90.10.8, skull length 70 mm. B. *Holodipterus gogoensis* ANU 49102, length of skull 130 mm. C. new genus B (Long in press), lower jaw. ANU 49103, width across base of postdentaries 30 mm. D. new genus A (Long in press). MV P221813, length of skull 115 mm. E. *Griphognathus whitei* WAM 86.9.651 length of skull 160 mm. F. *Gogodipterus paddyensis* WAM 70.4.250 maximum width of preserved palate 71 mm. G. *Pillararhynchus longi* lower jaw. ANU 49196 maximum width across jaws 30 mm. All photographs by J.A. Long.

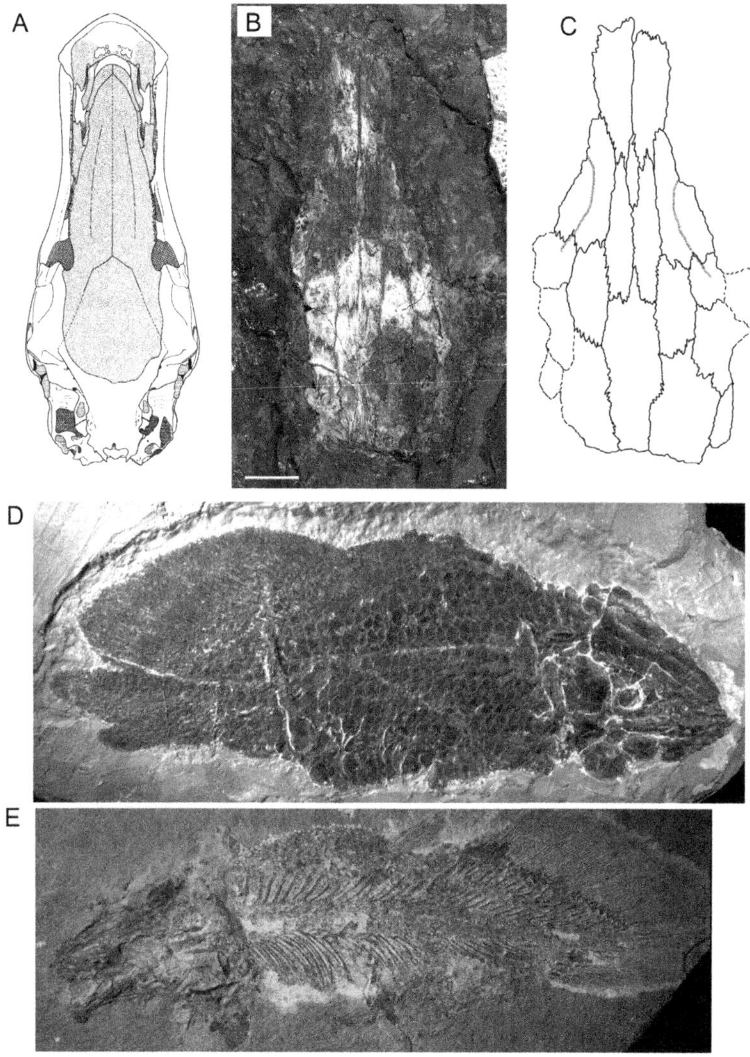

Fig. 9 Late Devonian lungfishes. A. Palate of *Griphognathus whitei*, a denticulated form (from Miles 1977). B, C. *Soederberghia groenlandica*, specimen ANSP (Academy of Natural Sciences, Philadelphia) 20902, photograph and interpretive drawing respectively. (Photograph, E. B. Daeschler). Scale bar 10 mm. D. *Scaumenacia curta*. Specimen number not traced. E. *Fleurantia denticulata* BMNH P.24745-P24736 length of specimen 215 mm. D, E, Photographs M. Arsenault, Parc de Miguasha, from Long 1995).

evolution towards enlargement of the second dorsal fin and its amalgamation with the caudal fin. *Fleurantia* still had a separate first dorsal, with an elongate second dorsal, whereas in *Scaumenacia* the first dorsal, though separate from the elongate second, was long and low and appears to lack any supports from the vertebral column (Cloutier 1996). The situation seems to have been comparable to that at Mount Howitt: *Scaumenacia* was a tooth plated form, whereas *Fleurantia* was a denticulated form, but the two genera had similar body plans. A detailed bed-by-bed study of the deposits at Miguasha was made by Cloutier *et al.* (1996), who showed that *Scaumenacia* occurs throughout the whole sequence of the Escuminac Formation, and is one of the commonest taxa at the locality. By contrast, *Fleurantia* occurs only in the lower parts of the sequence and is relatively rare. There appear to be no obvious correlations between the faunal assemblages and the lithostratigraphic units found there: the entire sequence may represent a brackish or marginal marine environment (Cloutier *et al.* 1996). These two genera may have had specific requirements that are not reflected in the preserved environment.

Devonian lungfishes were first found in East Greenland, in Famennian age deposits, in many cases alongside fossils of Devonian tetrapods. Four genera are known: *Soederberghia*, *Nielsenia*, *Jarvikia* and *Oervigia* (Bendix-Almgreen 1976). *Soederberghia* is the best known, and has congeners widely distributed throughout the world in similar age deposits (Figure 9B, C). For example, it is also known from the Catskill formation in Pennsylvania USA (Friedman and Daeschler 2006), from Belgium (Clement and Boisvert 2006), and from two localities in Australia: the spectacular fossil fish locality of Canowindra (Ahlberg *et al.* 2001), and the Jemalong Quarry that has also yielded tetrapod remains (Campbell and Bell 1982). Thus this genus is one associated with tetrapods in Greenland, Pennsylvania and Australia. It was a long snouted, denticulate form that could grow to a large size, and is one of the youngest genera to be represented by ossifications of the braincase region, though even it shows some aspects in which reduction has begun. It is known from postcranial remains that were quite well ossified, including long, curved ribs and substantial ring-centra in which the notochordal pit was sometimes completely obliterated. *Soederberghia* shows adaptations of the palate and postcranium interpreted as being associated with air-breathing — a cranial rib and a parasphenoid with a long posterior stem (Ahlberg *et al.* 2001).

Of the other Greenland genera, the tooth plated *Oervigia* and the denticulated *Jarvikia* were both long snouted, and resembled *Scaumenacia* and *Fleurantia* respectively. The short snouted *Nielsenia* is known from only a single specimen.

Other Famennian localities have yielded lungfish specimens, one of the most prolific being the Andreyevka-2 locality near Tula in Russia. Here, thousands of disarticulated bones especially of tooth plates of juvenile lungfish have been recovered by acid-digestion of the limestone, allowing growth series to be reconstructed from the great size-range of specimens available (Krupina and Reisz 1999). This genus, *Andreyvichthys*, is another lungfish found in association with tetrapod material (Figure 1B).

Orlovichthys, also from the Famennian of Russia, the Orel region, is known on the basis of an almost complete skull with dentition. It had a relatively narrow skull, with dentition organised into tooth plates (Krupina *et al.* 2001). *Adelargo* from the late Famennian of Australia is known from isolated elements, tooth plates, bones and scales, and a partial skull roof (Johanson and Ritchie 2000).

Two Famennian genera from Scotland are worth a mention. *Rhynchodipterus* from Rosebrae Quarry near Elgin has been considered a close relative of *Griphognathus* and *Soederberghia*, but this remains to be adequately tested. It has a fairly primitive appearance in terms of fin morphology, and is long snouted with a somewhat duck-like skull profile. Recently, the single specimen has been CT scanned, and work in progress by Friedman and Coates should help clarify its anatomy and relationships. *Phaneropleuron* from Dura Den has not been fully described. Huxley (1861) gave lithographs of a number of specimens, and Westoll (1949) suggested that it was like *Scaumenacia*. Thomson (1969a) reconstructed the body outline as having combined dorsal and caudal fins, more like *Uronemus*. Both *Rhynchodipterus* and *Phaneropleuron* require redescription, and could throw important light on lungfish postcranial evolution.

CARBONIFEROUS AND PERMIAN LUNGFISHES

After the Late Devonian, the taxonomic diversity of the lungfishes went into a significant decline, and there are many fewer genera known from the Carboniferous to the Recent than there are in the Devonian. However the timing of this decline is not altogether clear: the Early Carboniferous record is particularly sparse but this may represent a taphonomic artefact. The marked by a drop in the morphological diversity shown by the dipnoans in the Carboniferous is characterized by the loss of all long-snouted lungfish, and all lungfish without ridged tooth plates except one, *Conchopoma*. Thus it may be that some ecological niche previously occupied by long-snouted lungfish became unavailable at the Devonian–Carboniferous boundary. Similarly, by the Carboniferous, there seems to be an almost complete loss of endoskeletal ossification, and there are no known fossil lungfish from the post-Devonian which preserve the braincase or more than a few elements of the endoskeleton. This transition from morphological diversity to conservatism is still poorly understood.

For the Carboniferous, much of the lungfish record is known from the United Kingdom and, as for most freshwater and terrestrial fauna, the fossil record is sparse in the earliest parts of the era. Only isolated bones are known from the Tournaisian (eg. a rib from Dumbartonshire, Clack and Finney 2005)). The oldest Carboniferous taxon is probably *Ctenodus romeri* from the Viséan (Arundian) of Berwickshire (Thomson 1965). It is a tooth plated lungfish known only from its dentition, although there is some isolated cranial material probably attributable to this species (T. R. Smithson, pers. comm.).

Ctenodus is the most common Early Carboniferous (= Mississippian) genus known (Figure 4), and is distributed throughout the Carboniferous, although there does not seem to be any temporal overlap between the five apparently valid species. *Ctenodus romeri* from the Viséan, and *Ctenodus interruptus* (Barkas 1869) from the Viséan/Namurian boundary, are only known from tooth plates (Figure 1). *Ctenodus murchisoni* (Ward 1890), from the Westphalian C of the Staffordshire coal fields, has a single skull roof associated with it. *Ctenodus cristatus* (Agassiz 1838), from the Westphalian, and an un-named species from the Early Carboniferous (Sharp 2007) both have a not inconsiderable amount of skeletal material preserved, although little of it is articulated.

Ctenodus is an interesting genus because there is a progressive change from tooth plates which are similar in size to those of *Dipterus*, to being of extremely large size in the Late Carboniferous species, implying an increase in overall body size. Tooth plates of *C. murchisoni* reach up to 10cm in length, and have up to 23 tooth ridges, making them some of the largest tooth plates known. *Ctenodus* exhibits a variety of morphologies of tooth plate, within the framework of the ridged plate, that is quite striking. Were articulated specimens to be found in association with all these tooth plate morphologies, it is uncertain that they would be retained within the same genus.

Also from the Early Carboniferous are the genera *Straitonia* (Thomson 1965) and *Uronemus* (*Ganopristodus* of Schultze 1992a), both genera found in Scotland. *Straitonia* is a monospecific taxon from the Asbian (D1) of the Viséan, and is an almost complete specimen preserved in a nodule. It is the most completely preserved British Carboniferous lungfish and shows a postcranial skeleton that seems to demonstrate a dorsal fin fused with a diphycercal caudal fin, and no independent anal fin. *U. splendens* is known from remains of the skull roof and exhibits a unique form of dermal ornament, more akin to that of tetrapods than lungfish. There seems to be relatively little support for the union of this with another species, *U. lobatus*, within the same genus and it is desirable that the taxonomy of this genus be addressed.

The genus *Tranodis* is known from several localities in the late Mississippian of North America (Thomson 1965; Schultze and Bolt 1996). Recently numerous specimens of a *Tranodis*-like lungfish have been recovered from a site in Hancock County, Kentucky, also of late Mississippian age (Garcia *et al.* 2006a, b). This locality is also notable for the variety of the tetrapods that are preserved there. Many of the lungfish specimens are preserved in situ in carbonate concretions representing burrows at the top of a shale horizon in the upper part of the exposure, and are the oldest known such burrows. They are considered to be aestivation burrows by the authors. This part of the sequence at Hancock is intepreted as representing a small ox-bow lake in a floodplain environment, with intermitant connection to a larger lake or river.

Another Early Carboniferous form is *Delatitia* from the Mansfield Basin in Australia. It consists of a partial skull roof and tooth plates and bears some resemblance to *Ctenodus*, though it retains some primitive features of the skull bone pattern (Long and Campbell 1985).

The most well known and widespread Late Carboniferous (= Pennsylvanian) taxon, *Sagenodu*, is found in the United Kingdom as well as the Permian of the United States and the Czech Republic. A tooth plated genus, *Sagenodus* (Owen 1867) has a stratigraphic distribution of around 60 million years and is widespread in Europe and North America (Figure 4). The British species of *Sagenodus*, *S. inaequalis*, was well described by Watson and Gill (1923) and the American taxa were comprehensively addressed by Schultze and Chorn (1997). Almost all of the morphology of *S. copeanus* is known, from many disarticulated elements and a single articulated specimen, from the Hamilton Quarry in Kansas (Figure 10).

Sagenodus is found, along with *Ctenodus*, at the well known tetrapod locality of Newsham in Northumberland, UK, and its remains exist there in abundance. Schultze and Chorn (1997) have described *Sagenodus* as 'The beginning of modern lungfish', noting that it appears structurally intermediate between Devonian and post-Paleozoic lungfish. In addition they noted that the genus must be tolerant of a range of salinities because it is known from shallow marine to freshwater deposits. While *Sagenodus* is only known from one species in the UK, there are a number of species present in North American deposits and its range extends to the end of the Early Permian.

Fig. 10 *Sagenodus copeanus* Specimen KUVP (Kansas University Vertebrate Paleontology) 84201 part and counterpart. (Photographs, E. L Sharp, H.-P. Schultze). Scale bar 10 mm.

The generic diversity of lungfishes in the Permian is rather sparse, but two genera are known from many species. *Conchopoma* first appeared in the Late Carboniferous, with several species named from that Period, but species have also been described from the Early Permian. *C. gadiforme* from Europe is the best known of these (Schultze 1975). *Conchopoma* is the last-surviving genus of non-tooth plated lungfishes, and has cone-shaped individual teeth on the palate (Marshall 1988). It is best known from the Late Carboniferous of Mazon Creek, where almost complete skeletons have been found. It occurs alongside three other genera of lungfish including *Megapleuron*, that also persisted into the Early Permian.

Gnathorhiza is by far the most speciose of the Permian lungfish and is known from many localities in the United States, some preserving skeletal material and some only tooth plates (Olson and Daly 1972). *G. bothrotretus* is known from quite well preserved skeletons discovered within burrow-infills from a locality in New Mexico. This locality actually preserved large numbers of burrow-infills, although only four contained skeletal material. Two inferences were drawn from this. First, the burrows had been made by aestivating animals, indicating that the climate experienced seasonal drought. Secondly, that the mortality rate among those animals was quite low, given that only four skeletons remained among hundreds of burrow-infills (how many of these burrows were investigated is not stated (Berman 1976)). It stands in contrast to the discovery of the *Tranodis* specimens, many of which were found in articulation within their burrows. *Monongahela* is another genus named from the Permian (Marshall 1986), though it is only known from tooth plates and partial lower jaws and has been synonymized with *Palaeophichthys*, a genus known from the Middle Pennsylvanian (Schultze 1994). The synonomy was disputed by Kemp (1998b).

Figure 11 shows the distribution of genera of lungfishes, mentioned in the text, throughout the Devonian and Carboniferous arranged by stage, to illustrate the fall in diversity at the end of the Devonian. Readers should note that some genera survived through more than one stage and that the positions of genera within stages are not indicated.

MESOZOIC AND CENOZOIC LUNGFISHES

Subsequent to the end of the Paleozoic, remains of lungfishes decline in abundance. They consist largely of tooth plates, though about a dozen are known from skull roofs and a few from more complete remains. These include *Ariguna* and *Gosfordia* from the Early Triassic of Australia, two of the few post-Paleozoic forms known from complete fish (Kemp 1994b; Ritchie 1981). Mesozoic and Cenozoic lungfishes were reviewed by Schultze (2004), to whose work the reader is referred, and from which the summary presented here is largely drawn. More recently the most significant find has been of articulated skull remains and a limited amount

CARBONIFEROUS		Late Carboniferous 323mya	*Conchopoma, Gnathorhiza* *Ctenodus, Sagenodus*
CARBONIFEROUS 354mya		Early Carboniferous	*Ctenodus, Uronemus, Straitonia*
DEVONIAN	Late	Famennian	*Soederberghia,* other Greenland forms *Orlovichthys, Andreyevichthys* *Rhynchodipterus, Phaneropleuron*
DEVONIAN	Late	Frasnian 370mya	*Fleurantia, Scaumenacia* *Chirodipterus, Griphognathus,* *Holodipterus,* other Gogo forms
DEVONIAN	Middle	Givetian	*Howidipterus, Barwickia* *Dipnotuberculus, Eoctenodus* *Dipterus, Pentlandia*
DEVONIAN	Middle	Eifelian 391mya	*Stomiahykus, Iowadipterus*
DEVONIAN	Early	Emsian	*Sorbitorhynchus, Tarachomylax* *Speonesydrion* *Dipnorhynchus*
DEVONIAN	Early	Pragian	*Uranolophus, Melanognathus* *Ichnomylax* *Diabolepis, Youngolepis*
DEVONIAN 416mya	Early	Lockhovian	

Fig. 11 Devonian and Carboniferous timescale showing distribution of lungfish genera. (Note that position within stages does not necessarily indicate exact stratigraphic level of the taxon within that stage.)

of postcranial material from the genus *Ferganoceratodus* in Late Jurassic deposits in Thailand (Cavin *et al.* 2007) (Figure 12). This has added to the discussion of lungfish biology and relationships. It retains a bony snout, the only post-Devonian genus known to do so.

By the Mesozoic, lungfish dentitions had stabilized to the typical ridged tooth plates seen in modern genera, though they varied in the number and proportions of the ridges. These are often diagnostic as to genera and species, and show that there was quite a large array of genera and species during these later periods, but their morphological diversity was much more limited than in the Paleozoic (eg. Kemp 1993, 1994b, 1997, 1998a). Cavin *et al.* (2007) illustrated some of the variability to be found among them. Postcranial skeletons had stabilized to the long combined dorsal and caudal fins seen first among Carboniferous genera such as *Sagenodus*, and paired fins were long and narrow. Reduction of the osssification of the skull and postcranial skeleton also proceeded apace. The morphology is exemplified by the Triassic genus *Gosfordia* (Ritchie 1981) (Figure 12).

Taking their diversity as measured by the number of named taxa, there seems to be no obvious trend towards reduction. Taking that information at face value, there have been peaks during the Early and mid-Triassic, the Early and mid-Cretaceous, and the Late Cenozoic. Schultze (2004) has interpreted at least part of this pattern as attributable to the presence of extensive freshwater sequences from the Triassic compared to the prevalence of marine deposits from the Jurassic. At the Cretaceous/Tertiary (K/T) boundary, the number of taxa was cut by half, but five genera persist into the Cenozoic.

Most post-Paleozoic lungfish are from freshwater deposits, though not all: Schultze (2004) noted marine occurrences in Middle Triassic deposits, though possible Cretaceous marine records may be reworked (Kemp 1993; Cavin *et al.* 2007). Global distribution shows a trend towards restriction into southern continents. Triassic genera are found in Europe, Asia, southern Africa, Australia, and South America, though not, so far, in North America, despite being fairly common there in the Paleozoic. Asian and North American Jurassic deposits have yielded lungfish remains, though surprisingly, Australian ones have not. This may be a taphonomic effect in that the sediments are too acidic to preserve the remains (A. Kemp pers. comm. 2008). The last record of a European lungfish comes from the Middle Jurassic of England. In the Cretaceous they are found in South America, northern Africa, Madagascar, and once more, and most abundantly, in Australia. Cenozoic examples have been found in northern Africa, but otherwise they seem to have been restricted to South America and Australia, which is more or less their modern distribution. Cavin *et al.* (2007) explored this pattern based on their new phylogenetic analysis that suggested a deep split between lepidosirenids and *Neoceratodus*, inferring that vicariant events better explained the known distributions than did dispersal events.

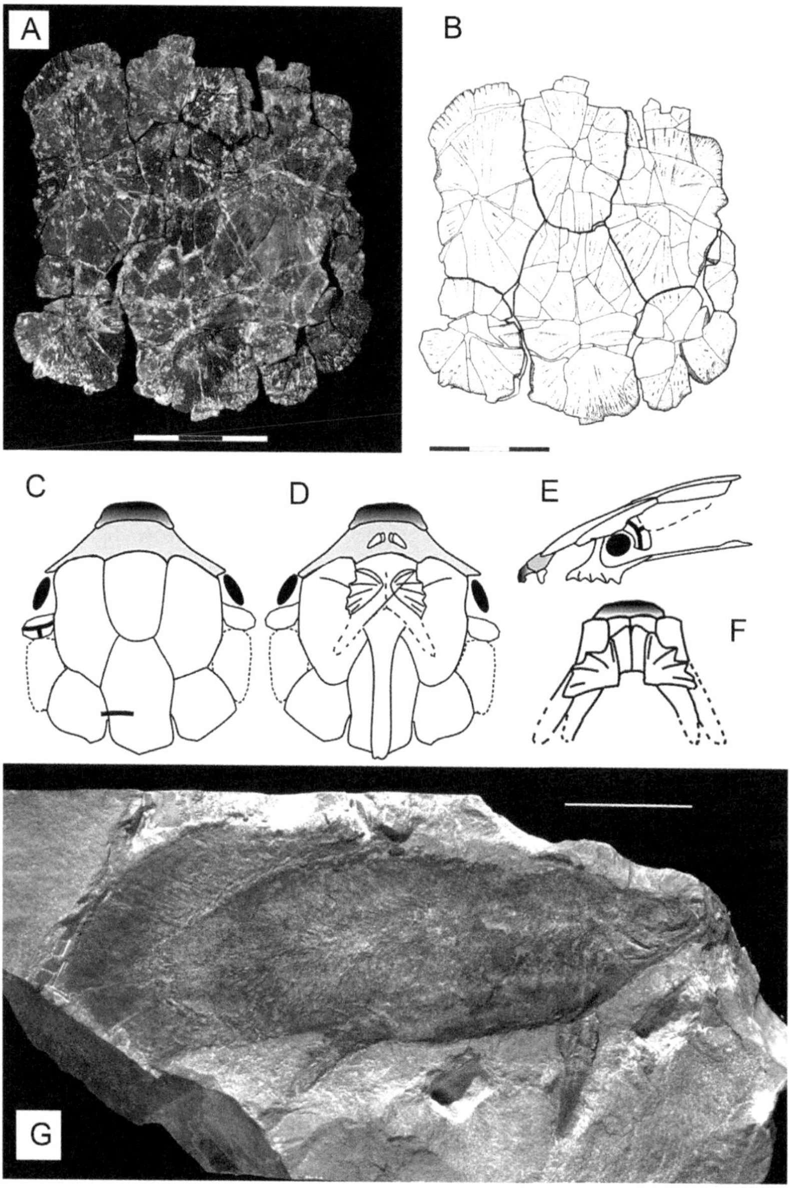

Fig. 12 Mesozoic lungfishes. A–F. *Ferganoceratodus martini*. A. Photograph and interpretive drawing respectively TF (Thai Fossils, Sahat Sakhan Dinosaurs Research Centre) 7712. Scale bars 30 mm. C, D. Reconstruction of skull in dorsal and ventral view respectively. E. Skull reconstruction in left lateral view. F. Lower jaw in dorsal view. A–F from Cavin *et al.* (2007). G. *Gosfordia truncata*, AMF 60621. (Photograph, J. A. Long). Scale bar 100 mm.

The extant Australian lungfish was initially named *Ceratodus*, on the basis of the similarity of its tooth plates to fossil forms from the Triassic. The extant form was given its present name in 1876 (de Castelnau 1876). As more '*Ceratodus*' tooth plates were found, discrimination between morphologies became more precise. Many '*Ceratodus*' tooth plates were assigned to new genera such as *Ptychoceratodus*, *Asioceratodus*, *Microceratodus* and *Archaeoceratodus* (Schultze 1981; Kemp 1992, 1993, 1997, 1998a). Despite this, *Ceratodus* itself still appears to be the longest-surviving form, with Early Triassic to end-Paleogene species being recognized. Neither Schultze (2004) nor Cavin *et al.* (2007) consider *Ceratodus* to be particularly closely related to *Neoceratodus*. *Neoceratodus* as a genus, in fact the species *N. forsteri*, is now traced back to the Early Cretaceous (Kemp 1997; Kemp and Molnar 1981), *Protopterus* back to the mid-Cretaceous, and *Lepidosiren* back to the Late Cretaceous (Schultze 2004).

Evolutionary History of Lungfishes

Fossil lungfishes, especially the rich Paleozoic records, throw light on the evolutionary history, biology and ecology of the group that would otherwise be impossible to infer. Despite the fragility of our understanding of their phylogeny, several significant 'trends' can be seen. Many of these were first noted in detail by Westoll (1949), but have been brought up to date over subsequent decades. Many areas of interest are not without their controversy, and many issues remain unresolved today.

Among the more consistent and striking observations is the reduction of ossification of both the dermal and endochondral skeleton through their geological record, though this is paralleled in other groups such as coelacanths. Early lungfishes retained a heavy cosmine covering over skulls and scales, a feature in common with other early sarcopterygians. It may have been resorbed periodically in several Devonian forms, and was lost completely by all post-Devonian taxa. The number of skull bones themselves became reduced, from the complex mosaic in *Dipnorhynchus* to the remnants seen in modern forms, with reduction progressing from front to back. Lateral lines came to run within the skin rather than through the skull bones. With few exceptions, *Ferganoceratodus* being one such, only the more posterior skull bones are retained in post-Paleozoic lungfish. Scales too not only lost their cosmine, but also their bone, causing Huxley (1861) to comment of the Famennian *Phaneropleuron* that they were 'exceedingly thin', 'containing very little bony matter'. That condition is not so different from *Neoceratodus* today, though even in *Protopterus*, some mineralization is retained in the scales (Zylberberg 1988). Other changes to the skull have been observed: enlargement of the orbit, shortening of the cheek, and reduction then loss of ossification of the braincase. These, along with other changes to the postcranial skeleton, such as loss of lepidotrichia, restriction of midline fins to the continuous dorsal and caudal fin, and reduction in ossification of the vertebral column have been suggested as resulting from heterochronic processes such as paedomorphosis (Bemis 1984; Joss and Johanson 2006). This may explain some of these observed sequential changes, but does not explain the gradual but complete limitation towards the ridged tooth plate pattern from the varied dental morphologies seen in early forms. The presence of substantially ossified ribs in many Carboniferous lungfish taxa suggests a delay in the trend towards reduction in ossification seen in the rest of the postcranial skeleton.

Loss of cosmine presents interesting problems of its own. Cosmine has been interpreted as intimately linked with an electrosensory or lateralis-derived system housed within the pore canals characteristic of that tissue (Thomson 1975; Northcutt and Gans 1983; Gans and Northcutt 1983; Cheng 1989). It has been shown fairly convincingly that comparable structures are seen in the snout of *Neoceratodus*, though the relationship to a sensory system has been disputed. Bemis and Northcutt (1992) interpreted the pore canal system as associated with a complex cutaneous vascular system involved in the deposition of mineralized tissues. Cavin *et al.* (2007) showed that *Ferganoceratodus*, the only post-Devonian lungfish with a 'hard snout' (though it lacked cosmine) nonetheless showed bony structures comparable to those of Devonian forms, and suggested it as an intermediate stage between Devonian forms and *Neoceratodus*. They agreed with Bemis and Northcutt (1992) that the system was not sensory, but was a system for deposition of mineralized tissue: this raises the question of why it was retained in *Neoceratodus*, which is poorly ossified throughout its skeleton.

The evolution of the postcranial skeleton has elicited comments from many authors, from Westoll (1949) onwards (eg. Long 1993, 1995; Ahlberg and Trewin 1995) (Figure 13). It is a manifest fact that whereas early lungfishes show a primitive sarcopterygian pattern, exemplified by *Dipterus*, at the other end of the spectrum from *Sagenodus* in the Carboniferous to the modern forms, the second dorsal, caudal and possibly anal fins amalgamate, the tail becomes symmetrical rather than heterocercal, and the paired fins elongate. Intermediate morphologies are shown by genera like *Fleurantia* and *Scaumenacia*. Further studies of genera such as *Pentlandia* or *Ctenodus* may help to understand aspects of postcranial evolution, though, until the phylogeny of early lungfish is sufficiently well understood, such questions will remain hard to answer. Why this derived morphology became established at the expense of more flexible patterns is speculative, but may be linked to paedomorphic processes (Bemis 1984; Long 1990), and/or environmental influences that also encouraged air-breathing (Long 1993). Recently discovered hatchlings from Devonian lungfish may go some way to illuminating this question (Newman and den Blaauwen 2008).

There has also been discussion about the inflation in genome size seen in modern lungfishes in comparison with almost all other osteichthyans (salamanders have acquired a similar feature, thought by most researchers today to be convergently derived). Increase in genome size is reflected in cell size, and so is amenable in a limited way to study in fossils. Thomson (1972) was the first to discuss this idea,

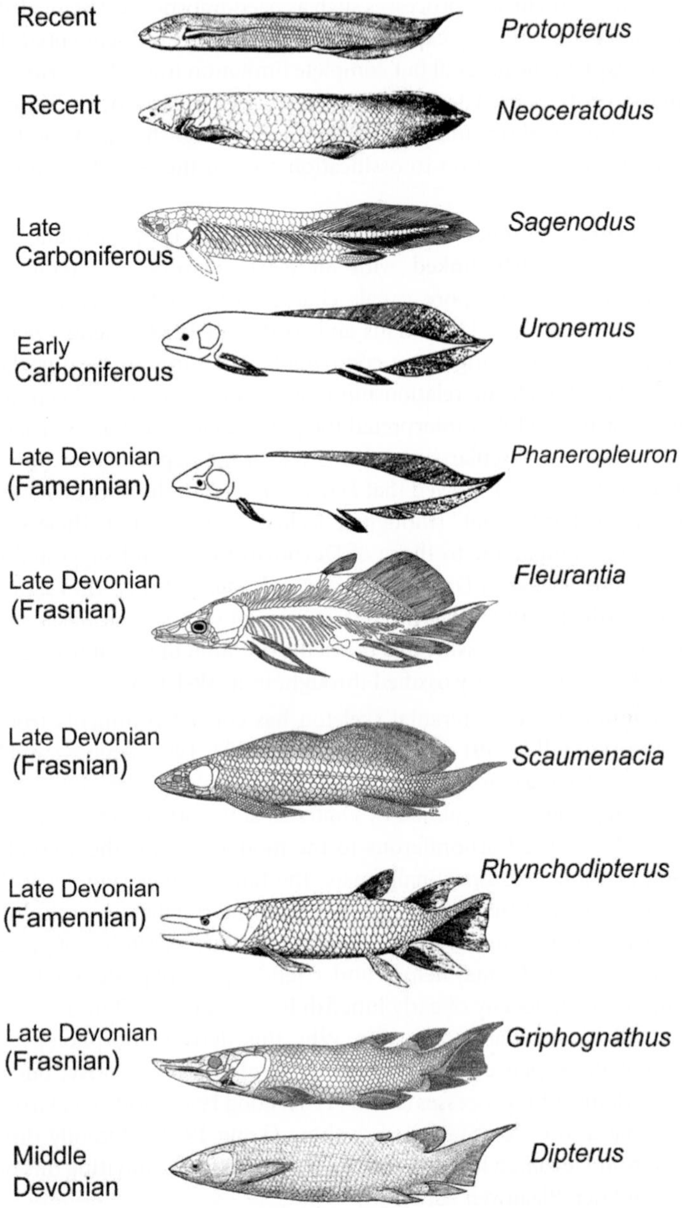

Fig. 13 Evolution of lungfish body form. *Dipterus* from Ahlberg and Trewin (1995); *Griphognathus* from Campbell and Barwick (1988); *Rhynchodipterus* from Jarvik (1980); *Fleurantia* from Cloutier (1996); *Scaumenacia* from Campbell and Barwick (1988); *Phaneropleuron* from Jarvik (1980); *Uronemus* from Jarvik (1980); *Sagenodus* from Schultze and Chorn (1997); *Neoceratodus* from Jarvik (1980); *Protopterus* from Jarvik (1980).

showing that cell size (ie. volume) was smallest in Devonian taxa, increased in Carboniferous taxa, again in Permian ones, and further again in Mesozoic forms. Thus the increase was not a sudden one-off event. Bemis (1984) pointed out that in salamanders, increase in genome size was linked to paedomorphosis (specifically, neoteny), and inferred that the same may be true for lungfish. Cell size is related to the rate of cell division, so that growth is retarded, and may result in neotenic species. Joss and Johanson (2006) pointed out that small cell size is characteristic of vertebrates showing both metamorphosis and direct development, and suggested that cell size increase might be related to loss of metamorphosis in post-Devonian lungfish, with retention of a *Neoceratodus*-type hatchling morphology into the adult. Unfortunately, since the hatchlings or juveniles of post-Devonian lungfish are rarely found in the fossil record, the existence or not of metamorphosis in these taxa is unlikely to be testable. Thomson (1972) further suggested that increase in cell-size may be linked to loss of diversity through the history of lungfishes. Though accepting the general principle of paedomorphism in lungfishes, Ahlberg and Trewin (1995) were skeptical that all the apparently paedomorphic changes seen in the morphology of the group were causally linked, noting that a robust phylogeny was the first requirement to test the hypothesis.

Another area of long-standing controversy concerns the morphology of the dentition in the Devonian forms. One view of the phylogeny divides the taxa into three separate lineages: denticulate forms, dentine plated forms, and those with true tooth plates, each of the dental types evolving as separate radiations (eg. Campbell and Barwick 1988, 1990). Other views of the phylogeny suggest that this scheme is unparsimonious when other skeletal information is taken into account, and the contrasting dentitions are seen as having arisen multiple times in several different lineages (eg. Schultze and Marshall 1993).

The radiating pattern of ridges on the tooth plates of post-Devonian forms grows by addition at the lateral ends of the ridges, without any shedding of teeth such as is found in other osteichthyans (Smith and Krupina 2001). In forms such as *Dipterus*, individual protuberances (teeth) can be identified, as they can in juveniles of *Neoceratodus*. In adults of *Neoceratodus*, they have become amalgamated into continuous ridges. Studies of growth series of both *Neoceratodus* and *Andreyevichthys* have shown that the same developmental pattern is observed in both showing that it is a very ancient mechanism, virutally unchanged since the Late Devonian (Krupina and Reisz 1999; Reisz and Smith 2001; Smith and Krupina 2001, Kemp 2002a). Work on the development of lungfish dentition has tended to suggest that, though it shows unique features compared to other osteichthyans, once established, only a limited number of processes are required to produce the varying patterns. Ahlberg *et al.* (2006) proposed a scheme whereby four different pterygoid dentitions could be generated by manipulating just two ontogenetic processes: tooth addition and tooth resorption (Figure 14A). They assumed that sheet dentine is added at the lateral pterygoid margin and that resorption is followed by the deposition of a denticle field. The idea that the dental morphologies fall along spectra of types is suggested for example by the differing morphologies seen within *Holodipterus*, and the fact that denticles lie between the toothed ridges in *Andreyevichthys* (Ahlberg *et al.* 2006) (Figure 1).

Little work has addressed questions of the functional significance of Paleozoic lungfish tooth plate morphology. Questions concerning the significance of numbers and spacing of ridges, or presence or absence of cusps would be hard to address because of the limited availability of modern analogues. Study of microwear might prove difficult because most Carboniferous tooth plates have been prepared using mechanical means which damage the surface of the enamel. However, tooth plate wear patterns and pathologies from Cenozoic forms have been shown to record environmental conditions, with increased numbers and types of pathologies occurring in populations under climatic or other stress. Different populations of lungfishes from the Cenozoic of Australia show some that were healthy, some that were aging and showed no recruitment, and some in which attrition or disease were prevalent (Kemp 2005).

Because of the exceptional preservation of some Devonian lungfishes, some unexpected aspects of their biology can be ascertained. For example, braincases of Gogo species, such as *Chirodipterus*, allow details of the vestibular system to be described (Miles 1977). Semicircular canals and their ampullae have perichondrally lined walls, and though the saccular region is less well ossified, the utricular region is discernible. It shows a derived character of lungfish, seen also in *Protopterus* (Retzius 1881) and *Lepidosiren* (J.M. Jørgensen pers. comm., see Chapter 19): a relatively enlarged utricular sac (Figure 14B). This character

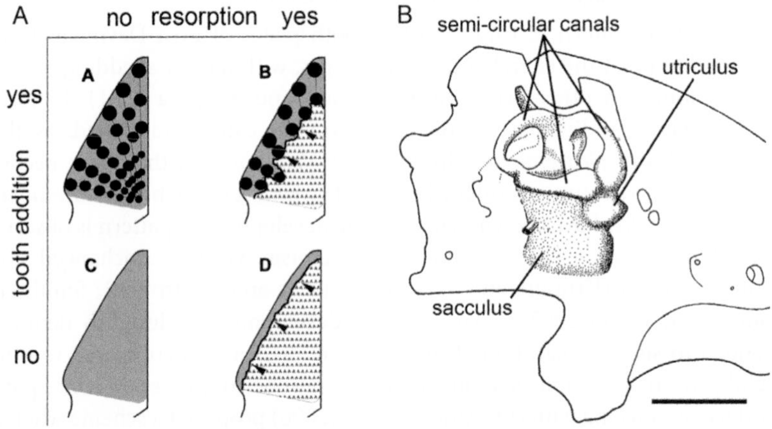

Fig. 14 A. Processes governing tooth plate development (from Ahlberg *et al.* 2006). B. Reconstruction of inner ear of *Chirodipterus australis* showing large utricular chamber (from Miles 1977). Scale bar 10 mm.

may therefore date back to an early stage in their evolutionary history, though its signficance is uncertain.

Details of the nasal capsules, blood vessels of the snout, and of the lower jaw have been reconstructed in detail for some taxa such as *Griphognathus* (Miles 1977), *Dipnorhynchus* (Campbell and Barwick 1982b, 2000), *Speonesydrion* (Campbell and Barwick 1984) and *Holodipterus* (Pridmore *et al.* 1994). Dental caries has been reported in Paleozoic lungfish (Kemp 1991) and many other pathologies, such as abrasion, attrition, spur and step wear, malocclusion, hyperplasia and abscesses are seen in Cenozoic examples (Kemp 2005).

The shift in palaeoenvironments of fossil lungfish from essentially marine to exclusively freshwater through their history is well documented, though details are not without their controversies. The earliest forms such as *Diabolepis* (Chang 1995) and *Dipnorhynchus* (Campbell and Barwick 1999), as well as all the Gogo lungfishes, are found in marine limestones. Through the Devonian, lungfishes seem progressively to invade marginal marine, estuarine and even fully freshwater environments. For example, in the mid-Devonian, *Dipterus* occurs in the Orcadie basin, basically a lake environment, though with occasional marine incursions (Trewin 1986). The Givetian forms *Howidipterus* and *Barwickia* are amongst the earliest forms found in fully freshwater environments (an intermontane lake). The Late Devonian *Soederberghia* occurs in freshwater flood-plain pond environments in the Catskill Formation of Pennsylvania, East Greenland and Australia (Ahlberg *et al.* 2001; Cressler 2006), though Ahlberg *et al.* (2001) suggest some continuity of the flood-plain habitat with marginal marine environments. The type of environment represented at Miguasha has been controversial. Prichonnet *et al.* (1996) suggested that it varied from fully marine, through marginal marine to freshwater at different times, and included horizons that reflect periods of stagnant conditions, whereas Cloutier *et al.* (1996) interpreted it as brackish to marginal marine throughout.

By the Carboniferous, most lungfishes are thought to have been freshwater, inhabiting coal swamp environments. However, there is some controversy over whether all coal swamp facies represent freshwater environments or whether some may have been under periodic marine influence (Schultze 1995). Recent work by one of the authors (ELS) has shown that among British Carboniferous lungfishes, *Ctenodus* seems to have been marine-intolerant. This contrasts with *Sagenodus* which is found from deltaic to more fully marine environments, and may have been euryhaline (Schultze and Chorn 1997). By the Permian the majority of lungfish were freshwater dwellers, but the environment inhabited by *Gnathorhiza* is disputed. Schultze and Chorn (1997) claim that it was tolerant of marine conditions and found in nearshore deposits. Evidence of marine dwelling lungfishes diminishes steadily through the Mesozoic, and by the Cenozoic, lungfishes were exclusively freshwater animals.

During the latter part of the Late Devonian and the Carboniferous, lungfish remains are often found associated with deposits that yield tetrapods, suggesting that they inhabited similar environments. In fact the presence of lungfish elements has been used to provide potential clues to the presence of Paleozoic tetrapods (T.R. Smithson, S.P. Wood, pers. comm. 2007).

Interpretation of their environment impinges on another topic of lungfish biology, that of aestivation (see below). Specimens of *Gnathorhiza* have been found embedded in burrows, which has often been taken to imply that they were aestivating. Schultze and Chorn (1998) suggest on the other hand that they burrowed to avoid high tides in a marginal marine situation, however this seems an unlikely explanation for the Hancock specimens of *Tranodis*. An alternative explanation of the 'burrows' might be as breeding nests. Modern *Protopterus* is known to build tubular or blind-ending burrows in which males, or less usually females, guard the eggs (Greenwood 1987; Mlewa *et al.*, this volume). Some types of nests resemble, though are smaller than, dry season aestivation burrows in *P. aethiopicus* (Greenwood 1987). The burrows of *Tranodis* are orientated sub-vertically to the bedding planes (Garcia *et al.* 2006b) and are described as 'banana-shaped' (G. Storrs, pers. comm. 2008).

Another topic of great interest is the development of air-breathing within the lungfish lineage. The earliest lungfishes were marine, and did not show any of the specializations for air breathing that are found in post-Devonian forms. The assumption was that marine forms are unlikely to have required any air-breathing facility, and that deep water dwellers are unlikely to have exploited it. However, it has recently been shown that even Gogo underwent periodic anoxic events (Trinajstic *et al.* 2007), though the relationship between them and where the lungfishes were found has not been established. The fact that *Griphognathus* was not exclusively a deep-water marine form is telling. Some air-breathing capacity is generally accepted among many paleoichthyologists as well as physiologists as being an ancient trait of osteichthyans (Schultze and Chorn 1997; Perry *et al.* 2001), so that even the earliest lungfishes probably had some capability in that respect.

Air-breathing adaptations in lungfish have been studied by Campbell and Barwick (1988) and Long (1993) based on well-preserved cranial and postcranial elements of the Gogo and Mount Howitt (Frasnian) lungfishes respectively. Campbell and Barwick (1988) showed how the buccopharyngeal chamber increased in size by the forward movement of the quadrates relative to the back of the skull, accompanied by a corresponding increase in length of the posterior parasphenoid stalk. They also showed how the posterior margins of the tooth plates and the region of the parasphenoid between them allowed the tongue to form a buccal seal in modern forms, reflected in fossil ones with similar palatal morphologies. They pointed out modifications of the hyobranchial system and shoulder girdle that developed enhanced capacity for rotation in association with enlargement of the gape for air-gulping. These observations were augmented by Long (1993), who added a cranial rib to the above list of features: this structure helps to anchor the pectoral girdle during air-gulping (Bishop and Foxon 1968). He found no evidence of a cranial rib in Gogo lungfishes, but did so in *Howidipterus* and *Barwickia*, from which he inferred some degree of air-breathing to be present. *Dipterus* was once thought to have had a cranial rib (Ahlberg and Trewin 1995), though this is no longer considered to be the case. However, den Blaauwen *et al.* (2005) suggested that nevertheless it showed some of the features of the palate suggesting that it was capable of some degree of air breathing though it did not have the complete suite of features found in later forms. Many of these features, including a cranial rib, have recently been described in the Gogo lungfish *Rhinodipterus* (Clement and Long 2010), suggesting that even some marine forms were air breathing.

It should also be borne in mind that the same adaptations as are used in air-gulping in lungfishes are also employed in their suction feeding mechanisms. It would be hard to disentangle evolution of the adaptations for these functions from each other. Early lungfish may have developed them first in connection with suction feeding, only later to co-opt them into air-gulping, so it may not be valid to infer air-breathing in the earliest forms (Sharp 2007).

The association of many later Devonian lungfishes with tetrapods has also been taken to suggest that both groups may have lived in relatively anoxic waters in which air-breathing might have been an advantage. It has been noted that during this time, it was the long-snouted denticulate forms that were more closely associated with tetrapods. Some of these, such as *Soederberghia*, possessed cranial ribs and an elongate parasphenoid stalk (Ahlberg *et al.* 2001), suggesting that they were air-breathers.

The similarity of the long curved trunk ribs of Carboniferous lungfish to those of tetrapods might lead to speculation that they may have been involved in costal ventilation, as they are in amniotes and their stem group. However, experiments on lungfish breathing suggests that they gulp air using the orobranchial and parabranchial chambers (Thomson 1969). They do not use hypaxial muscles for expiration, though they do have three hypaxial muscle layers, in contast to two as found in actinopterygians and four as in tetrapods. Apparently, the additional layer is used in locomotion but not in breathing, at least in modern forms (Brainerd *et al.* 1993). Modern lungfish have very reduced ribs, along with general reduction of the ossified endoskeleton, which leaves open the possibility that the strongly ossified and cylindrical rib-cage of fossil forms was used differently from that in recent ones, though was most likely used in locomotion. By the Carboniferous, most of the air-breathing adaptations seen in modern lungfishes were in place, and the co-occurrence of tetrapods and lungfish is continued.

Supposed lungfish burrows noted in the Devonian of the Catskills have been suggested as early evidence of aestivation in lungfishes. However, the burrows were probably misidentified (Friedman and Daeschler 2006), and could be tree-stump infills, to judge from the published photographs. The earliest hard evidence for burrows are thus probably those of *Tranodis* from the Mississippian of Kentucky.

Neither Schultze and Chorn (1997) nor Cavin *et al.* (2007) were convinced that the burrows containing *Gnathorhiza* necessarily implied a capacity to aestivate. Part of the reason for this is that among modern lungfish, only *Protopterus* and *Lepidosiren* are certainly known to aestivate in totally dry conditions, and *Gnathorhiza* is remote phylogenetically from these taxa. Thus they argue that aestivation could be i) a character that evolved among Paleozoic lungfish and was lost by *Neoceratodus*, or ii) evolved in parallel within several lineages, or iii) was absent in *Gnathorhiza*. Cavin *et al.* (2007) did allow the possibility that *Gnathorhiza* could burrow to survive short dry spells but perhaps not whole seasons. The inferences are challenged by the occurrence of the *Tranodis* burrows, assuming they were associated with aestivation. Cavin *et al.* (2007) also admit that even a robust phylogeny may not satisfactorily answer this question. There are many aspects of lungfish evolution and biology that may be answerable with future finds, though there will probably be others that remain a mystery.

Acknowledgements

Our thanks go to Hans-Peter Schultze for reading and improving the manuscript and for providing information and references and a photograph, and to Anne Kemp for help with Mesozoic and Cenozoic references. ELS was funded by the BBSRC (Biotechnology and Biological Sciences Research Council, UK) for three years of her PhD project. JL thanks Marius Arsenault for permission to use photographs of *Fleurantia* and *Scaumenacia*.

References

Agassiz, L. (1838). Recherches sur les poissons fossiles, 1833-1844, Neuchatel, Imprimerie de Petitpierre et Prince.

Ahlberg, P.E. (1991). A re-examination of sarcopterygian interrelationships, with special reference to the Porolepiformes. Zoological Journal of the Linnean Society of London 103: 241-287.

Ahlberg, P.E. and Trewin, N. (1995). The postcranial skeleton of the Middle Devonian lungfish *Dipterus valenciennesi*. Transactions of the Royal Society of Edinburgh: Earth Sciences 85: 159-175.

Ahlberg, P.E. and Johanson, Z. (1998). Osteolepiforms and the ancestry of tetrapods. Nature 395: 792-794.

Ahlberg, P.E., Johanson, Z. and Daescher, E.B. (2001). The Late Devonian lungfish *Soederberghia* (Sarcopterygii, Dipnoi) from Australia and North America, and its biogeographical implications. Journal of Vertebrate Paleontology 21: 1-12.

Ahlberg, P.E., Smith, M.M. and Johanson, Z. (2006). Developmental plasticity and disparity in early dipnoan (lungfish) dentitions. Evolution and Development 8: 331-349.

Barkas, T.P. (1868). *Ctenodus interruptus*. Scientific Opinion 2: 113-114.

Barwick, R.E., Campbell, K.S.W. and Mark-Kurik, E. (1997). *Tarachomylax*: A new Early Devonian dipnoan from Severnaya Zemlya, and its place in the evolution of the Dipnoi. Geobios 30: 45-73.

Bemis, W.E. (1984). Paedomorphosis and the evolution of the Dipnoi. Paleobiology 10: 293-307.

Bemis, W.E. and Northcutt, R.G. (1992). Skin and blood vessels of the snout of the Australian lungfish *Neoceratodus forsteri*, and their significance for interpreting the cosmine of Devonian lungfishes. Acta Zoologica 73: 115-139.

Bendix-Almgreen, S.E. (1976). Palaeovertebrate faunas of Greenland. In: A. Escher and W.S. Watt (eds). Geology of Greenland. Gronlands Geologiske Undersøgelse. Copenhagen, Denmark. pp. 536-573.

Berman, D.S. (1976). Occurrence of *Gnathorhiza* (Osteichthyes: Dipnoi) in aestivation burrows in the Lower Permian of New Mexico with a description of a new species. Journal of Paleontology 50: 1034-1039.

Bishop, I.R. and Foxon, G.E.H. (1968). The mechanism of air-breathing in the South American lungfish *Lepidosiren paradoxa*. Journal of Zoology 154: 263-271.

Brainerd, E.L., Ditelberg, J.S. and Bramble, D.M. (1993). Lung ventilation in salamanders and the evolution of vertebrate air-breathing mechansims. Biological Journal of the Linnean Society 49: 16-183.

Campbell, K.S.W. and Barwick, R.E. (1982a). A new species of the lungfish *Dipnorhynchus* from New South Wales. Palaeontology 25: 509-527.

Campbell, K.S.W. and Barwick, R.E. (1982b). The neurocranium of the primitive dipnoan *Dipnorhynchus sussmilchi* (Etheridge). Journal of Vertebrate Paleontology 2: 286-327.

Campbell, K.S.W. and Bell, M.W. (1982). *Soederberghia* (Dipnoi) from the Late Devonian of New South Wales. Alcheringa 6: 143-149

Campbell, K.S.W. and Barwick, R.E. (1984). *Speonesydrion*, an Early Devonian dipnoan with primitive tooth plates. Palaeo Ichthyologica 2: 1-48.

Campbell, K.S.W. and Barwick, R.E. (1988). Geological and palaeontological information and phylogenetic hypotheses. Geological Magazine 125: 207-227.

Campbell, K.S.W. and Barwick, R.E. (1990). Paleozoic dipnoan phylogeny: Functional complexes and evolution without parsimony. Paleobiology 16: 143-169.

Campbell, K.S.W. and Barwick, R.E. (1995). The primitive dipnoan palate. Journal of Vertebrate Paleontology 15: 13-27.

Campbell, K.S.W. and Barwick, R.E. (1998). A new tooth-plated dipnoan from the Upper Devonian Gogo Formation and its relationships. Memoirs of the Queensland Museum 42: 403-437.

Campbell, K.S.W. and Barwick, R.E. (1999). Dipnoan fishes from the Late Devonian Gogo Formation of Western Australia. Records of the Australian Museum 57: 107-138.

Campbell, K.S.W. and Barwick, R.E. (2000). The braincase, mandible and dental structures of the Early Devonian lungfish *Dipnorhynchus kurikae* from Wee Jasper, New South Wales. Records of the Australian Museum 52: 103-128.

Campbell, K.S.W. and Barwick, R.E. (2001). *Diabolepis* and its relationship to the Dipnoi. Journal of Vertebrate Paleontology 21: 227-241.

Campbell, K.S.W. and Barwick, R.E. (2002). The axial postcranial structure of *Griphognathus whitei* from the Upper Devonian Gogo Formation of Western Australia: Comparisons with other dipnoans. Records of the Australian Museum 21: 167-201.

Campbell, K.S.W. and Smith, M.M. (1987). The Devonian dipnoan *Holodipterus*: dental form variation and remodelling growth mechanisms. Records of the Australian Museum 39: 131-167.

Campbell, K.S.W., Barwick, R.E., Chatterton, B.D.E. and Smithson, T.R. (2002). A new Middle Devonian dipnoan from Morocco: Structure and histology of the tooth plates. Records of the Australian Museum 21: 39-61.

Cavin, L., Suteethorn, V., Buffetaut, E. and Tong, H. (2007). A new Thai Mesozoic lungfish (Sarcopterygii, Dipnoi) with an insight into post-Palaeozoic dipnoan evolution. Zoological Journal of the Linnean Society 149: 141-177.

Chang, M.M. (1995). *Diabolepis* and its bearing on the relationships between porolepiforms and dipnoans. In: M. Arsenault, H. Lelièvre and P. Janvier (eds). Studies on Early Vertebrates. Miguasha, Quebec: Bulletin du Muséum National d'Histoire Naturelle, Paris, France. pp. 235-268.

Chang, M.-M. and Wang, J.-Q. (1995). A new Emsian dipnorhynchid (Dipnoi) from Guangnan, southeastern Yunnan, China. Geobios 19: 233-239.

Cheng, H. (1989). On the tubuli in Devonian lungfishes. Alcheringa 13: 153-166.

Clack, J.A. and Finney, S.M. (2005). *Pederpes finneyae*, an articulated tetrapod from the Tournaisian of western Scotland. Journal of Systematic Palaeontology 2: 311-346.

Clement, G. and Boisvert, C.A. (2006). Lohest's true and false 'Devonian amphibians': Evidence for the rhynchodipterid lungfish *Soederberghia* in the Famennian of Belgium. Journal of Vertebrate Paleontology 26: 276-283.

Clement, A.M. and Long, J.A. (2010). Air-breathing adaptation in a marine Devonian Lungfish. Biology Letters (Published online doi:10.1098/rsbl.2009.1033).

Cloutier, R. (1996). Dipnoi (Akinetia: Sarcopterygii). In: H.-P. Schultze and R. Cloutier (eds). Devonian Fishes and Plants of Miguasha., Quebec, Canada. Verlag Dr Friedrich Pfeil. Munich, Germany. pp. 198-226.

Cloutier, R. and Ahlberg, P.E. (1996). Morphology, characters and the interrelationships of basal sarcopterygians. In: M.L.J. Stiassny and L. Parenti (eds). Interrelationships of Fishes II. Academic Press, London, UK. pp. 445-479.

Cloutier, R., Loboziak, S., Candilier, A.M. and Blieck, A. (1996). Biostratigraphy of the Upper Devonian Escuminac Formation, eastern Québec, Canada: A comparative study based on miospores and fish. Review of Palaeobotany and Palynology 93: 191-215.

Cressler, W.L. (2006). Plant paleoecology of the Late Devonian Red Hill locality north-central Pennsylvania, and *Archaeopteris*-dominated wetland plant community and early tetrapod site. In: S.F. Greb and W.A. DiMichele (eds). Wetlands through Time. Boulder, CO, USA: Geological Society of America Special Paper, 79-102.

Den Blaauwen, J.L., Barwick, R.E. and Campbell, K.S.W. (2005). Structure and function of the tooth plates of the Devonian lungfish *Dipterus valenncienesi* from Caithness and the Orkney islands. Records of the Australian Museum 23: 91-113.

De Castelnau, F. (1876). Remarques au sujet du genre *Neoceratodus*. Journale de Zoologie 5: 342-343.

Denison, R.H. (1968). Early Devonian lungfishes from Wyoming. Fieldiana, Geology 17: 353-413.

Denison, R. H. (1974). The structure and evolution of teeth in lungfishes. Fieldiana Geology 33: 31-58.

Forster-Cooper, C. (1937). The Middle Devonian fish fauna of Achanarras. Transactions of the Royal Society of Edinburgh 59: 223-239.

Friedman, M. (2007a). Cranial structure in the Devonian lungfish *Soederberghia groenlandica* and its implications for the interrelationships of 'rhynchodipterids'. Earth and Environmental Science Transactions of the Royal Society of Edinburgh 98: 1-20.

Friedman, M. (2007b). The interrelationships of Devonian lungfishes (Sarcopterygii: Dipnoi) as inferred from neurocranial evidence and new data from the genus *Soederberghia* Lehman 1959. Zoological Journal of the Linnean Society 151: 115-171.

Friedman, M. and Daeschler, E.B. (2006). Late Devonian (Famennian) lungfishes from the Catskill Formation of Pennsylvania, USA. Palaeontology 49: 1167-1183.

Gans, C. and Northcutt, R.G. (1983). Neural crest and the origin of vertebrates: A new head. Science 120: 268-274.

Garcia, W.J., Storrs, G.W. and Greb, S.F. (2006a). The Hancock County tetrapod locality: a new Mississippian (Chesterian) wetlands fauna from western Kentucky (USA). In: S.F. Greb and W.A. DiMichele (eds). Wetlands through Time. Geological Society of America Special Papers 399: 155-167.

Garcia, W.J., Storrs, G.W. and Greb, S.F. (2006b). Lungfish burrows from the Mississippian (Chesterian) of north-western Kentucky. Journal of Vertebrate Paleontology 26: 65A.

Graham Smith W. and Westoll, T.S. (1937). On a new long-headed dipnoan fish from the Upper Devonian of Scaumenac Bay, P.Q. Canada. Transactions of the Royal Society of Edinburgh 59: 241-268.

Greenwood, P.H. (1987). The natural history of African lungfishes. In: W.E. Bemis, W.W. Burggren and N.E. Kemp (eds). The Biology and Evolution of Lungfishes. Journal of Morphology, supplment 1 for 1986. pp. 163-180

Günther, C.A.L.G. (1871). Description of *Ceratodus*, a genus of ganoid fishes, recently discovered in rivers of Queensland, Australia. Transactions of the Royal Society of London 161: 377-379.

Huxley, T.H. (1861). Preliminary essay upon the systematic arrangement of the fishes of the Devonian epoch. Memoirs of the Geological Survey of the UK. Figures and descriptions illustrative of British organic remains: Scientific Memoirs 2: 421-460.

Jarvik, E. (1980). Basic Structure and Evolution of Vertebrates, Volume 1. Academic Press, London, UK.

Jarvik, E. (1996) The Devonian tetrapod *Ichthyostega*. Fossils and Strata 40: 1-206.

Johanson, Z. and Ritchie, A. (2000). A new Famennian lungfish from New South Wales, Australia, and its bearing on Australian-Asian terranes. Alcheringa 24: 99-118.

Joss, J. and Johanson, Z. (2006). Is *Palaeospondylus gunni* a fossil larval lungfish? Insights from *Neoceratodus forsteri* development. Journal of Experimental Zoology (Mol. Dev. Evol.) 306B: 1-9.

Kemp, A. (1991). Palaeopathology and lungfish tooth plates. In: M.M. Chang, Y. Liu and G. Zhang (eds). Early Vertebrates and Related Problems of Evolutionary Biology. Science Press, Beijing, China. pp. 441-464.

Kemp, A. (1992). New neoceratodont cranial remains from the Late Oligocene-Middle Miocene of Northern Australia with comments on generic characters for Cenozoic lungfish. Journal of Vertebrate Paleontology 12: 284-293.

Kemp, A. (1993). *Ceratodus diutinus,* a new fossil ceratodont from Cretaceous and Tertiary Deposits in Australia. Journal of Paleontology 67: 883-886.

Kemp, A. (1994a). Possible pathology in the snout and lower jaw of a Chinese Devonian dipnoan, *Sorbitorhynchus deleaskitus* (Osteichthyes: Dipnoi). Journal of Vertebrate Paleontology 14: 453-458.

Kemp, A. (1994b). Australian Triassic lungfish skulls. Journal of Paleontology 68: 647-654.

Kemp, A. (1995). Marginal tooth bearing bones in the lower jaw of the Recent Australian lungfish, *Neoceratodus forsteri* (Osteichthyes: Dipnoi). Journal of Morphology 225: 345-355.

Kemp, A. (1997). A revision of Australian Mesozoic and Cenozoic lungfish of the family Neoceratodontidae (Osteichthyes: Dipnoi) with a description of four new species. Journal of Paleontology 71: 713-733.

Kemp, A. (1998a). Skull structure in post-Paleozoic lungfish. Journal of Vertebrate Paleontology 18: 43-63.

Kemp, A. (1998b). On the generic status of *Palaeophichthys parvulus* Eastman 1908 and *Monongahela stenodonta* Lund 1970 (Osteichthyes: Dipnoi). Annals of the Carnegie Museum 67: 225-243.

Kemp, A. (1999). Ontogeny of the skull of the Australian lungfish, *Neoceratodus forsteri* (Osteichthyes: Dipnoi). Journal of Zoology 248: 97-137.

Kemp, A. (2002a). Growth and hard tissue remodeling in the dentition of the Australian lungfish, *Neoceratodus forsteri* (Osteichthyes: Dipnoi). Journal of Zoology 257: 219-235.

Kemp, A. (2002b). The marginal dentition of the Australian lungfish, *Neoceratodus forsteri* (Osteichthyes: Dipnoi). Journal of Zoology 257: 325-331.

Kemp, A. (2005). New insights into ancient environments using dental characters in Australian Cenozoic lungfish. Alcheringa 29: 123-149.

Kemp, A. and Molnar, R.E. (1981). *Neoceratodus forsteri* from the Lower Cretaceous of New South Wales, Australia. Journal of Paleontology 55: 211-217.

Krupina, N.I. and Reisz, R.R. (1999). Reconstruction of dentition in hatchlings of *Andreyevichthys epitomus,* a Late Famennian dipnoan from Russia. Modern Biology 24: 99-108.

Krupina, N.I., Reisz, R.R. and Scott, D. (2001). The skull and tooth system of *Orlovichthys limnatis,* a Late Devonian dipnoan from Russia. Canadian Journal of Earth Sciences 38: 1301-1311.

Long, J.A. (1987). A redescription of the lungfish *Eoctenodus* Hills 1929, with reassessment of other Australian records of the genus *Dipterus* Sedgwick & Murchison 1828. Records of the Western Australian Museum 13: 297-314.

Long, J.A. (1990). Heterochrony and the origin of tetrapods. Lethaia 23: 157-166.

Long, J.A. (1992). *Gogodipterus paddyensis* (Miles) gen. nov., a new chirodipterid lungfish from the Late Devonian Gogo Formation, Western Australia. The Beagle: Records of the Northern Territories Museum of Arts and Sciences 9: 11-20.

Long, J.A. (1993). Cranial ribs in Devonian lungfish and the origin of dipnoan air-breathing. Memoirs of the Association of Australasian Palaeontologists 15: 199-209.

Long, J.A. (1995). The Rise of Fishes. University of New South Wales Press, Sydney, Australia.

Long, J.A. (2005). Interrelationships of the holodontid lungfishes (Osteichthyes: Dipnomorpha) based on new material from the Upper Devonian Gogo Formation of Western Australia. Journal of Vertebrate Paleontology 25: 84A.

Long, J.A. (In press). Holodontid lungfishes from the Late Devonian Gogo Formation, Western Australia. In: X.Yu, J. Maisey and D.Miao (eds). Fossil Fishes and Related Biota: Morphology, Phylogeny and Palaeobiogeography–in Honour of Mee-Man Chang. Verlag Pfeil. Berlin, Germany. xx-xx

Long, J.A. and Campbell, K.S.W. (1985). A new lungfish from the Lower Carboniferous of Victoria, Australia. Proceedings of the Royal Society of Australia 97: 87-93.

Long, J.A., Campbell, K.S.W. and Barwick, R.E. (1994). A new dipnoan genus, *Ichnomylax*, from the Lower Devonian of Victoria, Australia. Journal of Vertebrate Paleontology 14: 127-131.

Long, J.A. and Clement, A.M. (2009). The postcranial anatomy of two Middle Devonian Lungfishes (Osteichthyes, Dipnoi) from Mt. Howitt, Victoria, Australia. Memoirs of the Museum of Victoria 66: 189-202.

Marshall, C.R. (1986). A list of fossil and extant dipnoans. Journal of Morphology, Supplement 1: 15-23.

Marshall, C.R. (1988). A large, well-preserved specimen of the Middle Pennsylvanian lungfish *Conchopoma edesi* (Osteichthyes: Dipnoi) from Mazon Creek, Illinois, USA. Journal of Vertebrate Paleontology 8: 383-394.

Miles, R.S. (1977). Dipnoan (lungfish) skulls and the relationships of the group: A study based on new species from the Devonian of Australia. Zoological Journal of the Linnean Society 61: 1-328.

Miller, A.E. (1930). Notes on the tail skeleton of *Lepidosiren paradoxa*, with remarks on the affinities of *Palaeospondylus*. Proceedings of the Zoological Society of London 1930: 782-789.

Mlewa, C.M., Green, J.M. and Dunbrack, R.L. (this book). The general natural history of the African lungfishes. Chapter 4. pp. 97-127.

Newman, M.J. and den Blaauwen, J.L. (2008). New information on the enigmatic Devonian vertebrate *Palaeospondylus gunni*. Scottish Journal of Geology 44: 89-91.

Northcutt, R.G. and Gans, C. (1983). The genesis of neural crest and epidermal placodes: A reintrepretation of vertebrate origins. Quarterly Review of Biology 58: 1-28.

Olson, E.C. and Daly, E. (1972). Notes on *Gnathorhiza* (Osteichthyes, Dipnoi). Journal of Paleontology 46: 371-376.

Owen, R. (1867). On the dental characters of genera and species, chiefly of the fishes from the Lower Main Seam and shales of coal, Northumberland. Transactions of the Ondontological Society of Great Britain 5: 323-375.

Perry, S., Wilson, R.J.A., Straus, C., Harris, M.B. and Remmers, J.E. (2001). Which came first, the lung or the breath? Comparative Biochemistry and Physiology. A 129: 37-47.

Prichonnet, G., Di Vergilio, M. and Chidiac, Y. (1996). Stratigraphical, sedimentological and paleontological context of the Escuminac Formation: Paleoenvironmental hypotheses.

In: H.-P. Schultze and R. Cloutier (eds). Devonian Fishes and Plants of Miguasha, Quebec, Canada. Verlag Dr Friedrich Pfeil. Munich, Germany. pp. 23-36.

Pridmore, P.A., Campbell, K.S.W. and Barwick, R.E. (1994). Morphology and phylogenetic position of the holodipteran dipnoans of the Upper Devonian Gogo Formation of northwestern Australia. Philosophical Transactions of the Royal Society of London. B 344: 105-164.

Reisz, R.R. and Smith, M.M. (2001). Lungfish dental pattern conserved for 360 myr. Nature 411: 548.

Reisz, R.R., Krupina, N.I. and Smith, M.M. (2004). Dental histology in *Ichnomylax karatajae* sp. nov., an Early Devonian dipnoan from the Taymire Penninsula, Siberia, with a discussion on petrodentine. Journal of Vertebrate Paleontology 24: 18-25.

Ritchie, A. (1981). First complete specimen of the dipnoan *Gosfordia truncata* Woodward from the Triassic of New South Wales. Records of the Australian Museum 33: 606-616.

Schultze, H.-P. (1975). Die lungenfisch-gattung *Conchopoma* (Pisces: Dipnoi). Senckenbergiana lethaea 56: 191-231.

Schultze, H-P (1981). Das Schadeldach eines ceratodontiden Lungenfisches aus der Trias Suddeutschlands (Dipnoi, Pisces). Stuttgarter Beiträge zur Naturkunde. Serie B 70:1-31.

Schultze, H.-P. (1992a). A new long-headed dipnoan (Osteichthyes) from the Middle Devonian of Iowa, USA. Journal of Vertebrate Paleontology 12: 42-58.

Schultze, H-P (1992b). Fossilium catalogus. Pars 131 Dipnoi. Kugler Publications, Amsterdam, The Netherlands.

Schultze, H.-P. (1994). *Palaeophichthys parvulus* Eastman, 1908, a gnathorhizid dipnoan from the Middle Pennsylvanian of Illinois, USA. Annals of the Carnegie Museum 63: 105-113.

Schultze, H.-P. (1995). Terrestrial biota in coastal marine deposits: Fossil-lagerstätten in the Pennsylvanian of Kansas, USA. Palaeogeography, Palaeoclimatology, Palaeoecology. 119: 255-273.

Schultze, H.-P. (2001). *Melanognathus*, a primitive dipnoan from the Lower Devonian of the Canadian arctic and the interrelationships of Devonian dipnoans. Journal of Vertebrate Paleontology 21: 781-794.

Schultze, H.-P. (2004). Mesozoic Sarcopterygians. In: G. Arratia and A. Tintori (eds). Mesozoic Fishes 3 – Systematics, Paleoenvironments and Biodiversity. Verlag Dr Friedrich Pfeil. Munich, Germany. pp. 463-492.

Schultze, H.-P. and Marshall, C. R. (1993). Contrasting the use of functional complexes and isolated characters in lungfish evolution. Memoirs of the Association of Australasian Palaeontologists 15: 211-224.

Schultze, H.P. Bolt, J.R. (1996). The lungfish *Tranodis* and the tetrapod fauna from the Upper Mississippian of North America. In: A.R. Milner (ed). Studies on Carboniferous and Permian Vertebrates. Palaeontological Association Special Papers in Palaeontology 52: 31-54.

Schultze, H.-P. and Chorn, J. (1997). The Permo-Carboniferous genus *Sagenodus* and the beginning of modern lungfish. Contributions to Zoology 67: 9-70.

Schultze, H.-P. and Chorn, J. (1998). Sarcopterygian and other fishes from the marine Upper Devonian of Colorado, USA. Mitteilungen aus dem Museum fur Naturkunde in Berlin. Geowissenschaftliche Reihe 1: 53-72.

Sharp, E. (2007). The systematics, taxonomy and phylogeny of the British Carboniferous lungfishes. PhD thesis, University of Cambridge (Unpublished).

Smith, M.M. and Chang, M.-M. (1990). The dentition of *Diabolepis sperratus* Chang and Yu, with further consideration of its relationships and the primitive dipnoan dentition. Journal of Vertebrate Paleontology 10: 833-834.

Smith, M.M. and Krupina, N.I. (2001). Conserved developmental processes constrain evolution of lungfish dentitions. Journal of Anatomy 199: 161-168.

Thomson, K.S. (1965). On the relationships of certain Carboniferous Dipnoi; with descriptions of four new forms. Proceedings of the Royal Society of Edinburgh, Section B (Biology) 69: 221-245.

Thomson, K.S. (1969). The biology of the lobe-finned fishes. Biological Reviews 44: 91-154.

Thomson, K.S. (1972). An attempt to reconstruct evolutionary changes in the cellular DNA content of lungfish. Journal of Experimental Zoology 180: 363-372.

Thomson, K.S. (1975). On the biology of cosmine. Bulletin of the Peabody Museum of Natural History 40: 1-59.

Thomson, K.S., Sutton, M. and Thomas, B. (2003). A larval Devonian lungfish. Nature 426: 833-834.

Trewin, N. (1986). Palaeoecology and sedimentology of the Achanarras fish bed of the Middle Old Red Sandstone, Scotland. Transactions of the Royal Society of Edinburgh: Earth Sciences 77: 21-46.

Trinajstic, K., Marshall, C., Long, J.A. and Bifield, K. (2007). Exceptional preservation of nerve and muscle tissues in a Late Devonian placoderm fish and their evolutionary implications. Biology Letters 3: 1-4.

Wang, S., Drapala, V., Barwick, R.E. and Campbell, K.S.W. (1993). The dipnoan species, *Sorbitorhynchus deleaskitus*, from the Lower Devonian of Guangxi, China. Philosophical Transactions of the Royal Society of London. B 340: 1-24.

Ward, J. (1890). The geological features of the north Staffordshire coal fields, their organic remains, their range and distribution: With a catalogue of the fossils of the Carboniferous system of north Staffordshire. Transactions of the North Staffordshire Institute of Mining and Mechanical Engineering 10: 189.

Watson, D.M.S. and Gill, E.L. (1923). The structure of certain Palaeozoic Dipnoi. Journal of the Linnean Society of London, Zoology 35: 163-216.

Westoll, T.S. (1936). On the structure of the dermal ethmoid shield of *Osteolepis*. Geological Magazine 73: 157-171.

Westoll, T.S. (1949). On the evolution of the Dipnoi. In: G.L. Jepsen, G.G. Simpson, and E. Mayr (eds). Genetic, Paleontology and Evolution. Princeton University Press, Princeton, USA. pp. 121-184.

Young, G.C. and Moody, J.M. (2002). A Middle-Late Devonian fish fauna from the Sierra de Perijá, western Venezuela, South America. Mitteilungen aus dem Museum für Naturkunde in Berlin, Geowissenschaftliche Reihe 5: 153-204.

Zardoya, R. and Meyer, A. (2001). Vertebrate phylogeny: Limits of infererence of mitochondrial genome and nuclear rDNA sequence data due to an adverse phylogenetic signal/noise ratio. In: P.E. Ahlberg (ed). Major Events in Early Vertebrate Evolution, Systematics Association Symposium Volume. Taylor and Francis, London, UK. pp. 135-155.

Zhu, M. and Schultze, H.P. (2001). Interrelationships of basal osteichthyans. In: P.E. Ahlberg (ed). Major Events in Early Vertebrate Evolution. Systematics Association Symposium Volume. Taylor and Francis, London, UK. pp. 289-314.

Zhu, M. and Yu, X. (2002). A primitive fish close to the common ancestor of tetrapods and lungfish. Nature 418: 767-770.

Zhu, M., Yu, X. and Janvier, P. (1999). A primitive fossil fish sheds light on the origin of bony fishes. Nature 397: 607-610.

Zylberberg, L. (1988). Ultrastructural data on the scales of the dipnoan *Protopterus annectens* (Sarcopterygii, Osteichthyes). Journal of Zoology 216: 55-71.

CHAPTER 2

Phylogeny of Lungfishes

Zerina Johanson[1,*] and Per E. Ahlberg[2]

[1]Department of Palaeontology, Natural History Museum, Cromwell Road, London, UK, SW7 5BD

[2]Subdepartment of Evolutionary Organismal Biology, Department of Physiology and Developmental Biology, Evolutionary Biology Centre, Uppsala University, Norbyvägen 18A, 752 36 Uppsala, Sweden

ABSTRACT

A great deal has happened in the field of lungfish phylogenetics over the two decades that have passed since the publication of *The Biology and Evolution of Lungfishes*. Three major contributory factors can be discerned: the application of cladistic methodology to ever larger data sets, the advent of molecular phylogenetics, and the conceptual impact of the node-based terminology of crown group, stem group and total group introduced by Jefferies. In this chapter we will review the effect of these factors on the development of the subject and present a brief overview of current opinion. We will not, however, be able to outline a neat set of consensus views; despite, or perhaps because of, the richness of the fossil data set, the deep phylogeny of the lungfishes is still the subject of much debate.

Keywords: lungfishes, Dipnoi, phylogeny, evolution, Sarcopterygii

INTRODUCTION

The Biology and Evolution of Lungfishes, published in 1987, provided the first computer-generated phylogenetic analysis of the lungfishes, presented by Charles Marshall (1987a). Later, Schultze and Marshall (1993) provided a more comprehensive analysis, including taxa from the Devonian to Recent, with 90 characters coded from the skull, dentition and postcranium. However, even before this, an attempt to apply phylogenetic principles to the question of

**Corresponding author:* E-mail: z.johanson@nhm.ac.uk

lungfish relationships had been made by Roger Miles. Miles (1977) presented a revised phylogeny of the Dipnoi, which differed from earlier phylogenies in that characters were clearly assessed as being primitive or derived, and principles of parsimony used to evaluate relationships within the group (i.e., with a minimum number of character changes allowed). Although Miles's resulting cladogram (Miles 1977: Fig. 157) left many taxa unresolved within the larger phylogeny, it was, as he noted, an advance over previous phylogenies because his hypotheses of character evolution and taxon relationships were explicit and could be tested by future researchers. In the years since Miles (1977) and Schultze and Marshall (1993), several new analyses have been performed, most involving computer-based parsimony programs. Interestingly, there has been disagreement among these trees, particularly with regard to the resolution of Devonian taxa. Nevertheless, there appears to be broad agreement regarding the identity of basal taxa as well as relationships among Carboniferous and younger taxa, but less agreement on the resolution of taxa between these nodes. Other researchers have examined relationships of the Dipnoi within the Sarcopterygii, most recently using EST sequences (express sequence tags; Hallström and Janke 2009). There seems to be broad agreement now that lungfish represent the extant sister taxon to the Tetrapoda. These agreements and disagreements are reviewed below, with a final word on potential directions for future research.

LARGE-SCALE PHYLOGENETIC STRUCTURE

In the broadest terms, the current state of the phylogenetic debate can be summed up by saying that the monophyletic status of the group Dipnoi is undisputed, but that its precise inclusiveness continues to be the subject of debate. Like any other clade, the Dipnoi can be viewed as a total group (all living and fossil forms that are more closely related to the living Dipnoi than to any other living or total group, e.g., the Tetrapoda; Jefferies 1979) that can be divided into a crown group (the living genera and their immediate relatives), and a stem group (all fossil members of the total group that fall outside the crown group). The debate about inclusiveness centres on the lowermost part of the stem group.

Because the lungfish crown group comprises only six living taxa (*Protopterus aethiopicus, P. amphibius, P. annectens, P. dolloi, Lepidosiren paradoxa* and *Neoceratodus forsteri*), only a limited number of phylogenetic problems can be addressed with molecular data. A recent analysis of the 16S rRNA gene of the mitochondrial genome (Tokita and Hikida 2005) has convincingly demonstrated the monophyly of *Protopterus* and its sister-group relationship to *Lepidosiren*. This is consonant with earlier molecular and morphological analyses showing *Neoceratodus* to be the sister-group to *Protopterus* + *Lepidosiren* (Miles 1977; Marshall 1987a; Hedges *et al.* 1993; Brinkmann *et al.* 2004). The general structure of the dipnoan crown group can thus be regarded as fully resolved; we return below to the presence and position of fossil forms within this crown group, and the dating of the nodes.

The sister-group relationship of the Dipnoi as a whole has been the subject of much debate, even though it has been clear for some decades (Miles 1977) that the Dipnoi form part of the Sarcopterygii and thus must be the sister group of the Actinistia (coelacanths), or the Tetrapoda (land vertebrates), or the Actinistia + Tetrapoda. In a phylogenetic analysis in *The Biology and Evolution of Lungfishes*, Schultze (1987) placed lungfishes as the sister group of coelacanths + tetrapods on the basis of skeletal characters. The same year, Panchen and Smithson (1987) used a combination of skeletal and soft-tissue characters to place lungfishes as the sister group of tetrapods. Ahlberg (1991) argued that Schultze (1987) had pushed the Dipnoi into an artificially basal phylogenetic position partly by scoring as absences certain characters that would be better treated as indeterminable. Ahlberg's (1991) analysis placed lungfishes and tetrapods as sister groups, and introduced the terms Dipnomorpha and Tetrapodomorpha for the lungfish and tetrapod total groups. The majority of subsequent morphological phylogenetic analyses have recovered a lungfish-tetrapod sister-group relationship (Cloutier and Ahlberg 1996; Zhu *et al.* 1999, 2001; Zhu and Yu 2002; Friedman 2007a; Friedman *et al.* 2007) although a few have placed lungfishes as the sister group of coelacanths + tetrapods (Zhu and Schultze 2001). A lungfish-tetrapod sister-group relationship is also weakly supported by analyses of mitochondrial DNA (Zardoya *et al.* 1998) and more robustly by nuclear DNA (Brinkmann *et al.* 2004). Most recently, expressed sequence tags, or ESTs, were used in an analysis of gnathostome relationships, resolving a lungfish-tetrapod sister group (Hallström and Janke, 2009). Together, these strands of evidence strongly suggest that the Dipnomorpha and Tetrapodomorpha are sister groups.

THE INCLUSIVENESS OF THE STEM GROUP AND THE ORIGIN OF LUNGFISH MORPHOLOGY

Numerous fossil sarcopterygians from the Middle and Late Palaeozoic have been identified unambiguously as "early lungfishes", or, in modern terminology, members of the dipnoan stem group (Marshall 1987b). These are characterized inter alia by autostyly, absence of an intracranial joint, absence of a maxilla, and a palatal bite between pterygoids and prearticulars where the dentition consists either of radially patterned tooth plates, dentine plates, or denticle fields (Campbell and Barwick 1987; Ahlberg *et al.* 2006). The earliest of these "early lungfishes", such as *Dipnorhynchus*, *Speonesydrion*, *Ichnomylax* and *Uranolophus*, are Early Devonian in age, dating from the late Pragian to Emsian (Clément *et al.* 2006).

In addition to the unambiguous stem lungfishes, most recent analyses include in the basal part of the dipnoan stem group a number of Devonian fossils that are less obviously lungfish-like: *Diabolepis*, *Youngolepis*, *Powichthys* and the porolepiforms. Of these, *Diabolepis* approaches closest to conventional lungfish morphology and is invariably placed in the most crownward position (Maisey 1986; Ahlberg 1991; Cloutier and Ahlberg 1996; Zhu *et al.* 1999, 2001; Zhu and Yu 2002; Friedman 2007a; Friedman *et al.* 2007). However, the inclusion of *Diabolepis* (and by implication *Youngolepis*, *Powichthys* and porolepiforms) in the dipnoan stem has been vigorously disputed by Campbell and Barwick in a series of papers (e.g., Campbell and Barwick 1987, 1990, 2001).

Diabolepis, which comes from the Early Devonian (Lochkovian) of Yunnan, China, is known from skull roofs with partial braincases, pterygoids with or without attached palatoquadrates, and lower jaws. It retains a number of primitive sarcopterygian or osteichthyan characters that are absent in more derived stem lungfishes: well-defined premaxillae are present, the palatoquadrate is not fused to the braincase, and the parasphenoid separates the pterygoids in the midline (Chang and Yu 1984; Chang 1995). Alongside these are found typical lungfish characters such as a B bone in the skull roof, radially arranged tooth rows on the pterygoids and prearticulars, and a short dentary with a lip fold. Campbell and Barwick (2001) critically reviewed the lungfish characters of *Diabolepis* in an attempt to show that it is not related to lungfishes. However, their analysis rests largely on character-by-character special pleading and we do not feel that it presents a compelling case. In agreement with current majority opinion we regard the inclusion of *Diabolepis* in the dipnoan stem group as well supported.

The situation is somewhat different with regard to porolepiforms, *Powichthys* and *Youngolepis*. Porolepiforms are an abundant and widespread group of Devonian lobe-finned fishes, ranging in age from Pragian to end-Famennian (latest Devonian) and known both from disarticulated material and complete bodies. In most regards they are not particularly lungfish-like; for example, they possess a maxilla, have a generalized sarcopterygian dentition of marginal teeth and fang pairs, and have a braincase divided by an intracranial joint. However, their pectoral fins are long and leaf-shaped, strikingly similar to those of many fossil lungfishes (e.g., *Dipterus*, Fig. 1) as well as the living *Neoceratodus*. The pelvic fins by contrast are short, asymmetrical and rounded. Ahlberg (1989) showed that the pectoral fin of the porolepiform *Glyptolepis* contains an archipterygial endoskeleton similar to that of *Neoceratodus*. Of particular interest is the fact that, as in all living lungfishes but in contrast to other sarcopterygians, there is no separate radius; instead, the humerus articulates distally with a single axial element, which in living lungfishes is formed from the fusion of radius and ulna (Johanson *et al.* 2007). Outgroup comparison with actinopterygians and chondrichthyans suggests that this pattern is derived relative to the shorter, more asymmetrical fin skeletons with separate radius and ulna that are found in other sarcopterygians (Ahlberg 1989; Friedman *et al.* 2007). The pelvic fin endoskeleton of *Glyptolepis* is shorter, asymmetrical and as far as we can tell comparable to the primitive sarcopterygian condition (Ahlberg 1989). The pectoral fin structure thus supports assignment of the porolepiforms to

Fig. 1 *Dipterus valenciennesi*, NHM P. 22187, showing leaf-like pectoral fin, also seen in living lungfish such as *Neoceratodus*, and stem dipnoans such as the Porolepiformes. Scale bars = 1 cm.

the dipnoan stem group whereas the pelvic fin is phylogenetically uninformative. The axial skeleton of porolepiforms (Ahlberg 1991) also resembles that of lungfishes, with a well-developed series of supraneural spines (Jarvik 1980: Fig. 339; Ahlberg and Trewin 1995; Campbell and Barwick 2002) that again appears to be a derived character within the Sarcopterygii (Ahlberg 1991).

Porolepiforms thus share a number of postcranial synapomorphies with undisputed stem and crown lungfishes. Unfortunately there is a complete absence of axial or appendicular skeletal material (apart from some shoulder girdles) for *Powichthys*, *Youngolepis*, *Diabolepis*, and most of the earliest and most primitive among the undisputed stem lungfishes. The Pragian genus *Uranolophus* is known from partly preserved postcrania, but the paired fins are unknown and the only information on the axial skeleton is provided by a short series of neural arches (Campbell and Barwick 1988). Only with the rather later and more derived *Dipterus* from the Middle Devonian (Eifelian-Givetian) do we get a reasonably detailed picture of the postcranial endoskeleton (Ahlberg and Trewin 1995). Like porolepiforms and *Neoceratodus*, *Dipterus* has a long archipterygial pectoral fin skeleton (Fig. 1) and well-developed supraneural spines. It also shows a porolepiform-like median fin architecture with a rather large posterior dorsal fin support carrying a branching array of radials posteriorly (Ahlberg and Trewin 1995). Comparable fin skeletons are found in the Late Devonian (Frasnian) genus *Griphognathus* (Campbell and Barwick 2002). The derived characters that appear to unite porolepiforms and lungfishes can thus be documented quite far down in the lungfish stem group, but not yet all the way to the bottom; this suggests that their status as synapomorphies should be treated with a small measure of caution, even though no conflicting data have been discovered.

Powichthys and *Youngolepis*, which are both of Lochkovian age, are known largely from braincases and lower jaws resembling those of porolepiforms (Jessen

1980; Chang 1982, 1991, 2004). A degree of caution is appropriate here as well, because some of the characters uniting these taxa are also found in the stem-group sarcopterygians *Psarolepis* and *Achoania* and are thus not dipnomorph synapomorphies (Yu 1988; Zhu *et al.* 2001). Like undisputed lungfishes, but unlike porolepiforms and non-dipnoan sarcopterygians, *Powichthys* and *Youngolepis* lack a mobile intracranial joint although traces of it questionably remain in the braincase in *Powichthys* (Chang 2004; Friedman 2007b). This causes them to occupy a higher position than the porolepiforms in the dipnoan stem group (Ahlberg 1991, Cloutier and Ahlberg 1996; Zhu *et al.* 1999, 2001; Zhu and Yu 2002; Chang, 2004; Friedman 2007a; Friedman *et al.* 2007).

If we accept the current consensus phylogeny as correct, this implies that the most primitive members of the lungfish stem group were generalized predatory lobe-finned fishes distinguished only by their lungfish-like pectoral fins. Later steps in the evolution of the stem group saw the consolidation of the braincase and later palatoquadrates into a single akinetic unit, the loss of the maxilla, and the switch from a marginal piercing bite to a palatal crushing bite. In *Diabolepis* the dermal skull bones still lie flat on the posterior part of the braincase, but in more derived stem lungfishes these bones are raised up on longitudinal cristae, creating space for the posterodorsal expansion of the jaw adductor muscles between the skull roof and braincase. All these cranial changes can be interpreted as coordinated aspects of a shift from soft-prey to durophagous feeding. However, very quickly some stem lungfish lineages developed more delicate jaws and denticulated dentitions, probably associated with a further feeding shift to soft-bodied prey such as worms. These forms are discussed in further detail below. Environmentally, *Powichthys*, *Youngolepis* and *Diabolepis* all derive from shallow or marginal marine environments (Jessen 1980; Chang 1982), while porolepiforms are known from both marginal marine and non-marine deposits. A marginal marine origin for the group thus seems likely. This contrasts with the fully marine environmental associations of the somewhat more derived "early lungfishes" of the Early Devonian, such as *Dipnorhynchus* and *Speonesydrion*, which typically occur together with stenohaline marine invertebrate groups like corals and articulate brachiopods.

PHYLOGENY OF DEVONIAN LUNGFISH: MAXIMAL DIVERSITY AT THE ORIGIN OF THE CLADE

The earliest undisputed dipnoans are early Devonian (Pragian-Emsian) in age and were numerous and geographically widespread by this time (Clack *et al.*, this volume). These include taxa such as *Uranolophus wyomingensis* (USA, Denison 1968; Campbell and Barwick 1988), *Ichnomylax kurnai* (Australia, Long *et al.* 1994), various species of *Dipnorhynchus* (Australia, Thomson and Campbell 1971; Campbell and Barwick 1982, 1999), *Sorbitorhynchus deleaskitus* (China, Wang *et al.*

1993), *Speonesydrion iani* (Australia, Campbell and Barwick 1984), *Tarachomylax opeki* (Russia, Barwick *et al.* 1997), *Jessenia concentrica* (Germany, Otto and Bardenheuer 1996), *Melanognathus canadensis* (Canadian Arctic, Jarvik 1967; Schultze 2001) and *Erikia jarviki* (China, Chang and Wang 1995). As summarised by Clément *et al.* (2006), indeterminate dipnoan remains have also been described from the Pragian of Vietnam and Idaho, and Lochkovian to Pragian of Russia. As noted above, *Diabolepis* is Lochkovian (earliest Devonian) in age, linking the origin of the dipnoan clade to the beginning of the Devonian.

These taxa also show a substantial amount of morphological diversity, particularly with respect to dentitions (Fig. 2). Dental morphology has formed the basis for

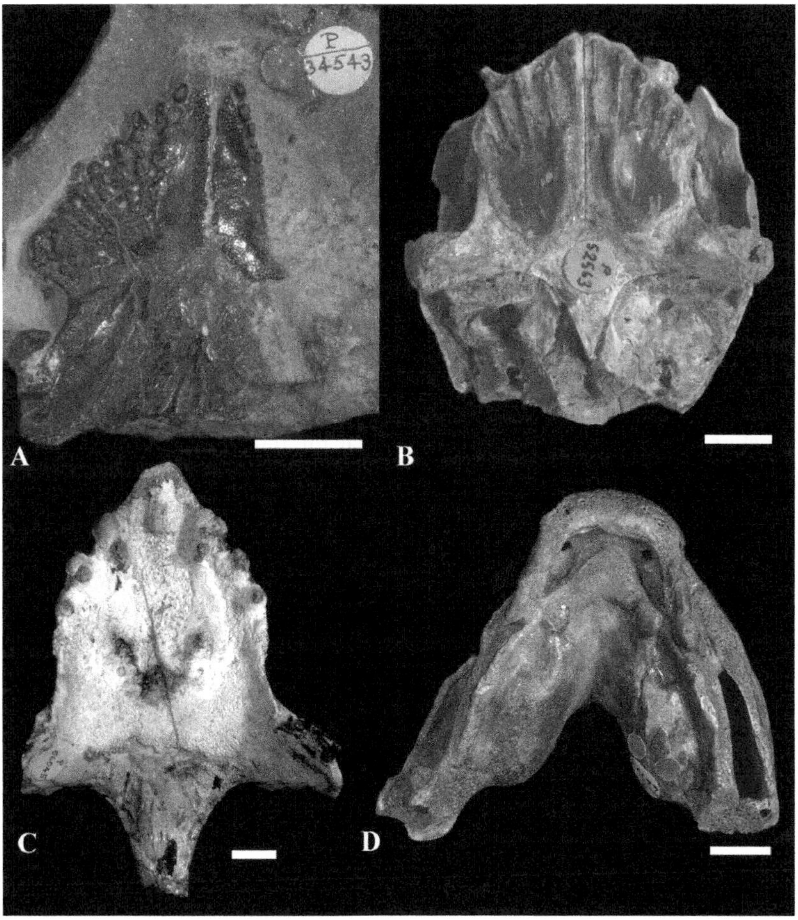

Fig. 2 Lungfish dentitions. A. *Dipterus valenciennesi*, NHM P. 34543 (toothplated dentition); B. *Chirodipterus australis* NHM P. 52563; C. *Holodipterus* sp. NHM P.56045; D. *Dipnorhynchus. sussmilchi* NHM P. 46773. Scale bars = 1 cm.

lungfish phylogenies produced by Campbell and Barwick (e.g., 1983, 1990), who continue to give precedence to functional feeding scenarios, dividing lungfish into tooth-plated, dentine-plated and denticulated lineages. As noted above, Marshall (1986) and Schultze and Marshall (1993) were among the first to apply more modern methods of phylogenetic analysis, involving the production of data sets and cladistic analysis of characters, to the Dipnoi. Marshall (1987a) incorporated characters from Miles's (1977) study of Devonian lungfish skulls into a data set that was then analysed with parsimony methods. Schultze and Marshall (1993) expanded on this database considerably, including Palaeozoic, Mesozoic and Cenozoic taxa. They found that *Dipnorhynchus*, *Speonesydrion* and *Uranolophus* were resolved as basal taxa. They also noted that while Mesozoic to recent taxa were phylogenetically stable, Palaeozoic taxa were not. Notably, both papers rejected the parsimony approach as 'not being a good criterion for choosing between the possible phylogenies for dipnoans' (Marshall 1987a: 151).

It is difficult to compare the cladograms figured in Schultze and Marshall (1993: Fig. 7), as they only present one of their 36 equal length trees, rather than a consensus of these trees. Nevertheless, there are broad patterns of similarity to more recent phylogenetic analyses (Schultze 2001; Ahlberg *et al.* 2006; Friedman 2007b). These recent analyses focused on Palaeozoic taxa, attempting to resolve the instability recognised by Schultze and Marshall (1993). Schultze (2001) primarily used the data set of Schultze and Marshall (1993), while Friedman (2007b) focused on characters of the neurocranium. Ahlberg *et al.* (2006) took a new approach to interpreting characters of the dentition, focusing on both pattern and process, stressing the independence and interplay of processes building dentitions and resorbing these dentitions (e.g., *Holodipterus*, Fig. 2C). By modifying these parameters, Ahlberg *et al.* (2006) showed that the morphologies of virtually all lungfish dentitions could be accounted for.

Broad phylogenetic patterns observed between recent analyses not only indicate regions of stability, but also highlight groups requiring further study. As noted, Schultze and Marshall's (1993) analysis resolved *Dipnorhynchus*, *Speonesydrion* and *Uranolophus* to the first basal nodes of the lungfish clade. Schultze (2001) recovered a similar result, but also resolved *Melanognathus* and *Jessenia* among the basal nodes (Fig. 3A). Ahlberg *et al.* (2006) resolved *Tarachomylax* and *Melanognathus* to the basalmost nodes. *Speonesydrion*, *Uranolophus*, *Dipnorhynchus* and *Ichnomylax* formed a clade at the next most inclusive node (with respect to the crown group), while *Jessenia* was resolved to the next node. Friedman's analysis (2007b) resulted in *Dipnorhynchus* and *Uranolophus* being resolved to basal nodes (Fig. 4A).

In other words, phylogenetic resolution of these early Devonian taxa broadly mirrors stratigraphic distribution. The Early Devonian taxa retain some of the plesiomorphic characters also present in *Diabolepis*, such as the B-bones meeting posterior to the midline I-bone, preventing the I-bone from reaching the rear margin of the skull (*Uranolophus*, *Dipnorhynchus*, *Jessenia*, *Speonesydrion*,

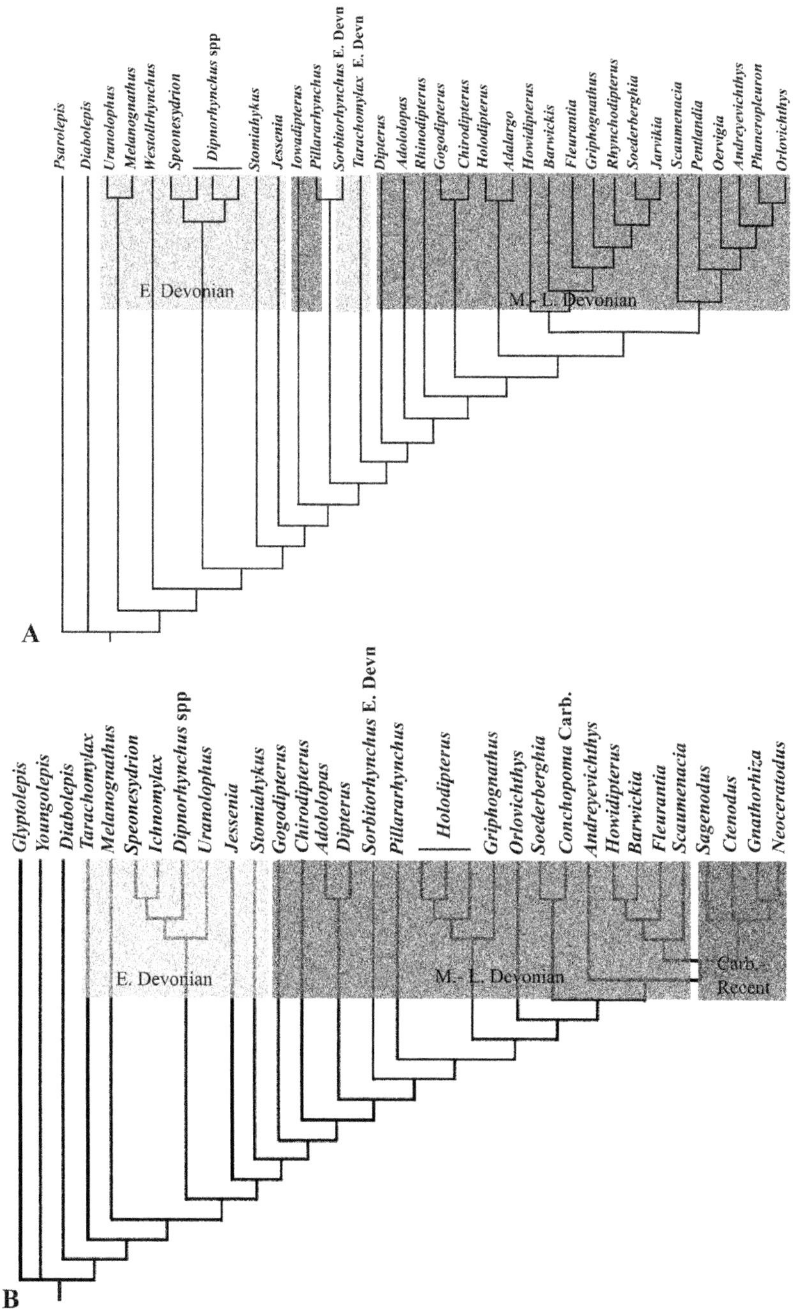

Fig. 3 Cladograms showing lungfish relationships. Note basal resolution of Early Devonian forms. A. modified from Schultze (2001: Fig. 5); B. modified from Ahlberg *et al.* (2006).

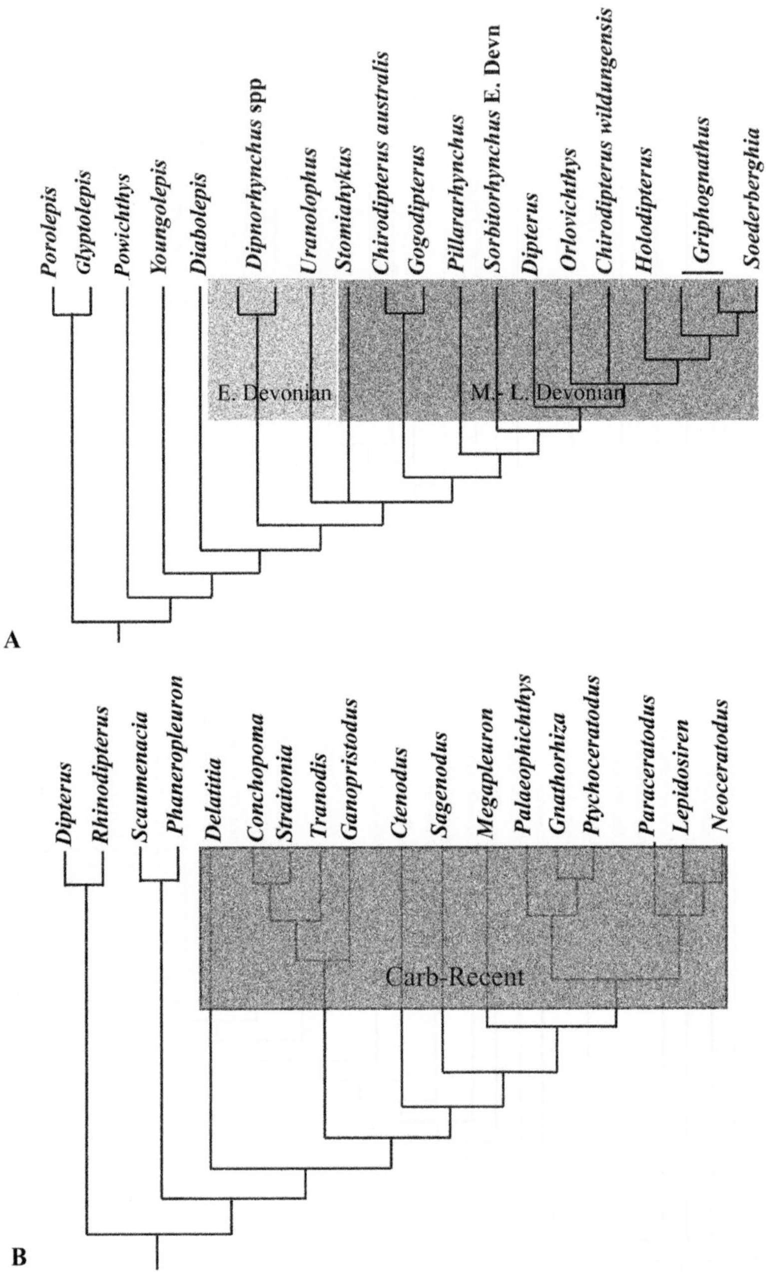

Fig. 4 Cladograms showing lungfish relationships. A. modified from Friedman (2007b: Fig. 18B), note basal resolution of Early Devonian forms. B. modified from Schultze and Chorn (1997), of Carboniferous and younger taxa.

Jarvikia) and a dentition comprising rows of teeth (*Speonesydrion*, *Melanognathus*, *Tarachomylax*). Other characters resolved as plesiomorphic within the Dipnoi include multiple D-bones (*Tarachomylax*, *Dipnorhynchus*, *Jarvikia*) and a large number of small bones at the front of the skull rather than paired E-bones (*Speonesydrion*, *Dipnorhynchus*, *Jarvikia*). However, the dentition-based groups of Campbell and Barwick are not recovered.

Correlation of stratigraphic age with phylogenetic resolution continues towards the lungfish crown group. Again, broadly, there is a region of the cladogram including Middle and Late Devonian taxa, and then Carboniferous and younger taxa in increasingly more crownward positions (Figs. 3, 4; also Schultze and Marshall, 1993). However, while there is disagreement as to the resolution of the Middle and Late Devonian taxa to particular nodes and the grouping of taxa into clades, there are also points of agreement among more recent analyses. The next nodes more crownward of the Early Devonian taxa include taxa such as *Chirodipterus australis*, *Gogodipterus paddyensis*, *Dipterus valenciennesi*, *Adololopas moyasmithae* and *Pillararhynchus longi* (see Clack, this volume). However, while Ahlberg *et al.* (2006) resolve *Chirodipterus* and *Gogodipterus* as more basal relative to *Dipterus*, *Adololopas* and *Pillararhynchus*, Schultze's (2001) analysis resolves *Pillararhynchus* and *Dipterus* to the more basal nodes. More crownward of this, both Schultze (2001) and Ahlberg *et al.* (2006) recover an association of *Howidipterus donnae*, *Barwickia downunda* and *Fleurantia denticulata*. *Scaumenacia curta* is resolved as sister taxon to these in Ahlberg *et al.* (2006), but not in Schultze (2001).

The more controversial nodes involve Devonian taxa previously assigned to the lungfish Family Rhynchodipteridae. This taxon was formalised by Miles (1977) to include *Rhynchodipterus*, *Soederberghia* and *Griphognathus*. This grouping was criticised by Ahlberg *et al.* (2001), who removed *Griphognathus*. *Griphognathus* and *Soederberghia* are retained as sister taxa in the phylogeny of Schultze (2001) and Friedman (2007b), but not in that of Ahlberg *et al.* (2006). The Family Holodipteridae, whose members show varying degrees of dental resorption and replacement by denticles (Fig. 2C), are closely related to *Griphognathus* or *Griphognathus+Soederberghia* in Long (2005), Ahlberg *et al.* (2006) and Friedman (2007b). In this instance, the dentition of *Griphognathus+Soederberghia*, consisting primarily of a denticulated surface, represents the endpoint of this resorption.

As noted above, when the dipnoan clade first evolves in the Early Devonian, there is a substantial range in dental morphology, although skull morphology is generally similar (there is reduction in the number of skull bones, particularly anteriorly, between early and Middle-Late Devonian taxa). This diversity of dental morphologies continues throughout the Middle and Late Devonian regions of the lungfish phylogeny until the Carboniferous. Ahlberg *et al.* (2006) explained this diversity as resulting from a decoupling of odontogenic (dentition building) and odontoclastic (dentition resorbing) processes in the Devonian, so that they could act independently, for example, at different rates. For example, *Griphognathus* and

Soederberghia show the predominance of odontoclastic, resorbing processes, while the dentition of taxa such as *Dipterus* represents the dominance of odontogenic processes, in this case tooth addition with little or no resorption.

Nevertheless, by the Carboniferous independence of these processes was lost, when taxa begin to show a stereotypical developmental pattern centered on the toothplated dentition. Here, new teeth are added along the lateral edges of the toothplate, to preexisting rows. Some changes involve the incorporation of teeth into developing ridges, but otherwise there is very little difference. Carboniferous taxa such as *Tranodus castrensis* and species of *Ctenodus* still have recognisable teeth within the ridges crossing the toothplate, but these are lost in taxa such as *Sagenodus* (Thomson 1965: pls. 1, 2; Schultze and Chorn 1997) At this time, overall body morphology also becomes more or less similar to living lungfishes, with loss of independent unpaired fins, a process first observed in the Devonian taxon *Phaneropleuron* (Ahlberg and Trewin 1995). Skull morphology continues to change, with a reduction of the number of bones present through to the present day and an overall loss of a mineralised or ossified snout (Cavin *et al.* 2007). The skull roof may represent the most variable part of the lungfish skeleton, and that most subject to evolutionary change, after the Devonian.

PHYLOGENY OF POST-DEVONIAN LUNGFISHES AND EVOLUTION OF THE CROWN GROUP

The phylogenetic relationships of post-Devonian lungfishes were evaluated by Schultze and Marshall (1993) and most recently by Schultze and Chorn (1997) and Cavin *et al.* (2007). As well as the evolutionary trends described above, overall ossification is reduced, such that very little is known about the postcranium, with most characters deriving from skull bones and teeth. Most taxa are known from the highly mineralised dentition, while skull bones are reduced in number and rarely found associated with teeth. As noted by Cavin *et al.* (2007), processes involved in reducing the number of bones in the skull are difficult to determine in fossil taxa, whether by bone loss or fusion of bones with each other. This makes determining bone homologies somewhat difficult, one example being homologies between the two or fewer midline skull bones characteristic of Mesozoic and younger lungfish with respect to the series of bones labeled A–E in Devonian lungfish. In an attempt to overcome this, Cavin *et al.* (2007) based their characters of the skull on numbers of bones and the topographical relationships of these to one another, without applying the labelling system used in Devonian lungfishes.

Despite these uncertainties, the recent results of Cavin *et al.* (2007) are of interest because of the resolution of certain fossil taxa within the lungfish crown group. The sister taxon to the crown group is *Microceratodus angolensis*, from the early Triassic of Angola. This node is characterised by the absence of paired bones in the midline of the skull, while the sensory canals are not enclosed in bony tubes within the skull bones. *Neoceratodus* is resolved to the base of the crown group, with the crown group being defined by the presence of two bones in the midline and mediolateral series of the skull, rather than three. There are several fossil taxa more closely related to the *Lepidosiren+Protopterus* grouping than *Neoceratodus*. The most basal of these is *Ferganoceratodus*, with two species including *F. martini* (Late Jurassic or Early Cretaceous of Thailand) and *F. jurassicus* (Middle Jurassic of Kyrgyzstan), with *F. martini* possessing a mineralised snout (Cavin *et al.* 2007). Taxa resolved to more derived nodes possess thin scales and palatines in contact with one another as well as very thin paired fins (*Ceratodus*, *Paraceratodus* and *Asiatoceratodus* and more derived taxa). Sister taxa to *Lepidosiren+Protopterus* include *Arganodus*, *Ptychoceratodus* and *Gosfordia*, all characterised by lower toothplates meeting in the midline.

Some of the more intriguing interpretations presented by Cavin *et al.* (2007), based on their phylogeny, are that *Neoceratodus* split from the rest of the crown group in the early Triassic, and so has a ghost lineage approximately 100 million years in length (the oldest record of *Neoceratodus forsteri* being Lower Cretaceous in age; Kemp and Molnar 1981). This is earlier than the vicariant event separating Australia from West Gondwana, which occurred in the mid-Jurassic, such that this event may not be related to the evolution of *Neoceratodus*. Many of the fossil taxa occupying nodes between these extant taxa are known from the Triassic, and Cavin *et al.* (2007) note that most cladogenetic events in this region of the phylogeny occur in the Triassic. Nevertheless, they were hesitant to identify this as a true biological radiation during the Triassic, or as the result of biases in the sedimentary record.

SUMMARY AND THE FUTURE FOR DIPNOAN PHYLOGENETIC ANALYSES

Marshall (1987a) and Schultze and Marshall (1993) were the first to apply computer-based parsimony programs to large character data sets, but came to the conclusion that parsimony was not adequate to the task of resolving lungfish relationships. Despite these difficulties, it is clear that phylogenies based on dentition morphologies and functional scenarios are also inadequate. Indeed, there is some consensus among recent phylogenetic analyses, which are broadly correlated to taxon age. Although Early Devonian taxa are placed at the basal nodes in the lungfish phylogeny, a better understanding and resolution of these taxa is crucial for determining evolution of characters in the more crownward nodes of the lungfish phylogeny. Middle and Late Devonian lungfish taxa remain difficult to resolve and should be a focus of new analyses; within this group of taxa there is agreement on the relationships of holodipterid lungfish and *Griphognathus*, representing a continuum of dentition resorption and replacement by a shagreen of denticles.

One area of potential research involves the changes seen in Carboniferous and younger taxa, whereby body morphology and that of the dentition become canalized to take on the general appearance of the extant lungfish *Neoceratodus forsteri*. For example, unpaired fins such as the dorsal and anal fins are either lost or merged with the caudal fin to form one continuous dorsal and ventral fin (Arratia *et al.* 2001). After the Carboniferous, variation in the dentition is largely lost and the toothplated dentition predominates. Living lungfish such as *Neoceratodus* are neotenic (juvenile form retained in the adult); whether this neoteny occurred in Carboniferous taxa and influenced body and dental morphology is a question that could be investigated further. However, skull morphology continues to evolve through the Carboniferous and Mesozoic, with reduction in the number of skull bones.

Acknowledgements

We would like to thank Jean Joss for inviting us to submit this chapter. Per Ahlberg acknowledges the support of the Swedish Research Council.

References

Ahlberg, P.E. (1989). Paired fin skeletons and relationships of the fossil group Porolepiformes (Osteichthyes: Sarcopterygii). Zoological Journal of the Linnean Society 96: 119-166.

Ahlberg, P.E. (1991). A re-examination of sarcopterygian inter-relationships, with special reference to the Porolepiformes. Zoological Journal of the Linnean Society 103: 241-287.

Ahlberg, P.E. and Trewin, N.H. (1995). The postcranial skeleton of the Middle Devonian lungfish *Dipterus valenciennesi*. Transactions of the Royal Society of Edinburgh: Earth Sciences 85: 159-175.

Ahlberg, P.E., Johanson, Z. and Daeschler, E.B. (2001). The Late Devonian lungfish *Soederberghia* (Sarcopterygii, Dipnoi) from Australia and North America, and its biogeographical implications. Journal of Vertebrate Paleontology 21: 1-12.

Ahlberg, P.E., Smith, M.M. and Johanson, Z. (2006). Developmental plasticity and disparity in early dipnoan (lungfish) dentitions. Evolution and Development 8: 331-349.

Arratia, G., Schultze, H.-P. and Casciotta, J. (2001). Vertebral column and associated elements in dipnoans and comparison with other fishes: development and homology. Journal of Morphology 250: 101-172.

Barwick, R.E., Campbell, K.S.W. and Mark-Kurik, E. (1997). *Tarachomylax*: A new Early Devonian dipnoan from Severnaya Zemlya, and its place in the evolution of the Dipnoi. Geobios 30: 45-73.

Brinkmann, H., Venkatesh, B., Brenner, S. and Meyer, A. (2004). Nuclear protein-coding genes support lungfish and not the coelacanth as the closest living relatives of land vertebrates. Proceedings of the National Academy of Sciences USA 101: 4900-4905.

Campbell, K.S.W. and R.E. Barwick. (1982). A new species of the lungfish *Dipnorhynchus* from the New South Wales. Palaeontology 25: 509-527.

Campbell, K.S.W. and R.E. Barwick. (1983). Early evolution of dipnoan dentitions and a new genus *Speonesydrion*. Memoirs of the Association of Australasian Palaeontologists 1: 17-49.

Campbell, K.S.W. and R.E. Barwick. (1984). *Speonesydrion*, an Early Devonian dipnoan with primitive toothplates. PalaeoIchthyologica 2: 1-48.

Campbell, K.S.W. and Barwick, R.E. (1987). Paleozoic lungfishes – a review. In: W.E. Bemis, W. Burggren and N.E. Kemp (eds.). The Biology and Evolution of Lungfishes. Journal of Morphology 1986 (Suppl. 1): 93-131.

Campbell, K.S.W. and Barwick, R.E. (1988). *Uranolophus*: a reappraisal of a primitive dipnoan. In: P.A. Jell (ed.). Devonian and Carboniferous Fish Studies. Memoirs of the Association of Australasian Palaeontologists 7: 87-144.

Campbell, K.S.W. and Barwick, R.E. (1990). Paleozoic dipnoan phylogeny: Functional complexes and evolution without parsimony. Paleobiology 16: 143-169.

Campbell, K.S.W. and R.E. Barwick. (1999). A new species of the Devonian lungfish *Dipnorhynchus* from Wee Jasper, New South Wales. Records of the Australian Museum 51: 123-140.

Campbell, K.S.W. and Barwick, R.E. (2001). *Diabolepis* and its relationship to the Dipnoi. Journal of Vertebrate Paleontology 21: 227-241.

Campbell, K.S.W. and Barwick, R.E. (2002). The axial postcranial structure of *Griphognathus whitei* from the Upper Devonian Gogo Formation of Western Australia: comparisons with other Devonian dipnoans. Records of the Western Australian Museum 21: 167–201.

Cavin, L., Suteethorn, V., Buffetaut, E. and Tong, H. (2007). A new Thai Mesozoic lungfish (Sarcopterygii, Dipnoi) with an insight into post-Palaeozoic dipnoan evolution. Zoological Journal of the Linnean Society 149: 141–177.

Chang, M.-M. (1982). The braincase of Youngolepis, a Lower Devonian crossopterygian from Yunnan, South-Western China. PhD Thesis, University of Stockholm Stockholm, Sweden.

Chang, M.-M. (1991). Head exoskeleton and shoulder girdle of *Youngolepis*. In: M.-M. Chang, Y.-H. Liu and G.-R. Zhang (eds.). Early Vertebrates and Related Problems of Evolutionary Biology, pp. 355-378. Science Press, Beijing, China.

Chang, M.-M. (1995). *Diabolepis* and its bearing on the relationships between porolepiforms and dipnoans. Bulletin Museum National d'Histoire naturelle, Paris 17 (Section C): 235-268.

Chang, M.-M. (2004). Synapomorphies and scenarios – more characters of *Youngolepis* betraying its affinity to the Dipnoi. In: G. Arratia, M.V.H. Wilson and R. Cloutier (eds.). Recent Advances in the Origin and Early Radiation of Vertebrates. Dr. Friedrich Pfeil. München, Germany. pp. 665-686.

Chang, M.-M. and Yu Xiaobo. (1984). Structure and phylogenetic significance of *Diabolichthys speratus* sp. nov., a new dipnoan-like form from the Lower Devonian of eastern Yunnan, China. Proceedings of the Linnean Society of New South Wales 107: 171-184.

Chang, M.-M. and Wang, J.-Q. (1995). A new Emsian dipnorhynchid (Dipnoi) from Guangnan; southeastern Yunnan; China. Geobios 28: 233-239.

Clément, G., Dupret, V., Goujet, D., Pernègre, V. and Roy, J.-C. (2006). First Devonian dipnoans (Vertebrata, Sarcopterygii) from Spitsbergen. Comptes Rendus Palevol 5: 893-890.

Cloutier, R. and Ahlberg, P.E. 1996. Morphology, characters and the interrelationships of basal sarcopterygians. In: M.L.J. Stiassny, L.R. Parenti and G.D. Johnson (eds.). Interrelationships of Fishes, Academic Press, San Diego, USA. pp. 445-479.

Denison, R.H. (1968). Early Devonian lungfishes from Wyoming, Utah, and Idaho, Fieldiana, Geology 17: 353–413.

Friedman, M. (2007a). *Styloichthys* as the oldest coelacanth: implications for early osteichthyan interrelationships. Journal of Systematic Palaeontology 5: 289-343.

Friedman, M. (2007b). The interrelationships of Devonian lungfishes (Sarcopterygii: Dipnoi) as inferred from neurocranial evidence and new data from the genus *Soederberghia* Lehman, 1959. Zoological Journal of the Linnean Society 151: 115-171.

Friedman, M., Coates, M.I. and Anderson, P. (2007). First discovery of a primitive coelacanth fin fills a major gap in the evolution of lobed fins and limbs. Evolution and Development 9: 329-337.

Hallström, B.M. and Janke, A. 2009. Gnathostome phylogenomics utilizing lungfish EST sequences. Molecular Biology and Evolution 26: 463-471.

Hedges, S.B., Hass, C.A. and Maxson, L.R. (1993). Relations of fish and tetrapods. Nature 363: 501-502.

Jarvik, E. (1967). On the structure of the lower jaw in dipnoans: with a description of an early Devonian dipnoan from Canada, *Melanognathus canadensis* gen. et sp. nov. Zoological Journal of the Linnean Society 47: 155-183.

Jarvik, E. (1980). Basic Structure and Evolution of Vertebrates, Volume 1. Academic Press, London, UK.

Jefferies, R.P.S. (1979). The origin of chordates: a methodological essay. In: M.R. House (ed.). The Origin of Major Invertebrate Groups. Systematics Association Special Volume 12. Academic Press, London, UK. pp. 443-447.

Jessen, H. (1980). Lower Devonian Porolepiformes from the Canadian Arctic with special reference to *Powichthys thorsteinssoni* Jessen. Palaeontographica A 167: 180-214.

Johanson, Z., Joss, J., Boisvert, C.A., Ericsson, R., Sutija, M. and Ahlberg, P. E. (2007). Fish fingers: digit homologues in sarcopterygian fish fins. Journal of Experimental Zoology (Molecular and Developmental Evolution) 308B: 757-768.

Kemp, A. and Molnar, R.E. (1981). *Neoceratodus forsteri* from the Lower Cretaceous of New South Wales, Australia. Journal of Paleontology 55: 211-217.

Long, J.A. (2005). Interrelationships of the holodontid lungfishes (Osteichthyes, Dipnomorpha) based on new material from the Upper Devonian Gogo Formation of Western Australia. Journal of Vertebrate Paleontology 25 (Suppl. 3): 84A.

Long, J.A., Campbell, K.S.W. and Barwick, R.E. (1994). A new dipnoan genus, *Ichnomylax*, from the Lower Devonian of Victoria, Australia. Journal of Vertebrate Paleontology 14: 127–131.

Maisey, J.G. (1986). Heads and tails: a chordate phylogeny. Cladistics 2: 201-256.

Marshall, C.R. (1986). Lungfish: Phylogeny and parsimony. Journal of Morphology 190: 151-162.

Marshall, C.R. (1987a). Lungfish: phylogeny and parsimony. In: W.E. Bemis, W. Burggren and N.E. Kemp (eds.). The Biology and Evolution of Lungfishes. Journal of Morphology 1986 (Suppl. 1): 151-162.

Marshall, C.R. (1987b). A list of fossil and extant Dipnoans. In: W.E. Bemis, W. Burggren and N.E. Kemp (eds.). The Biology and Evolution of Lungfishes. Journal of Morphology 1986 (Supplement 1): 15-23.

Miles, R.S. (1977). Dipnoan (lungfish) skulls and the relationships of the group: A study based on new specimens from the Devonian of Australia. Zoological Journal of the Linnean Society 61: 1-328.

Otto, M. and P. Bardenheuer. (1996). Lungfish with dipterid tooth-plates in the Lower Devonian of Central Europe. Modern Geology 20: 341-350.

Panchen, A.L. and Smithson, T.R. (1987). Character diagnosis, fossils, and the origin of tetrapods. Biological Reviews of the Cambridge Philosophical Society 62: 341-438.

Schultze, H.P. (1987). Dipnoans as sarcopterygians. Journal of Morphology 1986 (1987) (Suppl. 1): 39-74.

Schultze, H.-P. (2001). *Melanognathus,* a primitive dipnoan from the Lower Devonian of the Canadian arctic and the interrelationships of Devonian dipnoans. Journal of Vertebrate Paleontology 21: 781-794.

Schultze, H.-P. and Marshall, C.R. (1993). Contrasting the use of functional complexes and isolated characters in lungfish evolution. Memoirs of the Association of Australasian Palaeontologists 15: 211-224.

Schultze, H.-P. and Chorn, J. (1997). The Permo-Carboniferous genus *Sagenodus* and the beginning of modern lungfish. Contributions to Zoology 67: 9-70.

Thomson, K.S. (1965). On the relationships of certain Carboniferous Dipnoi; with descriptions of four new forms. Proceedings of the Royal Society of Edinburgh 64: 221-245.

Thomson, K.S. and Campbell, K.S.W. (1971). The structure and relationships of the primitive Devonian lungfishs—*Dipnorhynchus sussmilchi* (Etheridge). Bulletin of the Peabody Museum of Natural History 38: 1-109.

Tokita, M. and Hikida, T. (2005). Evolutionary history of African lungfish: a hypothesis from molecular phylogeny. Molecular Phylogenetics and Evolution 35: 281-286.

Wang, S., Drapala, V., Barwick, R.E. and Campbell, K.S.W. (1993). The dipnoan species, *Sorbitorhynchus deleaskitus*, from the Lower Devonian of Quangxi, China. Philosophical Transactions of the Royal Society London B 340: 1-24.

Yu, X. (1988). A new porolepiform-like fish, *Psarolepis romeri*, gen. et. sp. nov. (Sarcopterygii, Osteichthyes) from the Lower Devonian of Yunnan, China. Journal of Vertebrate Paleontology 18: 261-274.

Zardoya, R., Cao, Y., Hasegawa, M. and Meyer, A. (1998). Searching for the closest living relative(s) of tetrapods through evolutionary analyses of mitochondrial and nuclear data. Molecular Biology and Evolution 15: 506-517.

Zhu, M. and Schultze, H.P. (2001). Interrelationships of basal osteichthyans. In: P.E. Ahlberg (ed.). Major Events in Early Vertebrate Evolution. Systematics Association Special Volume 61: 289-314.

Zhu, M. and Yu, X. (2002). A primitive fish close to the common ancestor of tetrapods and lungfish. Nature 418: 767-770.

Zhu, M., Yu, X. and Janvier, P. (1999). A primitive fossil fish sheds light on the origin of bony fishes. Nature 397: 606-610.

Zhu, M., Yu, X. and Ahlberg, P. E. (2001). A primitive sarcopterygian fish with an eyestalk. Nature 410: 81-84.

CHAPTER 3

The Natural History of the Australian Lungfish *Neoceratodus forsteri* (Krefft, 1870)

Peter K. Kind*

Principal Scientist, Freshwater, Queensland Primary Industries and Fisheries, Primary Industries Building, 80 Ann St Brisbane, Queensland, Australia 4001

ABSTRACT

The Australian lungfish *Neoceratodus forsteri* is the only extant member of a rich lungfish fauna known primarily from Mesozoic and Cenozoic deposits in Australia. Present day lungfish populations are confined to a series of coastal river systems within the south-east corner of Queensland. However, fossil evidence confirms that *N. forsteri* previously occupied a more extensive distribution. Throughout their distribution, Australian lungfish inhabit river channels and tributary streams upstream of the tidal interface. The life-cycle is completed entirely within freshwater reaches with an annual spring/summer breeding season. Spawning activity is concentrated along shallow river margins, with eggs deposited in dense beds of aquatic plants or amongst the submerged rootlets of riparian vegetation. Developing embryos are subject to a range of threats including predation, fungal and bacterial infections, fluctuating water levels and physical damage. The incubation period varies according to temperature but generally extends for 23-30 days. The diet of *N. forsteri* changes over time reflecting a progression from larval to adult dentition. Recently hatched and juvenile lungfish feed largely on small invertebrates, while adults ingest large amounts of plant material as well as a range of invertebrates, small fish and tadpoles. Australian lungfish are predominantly nocturnal and occupy restricted home ranges around 1-1.5 km in length. Movements outside the home range are most commonly observed when individuals seek out suitable spawning habitat prior to the annual breeding season. However, longer movements can occur at any time and are frequently associated with instream flow events. At rest in well-

*E-mail: Peter.kind@deedi.qld.gov.au

aerated water *N. forsteri* rarely breathes air, respiration being supported almost entirely by the gills. Air breathing is used primarily to supplement short-term oxygen demand during periods of increased activity and is most commonly observed during courtship behavior and during flow events. The Australian lungfish is a slow-growing and long-lived species that reaches a maximum total length of approximately 1500 mm. However, the dominant size class in most surveys is 900-1000 mm. Sex ratio has been reported as approximately even, with largest size classes (>1200 mm) dominated by females. Males mature at a smaller size than females. The species exhibits low allelic diversity at allozyme and mitochondrial loci with little evidence of genetic differentiation between catchments within the distribution.

Debate over the conservation status of Australian lungfish has continued for over a century, primarily because juvenile lungfish are difficult to collect from the wild. This issue, coupled with the lack of a reliable method for ageing lungfish has hampered efforts to understand the population dynamics and true status of the species. Following European settlement, all the major rivers inhabited by *N. forsteri* have become progressively more regulated and degraded. Recent evidence indicates that successful spawning and recruitment rarely occurs in the still waters upstream of the numerous weirs and dams in south-east Queensland. These barriers are also restricting lungfish movements, raising concerns over the genetic structure of remaining populations. The species also faces other threats including introduced species, habitat alteration, stranding events and reduced water quality. Current management strategies are focussed on reducing adult mortality, improving connectivity between remaining populations and providing suitable habitat for successful spawning and recruitment.

Keywords: Australian lungfish (*Neoceratodus forsteri*), Queensland, *Ceratodus*

INTRODUCTION

"It will take considerable time before the Ceratodus is thoroughly understood; but it will come" (Welsby 1905).

The Australian lungfish drifted quietly into the scientific spotlight on the morning of January 18th, 1870. In a letter to the Editor of the *Sydney Morning Herald*, Head Curator and Secretary of the Australian Museum, Gerard Krefft, described two unusual specimens, which he had received from the Wide Bay/Burnett District of south-east Queensland. The specimens had been procured by Krefft's long time friend and prominent politician William Forster, a former Queensland resident who had acquired large tracts of land in the Burnett Catchment. During their long friendship Forster had tantalised Krefft with tales of fish eaten by squatters along the Burnett and known locally as 'fresh water salmon' (Whitley 1929). The specimens presented to Krefft were collected by Forster's cousin W.F. M'Cord of Coonambula, near Mundubbera.

Krefft's initial examination was hampered by a lack of internal organs, which had been removed for transport to Sydney. Despite this obstacle he recognised the significance of the specimens and described them as amphibians allied to fish from the genus *Lepidosiren*. More remarkably, he also recognised that tooth plates from the specimens bore similarities to fossil material from an extinct genus of fishes known as *Ceratodus* and previously described by Agassiz (1844) as selachians (sharks). Fulfilling a promise to William Forster, Krefft named the new species *Ceratodus forsteri* in his letter and thus provided the first scientific record of the Australian lungfish (Krefft 1870a). The letter was followed by a formal description including illustrations (Fig. 1), which was presented to members of the Zoological Society of London (Krefft 1870b), in which Krefft noted that:

"*It is strange that a curious creature like this, which was well known to the early settlers of the Wide Bay and other Queensland districts, should so long have escaped the eyes of those interested in natural history*".

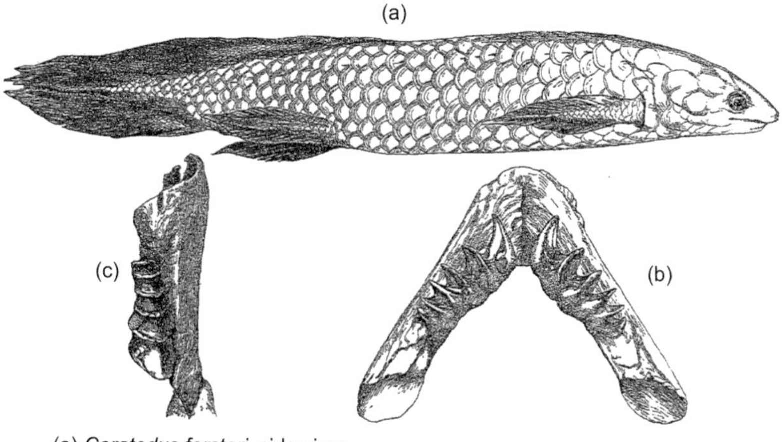

(a) *Ceratodus forsteri*, side view
(b) Lower jaw of ditto, from above
(c) Left ramus of lower jaw, seen from the front s. *Symphysis*

Fig. 1 Illustration of the Australian lungfish and tooth plate structures, reproduced from Krefft's original description of the species in 1870.

The first scientist to examine an intact Australian lungfish specimen was Krefft's German colleague Albert Günther, who published a thorough species description accompanied by detailed illustrations using one of Krefft's original specimens and two others procured from the Mary River (Günther 1871a). Günther dismissed Krefft's reference to the amphibians and placed the species decisively into the class Pisces. After careful examination of the teeth, Günther concluded that there was also no valid reason to separate recent and fossil forms of the genus *Ceratodus* noting that:

"*so well marked by peculiar characters, and the recent teeth so similar to those of certain extinct species, that we should be better justified in making generic distinctions among the fossil forms, than in separating the living from the extinct*".

When comparing the specimens Günther recorded what he believed to be "remarkable differences" between those procured from the Mary and Burnett Rivers. These differences centred on the size, shape and number of scales encircling the middle of the trunk. He proposed that the Mary River lungfish be named *Ceratodus miolepis*, a distinction subsequently dismissed to synonymy by Meyer (1875), who noted considerable natural variability in scale counts from Burnett River lungfish.

In 1876, Count F de Castelnau, the French Consul in Melbourne and a keen naturalist examined a lungfish specimen allegedly collected from the Fitzroy River, some 300 km north of the Burnett River system (de Castelnau 1876a). On the basis of perceived differences in the teeth and body proportions between his fish and other *C. forsteri* specimens, de Castelnau raised the genus *Neoceratodus* and named the species *Neoceratodus blanchardi.* Unfortunately, before a year had passed de Castelnau was forced to retract this description, recognising that he had in fact examined an immature Australian lungfish and failed to properly examine the tooth plates (de Castelnau 1876b).

Twenty years after Günther's description, Teller (1891) examined a fossil calvarium associated with typical ceratodont tooth plates and compared it to a skull from the recent fish. He noted significant differences between the specimens and reserved the genus *Ceratodus* for fossil forms, raising a new genus *Epiceratodus* for the recent forms. However, by the laws of priority, the genus *Neoceratodus* takes precedence for separating living and recent species and the extant Australian lungfish are now recognised as forming a single taxon *Neoceratodus forsteri.*

Reaction to the discovery of the Australian lungfish has been chronicled by Monteith (2001), who described subsequent events as one of the great sagas in Australian natural history. Bemis *et al.* (1987) surmised that such fervour in the scientific community has not been witnessed again, apart from events following the discovery of the coelacanth (*Latimeria chalumnae*) in 1938. The discovery breathed life into Australia's rich fossil lungfish fauna and triggered an immediate response from eminent scientists in Europe and Australia. In the latter years of the 19th century, a series of collectors converged on the rivers of south-east Queensland, all eager to obtain fresh specimens and record details of the species' distribution and behaviour. Such expeditions frequently frustrated the expectant ichthyologists but made significant contributions to documenting the flora and fauna of Queensland. In one notable example, Krefft's collector George Masters reportedly amassed a collection of more than 16,000 beetle specimens representing 1,100 species during a three month trip that yielded only 19 lungfish (Macleay 1871). Expeditions to search for lungfish also provided opportunities to study other Australian faunal curiosities such as platypus and echidna. W. H. Caldwell, who published on the development of lungfish eggs and larvae (and was able to settle a raging controversy in 1884 with proof that platypus are oviparous), reported that local aborigines collected more than 1300 echidna on his behalf during one visit to Gayndah alone (Caldwell 1885b).

Confirmation of a new extant lungfish fuelled the flames of debate surrounding concepts of evolution and raised hopes for the discovery of a true 'missing link'. Agassiz (1870) asserted that the Australian lungfish provided conclusive evidence for the concept of 'synthetic types', animals sharing physical characteristics from multiple taxonomic groups. Huxley (1880) made the observation that:

"*This wonderful creature seems contrived for the illustration of the doctrine of Evolution. Equally good arguments might be adduced for the assertion that it is an amphibian or a fish, or both, or neither*".

Enthusiasm for the idea of a missing link pervaded early literature describing the habits of the Australian lungfish, frequently reporting evidence that *N. forsteri* was able to emerge onto the land. In his original description Krefft (1870b) postulated that the species buried itself in mud during certain seasons. Within 12 months, Ramsey (1871) reported to the Zoological Society of London that *C. forsteri* probably emerged from the water at night to graze on terrestrial grasses, a notion ridiculed by Krefft (1874) and later retracted by Ramsey himself after observing lungfish in captivity (Ramsey 1876). Lumholtz (1889) perpetuated the myth by suggesting that lungfish foraged on shore during the night and spent the daylight hours sunning themselves on logs out of the water.

The excitement surrounding the discovery of the Australian lungfish and the furtive nature of the species have combined to create a mystique that has persisted into the present. Despite more than a century of consistent scientific attention, the Australian lungfish remains poorly understood by locals and scientists alike. Until recently, studies of *N. forsteri* have focussed predominantly on aspects of anatomy, physiology, embryology or endocrinology. Surprisingly little is known of the species daily life, behaviour or the true status of the current lungfish populations. Despite this the species has widespread social significance in the Wide Bay and Burnett regions. In 1922, a railway siding and settlement on the upper Burnett River was named Ceratodus after the species and according to O'Connor (1898) the township of Theebine on the Mary River owes its name to 'Teebine' or 'Teevine' an aboriginal name for lungfish. The water pumping station in Mundubbera bears a mural depicting lungfish, as do electricity poles in nearby Gayndah. Live lungfish can also be viewed in a display at the local Gayndah Museum and in Bundaberg a metallic lungfish sculpture has been erected in the town centre (Fig. 2).

While modern day locals often refer to lungfish as 'lungies' or 'cerats', a number of other common and aboriginal terms exist for the species. At the time of their discovery, the name 'Burnett River Salmon' was widely used for lungfish, reflecting the pink colour of their flesh. In addition to Teebine, other aboriginal terms have been recorded including 'Djellah' and 'Dala' and a review of historic literature will reveal other common names including mudfish, barramoondi, barramundi and red fish.

Fig. 2 Lungfish sculpture in the City of Bundaberg, Queensland, Australia.

The paucity of information surrounding lungfish was highlighted in the mid 1990s by a series of major dam proposals, designed to service the growing water demands of a rapidly expanding region. These proposals renewed interest in the status of lungfish and has triggered renewed scientific efforts to understand the species' ecology. Successive dam proposals on the Burnett River have been met with protests and outcry, which have so far failed to halt their progress. In 2009 the fate of the Traveston Dam proposal on the Mary River remains unresolved and will ensure that debate over lungfish will continue into the future.

DISTRIBUTION

Fossil evidence confirms that *N. forsteri* previously occupied a distribution in Australia extending to at least Lightning Ridge in New South Wales (Kemp and Molnar 1981), and Chinchilla in Queensland (Kemp 1982). However, this distribution has contracted considerably over time and lungfish now only occur in a series of adjacent coastal river systems within south east Queensland, Australia. The geographic boundaries of the species' distribution and the origin of some present day lungfish populations have been subject to ongoing debate for more than a century. The original specimens described by Krefft (1870a) were collected from the Burnett River or one of its major tributaries upstream from the township of Gayndah. Many of the early lungfish collectors reported difficulty obtaining specimens of *N. forsteri.* Juveniles were particularly scarce despite the use of destructive sampling techniques such as lime and dynamite to search for them (Illidge 1893; Bancroft 1911). Ramsey (1871) reported that the distribution of Australian lungfish extended beyond the Burnett and Mary into the Fitzroy, Dawson and Burrum Rivers, claims that are now widely accepted as untrue. After an initial flurry of questionable information, most collectors concentrated on the Burnett and Mary River systems. In time, authors such as Semon (1899) and Bancroft (1911) concluded that Australian lungfish were probably confined to these two catchments at the time of their discovery.

Concern over the apparent lack of juvenile lungfish was heightened by reports of early settlers using explosives to catch adult lungfish and then eating them in large numbers (Krefft 1870b; O'Connor 1898). In response to these concerns the Royal Society of Queensland commissioned a translocation program in which adult lungfish collected from the Mary River at Miva were distributed into a series of nearby rivers and still water bodies (see O'Connor 1902 quoted in Welsby 1905). Of 109 lungfish originally collected for the translocation program, 11 escaped from a holding pond and 29 died. The remainder were distributed to the North Pine River, a lagoon near the Albert River, a dam on the upper Brisbane River, Enoggera Reservoir (on a tributary of the Brisbane River), the Condamine River, the Coomera River and a pond in the Brisbane Botanic Gardens (O'Connor 1897). A small number of lungfish were also translocated into Blue Lake and Eighteen Mile Swamp on North Stradbroke Island off the Queensland coast (Thompson 1975; Kemp 1990). Bancroft (1933) documented his unsuccessful attempts to establish a lungfish hatchery at Brown Lake on the same island. Bancroft had argued that the paucity of small fish in early collections was a result of high predation levels on the mainland. He proposed that establishing a breeding colony on North Stradbroke would provide a ready source of fish to be released into mainland rivers. The translocation program and proposed captive breeding colony reflect the high level of concern about low juvenile numbers at the time. No satisfactory explanation has ever been found for the difficulty of collecting small lungfish. The most likely explanation remains that successful recruitment events are sporadic and perhaps uncommon (see later discussion).

Translocations of lungfish met with varied success but were responsible for establishing some of the now more familiar lungfish populations. After several years of studying lungfish, Kemp (1987) argued that present day lungfish populations in the Brisbane and North Pine Rivers are unlikely to have expanded from such small numbers of founders and therefore these two rivers probably form part of the historic distribution. The proposal is inconsistent with a general lack of historic lungfish knowledge among locals in the Brisbane Rivcr Catchment, but plausible given the cryptic and largely nocturnal habits of the species. A genetic study subsequently published by Frentiu *et al.* (2001) was unable to resolve this ongoing debate, but concluded that a rare haplotype shared by lungfish from the Brisbane and Mary Rivers is consistent with details of the translocation program described previously. In addition to the various sanctioned translocations, there have been numerous illegal and unreported movements of lungfish. These translocations have further clouded the actual natural distribution of the species. Recent evidence indicates that illegal and incidental translocations of lungfish are continuing to occur. In one example, eight lungfish were rescued from an isolated lagoon in the coastal Brisbane suburb of Redcliffe in 2005. The origin of these individuals remains unknown.

Kemp (1995) reported that the present day distribution of Australian lungfish extends to sections of the Burnett, Mary, North Pine, Brisbane, Logan, Albert and Coomera Catchments, as well as Enoggera and Gold Creek Reservoirs, Lake Manchester and possibly the Condamine River (Fig. 3). Johnson (2001) questioned the inclusion of the Condamine and Coomera Rivers, Lake Manchester and Gold Creek Reservoir stating that no recent museum records exist for these waterways. Ogilby (1917) reported that a lungfish measuring 495 mm had been received by the Queensland Museum from the Coomera River, confirming the persistence at that time. As recently as 2007, a small number of lungfish were collected in the lower Coomera River by scientists from Queensland Primary Industries and Fisheries (QPI&F) confirming their ongoing presence in this catchment. Likewise, biologists working for Brisbane City Council relocated several lungfish ranging from 66-100 cm in length from Lake Manchester into the Brisbane River when the lake was drawn down to supplement other domestic water supplies in 2007 (GHD 2007). Reliable reports of lungfish sightings also exist from Gold Creek Reservoir (QPI&F unpublished information) and Kemp (1995) reported that lungfish have been collected in sections of the Condamine River. The introduction of lungfish to Stradbroke Island almost certainly failed due to unsuitable water quality in the lakes other factors.

In recent years there has been growing evidence of a decline in lungfish numbers in the Enoggera Reservoir. Kemp (1987) reported that the reservoir provides poor lungfish habitat, with steep sloping banks, poor macrophyte communities and acidic water. Lungfish spawning has not been recorded in Enoggera Reservoir since water hyacinth (a declared noxious weed in Queensland) was removed in 1974. Lungfish had previously used the rootlets of this species as a spawning substrate (Kemp 1987). Surveys conducted in 2007 (see Hydrobiology 2007) failed to locate any lungfish in the lake but noted numerous individuals stranded in pools downstream of the dam wall. It is likely that gradual emigration of adult lungfish from the reservoir during flow events has not been matched by successful recruitment.

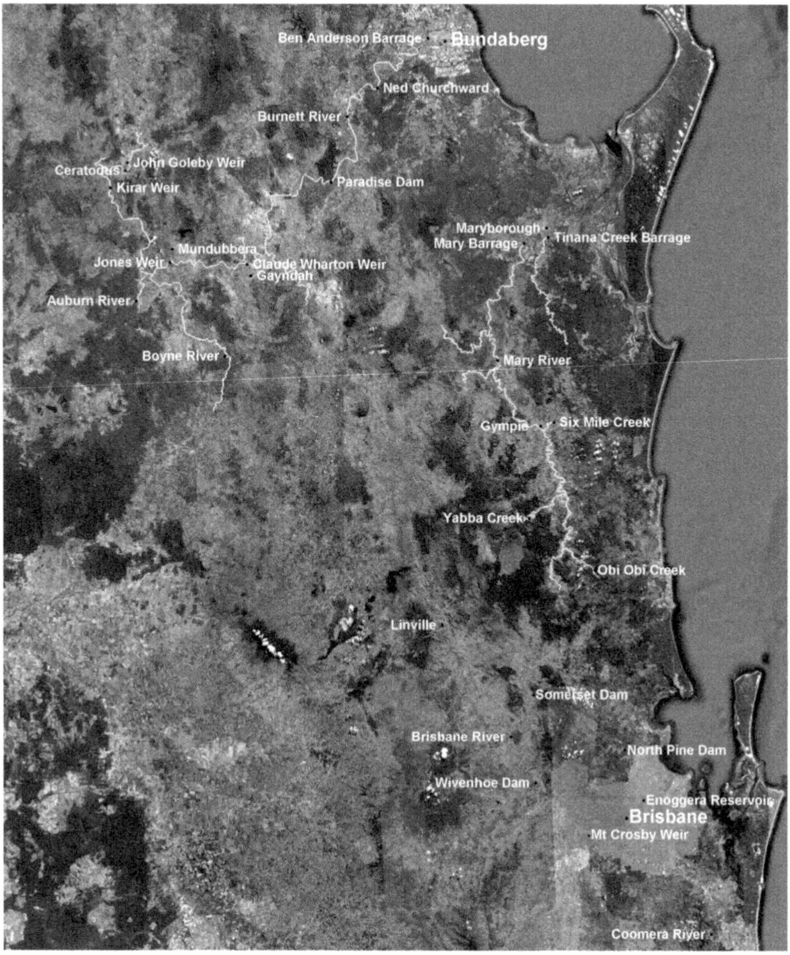

Fig. 3 Map of south east Queensland illustrating the presumed natural distribution of *N. forsteri* (yellow) and other catchments inhabited by the species (orange).
Color image of this figure appears in the color plate section at the end of the book.

Since their discovery, Australian lungfish specimens have been distributed to a large number of public aquariums, universities and zoos in Australia and overseas. The most celebrated of these exports is 'Grandad', a lungfish procured by the Shedd Aquarium in Chicago in 1933. This fish was fully grown on arrival and therefore is likely to be more than 80 years old. In the late 1990s, three commercial hatchery operators in Queensland were issued permits to breed lungfish for sale to the aquarium trade. One operator was successful in doing so and has exported juvenile Australian lungfish to collectors in several countries. These fish are individually tagged to assist with efforts to control the black market trade of lungfish.

Throughout their distribution Australian lungfish occupy a variety of habitats including tributary streams, the main river channel and still waters impounded by dams and weirs. Their existence in south east corner of Queensland coincides with the most densely populated areas in the state and one of the fastest growing regions in Australia. The Brisbane, North Pine, Mary and Burnett Rivers are regulated by dams and weirs to provide water for domestic supply, agriculture, forestry and mining. The Burnett River is the most heavily regulated system with more than 25 barriers dotted along the main river channel and its tributaries. The Mary River is less regulated with a small number of dammed tributaries and a tidal barrage in the lower reaches of the main river channel. The Brisbane River supplies water to the City of Brisbane and is dammed by two large water storages in addition to other balancing storages and smaller weirs. Likewise, the North Pine Dam forms a major barrier on the North Pine River as part of the water delivery program to northern Brisbane suburbs. In 2006, the Queensland Government announced the development of a "water grid" to counter serious drought, which had seen water supplies dwindle to crisis levels. The proposed grid system includes plans to create more dams in the south east corner of Queensland, linked by pipelines to provide maximum flexibility in water delivery.

The Burnett River lungfish populations have been surveyed extensively since 1995 in a series of ongoing studies by government scientists and other researchers. These surveys were commissioned as part of environmental measures associated with the construction of three major weirs and dams in the river channel between 1997 and 2005. Surveys conducted from 1997-2000 confirmed that lungfish occupy the main channel of the Burnett River from the tidal barrage at Adopted Middle Thread Distance (AMTD) 25.9 km, upstream to at least AMTD 334.0 km and probably beyond (Brooks and Kind 2002). Lungfish are present in major tributaries of the Burnett River including in the Boyne River upstream to the wall of Boondooma Dam, the Auburn River to Auburn River Gorge and Barambah Creek to the Barambah Gorge. Fish stocking surveys in Cania Dam (on Three Moon Creek) have yielded lungfish that were apparently translocated from drying waterholes in the upper Burnett (Alex Hamlyn, QPI&F unpublished data). An isolated population has recently been confirmed in Splitters Creek, downstream of the Burnett River tidal barrage. These fish have evidently moved into the Burnett River estuary during flood events and sought refuge in the headwaters of this minor tributary as high salinity levels re-establish below the barrage when flows recede (QPI&F unpublished information). In the main channel of the Burnett River, the majority of lungfish occur downstream from the township of Ceratodus. Brooks and Kind (2002) reported that Catch Per Unit Effort (CPUE) was highest in river reaches 50-100 km from the river mouth.

Surveys conducted by the QPI&F and other records indicate that lungfish are also widely distributed throughout the Mary River and its tributaries. In the main river channel, lungfish occur from the tidal barrage to at least the township of Conondale approximately 290 km upstream. Little is known about the relative abundance of *N. forsteri* between these two points. Lungfish are also common in several tributaries including the Tinana/Coondoo Ck system, Obi Obi Ck, Six Mile Ck, Yabba and Little Yabba Ck, and Wide Bay Ck (Simpson 1994; Kind 2002). Individuals have been observed in Borumba Dam on Yabba Creek and some other minor tributary streams (QPI&F unpublished data). Lungfish in the Tinana/Coondoo Creek complex have been isolated from the main channel population since construction of the tidal barrage at AMTD 55 km in 1979. Because Tinana Creek discharges downstream of this point, lungfish would need to negotiate estuarine sections of the lower Mary River but also the fish ladder at the tidal barrage to mix with main channel populations. This scenario is possible during major flow events, but highly unlikely.

The most widely recognised present day population of Australian lungfish occurs in a highly regulated section of Brisbane River between Wivenhoe Dam and the Mt Crosby Pumping Station. Wivenhoe Dam was constructed in 1985 to complement the nearby Somerset Dam on the Stanley River. Both dams were originally built as flood mitigation measures for the Brisbane River, but have become increasingly important as water storages, supplying the City of Brisbane and other key regional centres such as Ipswich, Logan, Redcliffe and northern sections of the Gold Coast. Water releases from Somerset Dam supplement Wivenhoe Dam, which in turn releases water to a treatment plant and pump station at Mt Crosby some 60 km further downstream. This regulated section enjoys constant steady flows and is heavily utilised for recreational pursuits such as swimming, canoeing, fishing and camping. Water in these reaches is generally very clear and lungfish are easily observed both from the bank and by snorkelling. The area has been the focus of long-term studies of lungfish breeding and biology (e.g. Kemp 1984, 1987, 1993). Lungfish also occur in smaller numbers downstream of the Mt Crosby Weir to the tidal limit of the Brisbane River. Upstream of Wivenhoe Dam, lungfish inhabit Lake Wivenhoe, Lake Somerset and upper reaches of the Brisbane and Stanley Rivers. The distribution and abundance of lungfish and particularly, the reliability of recruitment events in these areas is poorly understood. With such a long-lived species, a population of adults could survive for decades, conveying the impression of a healthy population even if no successful recruitment was occurring.

In the North Pine River, lungfish populations are known both above and below the North Pine Dam. When flow events necessitate water releases from the dam, lungfish gather immediately below the dam wall and subsequently become stranded in large numbers when flows recede. Lungfish moving through the outlet works enter lower reaches of the North Pine River where they are restricted by increasing salinity to a short section upstream of the tidal influence. Fish kills involving lungfish have been observed during king tides in pools near the tidal interface. A similar situation exists in and below the Enoggera Reservoir, where lungfish moving downstream in flow events occupy pools in the stream below the reservoir wall to the tidal limit. Lungfish populations in the Coomera and Albert Rivers are poorly understood, with only a few recent records to confirm their existence. In the Coomera River, adult fish have been collected immediately upstream of a small weir that separates freshwater from the estuary. However, apparently suitable habitat is available in several river reaches upstream of this point. There have been no surveys conducted to determine whether the fish that were captured are the remnants of a declining population or part of a viable breeding population that extends throughout the catchment.

EXTERNAL MORPHOLOGY

Detailed descriptions of lungfish anatomy have been provided by several authors including Krefft (1870b), (Günther 1871a), Kemp (1987) and Lambkin (1987). The following description is largely drawn from these references.

Australian lungfish are characterised by a robust elongate body covered in overlapping layers of large, thick scales (Fig. 4). Detailed information on body size is provided in subsequent sections. The upper body surface varies from light shades of olive-green to brown or almost black, with a series of small patches of darker pigment scattered in random patterns. These patches are generally more obvious and widespread in juvenile specimens. In contrast, the underbelly is coloured in various shades of orange, yellow, pink or cream. Dean (1912) suggested that the underbelly colour is brighter during the spawning season. There are also rare examples of individuals that are predominantly yellow or orange in colour on both dorsal and ventral surfaces. The body surface is protected by a viscous transparent epidermal layer. Despite reports to the contrary, there are no visible external features that reliably distinguish the sex of adult Australian lungfish. Characters such as changes in underbelly colour and the size or shape of the cloaca have all proved to be unreliable indictors when verified by internal examination of the gonads (Brooks and Kind 2002). During the spawning season, running ripe individuals can be easily stripped to collect eggs or milt thus determining sex. Outside of this period, sex can only be determined only by internal examination of the gonads.

The head is somewhat conical but flattened to produce a broad dorsal surface. Two unusually large scales protect a thin section of the skull in this area (Krefft 1870b). The snout and head exhibit clusters of sensory pits or 'ampullary organs' used for detecting weak electric fields emitted by potential prey items (Watt *et al.* 1999). The eyes and mouth are small relative to the overall body size and the jaws have only limited gape. The eyes usually appear brown or shades thereof. Dentition in adult specimens consists of complex crushing tooth plates (Krefft 1870b), described in detail by Kemp (1977). The dentition of juvenile lungfish consists of sharp cones used to grasp and hold live prey items. Gradual wearing of these cones is matched by changes in the diet to include a wider range of prey and eventually plant material (Kemp 1977, 1987).

Four pairs of functional gills are protected by fleshy opercula, which close over the base of the pectoral fins. The pectoral and pelvic fins are frequently described as 'flippers' or 'paddles' reflecting their roughly diamond shape and smooth surface. These fins act to brace the body against the substrate and are used to manoeuvre the fish during foraging (Dean 1906, 1912). The dorsal fin emerges approximately half way along the length of the body and joins the caudal and anal fins to outline the broad flattened tail. All the fins are covered by scales smaller than those on the body. Unlike most teleosts there are no fin spines or other sharp surfaces on the body or gill covers.

Juvenile Australian lungfish appear at first glance to be miniatures of adult specimens. However, clear differences become apparent on closer inspection. Juveniles exhibit a more pronounced and slightly rounded head (Kemp 1987). The dark spots on the body are frequently clustered giving a mottled appearance and the sensory pores are more prominent on the head and snout. As noted by Kemp (1987), the dorsal fin arises further forward, close to the back of the head.

Fig. 4 Australian lungfish exhibiting distinctive external features described in the text. This individual was collected as part of ongoing monitoring programmes in the Burnett River.

LIFE CYCLE & REPRODUCTION

Australian lungfish complete their life cycle entirely in freshwater and can be bred in captivity. The annual breeding season for *N. forsteri* can commence as early as July and continue until at least January (Kemp 1987; Brooks 1995). However, the majority of spawning occurs between August and December with some variability evident between catchments (Bancroft 1928; Grigg 1965; Kemp 1984, 1986, 1993; Brooks 1995; Brooks and Kind 2002). Suitable breeding habitat is often widespread during this period after low, stable winter flows have provided opportunities for aquatic plant communities to establish and flourish.

The exact triggers for spawning in *N. forsteri* have never been fully resolved. Kemp (1984) noted that the onset of spawning coincides with increasing day length and that spawning activity usually ceases before the arrival of summer rainfall and subsequent flow events. However, other triggers such as temperature, river flow and water quality may also exert influences that are not yet fully understood. Captive breeding of lungfish has been achieved as far south as Sydney in New South Wales, where spawning typically occurs from October to December (Joss and Joss 1995).

Lungfish courtship and spawning behaviour has been described several times and can be easily recognised. Prior to spawning, individuals gather in patches of suitable habitat forming aggregations that may number up to 100 individuals (Brooks 1995). The shallow water habitat lends itself to viewing these spectacles from elevated river banks. During the initial stages of courtship, individuals separate into pairs or small groups and exhibit repetitive circling behaviour, described by Grigg (1965) as 'follow the leader'. Groups typically contain only one female but can include up to eight males (Brooks 1995). Courtship is usually accompanied by frequent air breathing as individuals compensate for short-term oxygen deficits generated by their increased activity (Grigg 1965; Kind *et al.* 2002). Females swim through patches of macrophytes in shallow water, typically close to the banks. Once a female begins to deposit eggs into suitable spawning habitat, the accompanying male(s) exhibit increased interest, often nudging the female's cloaca and jostling for positions close to her (Grigg 1965; Kemp 1987; Brooks 1995). Attendant males fertilise the eggs as they emerge after which the female moves off and the circling behaviours recommence. The negatively buoyant eggs have an outer membrane that remains sticky for a short period allowing the eggs to adhere to the macrophtyes (Kemp 1987). Spawning events may commence at any time of the day and often continue for several hours. No parental care is evident after the eggs have been deposited.

Spawning activity occurs in shallow glides and along river margins in close proximity to dense macrophyte beds or partially submerged riparian vegetation. Early researchers reported collecting lungfish eggs from 'weeds' growing in shallow water along the banks (Caldwell 1885a; Illidge 1893; Semon 1899). Kemp (1984) collected viable eggs from submerged macrophytes and tree roots in the Brisbane River and from root masses of the noxious weed water hyacinth in Enoggera Reservoir. Some apparently suitable macrophyte beds did not contain lungfish eggs leading to a conclusion that lungfish may be specific in their choice of spawning sites. Following a series of similar collections, Kemp (1993) noted that lungfish spawning is largely restricted to submerged plants that occur in shallow water, have dense growth forms and contain food items such as algae, protozoa, small molluscs and crustaceans. This led to the hypothesis that microhabitat characteristics at the spawning site may be more important than the specific macrophyte species involved. Subsequent studies such as Brooks (1995) and Brooks and Kind (2002) have expanded on this idea and modelled empirically the characteristics of suitable lungfish spawning habitat. These studies have demonstrated that lungfish spawning occurs in a diverse range of aquatic, semi-aquatic and submerged terrestrial plant species. There is a strong positive correlation between macrophyte density and the intensity of lungfish spawning. A large proportion of lungfish eggs collected by Brooks and Kind (2002) were located in river sections where macrophytes covered 90% or more of the available substrate. Water flow is not mandatory in spawning areas but influences the depth at which eggs are deposited (Kemp 1993; Brooks 1995; Brooks and Kind 2002). In still water, up to 90% of eggs are deposited at depths < 200 mm, whereas in flowing water highest egg densities occur between 400-600 mm (Brooks and Kind 2002). The unifying features of lungfish spawning sites are that they provide protection from predators, suitable dissolved oxygen levels for developing embryos and abundant food supplies for recently hatched lungfish (Kemp 1984; Brooks 1995; Brooks and Kind 2002).

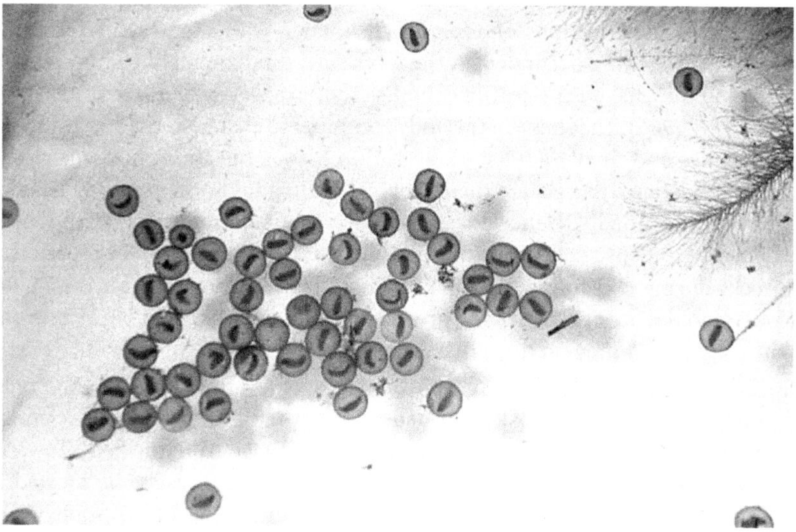

Fig. 5 Eggs of the Australian lungfish containing larvae at advanced stages of development. These eggs were spawned at a commercial fish hatchery producing lungfish for the aquarium trade (Photo by Gordon Hides).

Fertilised lungfish eggs are hemispherical when released, green or brown in colour and 2-3 mm in diameter (Fig. 5). The sticky outer layer mentioned earlier is one of four translucent membrane layers that protect the developing embryo (Kemp 1987, 1994). The eggs swell on contact with water until the spherical 'jelly package' reaches a final diameter of approximately 10 mm. The egg package quickly becomes opaque as detritus in the water column clings to the sticky outer surface.

Developing lungfish embryos are subject to a range of potentially lethal environmental stressors and predators. Kemp (1994) reported that a small proportion of eggs are infertile or suffer mortal physical damage. Other embryos exhibit developmental abnormalities or suffer damage when egg membranes fracture during hatching. In most cases where the egg or embryo is damaged, bacterial or fungal infection follows. The rate of infection varies between locations and can exceed 30% (Kemp 1994; Brooks and Kind 2002; Kind *et al.* 2005). Because the eggs are laid in shallow water, even small fluctuations in water level can expose the eggs or otherwise inundate them to levels where oxygen uptake is compromised.. Eggs that detach from macrophytes and settle into the substrate have reduced chance of success as the surface area for oxygen uptake is reduced under these conditions. Kemp (1981) reported that extreme temperatures (<10°C or >30°C) can also be lethal to cleaving eggs. The extent of predation on early lungfish life history stages is poorly understood. However, a range of potential predators including fish, birds, invertebrates and even other lungfish are believed to prey on lungfish eggs, larvae and juveniles.

The incubation period for lungfish embryos varies according to temperature but generally extends for 23-30 days (Kemp 1987). Fully developed embryos resemble miniature versions of adult lungfish albeit with slightly different body proportions. Hatching occurs when the larvae are approximately 10 mm long, at which time they have thin 'leaf like' bodies that exaggerate the size of the head (Kemp 1982, 1986). The shape of the body continues to change as the larvae mature into juveniles and finally adults. There is no obvious metamorphosis or clearly discernable changes that define their progression to adulthood. Recently hatched lungfish are essentially defenceless and avoid predation only by lying motionless on their side in dense cover. The hatchlings do not begin feeding for a period of 2-3 weeks during which they rely on remaining yolk supplies for nutrition. Once the yolk globules are exhausted, the juveniles right themselves and begin using an ambush strategy to prey on tiny crustaceans and molluscs (Kemp 1977).

Juvenile Australian lungfish are encountered only infrequently in the wild. Ever since the species was first described by Krefft (1870a), concerned scientists have pondered the paucity of juvenile specimens collected from the field. Towards the end of the annual spawning season, late stage eggs can easily be located in the spawning habitat using push nets or searching in patches of macrophytes by hand (Kemp 1984, 1986, 1993; Brooks 1995). Brooks (1995) also collected recently hatched larvae from the same habitat. After this point, lungfish rarely appear in targeted surveys until they are approximately 300 mm in length (Brooks and Kind 2002). The early collectors such as Bancroft (1911) were unable to locate juveniles, despite employing destructive methods such as dynamite and lime. Even modern techniques such as backpack electrofishing have not yielded consistent catches of young fish. Some researchers such as Illidge (1893) and Bancroft (1911) believed that juvenile lungfish buried themselves in the mud, a suggestion now widely accepted as unlikely. Records collated by Kemp (1987) provided evidence of sporadic peaks in recruitment. In a long-lived, slow-growing fish with low adult mortality this may be sufficient to maintain populations.

Juvenile lungfish held in breeding facilities actively avoid bright light (Joss pers comm.). Yearling lungfish filmed in above ground tanks by Kind (2002) invariably sheltered in the most complex habitat made available to them and especially in areas where the surface was shaded. In the wild, it is likely that juveniles exhibit a similar affinity for complex instream habitat (Brooks and Kind 2002). Records of juveniles collected in the field have typically been taken from dense cover in water < 300 mm deep. On the basis of available evidence it seems likely that juveniles remain in their natal habitat for long periods.

Because so few specimens have been collected from the wild, natural growth rates of juvenile lungfish are poorly known. In captive situations, growth rate appears to vary according to husbandry techniques and feeding regimes. Laboratory studies such as Kemp (1981) reflect extremely slow growth in keeping with small enclosures and individual feeding. At the other extreme, commercial hatchery operators have reported that fish reared in 'green water' conditions can attain lengths of 300 mm in 12 months and 450-550 mm in 24 months (Gordon Hides pers comm.). It is unlikely that either of these scenarios accurately reflect growth rates that would be achieved in the wild. Wild fish are subject to environmental stressors such as flow events, interspecific and intraspecific competition and changes in resource availability. Captive specimens receive regular food portions and lead a largely sedentary existence.

FOOD AND FEEDING

The diet of *N. forsteri* changes over time reflecting a progression from larval to adult dentition (Kemp 1977). Recently hatched lungfish possess isolated conical tooth cusps, which are used to catch and hold tiny crustaceans and worms, occasionally supplemented by filamentous algae (Kemp 1977, 1995). During this period, they are essentially ambush predators. These cusps fuse over time to form a series of ridges and eventually the crushing tooth plates that are characteristic of the adult stage (Kemp 1977). The digestive system of *N. forsteri* is also unusual. The thick, straight intestine contains a complicated spiral valve, which begins in the prepyloric or gastric region (Gunther 1871; Rafn and Wingstrand 1981). In addition, the pancreas contains numerous Islets of Langerhans, similar to those of tetrapods (Rafn and Wingstrand 1981).

Adult Australian lungfish may be broadly described as opportunistic omnivores. Individuals held in captivity will eat a range of live food items, meat, fruit and vegetables, dried pet food, filamentous algae and various other plant parts (Kemp 1987). In the wild, individuals forage amongst macrophytes, along the banks and on the river bed. Watt *et al.* (1999) demonstrated that *N. forsteri* are capable of locating hidden prey items using ampullary organs scattered in clusters around the head. Large quantities of plant material and substrate particles are ingested during feeding, much of which passes through the digestive system in a relatively intact state. In his original description of the species Günther (1871a) described the gut contents of two specimens as "crammed full of more or less masticated leaves of various plants". In some cases this includes terrestrial grasses that have been recently inundated, an observation which historically fuelled proposals that the species could forage on land. There are also reports of wild lungfish ingesting the fruits and seed pods of terrestrial plants such as *Eucalyptus sp.*, *Ficus sp.*, and prickly pear (Spencer 1892; Whitley 1927; G Hafner pers comm.). Food items such as small molluscs, crustaceans and worms are gleaned from the plant material and broken down by the crushing action of the mouthparts (Kemp 1987). It is also reported that lungfish will prey on tadpoles and small fish if the opportunity arises (Kemp 1995). This suggestion is supported by reports of anglers catching lungfish using artificial lures and flies. Accidental ingestion is evident in that lungfish eggs have been recovered from the faeces of lungfish surveyed during the spawning season. It is likely that the eggs are ingested accidently as adults graze in shallow macrophyte beds also utilised for spawning. Interestingly, largely intact cane toad skins were retrieved during 2006 from the faeces of lungfish after large numbers of toads died during a 'cold snap' in the Burnett River (QPI&F unpublished data).

POPULATION ECOLOGY

Adult Australian lungfish are among the largest of Australia's freshwater fish exceeded only by the Barramundi *Lates calcarifer*, the Murray Cod *Maccullochella peeli peeli* and the Freshwater Sawfish *Pristis microdon*. Even so, they rarely if ever reach the length of six feet (1.8 m) described by authors such as Krefft, (1870b) and Spencer (1892). A sample of almost 3000 lungfish captured and released by Brooks and Kind (2002) in the Burnett River included individuals ranging in total length (TL) from 345-1420 mm, with a mean TL of 906 mm. The length-weight relationship for these fish is illustrated in Fig. 6. The heaviest fish in this collection was 25 kg. Likewise, 109 fish taken from the Mary River by O'Connor (1897) ranged from 825-1125 mm. A recent 'snapshot' survey of lungfish in the North Pine, Brisbane and Mary River systems (Hydrobiology 2007), yielded similar results. This survey caught 245 lungfish from the Brisbane River ranging from 405-1175 mm TL, 102 from the North Pine River ranging from 675-1205 mm TL and 46 from the Mary River reaching 476-1220 mm TL.

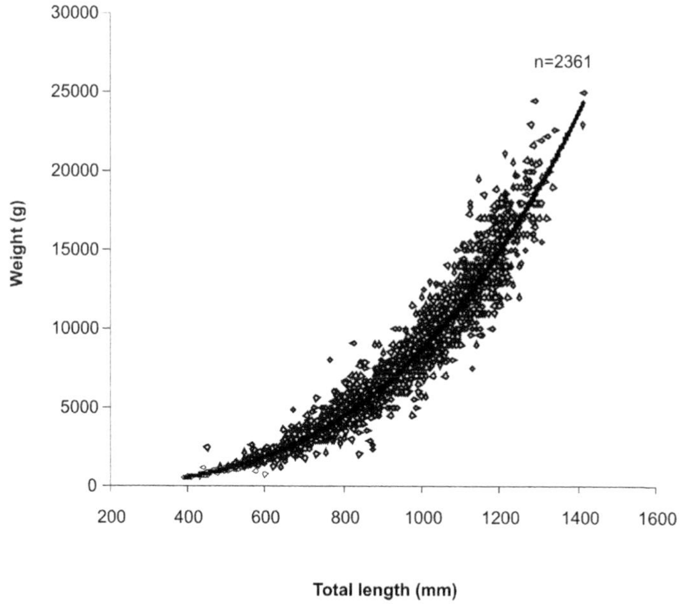

Fig. 6 Relationship between total length and weight for all lungfish collected by Brooks and Kind (2002) (n = 2361). The relationship is described by the formula $W = 9.96 \times 10^{-6} \times L^{2.98}$.

Notwithstanding some uncertainty in relation to juvenile growth rates, *N. forsteri* is clearly a slow-growing species across the bulk of its life span. Mark/recapture data from the Burnett River indicates that growth may be as slow as 5 mm/yr after individuals reach sexual maturity (Brooks and Kind 2002). There is some evidence of resource mediated variability in growth patterns across the range of *N. forsteri*. Lungfish collected from the lower Mary River are considerably heavier than fish of equal length in the upper Mary River, Burnett River and Brisbane River (QPI&F unpublished data). Likewise (Kind *et al.* 2005) reported that lungfish resident in isolated pools below Claude Wharton Weir exhibited generally poorer condition than fish from other sections of the Burnett Catchment.

Internal examination of 586 lungfish in the Burnett River revealed that the sex ratio was approximately 50:50 and that 80% of individuals > 1200 mm were female (Brooks and Kind 2002). Males appear to mature at an earlier age than females. The average length of sexually mature males collected by Brooks and Kind (2002) was 767 mm compared to 834 mm for mature females. Growth curves based on mark/recapture data predict that lungfish of these lengths would be 15 and 20 years old respectively (Fig. 7). In both sexes, 100% of individuals >1000 mm were sexually mature. In keeping with its slow growth, *N. forsteri* is likely to be a long-lived species and they may reach 50+ years in the wild (Brooks and Kind 2002). However, they may not receive the dietary variety available in the wild.

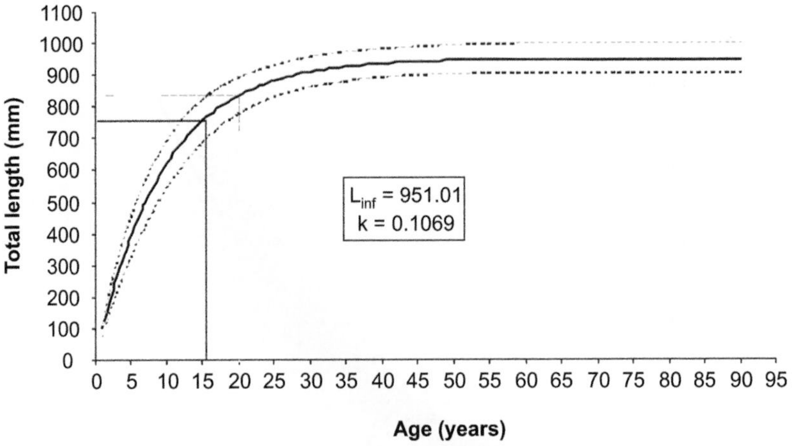

Fig. 7 Von Bertalanffy growth curve and 95% confidence intervals calculated from lungfish mark/recapture data collected by Brooks and Kind (2002) using the method of Francis (1988). Estimates of average age at maturity for males (solid vertical line) and females (broken vertical line) are presented based on average lengths at maturity of 767 mm and 834 mm for males and females respectively.

The genetic structure of Australian lungfish populations was examined by Frentiu *et al.* (2001) using tissue samples collected from the Brisbane, Mary and Burnett River Catchments. These authors reported low allelic diversity at allozyme and mitochondrial loci and minimal genetic differentiation between catchments within the species distribution. Low levels of genetic variability were unexpected given the abundance of lungfish, geographic separation of river drainages and the long evolutionary history. The results were interpreted as evidence of possible population bottlenecks, where populations were severely reduced during historic periods of aridity and subsequently recovered from a small number of survivors.

RESPIRATORY PHYSIOLOGY

Undoubtedly the most notable feature of *N. forsteri* is its bimodal respiratory system. Australian lungfish are facultative air-breathers, with well-developed gills on all branchial arches and a dorsally situated, lung (Gunther 1871a; Grigg 1965a; Johansen and Hansen 1968). At rest in well-aerated water *N. forsteri* rarely breathes air, respiration being supported almost entirely by the gills (Lenfant *et al.* 1966). This differs markedly from other extant lungfishes, which must breathe air to survive.

Until recently opinions were divided regarding the functional significance of the lung. The commonly accepted explanation was that the lung evolved to counter low levels of oxygen during periods of drought. However, several authors have offered alternative hypotheses. Spencer (1892) argued that the lung would be most useful when water is turbid, such as during periods of flooding. Bancroft (1918) suggested that the lung might be most useful when fish are sick or suffering stress. Grigg (1965b) measured oxygen levels in the Burnett River and concluded that lungfish in that river would only be exposed sporadically to low oxygen levels. He was able to show experimentally that the rate of aerial breathing increased when individuals were forced into exercise and concluded that they would use their lungs routinely as an adjunct to aquatic respiration when activity demanded increased oxygen supply. This is not to say that lungfish do not also make use of the lung when when oxygen levels are low, and the physiology accompanying exposure to low oxygen has been explored many times (see Lenfant *et al.* 1966; Johansen *et al.* 1967; Ftitsche *et al.* 1993). More recently, Kind *et al.* (2002) examined the effects of moderate and severe hypoxia on air breathing frequency and respiratory properties of the blood. This study concluded that *N. forsteri* employs at least three mechanisms in combination to preserve oxygen delivery during periods of hypoxia; increased branchial ventilation, increased air breathing and increased haemoglobin oxygen affinity, the latter being mediated by altering levels of organic phosphate within the red cells. Air breathing observed during the early onset of hypoxia was consistent with the interpretation that aerial breathing supplements short-term oxygen demand during periods of increased activity (Fig. 8). Frequent air breathing observed during courtship and spawning behavior also support this explanation.

Fig. 8 Aerial breathing is used to supplement oxygen supply during periods of increased activity (Photo by Gordon Hides).

Color image of this figure appears in the color plate section at the end of the book.

HABITAT USE AND MOVEMENT PATTERNS

Until recently, the manner in which Australian lungfish move through and exploit the riverine environment was inferred largely from field observations and studies examining spawning behaviour (e.g. Bancroft 1928; Grigg 1965; Kemp 1984, 1986, 1993; Brooks 1995). On the basis of angling experiences, early collectors such as Masters (1871) (quoted in Monteith 2001) and Spencer (1892) proposed that lungfish were most often encountered in deep pools, and were rarely seen in shallow water. Other authors such as Lambkin (1987) and more recently Kemp (1986, 1995) have reiterated this conclusion on the basis of field observations. Kemp (1995) added that lungfish congregate amongst logs, aquatic plants and in underwater caves, preferring depths between 3 and 10 metres.

Application of radio telemetry and mark/recapture techniques in recent years has furthered our understanding of the way lungfish use their environment (Kind 2002, Brooks and Kind 2002). These studies have highlighted the importance of overhanging riparian vegetation and complex shallow-water habitat, particularly during early life-history stages, contradicting the notion that deep pools are the preferred habitat. Kind (2002) demonstrated that Australian lungfish in the Mary River were strongly associated with instream wood piles, macrophyte beds and overhanging vegetation, utilising depths between 2.0 and 3.0 m far in excess of their availability. Patches of open water and deep sections were largely avoided by the tagged individuals. Some common aquatic macrophyte species were also utilised more frequently than predicted from the extent of their coverage, particularly in sections where submerged macrophyte beds were associated with species such as water lilies. Year old lungfish also showed strong associations with habitat that included underwater structure and surface shading. Kind (2002) observed that lungfish breeding grounds, nursery areas and adult foraging areas shared similar habitat features and frequently coincided or overlapped. All three were unified by the presence of submerged aquatic plants highlighting the importance of shallow water habitat to the species.

Apart from an assertion by Semon (1899) that lungfish forage by day and night, it has long been accepted that *N. forsteri* is a predominantly nocturnal species (Dean 1906; Longman 1928; Kemp 1987). Grigg (1965b) demonstrated increased nocturnal activity using a paddle wheel and counter wheel system in tanks containing captive lungfish. Kind (2002) observed a similar pattern in radio tagged wild fish, noting that tagged individuals were relatively inactive between 11.00am and 3.00pm. Outside of this period movements were more variable with peaks in the early morning and late afternoon. The most distinct daily movements were observed in the early morning as individuals returned to daytime refuges (Kind 2002). Captive juveniles followed a similar daily rhythm, leaving daytime shelters in the late afternoon to forage during the night (Kind 2002).

Kemp (1987) stated that in her experience, lungfish are a generally sedentary species, seemingly reluctant to move even during times of flood. Lambkin (1987) provided anecdotal support for this notion, documenting repeated sightings of distinctively marked lungfish that appeared to reside in a short stretch of the Burnett River. In contrast other authors have inferred that Australian lungfish may undertake regular longitudinal movements in response to triggers such as increased water flow or temperature. Masters (quoted in Monteith 2001) reported that lungfish in the Burnett River were distributed widely during floods but retreated to deep waterholes when the water levels receded. Ramsey (1871) noted that lungfish are easiest to catch during spring and suggested that they become move active with the onset of the spawning season. This suggestion has since been supported by Kemp (1984, 1986). A series of interviews and anecdotal accounts collated by Brooks (1995) revealed that lungfish regularly congregate below instream barriers on the Burnett River, particularly during flow events. This was interpreted as evidence that the fish were attempting to move upstream in response to elevated flow, a common trigger for migration in many Australian freshwater fish (Northcote 1984).

Recent radio telemetry and mark/recapture programs have demonstrated that Australian lungfish undertake largely localised movements around a distinct home range (Kind 2002). The home range is typically 1-1.5 km in length and centred on a small number of regular refuges where individuals shelter during the day. Daytime refuges include instream features such as macrophyte beds, log piles or submerged tree branches. Nocturnal foraging is largely confined to sections of the "home pool" or adjacent pools nearby. Lungfish moving between pools during nightly foraging events have been recorded moving through glide and riffle sections as shallow as 12cm. Movements are more restricted during the winter months with individuals foraging close to their daytime refuges (Kind 2002).

In flowing river reaches, movements outside of the normal home range appear to be rare and were observed only in a small proportion of the population. In contrast, lungfish in impounded waters can be highly mobile. Lungfish spawning habits are well documented, and confirm that *N. forsteri* is not diadromous. Spawning occurs exclusively in freshwater, generally at depths less than 1.5 m (Bancroft 1928; Grigg 1965; Kemp 1984, 1986, 1993; Brooks 1995; Brooks and Kind 2002). Large, negatively buoyant eggs are deposited predominantly amongst aquatic macrophytes as well as in the roots of terrestrial vegetation protruding into the water and the rootlets of some floating aquatic plants (Caldwell 1885a; Illidge 1893; Kemp 1982, 1984, 1986, 1993; Brooks 1995; Brooks and Kind 2002). Although this type of habitat is common throughout the distribution of *N. forsteri*, freshwater rivers are dynamic and sometimes ephemeral environments.

During late autumn and winter, lungfish tagged in impounded reaches of the Burnett River by Brooks and Kind (2002) moved out of the impoundments to seek out suitable spawning habitat in shallow pools and glides. Individuals were subsequently recorded in sections upstream of the impoundments and in the lower reaches of some tributary streams. This pattern was observed in both radio tagged individuals and mark/recapture records. Movements up to 35 km were recorded by lungfish in this study (Fig. 9). Following the spawning period, return movements occurred on a staggered basis with individuals often utilising small flow events or the tail of larger flows to assist their downstream passage (Brooks and Kind 2002). Berghuis and Broadfoot (2004) reported that lungfish downstream of Ned Churchward Weir (on the lower Burnett River) also made upstream movements during minor flow events, taking advantage of increased connectivity to move between pools.

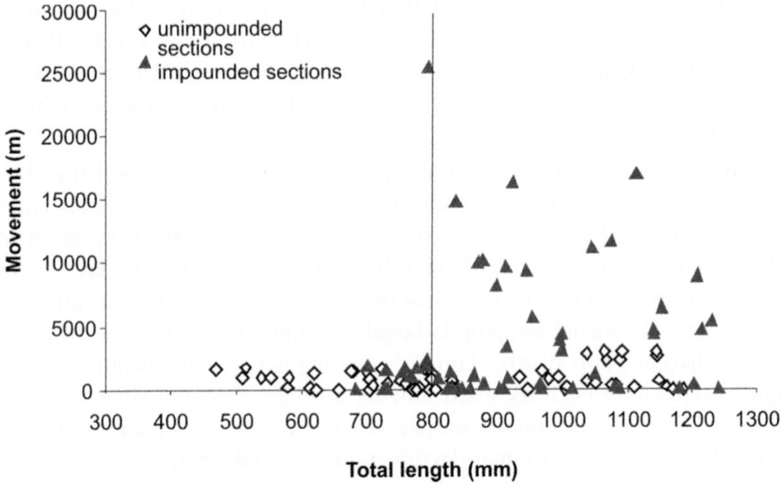

Fig. 9 Movements of recaptured lungfish from impounded and flowing sections of the Burnett River (n = 124). Vertical line at 800 mm represents the approximate threshold for sexual maturity (data from Brooks and Kind (2002)).

Color image of this figure appears in the color plate section at the end of the book.

Radio telemetry has also highlighted aspects of Australian lungfish movements during major flow events. Radio tagged lungfish being monitored by Kind (2002) experienced a major flood event in the Mary River during February 1999. These fish responded to rising water levels by retreating to shallow water along the river margins and eventually sought shelter in dense patches of inundated riparian vegetation and behind large tree trunks after the river broke its banks. When the water levels receded, all but one of the 13 fish had re-entered the river channel and occupied positions within their normal home range.

Many aspects of behaviour in *N. forsteri* remain unexplored. Brooks and Kind (2002) observed that many captured individuals exhibit fresh or healed crescent shaped wounds on the dorsal fin. The shape of these wounds is consistent with suffering bites from other lungfish. The frequency of these injuries was closely correlated to catch rates, suggesting that density related territorial behaviour may occur in the species. Observations also suggested that the size of the wounds was correlated to size of the fish. Aggressive behaviour is clearly evident in captive juveniles raising the possibility of dominance hierarchies among adults in feeding and for shelter areas.

CONSERVATION STATUS

There has been debate about the conservation status of Australian lungfish for more than a century. Generations of Wide Bay/Burnett residents have pondered the curiosity in their local waterways and formed various opinions on the significance and status of the species. Early settlers are known to have exploited the lungfish for food (Krefft 1870a; O'Connor 1898), a practice that persists to some extent despite its protected status. Some residents, particularly anglers consider lungfish to be a pest, which has expanded in numbers since the protection of the species, introduction of large-scale river regulation and damming. Others view the Australian lungfish as an endangered species teetering on the edge of extinction as the modern world closes in around it. Aboriginal groups, the traditional owners of the land have recently expressed strong interest in the ongoing conservation of lungfish, citing the benefits to other species and general river health as positive outcomes that will be associated with lungfish conservation. Eve Mumewa Doreen Fesl, Elder of the Gubbi Gubbi people has stated publicly that lungfish, known to the Gubbi Gubbi as 'Dala' have sacred (totemic) value. The Gubbi Gubbi did not kill or eat lungfish, but sought to ensure that they were protected from harm.

Many early collectors had difficulty obtaining specimens and soon realised that lungfish occupied a more restricted distribution than had been previously expected. In response to this, the apparent lack of juveniles and large numbers adult lungfish being removed for food by early settlers, the Royal Society of Queensland commissioned the translocation programs discussed earlier.

In recognition of the potential impacts of continued subsistence and recreational harvest, *N. forsteri* was listed as a protected species under the Queensland *Fish and Oyster Act 1914*. This status prevented the taking of lungfish by any means, a situation that remains current under the Queensland *Fisheries Act 1994*. This legislation requires that a permit be obtained to collect and hold a lungfish specimen for display or research and recreational harvest of the species is totally prohibited.

Debate over the conservation status of Australian lungfish was reignited in the mid 1990s by a proposal for a new weir to be erected near Wallaville on the Burnett

River (Fig. 10). The weir site included well known lungfish breeding grounds and the proposal created considerable debate about the impacts of expanding water infrastructure on resident lungfish populations. A short-term study of lungfish breeding in the area by Brooks (1995) highlighted the paucity of information available on aspects of lungfish ecology such as movement patterns, habitat requirements, population structure and genetics. Approval for the Ned Churchward (Walla) Weir project included a recommendation for long-term studies to assess the cumulative impacts of river regulation on lungfish populations, to be undertaken prior to further water infrastructure expansion. This recommendation was taken up and a four year study of Burnett River lungfish populations commenced in January 1997. Later that same year a nomination was raised to list *N. forsteri* as an endangered species under the Australian Government's (now repealed) *Endangered Species Protection Act 1992*. This nomination was rejected on the basis their numbers had not been reduced to such a critical level, or their habitat so reduced that the species was in immediate danger of extinction.

Fig. 10 Proposals for new dams in south east Queensland during the 1990s reignited debate about the conservation status of lungfish.

The four-year study of lungfish undertaken by Brooks and Kind (2002) concluded that impounded water bodies represent only marginal breeding habitat for Australian lungfish. Inundation of the river channel upstream of Ned Churchward Weir resulted in steep bank profiles that were unsuited to the proliferation of aquatic plants commonly used by lungfish for spawning. Despite intensive sampling effort, no evidence of breeding activity was observed within Ned Churchward Weir during the entire study. This finding was consistent with the observation that lungfish ceased breeding in Enoggera Reservoir after removal of water hyacinth from the lake margins (Kemp 1987). These findings coupled with discussions about substantial expansion of water infrastructure in the catchment prompted the nomination to list *N. forsteri* as a vulnerable species under the Australian Government *Environmental Protection and Biodiversity Conservation Act 1999* (the EPBC Act). This nomination was successful and *N. forsteri* was listed as a vulnerable species in August 2003.

In summary, the nomination was accepted on the basis that Australian lungfish were eligible for elevation to vulnerable status under categories 1 & 2 of the EPBC Act threatened species guidelines. A scientific panel considering the application accepted that marked declines in the quality and extent of lungfish breeding habitat resulted from the creation of impoundments. The decline was predicted to impact on lungfish populations over the next three generations. The listing took into account the restricted distribution of the species and the high likelihood of further water infrastructure development in the south east Queensland region. Few data were available to assess changes in the size of lungfish populations or to quantify the probability of extinction. This was acknowledged and meant that *N. forsteri* was ineligible for listing under categories 3-5. However, the overall conclusion was that the recent reduction in the extent and quality of breeding habitat was likely to result in unsustainably low recruitment levels, leading to a substantial reduction in population size within the next three generations.

In the Mary River catchment, Australian lungfish coexist with two other threatened aquatic species, the Mary River cod *Maccullochella peelii mariensis* and the Mary River Turtle *Elusor macrurus*. Both these species are listed as endangered under the EPBC Act. The Mary River Turtle is also listed as Endangered under the Queensland *Nature Conservation Act 1992* and Mary River cod are a protected species under the *Fisheries Act 1994*. In the Burnett River, the distribution of lungfish coincides with the recently described Southern Snapping Turtle *Elseya albagula*. This species is considered to be rare and has high conservation value. The successful implementation of recovery actions for lungfish including in-stream and riparian habitat rehabilitation, improved fish passage, improved management of water storages, provision of suitable flow regimes, maintenance of water quality, prevention of further disruptions to hydrology and management of introduced species would almost certainly benefit all of these species as well as other native aquatic species.

Australian lungfish have been protected from international trade since 1977 under Appendix II of the Convention on International Trade in Endangered Species (CITES). However, three commercial fish hatcheries have been granted permission to captive breed specimens for the aquarium trade, under strict conditions and one has been granted approval to export specimens. Despite its vulnerable status under the EPBC Act, *N. forsteri* is not a listed species under the Queensland *Nature Conservation Act 1992* or on the IUCN Red List of Threatened Species. The Australian Society for Fish Biology still lists lungfish as 'common and secure'. This situation highlights the difficulty of assessing the true conservation status of an abundant, long-lived and slow-growing species that occupies highly variable and disturbed habitat within a restricted distribution, especially when recruitment rates are difficult to assess.

THREATS TO AUSTRALIAN LUNGFISH AND MANAGEMENT MEASURES

Despite protection under State law in Queensland, fishing competition data, anecdotal reports and carcasses observed along the river banks confirm that some incidental capture of lungfish occurs in the recreational fishery. Recent reports from the Mary River Catchment indicate that a small number of residents may still be consuming lungfish or feeding carcasses to livestock and pets. The true extent of this practice remains unclear, however, the flesh of *N. forsteri* is generally considered to be unpalatable by modern day anglers. A market for lungfish exists within the aquarium trade, which is at least partially serviced by the three approved fish hatchery operators. No data are currently available on the post-release survival rate of lungfish taken by recreational anglers or the extent of the residual black market trade. However, it is unlikely that overfishing represents a significant threat to the Australian lungfish. The use of 'shrimp rakes' is also widespread within the distribution of *N. forsteri*. This bait collection method can be responsible for disturbance to aquatic plants necessary for lungfish breeding and may need revision in future reviews of Queensland legislation. Lungfish are also known to have been struck and killed by boats travelling at speed, including ski boats. This artificially increases adult mortality rates, which is a threat to recovery of the species when a fundamental conservation principle is protection of the breeding population. No data are currently available on the extent or impacts of this problem.

Barriers to fish movement such as dams, weirs and culverts occur widely throughout the distribution of *N. forsteri* and more large dams may be proposed to cater for ongoing population growth in southeast Queensland. Only a small number of the existing barriers incorporate working fish ladders or other fishways. Australian lungfish have highly specialised spawning requirements characterised by shallow water and submerged aquatic plants. Barriers to fish movement prevent

Fig. 11 Ned Churchward Weir, constructed on the Burnett River, Queensland in 1998 was fitted with a fish lock to provide for lungfish passage. The effects of this weir on lungfish and the operation of the fishway were the subject of considerable debate in the local community.

individuals from accessing suitable spawning habitat located in other parts of the catchment. By acting to effectively fragment lungfish populations, such barriers may also be having a significant effect on genetic structure within lungfish populations (Fig. 11).

Dams constructed on the Burnett River during the last decade have incorporated fishways of various designs, including vertical-slot fish ladders, fish locks and fish lifts. These devices have been designed to pass a large variety of fish species ranging from small-bodied fish, larvae and juveniles, to adult lungfish and other large predatory species. Regardless of the design, all these fishways are dependant on sufficient water flow to operate and can be severely compromised during periods of drought. Consequently, these fishways have met with mixed success and become the focus of intense debate over lungfish conservation. Water resource allocation is a complex and often controversial issue in the Wide Bay / Burnett area of Queensland where regional economies rely heavily on industries such as citrus, cattle and sugar cane.

Recent evidence has also highlighted that dams and weirs can lead to physical damage when lungfish inadvertently swim over impoundment walls. This problem was most recently seen at the North Pine Dam in June 2009 where large numbers of lungfish were injured during water releases following heavy rain. Radio-tagged lungfish have also been recorded moving downstream over weir walls in the Burnett and Mary Catchments (Kind 2002; Brooks and Kind 2002). When lungfish swim over tidal barriers separating freshwater and estuarine areas, evidence suggests that they rarely survive due to predation by sharks and salinity intolerance. At least one lungfish tagged by Brooks and Kind (2002) moved downstream and over the tidal barrage, entering the Burnett River estuary. This fish was never relocated and presumably perished. In the Burnett River plans are currently being considered for a lungfish specific fishway, to be operated for short periods following flow release events at the tidal barrage. Lungfish are also often rescued physically from the estuary by field officers operating the tidal barrage and returned to the freshwater.

Damming of rivers within the distribution of *N. forsteri* has also significantly altered natural flow regimes. The construction and operation of such water storage infrastructure, has altered the volume, frequency, duration and timing (seasonality) of riverine flows. It has effectively changed the aquatic environment upstream of barriers from lotic (flowing water) to lentic (still water, e.g. lakes). Permanently flooded riverine areas and fluctuating water levels make it difficult for macrophytes to establish within impounded areas. Altered flow can significantly impact on the growth and condition of macrophyte beds downstream of barriers. There is also evidence that flow regulation has led to increased fluctuations in water level, reduced volumes of water in river channels and therefore reduced habitat for fish, stranding of adult lungfish, exposure of breeding habitat including eggs and inundation of habitat leading to reduced habitat availability and changes in lungfish movement patterns (Fig. 12). Pumping of water directly from the rivers also occurs for agricultural use. Present water allocations in the Burnett River have occasionally resulted in the stranding of lungfish in drying pools. It is not known whether any lungfish are physically extracted from the river or injured by pumps however, significant losses of native fish have been reported in other catchments such as the Murray-Darling Basin when water is extracted for irrigation.

Fig. 12 Lungfish stranded downstream of a weir on the Burnett River after water releases were shut down.

Color image of this figure appears in the color plate section at the end of the book.

The impacts of stocked native fish and introduced pest fish on Australian lungfish are poorly understood. Native fish have been legally stocked into all the major catchments within the distribution of *N. forsteri* to enhance recreational fishing opportunities. Stocked species include highly predatory species such as Australian bass, golden perch, saratoga, silver perch and barramundi. Apart from the possibility of increased predation on lungfish eggs and juveniles, this practice may also be increasing the threat of competition for food and habitat, disturbance to spawning habitat and spread of disease and parasites. The noxious pest fish *Oreochromis mossambicus* (a species of tilapia) occurs widely in the Brisbane River Catchment. The species has also established in Boondooma Dam and other private dams within the Burnett River Catchment. Tilapia are highly and there is anecdotal evidence that tilapia disturb macrophyte beds when making their nests which could reduce the availability of suitable lungfish breeding habitat. Once established in flowing river systems, pest fish species such as tilapia are impossible to completely eradicate. In such instances information campaigns to prevent further spread and ongoing control measures based around physical removal have been implemented. Interactions between lungfish and such pest fish species remain poorly resolved. Floating and submerged aquatic weeds such as water hyacinth *Eichornia crassipes*, Fanwort *Cabomba caroliniana*, Elodea, Dense Water weed *Egeria densa* and Salvinia *Salvinia molesta*, are also widespread in south east Queensland. These weeds have reduced the quality of existing lungfish habitat by lowering dissolved oxygen levels, reducing water temperature and affecting breeding habitat by preventing light penetration and physically smothering existing native macrophytes. Some local projects have recently been established to provide ongoing control with mixed success. Secondary problems associated with ongoing aquatic weed control are also becoming evident, most notably a build up of decaying plant matter on the substrate in recently sprayed river sections where regulation has reduced flow events that would remove this decaying vegetation.

Juvenile and adult lungfish rely on complex underwater habitat, predominately macrophyte beds, woody debris and undercut banks for foraging and shelter (Kemp 1987; Kind 2002; Brooks and Kind 2002). In sections of south east Queensland, this habitat has been compromised by activities including livestock trampling, de-snagging, erosion, increased sedimentation and land clearing. Livestock watering is widespread throughout the distribution of the Australian lungfish. Livestock have been observed trampling lungfish eggs and breeding habitat while drinking. Loss of breeding habitat has the potential to impact on recruitment levels in the future. Brooks and Kind (2002) suggested that ongoing loss of spawning habitat and infrequent spawning in impounded waters may also have resulted in overcrowding within the remaining breeding grounds. Overcrowding may lead to a number of negative impacts on lungfish including increased stress, damage to eggs, increased predation on eggs and juveniles, and increased competition between hatched juveniles.

A reduction in water quality parameters such as dissolved oxygen, temperature, pH, salinity and waterborne toxins can impact on lungfish growth and egg development. Increased salinity is an issue for many river catchments including the Burnett and Mary. Increased salinity could have major effects on lungfish due to their intolerance of saline conditions. The sub-lethal impacts of salinity and effects of salinity on early stages of lungfish life-history are also poorly understood. Increased sedimentation and turbidity due to erosion and runoff can affect river productivity and thereby reduce macrophyte growth that is important breeding habitat for lungfish. Increased suspended particles can lead to smothering of eggs and substrates essential for spawning, refuge and feeding. Anecdotal observations have indicated that no eggs are found in areas where macrophytes are covered by fine sediment.

In summary, current management arrangements effectively protect the Australian lungfish from overfishing but are less effective in preventing threats to remaining areas of suitable habitat. In a long-lived and slow-growing species such as *N. forsteri*, management strategies should focus on reducing adult mortality, improving connectivity between remaining populations and providing suitable habitat for successful spawning and recruitment.

References

Agassiz, L. (1844). Recherches sur les Poissons Fossiles. Du Genre *Ceratodus*. Vol III. Neuchâtel: Imprimerie de Petitpierre, pp. 129-136.

Agassiz, L. (1870). Letter on 'synthetic types'. Nature 3: 166-167.N4 17.

Bancroft, T.L. (1911). On a weak point in the life history of *Neoceratodus forsteri* Krefft. Proceedings of the Royal Society of Queensland 23: 251-256.

Bancroft, T.L. (1918). Some further notes on the life history of Ceratodus *Neoceratodus forsteri*. Proceedings of the Royal Society of Queensland 30: 91-94.

Bancroft, T.L. (1928). On the Life history of *Ceratodus*. Proceedings of the Linnean. Society of N.S.W. 53: 315-317.

Bancroft, T.L. (1933). Some further observations on the rearing of *Ceratodus*. Proceedings of the Linnean Society of N.S.W. 58: 467-469.

Bemis W.E., Burggren, W.W. and Kemp, N.E. (1987). The Biology and Evolution of Lungfishes. Alan R. Liss, Inc. New York.

Berghuis, A.P. and Broadfoot, C.D. (2004). Downstream fish passage of fish at Ned Churchward Weir. Report prepared by Queensland Department of primary industries for the Department of State Development.

Brooks, S. (1995). Short-Term study of the breeding requirements of lungfish (*Neoceratodus forsteri*) in the Burnett River with specific reference to the possible effects of the proposed Walla Weir. Unpublished report for the Queensland Department of Primary Industries.

Brooks, S.G. and Kind, P.K. (2002). Ecology and demography of the Queensland lungfish (*Neoceratodus forsteri*) in the Burnett River, Queensland—with reference to the impacts of Walla Weir and future water infrastructure development. Final Report May 2002. Department of Primary Industries Report QO02004. Prepared for the Department of Natural Resources and Mines.

Caldwell, W.H. (1885a). The eggs and larvae of *Ceratodus*. Journal and Proceedings of the Royal Society of N.S.W. 18: 138.

Caldwell, W.H. (1885b). On the development of the Monotremes and *Ceratodus*. Journal and Proceedings of the Royal Society of N.S.W. 18: 177-122.

Dean, B. (1906). Notes on the living specimens of the Australian lungfish, *Ceratodus forsteri*, in the Zoological Society's collection. Proceedings of the Zoological Society of London 1906: 168-178.

Dean, B. (1912). Additional notes on the living specimens of the Australian lungfish, (*Ceratodus forsteri*) in the Zoological Society's collection. Proceedings of the Zoological Society of London 1912: 607-612.

De Castelnau, F. (1876a). Mémoire sur les poissons apelés barramundi par les aborigenés du Nord-Est de l'Australie. Journal of Zoology 5: 129-136.

De Castelnau, F. (1876b). Remarques au sujet du genre *Neoceratodus*, Journal of Zoology 5: 342-343.

De Vis, C.W. (1885). *Ceratodus* Post-Pleistocene. Proceedings of the Royal Society of Queensland 1: 40-43.

Fritsche, R., Axelsson, M., Franklin, C.E., Grigg, G.C., Holmgren S. and Nilsson, N. (1993). Respiratory and cardiovascular responses to hypoxia in the Australian lungfish. Respiratory Physiology 94: 173-187.

Frentiu, F.D., Ovenden, J.R. and Street, R. (2001). Australian lungfish (*Neoceratodous forsteri*: Dipnoi) have low genetic variation at allozyme and mitochondrial DNA loci: a conservation alert. Conservation Genetics 2: 63-67.

GHD (2007). Lake Manchester Dam Flood Security Upgrade. Unpublished Newsletter.

Grigg, G.C. (1965a). Studies on the Queensland lungfish, *Neoceratodus forsteri* (Krefft) I. Anatomy, histology, and functioning of the lung. Australian Journal of Zoology 13: 243-253.

Grigg, G.C. (1965b). Studies on the Queensland lungfish, *Neoceratodus forsteri* (Krefft). III. Aerial respiration in relation to habits. Australian Journal of Zoology 13: 413-421.

Grigg, G.C. (1965). Spawning behaviour in the Queensland Lungfish, *Neoceratodus forsteri*. Australian Natural History 15: 75.

Günther, C.A.L.G (1871a). Description of *Ceratodus*, a genus of gadoid fishes, recently discovered in rivers of Queensland, Australia. Transactions of the Royal Society of London 161: 511-792.

Günther, C.A.L.G (1871b). The new gadoid fish (*Ceratodus*) recently discovered in Queensland. Nature, London 4: 406-408, 428-429, 447.

Huxley, T.H. (1880). On the application of the laws of evolution to the arrangement of the Vertebrata and more particularly of the Mammalia. Proceedings of the Zoological Society of London 1880: 649-662.

Hydrobiology (2007). Lungfish Study. Survey of Lungfish in the Brisbane, North Pine and Mary River Systems. Hydrobiology Pty Ltd, Environmental Services.

Illidge, T. (1893). On *Ceratodus forsteri*. Proceedings of the Royal Society of Queensland 10: 40-44.

Johnson, J.W. (2001). Review of the draft lungfish scientific report 4 July 2001. Unpublished report.

Joss, J. and Joss, G. (1995). Breeding Australian lungfish in captivity. Fish Symposium 95, Austin, Texas, (USA).

Kemp, A. (1977). The pattern of tooth plate formation in the Australian lungfish *Neoceratodus forsteri* Krefft. Zool. Journal of the Linnean Society 60: 223-258.

Kemp, A. (1981). Rearing of embryos of the Australian lungfish *Neoceratodus forsteri* under laboratory conditions. Copeia 4: 776-784.

Kemp, A. (1982). The embryological development of the Australian Lungfish, *Neoceratodus forsteri* (Krefft). Memoirs of the Queensland Museum 20: 553-597.

Kemp, A. (1984). Spawning of the Australian Lungfish, *Neoceratodus forsteri* (Krefft) in the Brisbane River and in Enoggera Reservoir, Queensland. Memoirs of the Queensland Museum 21: 391-399.

Kemp, A. (1987). The biology of the Australian Lungfish, *Neoceratodus forsteri* (Krefft 1870). Journal of Morphology Supplement 1: 181-198.

Kemp, A. (1993). Unusual oviposition site for *Neoceratodus forsteri* (Osteichthyes:Dipnoi). Copeia 1993: 240-242.

Kemp, A. (1995). Threatened fishes of the world: *Neoceratodus forsteri* (Krefft, 1870). (Neoceratodontidae). Environmental Biology of Fishes 43: 310.

Kemp, A. and Molnar, R.E. (1981). *Neoceratodus forsteri* from the lower Cretaceous of New South Wales, Australia. Journal of Palaeontology 55: 211-217.

Kemp, A. (1990). A relic from the past - The Australian lungfish. Wildlife Australia (Autumn). 10-11.

Kind, P.K. (2002). Movements patterns and habitat use in the Queensland lungfish *Neoceratodus forsteri* (Krefft, 1870). PhD Thesis, University of Queensland, Australia.

Kind, P.K., Grigg, G.C. and Booth, D.T. (2002). Physiological responses to prolonged aquatic hypoxia in the Queensland lungfish *Neoceratodus forsteri*. Respiratory Physiology and Neurobiology 132(2): 179-190.

Kind, P.K., Brooks, S.G. and Piltz, S.A. (2005). Burnett River Dam baseline lungfish monitoring. Report prepared for Burnett Water Pty Ltd May 2005.

Krefft, G. (1870a). Letter to the Editor. *Sydney Morning Herald*. January 17, 1870. p5.

Krefft, G. (1870b). Description of a giant amphibian allied to the genus *Lepidosiren* from the Wide Bay district, Queensland. Proceedings of the Zoological Society of London 1870: 221-224.

Krefft, G. (1874). Fossil tooth of *Ceratodus palmeri*. Nature, (London). 9: 293.

Lambkin, L. (1987). *Ceratodus*, the Australian Lungfish. Vantage Press, New York, USA.

Lenfant, C., Johansen, K. and Grigg, G. C. (1966). Respiratory properties and pattern of gas exchange in the lungfish *Neoceratodus forsteri* (Krefft). Respiratory Physiology 2: 1-21.

Longman, H. (1928). Discovery of juvenile lung-fishes with notes on *Epiceratodus*. Memoirs of the Queensland Museum 9: 161-173.

Lumholtz, C. (1889). Among Cannibals. An account of four years travel in Australia and of camp life with the aborigines of Queensland. John Murray, London.

Macleay, W.J. (1871). Notes on a collection of insects from Gayndah. Transactions of the Entomological Society of N.S.W. 2: 79-158, 159-205, 239-318.

Macleay, W.J. (1883). Notes on a collection of fishes from the Burdekin and Mary Rivers, Queensland. Proceedings of the Linnean Society of N.S.W. 8: 199-213.

Meyer, A.B. (1875). *Ceratodus forsteri* and *Ceratodus miolepis*. Annals and Magazine of Natural History 15: 368.

Monteith, G.B. (2001). C.T. White Memorial Lecture for 2000. Gayndah, Lungfish and Beetles - A Fishy Tale. The Queensland Naturalist 39: 5-32.

Northcote, T.G. (1984). Mechanisms of fish migration in rivers. In. Mechanisms of migration in fishes. (McCleave, J.D., Arnold, G.P., Dodson, J.J. and W.H. Neill eds.). pp. 371-375. Plenum Publishing Corporation, New York.

O'Connor, D. (1897). Report on preservation of *Ceratodus*. Proceedings of the Royal Society of Queensland 12: 101-102.

O'Connor, D. (1898). Report of live *Ceratodus forsteri*. Proceedings of the Zoological Society London 1898: 586.

Ogilby, J.D. (1917). Ichthyological notes (No. 4). Memoirs of the Queensland Museum 6: 97-105.

Ramsey, E.P. (1871). Letter read on *Ceratodus forsteri*. Proceedings of the Zooogical Society of London 1871: 7-8.

Ramsey, E.P. (1876). On the habits of *Ceratodus*. Proceedings of the Zooogical Society of London 1876: 698-699.

Semon, R. (1899). In the Australian Bush. London, Macmillan and Company.

Simpson, R. (1994). An investigation into the habitat preferences and population status of the endangered Mary River cod (*Maccullochella peelii mariensis*) in the Mary River system, south-eastern Queensland. Department of Primary Industries - information series. Report QI94011.

Spencer, W.B. (1892). A trip to Queensland in search of the *Ceratodus*. Victorian. Naturalist 9: 16-32.

Stuart, I.G. and Berghuis, A.P. (2002). Upstream passage of fish through a vertical-slot fishway in an Australian subtropical river. Fisheries Management and Ecology 9: 111-122.

Thompson, J.M. (1975). A place for fishes. Proceedings of the Royal Society of Queensland 86: 25-27.

Teller, F. (1891). Űber den Schädel eines fossilen Dipnoers *Ceratodus sturii*. Geol. Reichsanst. Wien, Abh. 15: 1-38.

Watt, M., Evans, C.S. and Joss, J.M.P. (1999). Use of electroreception during foraging by the Australian lungfish. Animal Behaviour 1999, 58: 1039-1045.

Welsby, T. (1905). Schnappering and Fishing in the Brisbane River and Moreton Bay Waters. Outridge Printing Co., Brisbane.

Whitley, G.P. (1929). The discovery of the Queensland lungfish. Australian Museum Magazine 3: 363-364.

CHAPTER 4

The General Natural History of the African Lungfishes

Chrisestom M. Mlewa[1], John M. Green[2,*] and Robert L. Dunbrack[2]

[1]Department of Biological Sciences, Pwani University College, P.O. Box 195-80108, Kilifi, Kenya

[2]Department of Biology, Memorial University, St. John's, NL, Canada A1B 3X9

ABSTRACT

The four African lungfishes: *Protopterus aethiopicus*, *P. amphibius*, *P. annectens* and *P. dolloi*, typically inhabit fringing weedy areas of lakes and rivers, where dissolved oxygen levels are low, daytime temperatures are high, and seasonal drying is common; conditions they are adapted to because of their broad generalist diet, wide tolerances to variations in temperature and salinity, parental care of eggs and young, aerial respiration, and the ability to aestivate. Collectively they have a broad geographical distribution but sympatric populations are probably rare and are not well documented. The adaptations that enable protopterids to inhabit harsh environments may also be important in the re-colonization of areas subject to local extinction, and in their competition for resources with actinopterygian fishes. In this chapter we discuss these adaptations against what is known about the natural history of African lungfishes, an approach limited by the lack of comparative data for the four species.

Growth in sub-adult African lungfishes appears to be rapid and has been estimated at 0.3% body mass per day for a 1000 g *P. aethiopicus* in Lake Baringo, Kenya. Males may become mature at a larger size than females, possibly because they must construct and defend nest sites, and provide care to eggs and young. Female *P. aethiopicus* in lake populations only move to inshore spawning areas when ready to spawn and return to open waters soon after spawning, which probably accounts for skewed sex ratios in favor of mature females reported in open waters. Spawning is likely seasonal in most African lungfish populations but appears to be less so in lake populations of *P. aethiopicus*. Age at

*Corresponding author: E-mail: jmgreen@mun.ca

maturity for female *P. aethiopicus* is approximately 3 years. The instantaneous mortality rate for the same species in an exploited population (Lake Baringo) is estimated at 0.4 year^{-1}. There are no comparative data for other protopterid species or populations.

Ultrasonic tracking of this lake population of *P. aethiopicus* recorded daily movements of up to 5.2 km and indicated that this species is active throughout the diel cycle. Some individuals had large (5.8-19.8 km^2) home ranges which they occupied for 8 to 18 weeks. Prey availability appears to be the most important factor influencing movement in open waters but intra-specific aggression and the presence of predators may also influence the use of space. Individuals are capable of long linear movements in very turbid water suggesting a well developed navigational ability. While laboratory studies have shown that protopterids are obligate air breathers, a recent field study suggests that in well oxygenated water wild *P. aethiopicus* can meet their oxygen demands solely through aquatic respiration, thereby reducing both their metabolic costs and exposure to aerial predators. Although this result needs to be verified, it illustrates the importance of a thorough knowledge of lungfish natural history for understanding lungfish biology.

Keywords: ecology, behavior, distribution, growth, reproduction

INTRODUCTION

The four African lungfish species, the West African lungfish *Protopterus annectens* (Owen 1839), the marbled lungfish *P. aethiopicus* (Heckel 1851), the slender lungfish *P. dolloi* (Boulenger 1900), and the gilled lungfish *P. amphibius* (Peters 1844) are the only extant members of the family Protopteridae. The protopterid lungfishes are placed with the South American lungfish *Lepidosiren paradoxa* (Fitzinger 1837) (Lepidosirenidae) in the suborder Lepidosirenoidei and with the more distantly related Australian lungfish *Neoceratodus fosteri* (Krefft 1870) (Ceratodontidae) in the order Ceratodontiformes, which also includes a number of fossil forms, some dating back to the late Permian or early Triassic (Nelson 2006). Although the Ceratodontiformes are generally accepted as the closest living relatives of the Tetrapods (Rosen *et al.* 1981; Forey 1986; Brinkman *et al.* 2004), other extinct non-dipnoan sarcopterygian groups are thought to be more closely related to the tetrapods (Schultze 1986; Clack 2002). The earliest lungfishes may have been largely marine (Miles 1977; Schultze and Chorn 1998), but the ceratodont lineage appears to have been a freshwater one since the Permian (Campbell and Barwick 1986; Cavin *et al.* 2007) and all extant ceratodonts are primary freshwater fishes. Within the Ceratodontiformes the split between the *Neoceratodus* and *Protopterus*/*Lepidosiren* lineages predates the breakup of Gondwana in the Cretaceous (Cavin *et al.* 2007), with the later split between the South American and African lungfishes possibly coinciding with the final breakup of Gondwana in the Late Cretaceous (Lundberg 1993; Tokito *et al.* 2005). A continuous fossil record of protopterid lungfishes in Africa exists throughout the Cenozoic (Stewart 2001), although only recently have molecular techniques been used to resolve the relationships among the extant species (Tokito *et al.* 2005) (Figure 1).

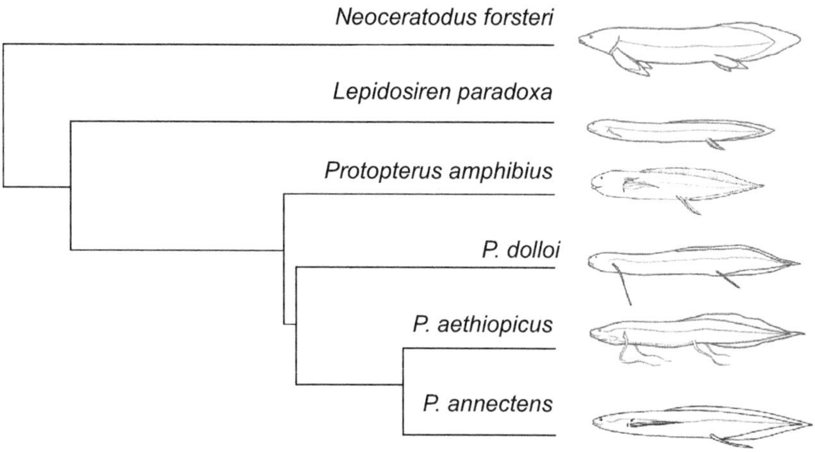

Fig. 1 Phylogram based on Tokito *et al.* 2005 showing relationships among extant dipnoan fishes. General body forms are indicated but not relative sizes. Adult *P. annectens* and *P. amphibius* have external gills, but they are much larger in the latter. See text for other morphological differences between species.

Nelson (2006) has suggested that lungfish of the genus *Gnathoriza*, which have been found preserved within their aestivation burrows in mid-continental North American Permian lake sediments and used as key indicators of seasonally arid climates (Carlson 1968; Berman 1976), were possibly early lepidosirenoids and evidence suggests that the common ancestors of the extant lungfishes were freshwater forms that could survive periods of drought in burrows, possibly through aestivation (Cavin *et al.* 2007). It is thus likely that lungfish, morphologically and behaviorally similar to extant protopterids and capable of aerial respiration, have inhabited comparable seasonal environments for more than 250 million years. Although the behavioral and physiological traits of the extant lepidosirenoid lungfishes associated with their use of seasonal freshwater environments were acquired after the appearance of the tetrapods, they have nonetheless been of continuing interest for studies of tetrapod evolution as they may be the best living analogues of the earliest proto-tetrapods, a group generally believed to have occupied shallow water environments and capable of aerial respiration (Clack 2002).

The survival of the protopterid and lepidosirenid lungfishes is remarkable in light of the decline in the relative diversity of sarcopterygian fishes since their initial radiation in the Paleozoic (Carroll 1988), a decline that has been attributed to competition from more derived actinopterygians (Helfman *et al.* 1997). Although sarcopterygians were ancestral to the tetrapods and were an important component of Paleozoic fish faunas (Carroll 1988), presently sarcopterygian fishes make up only six of the approximately 10,000 freshwater fish species, an assemblage overwhelmingly dominated by the teleosts (Lundberg *et al.* 2000). The persistence of lepidosirenoid lungfishes despite the presence of a diverse array of sympatric teleost competitors must be due in large part to the strongly seasonal environments they generally occupy. Protopterid and lepidosirenid lungfishes are typically found in weedy, marginal areas, where dissolved oxygen levels are low, daytime temperatures are high, and seasonal drying is common (Greenwood 1986). They display a suite of behavioral and physiological adaptations to these harsh conditions including aerial respiration, aestivation, extensive parental care of eggs and larvae in burrows, a broad generalist diet, and wide tolerances to a range of environmental conditions. *Neoceratodus*, on the other hand, occupies less seasonal environments than the lepidosirenoids and is correspondingly less specialized (it is not an obligate air breather, does not aestivate, and has no parental care). Its persistence in Australia may be somewhat less remarkable because there were no teleosts in Australia prior to its separation from the rest of Gondwana. Current endemic freshwater Australian teleosts are derived from more recently arrived anadromous and marine groups and there are no native primary freshwater ostariophysan fishes, the dominant freshwater fishes elsewhere (Lundberg *et al.* 2000).

The first description of an extant protopterid was of the West African lungfish *P. annectens* by Owen (1839) based on samples collected at MacCarthy Island in the lower Gambia River by Thomas Weir in 1835, two years before the discovery and scientific description of the South American lungfish *Lepidosiren paradoxa*. Owen's original designation of *Protopterus anguilliforms* was changed to *Lepidosiren annectens* in the published description after he noted his specimen's similarity to the then recently described South American species. Descriptions of two additional protopterid species: *P. amphibius* (Peters 1844) and *P. aethiopicus* (Heckel 1851) followed shortly based on collections from northern Mozambique and the White Nile, respectively. *Protopterus dolloi* was described somewhat later from material collected from the Congo River.

The finding of living ceratodontids in the 19th century generated considerable interest, particularly after the publication of *The Origin of Species*. They seemed to be intermediate in many respects between the fishes and tetrapods and thus may have been closely related to the immediate tetrapod ancestors (Conant 1986). Although many collections of protopterids were made in the 19th and early 20th centuries, most work concentrated on comparative morphology and observations of protopterid natural history were rare or anecdotal and primarily concerned aestivation. Jardine (1841), for example, described an aestivation cocoon of *P. annectens* supplied by Weir that had been excavated from dry mud and Peters original description of *P. amphibius* noted that the species resides in cavities formed in the earth during the dry season (Gray 1845).

Other early contributions to the natural history of *Protopterus* included information on parental care and reproductive biology (Budgett 1899, 1901; Greenwood 1958; Brien *et al.* 1959), aestivation and aerial respiration of adults and juveniles (Smith 1930, 1931), fasting endurance and metabolism (Smith 1931), and larger works on natural history of *P. annectens* (Johnels and Svensson 1954) and *P. aethiopicus* (Curry-Lindahl 1956). Much of this early work is summarized in Greenwood (1986). Unfortunately, Greenwood's (1986) observation that remarkably little is known of protopterid natural history remains largely true, as is demonstrated by the small amount of new information on this group contained in two recent texts on African freshwater fishes (Leveque 1997; Skelton 2001).

This chapter reviews the current state of knowledge on the natural history of the extant protopterids, highlighting, where possible, adaptations specifically related to their use of seasonal environments and concentrating on studies carried out since the review by Greenwood (1986). Much of this recent work has focused on *P. aethiopicus,* which is easily the most studied of the four *Protopterus* species.

GENERAL APPEARANCE OF AFRICAN LUNGFISHES

African lungfishes are easily recognized by their elongate bodies, filamentous pectoral and pelvic fins, and the fusion of their dorsal, caudal and anal fins into one continuous diphycercal tail (Bemis *et al.* 1987) (Figure 1). The anal opening is conspicuously off-set from the midline. *Protopterus* larvae possess external gills (Figure 2), a trait shared by *Lepidosiren*, but not *Neoceratodus.*

As adults, *P. amphibius* are the most easily identified as they retain three large external gills on each side of the head. These larval characteristics are lost in juveniles and adults of *P. aethiopicus* and *P. dolloi* (Greenwood 1986). *Protopterus amphibius* has a long skull (more than 30% distance snout to vent), broad membranes on its pectoral fins, and a uniformly blue or slaty-green body color, with small or inconspicuous black spots (Boulenger 1909; Trewavas 1954). *Protopterus annectens* also retains external gills in the adult but these are considerably smaller than those in *P. amphibius.* For example, they were not noted in Owen's (1839) original description, although they are clearly illustrated (but misidentified as components of the pectoral fin) in Jardine's (1841) description of the same specimens. *Protopterus annectens* also has broad membranes on its

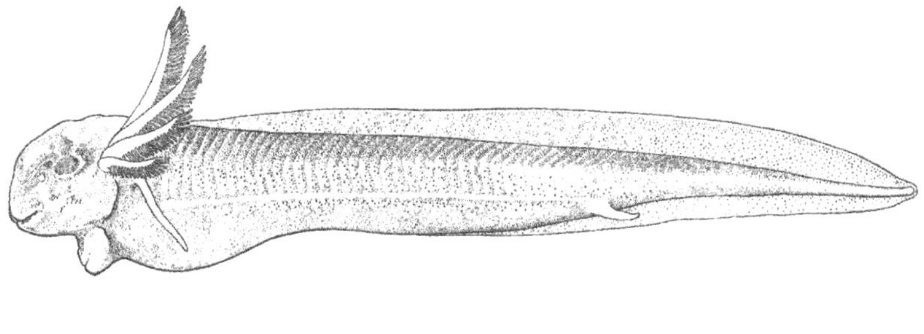

Fig. 2 Stage II *Protopterus dolloi* larva (after Brien 1959). Cement from a gland located on the anterior ventral surface enables them to adhere to the substrate.

Table 1 Morphological characteristics that distinguish *Protopterus* species

Species	*Number of ribs*	*HL/AL(%)*	*DL/AL (%)*	*External gills*
Protopterus aethiopicus	38-39	24-29.6	62.1-67.1	Absent in adult
Protopterus amphibius	27-30	33.2	45-56	Present in adult
Protopterus annectens	32-37	22.8-28.2	51-57.5	Present in adult
Protopterus dolloi	47-55	16.2-19.6	63.7-66	Absent in adult

Numerical data from Poll (1961). Abbreviations: DL, distance from snout to anterior base of dorsal fin; HL, head length; AL, distance from snout to anterior base of anal fin.

pectoral fins. *Protopterus dolloi* is relatively slender with brown coloration and can be distinguished from *P. aethiopicus* by the latter's more cylindrical body and larger adult size (Bailey 1994). Morphometric traits that distinguish the four species are given in Table 1.

While protopterids are morphologically similar to *Lepidosiren*, they have an additional gill arch (5 vs 4) and lack the hypervascularized pelvic fins that *Lepidosiren* males develop during the spawning season. Unlike protopterids and *Lepidosiren*, *Neoceratodus* has robust flipper-like pectoral and pelvic fins (Figure 1), larger scales, a more laterally compressed body, single rather than paired lungs, and no external gills in the larval stage (Thomson 1969; Kemp 1986; Nelson 2006).

DISTRIBUTION

The four *Protopterus* species are endemic to the African zoogeographic region (Roberts 1975) and their collective ranges cover a large part of the African continental landmass (Greenwood 1986) (Figure 3). *Protopterus aethiopicus* has an extensive native range in eastern and central Africa encompassing the upper tributaries of the Congo and Nile basins, and major rivers and lakes in the Rift Valley including Lakes Victoria, Tanganyika, Albert, Edward, George, and Kyoga. This species has been the subject of two reported and successful introductions. It was translocated (*sensu* Shafland and Lewis 1984) to Lake Baringo, Kenya in the mid 1970s where it is now the subject of an artisanal fishery (Mlewa and Green 2006). In 1988 and 1990 *P. aethiopicus* from Lake Edward in Uganda were introduced into Lake Muhazi in Rwanda where it is now apparently well established and being caught by fishers (Micha and Gashagaza 2002). *Protopterus annectens* is widely distributed in western Africa and is also found in the Zambezi and Limpopo River basins of southeast Africa. *Protopterus dolloi* is primarily found in the Congo Basin, whereas *P. amphibius* has a restricted but disjunct range in East Africa, appearing to be absent from the rift lakes (Okeyo 2006). At least two other species show disjunct distributions. *Protopterus annectens* is found in the Congo drainage basin, in a number of unconnected drainage systems in north-west Africa, and in the Nile and the Zambezi. The northern (across West

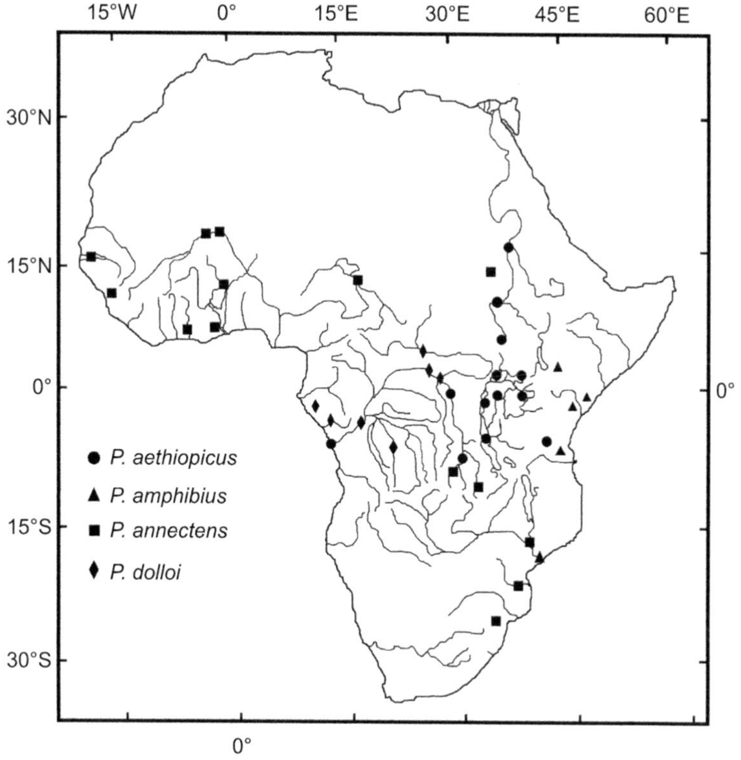

Fig. 3 The natural distribution ranges of the four African lungfish species. Redrawn from Greenwood (1986).

Africa to the Sudd area of the upper Nile) and southern African populations are recognized as the subspecies *P. annectens annectens* and *P. annectens brieni*, respectively (Poll 1961; Eschmeyer 1998; Skelton 2001). *Protopterus aethiopicus* is found in both the Congo and Nile basins. These disjunct distributions may have arisen due to periodic shifts in headwater drainage patterns, primarily during the period preceding the uplifting and rifting of the African continent (Stewart 2001). However, the ability of lungfishes to occupy the shallow swampy areas often found at higher elevations may allow dispersal between drainage basins during seasonal rainy periods, thus allowing them to re-colonize areas where populations are subject to periodic local extinction. *Protopterus annectens*, *P. aethiopicus*, and *P. dolloi* co-occur in the Congo basin (Greenwood 1986) and both *P. annectens* and *P. amphibius* have been reported from the lower Tana River in Kenya (Trewavas 1954). There are also reports of co-occurrence of *P. annectens* and *P. aethiopicus* in Lake Edward, Uganda (Johnels and Svensson 1954; Curry-Lindahl 1956). Thus sympatric populations of some protopterid species likely exist, although area-specific faunal lists generally include only one lungfish species (Gosse 1984).

These sympatric populations should permit comparisons of habitat use between allopatry and sympatry that might reveal important information on lungfish natural history, including species-specific habitat differences, not evident from studies of allopatric populations.

SIZE AND GROWTH

African lungfishes are among the largest fishes throughout their ranges, with mature males of *P. aethiopicus,* the largest of the four species, reaching up to 180 cm total length (Greenwood 1986). Maximum reported lengths of the other species are 130 cm, 82 cm, and 44 cm for *P. dolloi*, *P. annectens*, and *P. amphibius*, respectively (Greenwood 1986). The report by Webb *et al.* (1981, as cited in Oniye *et al.* 2006) that *P. annectens* reaches a length of 200 cm is likely in error.

Large adult size in fishes has been related to a number of factors including diet and lower predation rates (Dunbrack and Ware 1987; Bonnet *et al.* 2000). However, selection on body size related to diet or size dependent predation rates would not be unique to lungfishes and their large body size may also be influenced by more specific features of their life history, such as a requirement for increased fasting endurance and reduced relative rates of water loss during aestivation, correlates of large body size because of the decrease in mass specific metabolic requirements with increasing body mass (Dunbrack and Ramsay 1992).

Johnels and Svensson (1954) are the only authors to describe what they considered to be unequivocal differences in external morphology between male and female African lungfishes. They described the snout of *P. annectens* males above the length of about 250 mm as being rounder and shorter than that of females; and the limit between the head and body as being well defined in males, because of their broader heads than bodies, but not in females. Such differences have not been reported for the other African lungfishes, but males of *P. aethiopicus* may reach larger size than females, with data from the Lake Baringo fishery indicating maximum lengths of 140 cm and 130 cm for males and females respectively (Mlewa and Green 2004). Although these values may approach the asymptotic lengths in this population, the absence of larger fish may also reflect the high mortality rates associated with exploitation (Mlewa 2003). Larger male size was also reported for the *P. aethiopicus* population in Mwanza gulf where males grow to a total length of 156 cm compared to 146 cm for females (Mosille and Mainoya 1988) and the largest male *P. annectens* from Jachi Dam, Nigeria were also slightly larger (~6 cm) than the largest females (Oniye *et al.* 2006). The slightly larger maximum size of males suggests that reproductive success may be more strongly dependent on body size in males than females causing males to allocate more energy to growth and less to reproduction than females (Mlewa and Green 2004).

There may be a small difference in the size at first maturity of male and female *P. aethiopicus*. In Lake Baringo, males and females mature at 82.0-88.0 and 70.0-

76.0 cm total length (TL), respectively (Mlewa and Green 2004). Males are also larger at first maturity in northern parts of Lake Victoria (Okedi 1971), but the converse is true in the Mwanza gulf (Mosille and Mainoya 1988). The reason for these differences is not known, however larger male size at sexual maturity could be related to their construction and defense of nest sites, and the care they provide for eggs and larvae (Mlewa and Green 2004).

Growth rate data for African lungfishes are scarce. Baer *et al.* (1992) reported a specific growth rate in mass of 1.52% day^{-1} for *P. amphibius* under culture conditions with *ad libitum* feeding. The only information on the growth of *Protopterus* species under natural conditions is a study by Dunbrack *et al.* (2006) which used mark-recapture data to determine the mass specific growth rate for sub-adult *P. aethiopicus* in Lake Baringo. Their empirical growth equation gives a daily growth rate of 0.3% per day for a 1000 g fish. This is comparable to growth rates for other fishes sympatric with *P. aethiopicus*, including the Nile perch *Lates niloticus* Linn. (Dunbrack *et al.* 2006) (Figure 4). Their data indicated that the mass at sexual maturity for females (3000 g in Lake Baringo) was reached after

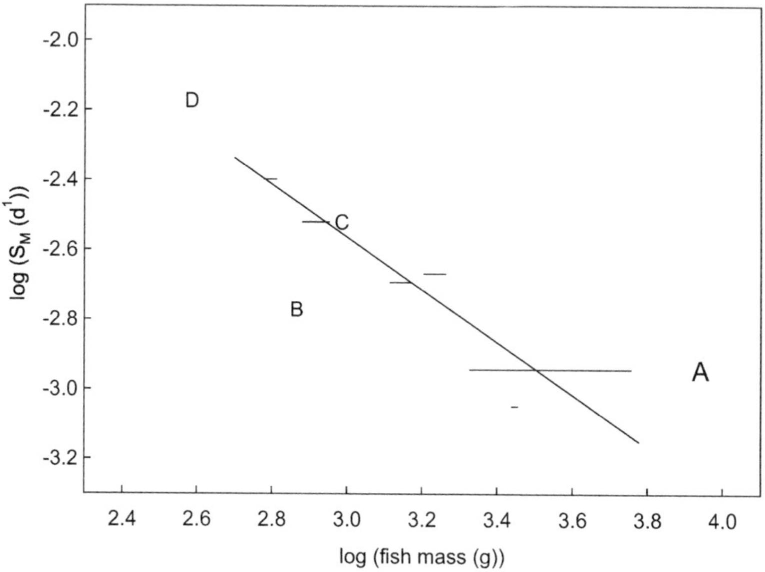

Fig. 4 The relationship between mass specific growth rate (S_M; day^{-1}) and body mass (g). Both axes are log_{10} transformed. Each short horizontal line is an S_M value calculated from mark-recapture data for a single fish. Single capital letters are S_M values for four fish species sympatric with *P. aethiopicus*: A, nile perch *Lates niloticus* (L.); B, African bonytongue *Heterotis niloticus* (Cuvier); C, semutundu *Bagrus docmak* (Forsskal); D, African tetra *Hydrocynus forskahlii* (Cuvier) (after Dunbrack *et al.* 2006).

approximately 3 years, the only estimate of age at maturity for any *Protopterus* species.

Larval growth in protopterids is initially rapid but, according to Johnels and Svensson (1954), slows considerably once exogenous feeding begins in post-absorptive larvae. A similar pattern is seen in *Lepidosiren* although the yolk dependent period is longer and active feeding begins at a larger size; probably due to the larger egg size of *Lepidosiren* (Johnels and Svensson 1954) (Figure 5). However, the slow growth of post-absorptive larvae described above may have been due to their rearing conditions since other data in the same paper suggest, as expected, that growth rate increases following commencement of exogenous feeding. *Protopterus annectens* juveniles grow for about three months before aestivating for the first time at which point their lengths vary from 70 to 120 mm (Johnels and Svennson 1954). There are few data on early juvenile growth in other species but they would be expected to be similar.

Juvenile *P. aethiopicus*, inhabiting large bodies of water such as lakes Victoria or Baringo, appear to undergo an ontogenetic habitat shift from the shallow lake

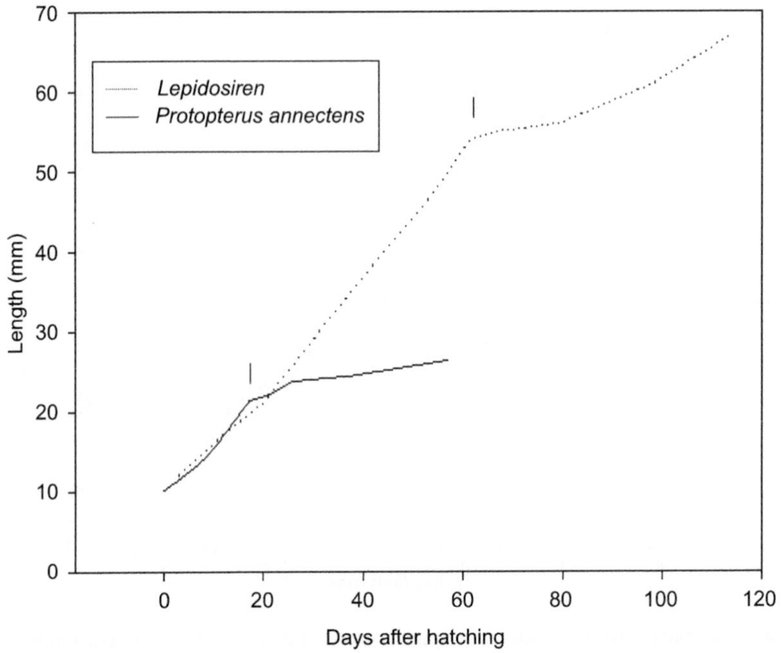

Fig. 5 Comparison of growth rate of larval *Lepidosiren paradoxa* and *Protopterus annectens*. Vertical bars indicate time at which all yolk has been reabsorbed. (Data from Johnels and Svensson 1954.)

margins to the open lake. Goudswaard *et al.* (2002) noted that 200-400 mm *P. aethiopicus* in Lake Victoria were "most abundant within the swamps, particularly swamp lagoons", suggesting a migration to open waters at approximately 400 mm TL. Smith (1931) collected large numbers of juvenile *P. aethiopicus* between 70 and 350 mm in length from the roots of floating papyrus islands in Lake Victoria and noted that specimens smaller than 300 mm were never caught in nets set in the open lake waters. In Lake Baringo, only *P. aethiopicus* longer than 360 mm were caught in open waters by long-line fishers (Mlewa and Green 2004). The period required to reach this size, presumably spent in the weedy lake margins, is between four and five months once active feeding begins outside the nest (Dunbrack *et al.* 2006) (Figure 5), which occurs approximately 50 days following egg fertilization (Greenwood 1986).

Once maturity is attained, growth rates are dependent upon the proportion of potential growth allocated to reproductive mass (Dunbrack and Ware 1987). In Lake Baringo, *P. aethiopicus* females appear to have a low initial reproductive effort, suggesting that growth may remain high during the early adult period (Dunbrack *et al.* 2006). However, gradual increases in reproductive effort with body mass following maturity (see below) precludes estimating age relative to size of older females. Length exponents of length–weight relationships for *P. aethiopicus* range from 3.4 to 3.5 (Mosille and Mainoya 1988; Mlewa and Green 2004), indicating positive allometric growth; the typical pattern for most fishes (Wooton 1990). Mlewa and Green (2004) found no significant difference between the length exponents for male and female *P. aethiopicus* in Lake Baringo. Positive allometric growth has also been reported for *P. annectens* (Otuogbai *et al.* 2001; Oniye *et al.* 2006).

FOOD AND FEEDING

Protopterids rely primarily on olfaction and sensory receptors on their filamentous paired fins, rather than vision, to locate prey (Otuogbai *et al.* 2001). Curry-Lindahl (1956) reported the presence of tastebud-like structures on both the pelvic and pectoral fins of *P. aethiopicus* and similar structures have been observed on *P. annectens* (Johnels and Svensson 1954; Otuogbai 2001). They also possess electroreceptor organs (Jorgensen 1984, 2005; Bullock *et al.* 2005) as does *Lepidosiren* (Roth 1973) and *Neoceratodus* (Jorgensen 1984). Watt *et al.* (1999) described the use of electroreceptors to locate prey in *Neoceratodus* and it would not be surprising to find similar capability among the protopterids. Considering that their diets are usually dominated by non-motile prey, protopterids must primarily be active foragers rather than sit and wait predators.

Unlike the South American and Australian lungfish which may consume considerable plant material (Kemp 1986), all four African lungfishes are more carnivorous in their food habits, consuming a wide array of prey including mollusks, small fish, crustaceans, aquatic insects, and worms (Curry-Lindahl 1956; Corbet 1961). The anoxic, seasonal habitats of most protopterids do not support large populations of resident fish or invertebrate prey species, leading to the expectation that they should be generalist feeders able to exploit a wide variety of food. Based on the limited information available this seems to be the case. Juvenile protopterids consume primarily insects, small mollusks, and crustaceans while adults consume larger prey including fish (Greenwood 1986; Mlewa and Green 2004). Otuogbai (2001) analyzed gut contents of *P. annectens* from a tributary of the River Niger and reported a "clear selective preference for diets of flesh such as other fishes, mollusks, insects, crustaceans and annelids".

Similar food items have been reported for *P. aethiopicus* in Lake Victoria (Corbet 1961; Okedi 1990; Pabari 1997). The introduced *P. aethiopicus* in Lake Baringo probably have a typically diverse diet as juveniles but are largely piscivorous as adults, feeding primarily on the Baringo tilapia, *Oreochromis niloticus baringoensis* Trewavas 1983. This piscivory is possibly related to a lack of benthic macro-invertebrates in the open water areas of the lake (Ssentongo 1995), the avoidance by adults of the shallow lake margin, and the extreme murkiness of the lake, which in the absence of visual cues may allow close encounters between predators and prey (Mlewa and Green 2004). Micha and Gashagaza (2002) also reported piscivory in *P. aethiopicus* introduced to Lake Muhazi, Rwanda.

The stomach contents of wild-caught protopterids generally indicate a varied diet characteristic of opportunistic foraging, although a laboratory test of diet choice in *P. annectens* gave different results. When offered a range of benthic prey, subjects showed a clear preference for snails (Daffalla *et al.*1985). This preference may seem inconsistent with the reported diets of wild fish, but it may only indicate that prey choice in protopterids is based on prey profitability (net energy in prey/handling time). In this case, foraging theory predicts that a preference for snails would be expressed if the rate at which they were encountered was sufficiently high that energy gain could be maximized by ignoring other prey (Stephens and Krebs 1986). In this diet choice experiment (Daffalla *et al.* 1985), snail densities and rates of consumption were high and even small lungfish (< 30 cm) consumed up to 200 snails per day. Based on these results it was suggested that protopterids could have value in controlling local populations of schistosomiasis snail vectors when snails reached high population densities (Daffalla *et al.* 1985).

Several workers have reported plant material in the diet of protopterids (e.g. Curry-Lindahl 1956; Otuogbai 2001; Micha and Gashagaza 2002; Mlewa and Green 2004; Oniye *et al.* 2006). The presence of plant material in the diet of most protopterids has led to their classification as omnivores (Curry-Lindahl 1956; Greenwood 1986; Oniye *et al.* 2006). However, some workers have attributed this to incidental ingestion while foraging on prey since the structure of the digestive tract of *P. aethiopicus* is thought to be adapted to a carnivorous diet (Amongi *et al.* 2001; Mlewa and Green 2004). Oniye *et al.* (2006) considered plant material to be an important component of the diet of *P. annectens* in Jachi Dam, Nigeria. They reported a low prey population at the time of their study and considered the omnivorous feeding habits to be a survival strategy for dealing with such conditions.

The dentition of the protopterids certainly reflects their predatory diet. Observations on captive individuals in aquaria show that they initially capture prey by suction, but then masticate it prior to swallowing using sharp incisors capable of cutting flesh and highly modified lateral continuously growing molariform tooth plates that can crush hard bodied prey (Greenwood 1986; Otuogbai *et al.* 2001). This mode of feeding contrasts with the suction feeding using lightly built kinetic jaws common in teleosts. The latter is effective for the rapid ingestion of small prey but is not well adapted for crushing hard bodied prey or cutting large prey into manageable sections. The versatile dentition of lungfishes is accompanied by powerful jaw adductor muscles housed within adductor chambers that are much larger than is typical for actinopterygians (Bemis and Lauder 1988; Carroll 1988). The power and efficiency of lungfish dentition is evident in the clean and fresh excision marks often found on the tails and fins of lungfish caught on long-lines, which bear the characteristic bite pattern of lungfish incisors (Mlewa and Green 2004). This cutting and crushing dentition is characteristic of all *Protopterus* species and is of ancient origin, having appeared in the Devonian (Carroll 1988). This dentition allows protopterids to process a wide range of prey, an important factor in their ability to inhabit harsh unproductive environments.

REPRODUCTION

The descriptions of nest structure and parental care given below are taken primarily from Johnels and Svensson (1954), Bouillon (1961), and Greenwood (1986), indicating the virtual absence of any recent information on this aspect of the natural history of African lungfishes. Spawning nests of *P. amphibius* have not been described. In the other three species males construct tunnel-like nests in shallow, weedy areas that are the site of spawning and also where they guard the eggs and larvae. Nest structure varies both within and among species but generally consists of tunnels with multiple entrances excavated from the soft substrate of mud and vegetation with eggs and larvae often occupying an expanded area at the base of the tunnel. Tunnel entrances may be submerged or exposed depending on water level, and additional sub-surface tunnels may connect the nest with adjacent bodies of water. Johnels and Svensson (1954) noted additional pathways in the bottom vegetation leading to subsurface tunnel entrances in *P. annectens*, possibly produced by nocturnal foraging as movements of males outside their tunnels were seldom observed during the daylight hours.

The start of the breeding season in *P. annectens* in the lower Gambia River is closely tied to the beginning of the wet season. As water levels rise, males emerge from their aestivation burrows and excavate breeding burrows in the mud of the temporary wet season marshes (Johnels and Svensson 1954). In lake populations of *P. aethiopicus* and *P. dolloi* nest burrows are constructed in permanent marshes (Bouillon 1961; Greenwood 1986). However, *P. dolloi*, unlike other protopterids, reproduces in the dry season in the Stanley Pool, Uganda, where males appear to use the same tunnels for both dry season refuges and reproduction (Brien *et al.* 1959). The permanent water in the Stanley Pool allows burrows to be excavated below the dry season water table and females are able to reach the burrows from the open lake waters through the submerged vegetation mat (Bouillon 1961). However, protopterid reproduction may not be entirely subterranean. For example, Greenwood (1986) described an exposed nest of *P. aethiopicus* that consisted only of a small area in shallow water cleared of aquatic vegetation.

Details of protopterid spawning behavior are poorly known. Males excavate the burrows and remain there for extended periods guarding eggs and larvae but it appears that females only enter burrows briefly to deposit eggs (Johnels and Svensson 1954; Bouillon 1961; Greenwood 1958). In both *P. annectens* and *P. aethiopicus* eggs and larvae at different developmental stages have been found in the same tunnel, suggesting that males may spawn with multiple females. However, this may not be the case for *P. dolloi* in the Stanley Pool where individual tunnels have only been found to contain larvae of the same developmental stage (Bouillon 1961). In *P. aethiopicus*, hatching of eggs takes approximately two weeks and parental care appears to continue at least to the end of yolk absorption. Greenwood (1958) observed one adult *P. aethiopicus* that stayed for over 2 months guarding eggs and young in a nest in the northern part of Lake Victoria.

While adult males are assumed to construct the nest and guard the eggs and larvae in all species, this has been verified only for *P. aethiopicus* and only for one adult (Greenwood 1958, 1986), highlighting the need for more field work. However, male, in contrast to female, parental care is far more common in fishes (Breder and Rosen 1966) and compelling indirect evidence indicates that female protopterids do not provide parental care. Female *P. aethiopicus* in both spawning and recently spawned condition have been observed in samples from open water in Lake Victoria (Greenwood 1958; Mosille and Mainoya 1988) and Lake Baringo (Mlewa and Green 2004). Brien *et al.* (1959) made similar observations in *P. dolloi*. This suggests that females move to spawning areas only when ready to spawn and return to open waters shortly thereafter (Greenwood 1958; Mlewa and Green 2004). This may also explain sex ratios skewed in favour of females among *P. aethiopicus* samples taken from the open waters of both lakes (Greenwood 1958; Okedi 1971; Mosille and Mainoya 1988; Mlewa and Green 2004). In Lake Baringo, skewed sex ratios were found only in lungfish longer than 70-80 cm (Mlewa and Green 2004), consistent with males periodically migrating to inshore spawning habitats after attaining sexual maturity. However, differential mortality between the sexes could also contribute to the skewed sex ratio, as guarding males are frequently captured in their nesting burrows by fishermen and would be more vulnerable to predation by Nile crocodiles (*Crocodylus niloticus*) in the shallow nesting areas (Mlewa and Green 2004). Oniye *et al.* (2006) reported a nearly equal sex ratio among 179 *P. annectens* from Jachi Dam Nigeria, however the maturity states of these fish were not given and fish were as small as 19 cm total length.

A number of factors associated with protopterid reproductive behaviour may have particular relevance to the persistence of protopterid lineages. The location of the tunnel nest in shallow, weedy, anoxic environments isolates developing lungfishes from most predation by actinopterygian fishes when they would be most vulnerable. The tunnel nest also provides a well hidden, cryptic, and easily defended refuge for eggs, larvae, and guarding adults, and is of sufficient depth to offer protection from low water levels. Parental care has obvious benefits in reducing predation on eggs and larvae but may have evolved initially because of the advantages of aerating eggs sequestered in the stagnant waters of tunnel nests. In the closely related South American lungfish, males are able to increase the oxygen content of the water in the nest by the reverse diffusion of oxygen from hyper-vascularized pelvic fins (Graham 1997). Such morphological structures do not exist in African lungfishes, but *P. aethiopicus* and *P. annectens* may increase dissolved oxygen levels for eggs and larvae by "tail lashing", a vigorous mixing of the surface waters of the tunnel nest with air using the posterior portion of the body (Greenwood 1986). Tail lashing would have obvious benefits for embryonic lungfishes and could also lower predation risk for larvae by reducing the frequency of surfacing for aerial respiration. In all protopterid species males probably aid the respiration of eggs and larvae by maintaining water movements through the tunnel system using undulations of their caudal fin, a behaviour that has been reported in both *P. annectens* and *P. dolloi* (Greenwood 1986; Bouillon 1961). *Protopterus dolloi* has also been reported to release inspired air via the opercula as bubbles that may increase the oxygen content of the water, or produce air pockets in the tunnel system which larvae can use for aerial respiration (Brien *et al.* 1959). The general characteristics of parental care are similar in *Protoperus* and *Lepidosiren*, whereas *Neoceratodus* is not reported to provide parental care, and the fertilized eggs adhere to aquatic vegetation during development (Kemp 1986).

The possible connection between the burrows constructed for reproduction and for avoiding desiccation, suggests that parental care in lungfishes may have a very long evolutionary history. Structures resembling lungfish aestivation burrows are known from the Devonian and similar structures containing fossilized lungfish from the Permian (Berman 1976) indicate a long history of adaptation to habitats subject to episodic drying. Although these fossil structures are similar to the aestivation burrows of extant lungfishes and not the more extensive tunnel nests, tunnels occupied by non-aestivating male *P. dolloi* during the dry season are subsequently used as nests sites (see above).

The onset of the reproductive period in *P. aethiopicus* generally coincides with seasonal precipitation maxima and swampland flooding (Greenwood 1986; Mosille and Mainoya 1988), however this may not be true of all populations. The finding of young *P. aethiopicus* (36-48 cm total length) throughout the year in Lake Baringo

(Mlewa and Green 2004), and the known rapid pre-reproductive rate of somatic growth in this population, is consistent with year round spawning. Year round reproduction is also suggested by the presence of two or more egg size classes in the ovaries of mature females (Greenwood 1958; Okedi 1971; Mosille and Mainoya 1988; Mlewa and Green 2004), indicating that in some populations eggs may be laid during a number of spawning bouts throughout all or much of the year.

Fecundity (pre-spawning egg number) in *Protopterus* may differ considerably among species. For example, *P. dolloi* is reported to have a maximum fecundity of less than 500 (Bouillon 1961) whereas in *P. aethiopicus* it may exceed 20,000. Fecundity also varies with body size within species. In Lake Baringo *P. aethiopicus* fecundity increased from 4000 to 16,000 over a body size range of 2 to 10 kg, producing a mass exponent for fecundity of 0.65 (Mlewa and Green 2004). A similar relationship has been described for *P. aethiopicus* from Lake Victoria (Okedi 1971; Mosille and Mainoya 1988) and likely applies to the other protopterids. Although fecundity increases with body mass, the allometric mass exponent of less than one indicates that the pre-spawning gonado-somatic index should decrease allometrically, with a mass exponent of approximately -0.35. It has not been possible to confirm this prediction empirically because of the variation among females in the proportion of eggs at different developmental states (Mlewa and Green 2004).

However, a negative relationship between body mass and egg production per unit mass does not necessarily imply that female lungfish decrease their reproductive effort (the proportion of total potential growth allocated to reproductive biomass) as they grow. The relationship between reproductive effort (RE) and body mass depends on the ratio of the mass exponents of fecundity and growth. The instantaneous yearly growth rate calculated by Dunbrack *et al.* (2006) for non-reproductive *P. aethiopicus,* is $dM/dt = 187.5M^{.246}$, where M is mass in g. The relationship between body mass and total potential yearly mass growth (TYG), derived from this expression, is approximately $TYG = k_1M^{.34}$, where k_1 is a constant and M is the mass at the end of the yearly growth period. An approximation of the weight exponent for the relationship between RE and body mass can then be obtained from the ratio of the expressions for mass specific fecundity and potential growth, in this case $W^{.65}/W^{.246}$ or $W^{.4}$. Thus RE increases with increasing body mass in this population as $k_2W^{.4}$, where k_2 is a constant. Increasing female RE with body mass is a general trend in fishes (Dunbrack and Ware 1987) and is partially responsible for the asymptotic growth of adults described by growth models such as the Von Bertalanffy growth equation.

There is no information on the body size relationship of RE in male protopterids but it is clear that males make a significantly smaller investment in reproductive biomass than females. For example, gonad mass in male *P. aethiopicus* is approximately ¼ that in females of comparable size (Mlewa and Green 2004). Despite the smaller investment in gametes by males, their indirect costs of reproduction, including the loss of potential growth during the nest guarding period, when male feeding rates are probably low; and increased predation on guarding males by predators such as crocodiles and fish eagles, probably exceed those of females.

Britton *et al.* (2006) reported that *P. aethiopicus* may have minimum doubling times greater than 14 years and would therefore exhibit low resilience and be slow to recover from local reductions in population size. However, their relatively high fecundity, combined with an extensive period of parental care and rapid somatic growth, suggest that protopterid populations could increase rapidly. The best data on the potential rate of increase for any protopterid comes from the translocation (*sensu* Shafland and Lewis 1984) of *P. aethiopicus* into Lake Baringo (Mlewa and Green 2006). Three individuals were introduced in 1975 and within 10 years this species had become an important component of the artisanal fishery, making up over 50% of total landings, with a reported catch in 1999 of 199 tonnes (Mlewa and Green 2006). The indicated doubling time of less than 1 year suggests that classifying protopterids as low resilience species may not be correct in all cases. Clearly, some protopterid populations can increase rapidly when there is little intraspecific competition, predation rates are low, and permanent water allows for year round growth and reproduction. Nonetheless, the success of the Lake Baringo introduction is still surprising considering the seasonal, unproductive environments usually associated with protopterids, and the suggestion, made above, that protopterids have been able to persist for hundreds of millions of years in their harsh environments because only here can they compete successfully with actinopterygians. It should be noted however, that such competition may have been largely absent in Lake Baringo because of its relatively low fish diversity (Mlewa and Green 2004; Okeyo 2006). The successful introduction of *P. aethiopicus* to Lake Muhazi and its later spread to Lake Mugesera (Micha and Gashagaza 2002) also seems to indicate the potential for rapid population growth in this species. Data on the population growth of other *Protopterus* species are not available, but potential rates of increase would be expected to be lower when there is a long annual period of aestivation, such as in *P. annectens* (Greenwood 1986).

MORTALITY

African lungfishes are known to occur in the open waters of large lakes (Greenwood 1986), but are more generally associated with shallows and swamps at the margins of lakes and rivers often subjected to seasonal drying. Thus they may avoid many potential predators because the swampy inshore areas are inaccessible to species not adapted to shallow or anoxic water (Chapman *et al.* 1996). Because of their large size natural predation rates on adult lungfishes may also be low (Greenwood 1986). However, predation by the even larger Nile perch introduced into Lake Victoria is considered to have contributed to the decrease in the relative abundance of *P. aethiopicus* and other indigenous fish stocks in open waters relative to marginal areas (Ogutu-Ohwayo 1990; Chapman *et al.* 1996; Goudswaard *et al.* 2002). This apparent habitat shift could arise either from a behavioural response to the presence of Nile perch in open water or differential mortality in the two habitats. An opposite effect on habitat use by *P. aethiopicus* could occur due to predation by the Nile crocodile. Such predation was observed by Greenwood (1986), and could explain the apparent avoidance of shallow inshore areas by adult non-nesting lungfish in Lake Baringo (Mlewa *et al.* 2005).

Several factors may tend to limit predation rates during the early life history. All *Protopterus* species have a protracted period of parental care when eggs and larvae are guarded by males within nests. Following this juvenile *P. aethiopicus* continue to avoid open water until they reach a length of approximately 40 cm (see above), thus escaping open water predators during a developmental stage that, in most fish species, is characterized by high rates of predation (Wooton 1990). Although most information on mortality is specific to *P. aethiopicus*, parental care and the use of shallow, weedy, low oxygen environments by both juveniles and adults of the other species must similarly reduce the threat posed by open water predators. However, the use of shallow water habitats, particularly those in the process of drying, may also expose protopterids to predators not encountered in open waters. In Lake Edward, for example, seasonal rains produce temporary pools that are rapidly colonized by a number of fish species, including *P. aethiopicus*. When these pools are cut off from the lake as they dry, the trapped fishes are taken by a variety of predators including leopards, hyenas, lions, eagles, and vultures (Curry-Lindahl 1956). Interestingly, although Curry-Lindahl (1956) observed large numbers of living *P. aethiopicus* in pools that had been recently isolated from the lake, he did not observe any predation on this species or find their remains among the fish carcasses discarded by predators. Despite this, pools in the last stages of drying did not contain any lungfish and it was suggested that they had avoided the fate of other species in the drying pools, primarily *Clarias* spp., by returning to the lake by tunneling or moving overland (Curry-Lindahl 1956), although this latter suggestion seems unlikely. Protopterids have been reported to be capable of snake-like terrestrial locomotion when escaping to the water after capture (Curry-Lindahl 1956; Bouillon 1961; Greenwood 1986), but purposeful overland movements have not been observed and do not appear to be part of their behavioural repertoire (Greenwood 1986).

In open waters protopterids, particularly smaller individuals, are probably eaten by a number of species. Curry-Lindahl (1956) observed fish eagles with prey resembling *P. aethiopicus*, possibly captured when surfacing for aerial respiration, as well as the remains of lungfish discarded along the shore of Lake Edward by nocturnal mammalian predators tentatively identified as hyenas, cerval cats, and leopards. Nonetheless it was suggested that the most important lungfish predators in Lake Edward were fish, primarily *Clarias* spp., although no supporting data were provided. While parental care, large size, and the use of habitats largely inaccessible to predators may allow protopterids to avoid or diminish potential sources of mortality, they do not enable them to avoid cannibalistic attacks from other lungfishes. Cannibalism has been reported for both *P. aethiopicus* (Curry-Lindahl 1956; Mlewa and Green 2004) and *P. annectens* (Johnels and Svensson 1954) and likely occurs in the other protopterid species as well.

Although there are no quantitative data on natural mortality rates for any *Protopterus* species there is one estimate of fishing mortality. Tag returns from the Lake Baringo fishery in 2001 (Mlewa 2003), indicated a mean monthly fishing mortality of 3.3%, giving an instantaneous mortality rate (q) of 0.4 $year^{-1}$ and a total yearly mortality of 33% (Dunbrack *et al.* 2006). The expected survival time for an adult exposed to this mortality rate is q^{-1}, or 2.5 years beyond its current age. However, this estimate of q is certainly low because it does not include non-returns of recaptured fish, tag loss, or non-fishing mortality (Dunbrack *et al.* 2006). The conservative expected life span therefore is probably less than 2 years for a female in the fished population after it attains sexual maturity. This is the only estimate of fishing mortality for African lungfishes, but declines in catches of *P. aethiopicus* in Lake Victoria between 1968 and 1976 and the failure of these populations to recover were probably the result of fishing pressure as they predated the documented impact of the introduced Nile perch (Goudswaard 2002).

Throughout their range African lungfishes are important food fishes in many small communities; exploited in minor commercial and subsistence fisheries (FishBase 2007). In West Africa *P. annectens,* commonly known as "cambona" are excavated during the dry season and their mud-bound cocoons can be found on sale in local markets in many parts of the region (Otuogbai *et al.* 2001). Among the Luo community on the eastern shores of Lake Victoria, even very small lungfish known locally as "dhuri" are harvested for food from inshore waters, whereas larger lungfish are reserved for distinguished guests (Malala 2005). In most lake fisheries African lungfishes (primarily *P. aethiopicus*) are landed as by-catch in gillnet fisheries but are also the primary target species of long-line fisheries in Lake Victoria, Lake Baringo and Lake Muhazi, Rwanda (Micha and Gashagaza 2002; Mlewa and Green 2006).

REGENERATION IN AFRICAN LUNGFISH

Incidences of partial fin excision in wild protopterid populations have been reported by several workers (e.g. Greenwood 1958; Okedi 1971; Goudswaard *et al.* 2002; Mlewa and Green 2004) and have been attributed to predator attack, including cannibalism (Curry-Lindahl 1958; Mlewa and Green 2004). In many cases excised fins were in the process of regeneration and indicate that protopterids have a well developed ability to regenerate both fins and body musculature. Based on laboratory studies in which fins were excised in different ways Conant (1972) found that regeneration started within 2 to 3 weeks. Complete restoration of cut sections occurred in 3 to 4 months with growth of regenerating tissue following a sigmoid curve. Conant (1970, 1972) also demonstrated that regenerating fins often formed branched structures such as those reported among wild lungfish in Lake Baringo and Lake Victoria (Mlewa and Green 2004). Regeneration has also been reported in the South American lungfish *L. paradoxa* (Kerr 1932). The advantage of regeneration is self-evident but it is a trait not generally well developed in actinopterygian fishes. However, fin regeneration only occurs when the damage affects the cartilaginous endoskeleton, which is considerably more extensive in protopterids than in typical actinopterygian fishes (Conant 1972).

ACTIVITY AND MOVEMENT

There are few data on diel activity patterns for any of the protopterids and none for the South American lungfish. Johnels and Svensson (1954) found higher breathing rates in captive *P. annectens* nestlings at night than during the day and concluded that this pattern would also be observed in the wild. From field observations in Lake Edward, Curry-Lindahl (1956) reported that *P. aethiopicus* remained largely inactive during the early part of the day, except for regular visits, at 15–25 minute intervals, to the surface to breathe; but became more active during the last hours of daylight. Greenwood (1986) noted that the tendency to surface regularly to breath, regardless of the time of day, could obscure diel patterns of activity. Ultrasonic tracking studies by Mlewa *et al.* (2005) found no significant difference in the distances moved by *P. aethiopicus* between the morning to late afternoon and late afternoon to morning periods. However, this technique was not able to distinguish the small-scale movements of benthic foraging. Mlewa and Green (2006) did report that long-line fishers in Lake Baringo were as likely to catch lungfish on freshly baited hooks fished during the day (08:00-15:00 hours) as those fished over night (15:00-08:00 hours) suggesting that foraging may occur throughout the day. If non-visual senses are used for foraging, the diel cycle would likely have little influence on foraging time unless factors such as predation risk or temperature were influenced by it.

The ultrasonic tracking studies conducted in Lake Baringo by Mlewa *et al.* (2005) provide the only data on spatial activity patterns of *P. aethiopicus*. Individual lungfish movements in the open waters of the lake were frequently restricted within an area of the lake (a home range) for periods ranging from 8 to 18 weeks followed by movements over a few days to a different part of the lake where another 'home range' was then used. Home range size varied from 5.8 to 19.8 km^2. The adaptive significance of this behaviour is uncertain but the occupation of non-breeding home ranges is generally thought to be related to the increased efficiency of foraging and predator avoidance possible in a familiar habitat (Lucas and Baras 2001). Overlap in the home ranges of two tracked lungfish suggests that they are not true territories i.e. defended areas, as suggested by Curry-Lindahl

(1956). The reason for the periodic changes in home range is unclear, but may have involved variation in food supply, the presence of predators, or competition with conspecifics. These ongoing movements appear to spread the effects of fishing over large areas of the lake, limiting local depletion. This may be a factor in the tendency of fishers in Lake Baringo to use fixed locations for their long-lines rather than move locations in response to local depletion.

How lungfishes navigate within large bodies of water such as Lake Baringo to stay within a home range or to make long linear movements has not been studied. Our impression from ultrasonic tracking and from observing tethered fish is that they can maintain fixed compass bearings in highly turbid water, suggesting chemical and/or electromagnetic cues may be involved in this behavior. Given what is known about orientation mechanisms in other fishes, either or both seem likely.

African lungfishes have been reported to migrate upstream (Greenwood 1986). *Protopterus aethiopicus* are now found in Lake Kichiritith (Taras-Wahlberg *et al.* 2003), about 6 km from Lake Baringo, and must have colonized this lake relatively recently from Lake Baringo via the Parkerra or Molo rivers. Movement between these lakes may now be a regular occurrence as evidenced by finding a lungfish tagged and released in Lake Baringo but recaptured in Lake Kichiritith five months later (Mlewa *et al.* 2005). Similarly, *P. aethiopicus* introduced to Lake Muhazi in Rwanda appear to have spread into other lakes in the same drainage system (Micha and Gashagaza 2002). Recent or ongoing movements between Lake Victoria and satellite lakes like Kanyaboli in Kenya, and Nabugoboo in Uganda would explain the similarity in genetic structure of these lungfish populations (Garner *et al.* 2006).

RESPIRATION

Respiration is the most widely studied aspect of protopterid biology. This is, in part, because their obligate use of aerial respiration and phyletic position make them the best living models for studies intended to clarify the role of obligate aerial respiration as a necessary intermediate step in the early evolution of the tetrapods (Graham 1997; Glass 2008). The view that African lungfishes are obligate air breathers is supported by compelling evidence from laboratory studies as well as anecdotal records of wild lungfish breaking the surface to breathe (Smith 1931; Curry-Lindahl 1956; Lenfant and Johansen 1968; McMahon 1970; Johansen *et al.* 1976; Babiker 1979; Greenwood 1986). However, only one study has attempted to quantify the extent of aerial respiration by wild African lungfish (Mlewa *et al.* 2007). In this study, the surfacing frequency of four *P. aethiopicus* was monitored in Lake Baringo using radio telemetry. It was possible to use this technique because of the high conductivity of Lake Baringo which caused radio signals to rapidly attenuate with depth. Radio tags could thus only be detected when the fish were close to the surface. As expected, radio tagged lungfish surfaced frequently to breathe immediately after their release, but, surprisingly, within 12 hours this surfacing behavior had ceased. The initial period of aerial respiration was attributed to higher oxygen demand as the fish recovered from handling stress. Unlike typical lungfish habitats that are characterized by low dissolved oxygen, open waters of Lake Baringo have relatively high oxygen levels (Oduor *et al.* 2003), suggesting that in well oxygenated waters *P. aethiopicus* may be infrequent air breathers, capable of meeting their oxygen demands entirely through aquatic respiration. Although these results appear to contradict the long held belief that protopterids are obligate air breathers, Mlewa *et al.* (2007) noted possible costs to aerial respiration in natural environments, not present in the laboratory, which may favour an increased dependence on aquatic respiration. These include the time and energy costs of transport between the bottom and the surface, exposure to aerial predators while surfaced, the interruption of foraging activities, and the increase in metabolic costs arising from periodic movements to a much warmer surface layer. In this context it is interesting to note that *P. aethiopicus* were regularly caught at depths greater than 30 meters in Lake Victoria prior to the introduction of Nile perch (Bergstrand and Cordone 1970) and it would be surprising if fish were surfacing regularly to breath from such depths. Despite these findings, the apparent absence of aerial respiration for such extended periods is not easily reconciled with the reported physiological constraints on protopterid respiration (Graham 1997), and it is hoped that these results will stimulate more work on the extent of aerial respiration by wild lungfish. Regardless of the study by Mlewa *et al.* (2007), protopterids can clearly meet their oxygen demands entirely from aerial respiration making them completely independent of the oxygen concentration in the water and giving them the ability to fully exploit habitats that most other fishes cannot. However, non-aestivating lungfishes remain dependent on water for respiration because carbon dioxide is exchanged primarily aquatically at the gills (Burggren and Johansen 1987; Graham 1997).

A similar bimodal respiration pattern is found in *Lepidosiren*; also considered an obligate air breather (Burggren and Johansen 1987). In contrast, *Neoceratodus* is a facultative air breather that requires aerial respiration only in poorly oxygenated water. Unlike *Protopterus* and *Lepidosiren* it has a single lung and like non-dipnoans retains holobranchs on all free branchial arches (Thomson, 1969).

AESTIVATION

All *Protopterus* species appear to have the ability to aestivate during periods of seasonal drying. Its use, however, varies considerably among species (Greenwood 1986) with *P. annectens* being most dependent on aestivation (Smith 1931; Johnels and Svensson 1954; Greenwood 1986). The following brief description of aestivation in *P. annectens* is summarized from Johnels and Svensson (1954). Prior to the total drying of the temporary marshlands adjacent to the lower Gambia

River, individuals excavate a short burrow with an enlarged lower portion which accommodates a 180° fold in their trunk so that both their snout and tail face upward. They continue to rise to the surface to breathe until their snouts are permanently exposed to the air by dropping water levels. As the surrounding mud dries, mucus secreted by the lungfish combines with the hardening mud to produce a waterproof cocoon that completely surrounds the lungfish except for a small respiratory opening above the snout. *Protopterus annectens* remain in their cocoons until these are submerged by rising water levels in the following wet season. *P. annectens* normally aestivate for periods of seven to eight months but captive lungfish have emerged from their aestivation cocoons after periods of up to seven years (Lomholt 1993), so that survival in the wild over the shorter periods of annual drought is probably high. *Protopterus dolloi* is not believed to aestivate in its natural environment (Brien *et al.* 1959) and reports of aestivation in *P. aethiopicus* and *P. amphibius* are anecdotal or problematic (Greenwood 1986). However, the former two species have been induced to aestivate under artificial conditions (Smith 1931; Brien *et al.* 1959; Greenwood 1986), producing cocoons similar to those described for *P. annectens*. Also, *P. dolloi* can be induced to enter an inactive state within a cocoon, but it does not burrow nor down-regulate its metabolism as the other protopterid species do (Perry *et al.* 2008). The word "terrestrialization" has been coined to describe this state to distinguish it from true aestivation (Perry *et al.* 2008).

Aestivation, in addition to allowing lungfish to permanently occupy seasonal habitats, may have adaptive value in avoiding high mortality rates and resource bottlenecks associated with the seasonal movements required of non-resident species. Reduced energy demands during aestivation also allow lungfish to attain large body size yet still grow and reproduce in habitats with extreme seasonal fluctuations in resource levels, and possibly attain higher population densities there than could be sustained by non-aestivating species. For example, Johnels and Svensson (1954) reported densities of aestivating large lungfish in dry seasonal swamps of 5 per m^2, with an even greater density of smaller specimens, and in Lake Chad, an increase in the relative catch of *P. annectens* following a severe drought in 1972-1973 was attributed to their use of aerial respiration and aestivation (Bukar and Gubio 1985). The ability of lungfishes to aestivate and thus permanently occupy habitats subject to periodic drying has also led to the suggestion that they could have potential as an aquaculture species in arid areas with only temporary seasonal rainwater ponds (Baer *et al.* 1992).

Although only *P. annectens* appears to aestivate habitually, both *P. dolloi* and *P. aethiopicus* have been reported to restrict their movements to larger water filled burrows during the dry season (Brien *et al.* 1959; Wasawo 1959). Dry season burrows are more extensive than aestivation burrows and more closely resemble tunnel nests, and in the case of *P. dolloi* serve both purposes. *Lepidosiren* aestivates in burrows resembling the dry season burrows of *P. dolloi*, but with out forming a cocoon (Kerr 1950; Harder *et al.* 1999). As might be expected in a riverine fish, *Neoceratodus* does not aestivate (Kemp 1986).

Although aerial respiration is a necessary prerequisite for aestivation, there are a number of other physiological adaptations associated with aestivation, including reduced metabolic rate, renal shutdown, and mechanisms limiting the harmful effects of fat and protein metabolism while fasting without access to water (Fishman *et al.* 1992). Many of these physiological responses are specifically related to the lack of water during aestivation, so would not be expected to be observed in lungfish occupying water filled burrows. However, when water in the surrounding environment is insufficient to allow foraging, lungfish would still experience a prolonged period of seasonal fasting similar in length to that required during aestivation. The ability of lungfishes to fast for prolonged periods without undue physiological stress, even in the absence of aestivation, is an adaptation that would certainly enhance their ability to survive in strongly seasonal environments. The fasting endurance of non-aestivating lungfish is no less impressive than is their ability to aestivate for extended periods. For example, a specimen of *P. aethiopicus* has been reported to have survived without food for 3.5 years in an aquarium, and apparently in good condition (El Hakeem 1979). Fasting in non-aestivating lungfish is accompanied by a reduction in metabolic rate and oxygen consumption similar to that observed in aestivation (Fishman *et al.* 1992), although the reduction is less rapid probably because metabolic wastes can still be excreted or diffused to the water. The possible role of large body size in fasting endurance has been noted above.

THREATS TO AFRICAN LUNGFISHES

African lungfishes are highly valued food fishes by many local communities within their natural distribution range where they support subsistence and small scale commercial fisheries. Consequently overexploitation, particularly of major populations in large lakes, such as Lake Victoria, poses a major threat to African lungfish (Goudswaard *et al.* 2002). In Nigeria aestivating *P. annectens* are dug out of their cocoons during the dry season when they "form sought-after delicacies for the local people" (Otuogbai *et al.* 2001). This mode of exploitation is reported elsewhere in Africa where lungfish aestivate (e.g. Smith 1930; Greenwood 1986). Exploitation of nesting-guarding males and aestivating lungfish, which may be primarily males, can be particularly detrimental to lungfish populations because the reproductive output of the population depends on the parental care provided by males (Goudswaard *et al.* 2002).

Like most other fishes, protopterid lungfishes are threatened by human alteration of aquatic habitats through such activities as dam construction, catchment modifications, and reclamation of wetlands for cultivation. The latter are particularly critical for lungfish because they are needed for spawning, and the loss of wetlands from large projects like the Yala swamp project in Kenya, pose significant threats to some populations. Although none of the four protopterid species is presently categorized as threatened or endangered (FishBase 2007), ongoing effects of exploitation and habitat loss indicate that there is need for more data to guide the formulation of rational policies for management and conservation of African lungfish populations.

SUMMARY

The four species of African lungfishes are remarkable for their persistence in the midst of competition from a diverse array of 3000 more derived actinopterygian fishes inhabiting the inland waters of Africa. This persistence can be attributed to adaptations to harsh environmental conditions acquired well before the appearance of teleosts fishes: aerial respiration and aestivation, the parental care of eggs and larvae, their relatively large size, and the morphological and sensory capabilities to capture and utilize a wide selection of prey. These traits apparently have been present in this group for several hundred million years. In this chapter we have discussed these adaptations against what is known about their natural history, an approach which was somewhat limited by a lack of comparative data for the four species.

Given the uniqueness of this group of fishes it is surprising that so little is known about their natural history. Relatively little new field data on protopterid fishes has been gathered during the last 20 years, and what new data have become available are mostly restricted to one species, *P. aethiopicus*, whose introduced population in Lake Baringo has been studied to some detail. However, the recent field work points to the need for more field studies on these remarkable fishes before their current habitats and populations are largely lost. It is only through a thorough knowledge of the natural history of these fishes that laboratory studies of their physiology and behaviour can be placed in an ecological context and the role of lungfish as models of how the evolution of tetrapods might have occurred can be evaluated.

Acknowledgements

We thank Green, A.A., Green, T.D. and Steele, D.H. for their comments on the manuscript; and R. Ficken for preparation of Figure 3. The Multimedia Laboratory at the Queen Elizabeth II Library, Memorial University provided assistance with Figures 1 and 2.

References

Amongi, T., Muwazi, R.T. and Adupa, I. (2001). The gastrointestinal tract of the African lungfish *Protopterus aethiopicus*: structure and function. Journal of Morphology 248: 202.

Babiker, M.M. (1979). Respiratory behavior, oxygen consumption and relative dependence on aerial respiration in the African lungfish (*Protopterus annectens*, Owen) and an air-breathing teleost (*Clarias lazera*). Hydrobiologia 65: 177-187.

Baer, S., Haller, R.D. and Freyvogel, T.A. (1992). Growth of the African lungfish, *Protopterus amphibius* Peter, in aquaculture. Aquaculture and Fisheries Management 23: 265-267.

Bailey, R.G. (1994). Guide to the fishes of the River Nile in the southern Sudan. Journal of Natural History 28: 937-970.

Bemis, W.E. and Lauder, G.V. (1988). Morphology and function of the feeding apparatus of the lungfish, *Lepidosiren paradoxa* (Dipnoi). Journal of Morphology 187: 81-108.

Bemis, W.E., Burggren, W.W. and Kemp, N.E. (1987). The Biology and Evolution of Lungfishes. Alan R. Liss Inc. New York, USA.

Bergstrand, E. and Cordone, A.J. (1970). Exploratory bottom trawling in Lake Victoria. East Africa Freshwater Fisheries Research Organization. 1969 Annual Report. pp. 42-52.

Berman, D.S. (1976). Occurrence of *Gnathorhiza* (Osteichthyes: Dipnoi) in aestivation burrows in the lower Permian of New Mexico with description of a new species. Journal of Paleontology 506: 1034-1039.

Bouillon, J. (1961). The lungfish of Africa. Natural History 70: 62-70.

Bonnet, X., Naulleau, G., Shine, R. and Lourdais, O. (2000). Reproductive versus ecological advantages to larger body size in female snakes, *Vipera aspis*. Oikos 89(3): 509-518.

Boulenger, G.A. (1909). Catalogue of the Freshwater Fishes in the British Museum (Natural History). Volume 1. London, UK.

Breder, C.M. and Rosen, D.E. (1966). Modes of Reproduction in Fishes. TFH Publ. Neptune City, NJ, USA.

Brien, P. (1959). Ethologie du *Protopterus dolloi* (Boulenger) et de ses larves. Annales de la Societe Royale Zoologique de Belgique 89: 9-48.

Brien, P., Poll, M. and Bouillon, J. (1959). Ethologie de les reproduction de *Protopterus dolloi* Blgr. Annales de la Societe Royale Zoologique de Belgique 71: 3-21.

Brinkman, H., Venkatesh, B., Brenner, S. and Meyer, A. (2004). Nuclear protein-coding genes support lungfish and not the coelacanth as the closest living relatives of land vertebrates. Proceedings of the National Academy of Sciences 101(14): 4900-4905.

Britton, R.J., Ng'eno, J., Lugonzo, J. and Harper, D. (2006). Can an introduced, non-indigenous species save the fisheries of Lakes Baringo and Naivasha, Kenya? Proceedings of the 11th World Lakes Conference, Volume 2, pp. 568-572.

Budgett, J.S. (1899). Observations on *Polypterus* and *Protopterus*. Proceedings of Cambridge Philosophical Society Biological Science 10: 236-240.

Budgett, J.S. (1901). On the breeding habits of some West African fishes, with an account of the external features in the development of *Protopterus annectens* and a description of the larva of *Polypterus laparadei*. Transactions of Zoological Society, London 16: 115-136.

Bukar, T.A. and Gubio, A.K. (1985). The decline in the commercially important species of fish and predominance of *Clarias lazera* in Lake Chad. Proceedings of the Annual Conference of the Fisheries Society of Nigeria.

Bullock, T.H., Hopkins, C.D., Popper, A.N. and Fay, R.J. (eds). (2005). Electroreception. Springer, New York, USA.

Burggren, W.W. and Johansen, K. (1987). Circulation and respiration in lungfishes (Dipnoi). In: W.E. Bemis, W.W. Burggren and N.E. Kemp (eds). The Biology and Evolution of Lungfishes. Alan R. Liss Inc. New York, USA. pp. 217-236.

Campbell, K.S.W. and Barwick, R.E. (1986). Paleozoic Lungfishes — A Review. Journal of Morphology Supplement 1: 93-131.

Carroll, R.L. (1988). Vertebrate Paleontology and Evolution. WH Freeman Inc., New York.

Carlson, K. (1968). The skull morphology and estivation burrows of the Permian lungfish, *Gnathoriza serrata*. Journal of Geology 76(6): 641-663.

Cavin, L., Suteethorn, V., Buffetaut, E. and Tong, H. (2007). A new Thai Mesozoic lungfish (Sarcopterygii, Dipnoi) with an insight into post-Palaeozoic dipnoan evolution. Zoological Journal of the Linnean Society 149: 141-177.

Chapman, L.J., Chapman, C.A., Ogutu-Ohwayo, R., Chandler, M., Kaufman, L. and Keiter, A.E. (1996). Refugia for endangered fishes from an introducted predator in Lake Nabugabo, Uganda. Conservation Biology 10(2): 554-561.

Clack, J.A. (2002). Gaining Ground: The Origin and Evolution of Tetrapods. Indiana University Press, Bloomington, USA.

Conant, E.B. (1970). Regeneration in the African lungfish, *Protopterus* I. Gross aspects. Journal of Experimental Zoology 74: 15-32.

Conant, E.B. (1972). Regeneration in African lungfish, *Protopterus* II. Branching structures. Journal of Experimental Zoology 181(3): 353-363.

Conant, E.B. (1986). A historical overview of the literature of Dipnoi: Introduction to the bibliography of lungfishes. Journal of Morphology Supplement 1: 5-13.

Corbet, P.S. (1961). The food of non-cichlid fishes in the Lake Victoria Basin, with remarks on their evolution and adaptation to lacustrine conditions. Proceedings of the Zoological Society, London 136: 1-101.

Curry-Lindahl, K. (1956). On the ecology, feeding behavior and territoriality of the African lungfish, *Protopterus aethiopicus* Heckel. Arkiv Zoologi 9: 479-497.

Daffalla, A.A., Elias, E.E. and Amin, M.A. (1985). The lungfish *Protopterus annectens* (Owen) as a biocontrol agent against schistosome vector snails. Tropical Medicine 88(2): 131-134.

Dunbrack, R. and Ware, D.M. (1987). Energy constraints and reproductive trade-offs determining body size in fishes. In: P. Calow (ed). Evolutionary Physiological Ecology. Cambridge University Press, Cambridge, UK.

Dunbrack, R.L. and Ramsay, M.A. (1992). The allometry of mammalian adaptations to seasonal environments: a critique of the fasting endurance hypothesis. Oikos 66: 336-342.

Dunbrack, R., Green, J.M. and Mlewa, C.M. (2006). Marbled lungfish growth rates in Lake Baringo, Kenya, estimated by mark recapture. Journal of Fish Biology 68: 443-449.

El Hakeem, O. (1979). A lungfish that survives three and a half years of starvation under aquatic conditions. Zoologischer Anzeiger 202(1-2): 17-19.

Eschmeyer, W.N. (1998). Catalog of Fishes. Volume 1. California Academy of Sciences, San Francisco, USA.

FishBase. (2007). Accessed at www.FishBase.org.

Fishman, A.P., Galante, R.J., Winokur, A. and Pack, A.I. (1992). Aestivation in the African lungfish. Proceedings of the American Philosophical Society 136(1): 61-72.

Forey, P.L. (1986). Relationships of lungfishes. Journal of Morphology Supplement 1: 75-91.

Garner, S., Birt, T.P., Mlewa, C.M., Green, J.M., Seifert, A. and Friesen, V.L. (2006). Genetic variation in marbled lungfish (*Protopterus aethiopicus*) in Lake Victoria and introduction to Lake Baringo, Kenya. Journal of Fish Biology 69 (Supplement b): 189-199.

Glass, M.L. (2008). The enigma of aestivation in the African lungfish *Protopterus dolloi* – Commentary on the paper by Perry *et al.* Respiratory Physiology and Neurobiology 160: 18-20.

Gosse, J.-P. (1984). Protopteridae. In: J. Daget, J.-P. Gosse, and Thys van den Audenaerde (eds). Checklist of the Freshwater Fishes of Africa CLOFFA, Volume 1 ORSTOM, MRAC, Paris.

Goudswaard, K.P.C., Witte, F. and Chapman, L.J. (2002). Decline of the African lungfish (*P. aethiopicus*) in Lake Victoria (East Africa). African Journal of Ecology 40: 42-52.

Graham, J.B. (1997). Air-breathing Fishes. Academic Press, New York, USA.

Gray, J.E. (1845). On a fish allied to *Lepidosiren annectens*. Annals and Magazine of Natural History 16: 348-350.

Greenwood, P.H. (1958). Reproduction in the East African lungfish *Protopterus aethiopicus*. Proceedings of the London Zoological Society 130: 547-567.

Greenwood, P.H. (1986). The natural history of the African lungfishes. Journal of Morphology Supplement 1: 163-179.

Harder, V., Souza, R.H.S., Severi, W., Rantin, F.T. and Bridges, C.R. (1999). Biology of tropical fishes. In: A.L.Val and V.M.F. Almeida-Val (eds), The South American Lungfish - Adaptations to an Extreme Habitat. INPA, Manaus, Brazil. pp. 99-110.

Helfman, G.S., Collette, B.B. and Facey, D.E. (1997). The Diversity of Fishes. Blackwell Science, Malden, MA, USA.

Jardine, W. (1841). Annals and Magazine of Natural History 7:21-26.

Johansen, K., Lomholt, J.P. and Maloiy, G.M. (1976). Importance of air and water breathing in relation to size of the African lungfish *Protopterus amphibius* (Peters). Journal of Experimental Biology 65: 395-399.

Johnels, A.G. and Svensson, G.S.O. (1954). On the biology of *Protopterus annectens* (Owen). Arkiv Zoologi 7: 131-164.

Jorgensen, J.M. (1984). On the morphology of the electroreceptors of two lungfish: *Neoceratodus forsteri* Krefft and *Protopterus annectens*. Videnskabelige Meddelelser fra dansk naturhistorisk. Forening 145: 77-85.

Jorgensen, J.M. (2005). Morphology of electroreceptive sensory organs. In: T.H. Bullock, C.D. Hopkins, A.N. Popper and R.R. Fay (eds). Electroreception. Springer Verlag, New York, USA. pp. 47-67.

Kemp, A. (1986). The biology of the Australian lungfish, *Neoceratodus forsteri* (Krefft 1870). Journal of Morphology Supplement 1: 299-303.

Kerr, J.G. (1932). Archaic fishes – *Lepidosiren, Protopterus, Polypterus* – and their bearing upon problems of vertebrate morphology. Jena Zeitschrift Fur Naturwissenschaften 67: 419-433.

Kerr, J.G. (1950). A Naturalist in the Gran Chaco. *Lepidosiren* Expedition. Cambridge Univ. Press, Cambridge, UK. pp. 169-229.

Lenfant, C. and Johansen, K. (1968). Respiration in the African lungfish *Protopterus aethiopicus*. I. Respiratory properties and normal patterns of breathing and gas exchange. Journal of Experimental Biology 49: 437-452.

Leveque, C. (1997). Biodiversity Dynamics and Conservation: The Freshwater Fish of Tropical Africa. Cambridge University Press, New York, USA.

Lomholt, J.P. (1993). Breathing in the aestivating African lungfish, *Protopterus amphibius*. Advances in Fish Research 1: 17-34.

Lucas, M.C. and Barras, E. (2001). Migration of Freshwater Fishes. Blackwell Science, Oxford, UK.

Lundberg, J.G. (1993). African-South American freshwater fish clades and continental drift: Problems with a paradigm. In: P. Goldblatt (ed). Biological Relationships between Africa and South America. Yale University Press, New Haven, CT, USA. pp. 156-199.

Lundberg, J.G., Kottelat, M., Smith, G.R., Stiassny, M.L.J. and Gill, A. (2000). So many fishes so little time: An overview of recent ichthyological discovery in continental waters. Annals of the Missouri Botanical Garden 87: 26-62.

Malala, R.O. (2006). A study of the nutritional status of fresh and traditionally preserved African lungfish, *Protopterus aethiopicus* Heckel 1851, from Lake Victoria and Lake Kanyaboli, Kenya. B.Sc. Dissertation, Moi University, Kenya.

McMahon, B.R. (1970). The relative efficiency of gaseous exchange across the lungs and gills of an African lungfish, *Protopterus aethiopicus*. Journal of Experimental Biology 52: 1-15.

Micha, J.-C. and Gashagaza, J.-B. (2002). Successful introduction of lung fish *Protopterus aethiopicus* in Muhazi Lake (Rwanda). Academie Royale des Sciences d'Outre-Mer Bulletin des Seances 48(3): 315-318.

Miles, R.S. (1977). Dipnoan (lungfish) skulls and the relationships of the group: A study based on a new species from the Devonian of Australia. Journal of the Linnean Zoological Society 61: 1-328.

Mlewa, C.M. (2003). Biology of the African lungfish *Protopterus aethiopicus* Heckel 1851, and some aspects of its fishery in Lake Baringo, Kenya. Ph.D Thesis, Memorial University, St. John's, Newfoundland, Canada.

Mlewa, C.M. and Green, J.M. (2004). Biology of the marbled African lungfish *Protopterus aethiopicus* Heckel in Lake Baringo, Kenya. African Journal of Ecology 42: 338-345.

Mlewa, C.M. and Green, J.M. (2006). Translocation of marbled African lungfish *Protopterus aethiopicus* (Teleostei: Protopteridae) and its fishery in Lake Baringo, Kenya. African Journal of Aquatic Science 31(1): 131-136.

Mlewa, C.M., Green, J.M. and Simms, A. (2005). Movement and habitat use by the marbled lungfish *Protopterus aethiopicus* Heckel 1851 in Lake Baringo, Kenya. Hydrobiologia 537: 229-238.

Mlewa, C.M., Green, J.M. and Dunbrack, R. (2007). Are wild African lungfish obligate air breathers? Some evidence from radio telemetry. African Zoology 42(1): 131-134.

Mosille, O.I.I.W. and Mainoya, J.R. (1988). Reproductive biology of the East African lungfish *Protopterus aethiopicus* in Mwanza Gulf, Lake Victoria. African Journal of Ecology 26: 149-162.

Nelson, J.S. (2006). Fishes of the World. 4th Edition, John Wiley & Sons, Inc. Hoboken, New Jersey, USA.

Oduor, S.O., Schagerl, M. and Mathooko, J.M. (2003). On the limnology of Lake Baringo (Kenya): I. temporal physico-chemical dynamics. Hydrobiologia 506: 121–127.

Ogutu-Ohwayo, R. (1990). The decline of the native fishes of lakes Victoria and Kyoga (East Africa) and the impact of introduced species, especially the Nile Perch, *Lates niloticus*, and the Nile tilapia, *Oreochromis niloticus*. Environmental Biology of Fishes 27: 81-96.

Okedi, J. (1971). Maturity, sex ratio and fecundity of the lung fish (*Protopterus aethiopicus* Heckel) from Lake Victoria. EAFFRO 1970 Annual Report, pp. 17-20.

Okedi, J. (1990). Observations on the benthos of Murchinson Bay, Lake Victoria, East Africa. African Journal of Ecology 28: 111-122.

Okeyo, D. (2006). On the distribution of fishes of the Kenya's Great Rift Valley drainage system. Discovery and Innovation 18(2): 141-159.

Oniye, S.J., Adebote, D.A., Usman, S.K. and Makpo, J.K. (2006). Some aspects of the biology of *Protopterus annectens* (Owen) in Jachi Dam near Katsina, Katsina state, Nigeria. Journal of Fisheries and Aquatic Sciences 1(2): 136-141.

Otuogbai, T.M., Ikhenoba, A. and Elakhame, L. (2001). Food and feeding habits of the African lungfish, *Protopterus annectens* (Owens) (Pisces: Sarcopterygii) in the flood plains of River Niger in Etsako east of Edo State, Nigeria. African Journal of Tropical Hydrobiology and Fisheries 10: 14-26.

Owen, R. (1839). A new species of the genus *Lepidosiren*. Proceedings of the Linnean Society of London 1: 27-32.

Owen, R. (1841). Description of *Lepidosiren annectens*. Transactions of the Linnean Society of London 18: 327-361.

Pabari, M. (1997). Aspects of the biology of *Protopterus aethiopicus* (Heckel 1851) in the lower Nyando River. MSc Thesis, Middle East Technical University, Ankara, Turkey.

Perry, S.F., Euverman, R., Wang, T., Loong, A.M., Chew, S.F., Ip, Y.K. and Gilmour, K.M. (2008). Control of breathing in African lungfish (*Protopterus dolloi*): A comparison of aquatic and cocooned (terrestrialized) animals. Respiratory Physiology and Neurobiology 160: 8-17.

Poll, M. (1961). Revision systematique et raciation geographique des Protopteridae de l'Afrique Centrale. Annals de l'Musee Royale Afrique Centrale Ser Quarto Zoologi 103: 1-50.

Roberts, T.R. (1975). Geographical distribution of African freshwater fishes. Journal of the Linnaean Zoological Society 57: 249-319.

Rosen, D.E., Forey, P.J., Gardiner, B.G. and Patterson, C. (1981). Lungfishes, tetrapods, paleontology and plesiomorphy. Bulletin of the American Museum of Natural History 167: 159-276.

Roth, A. 1973. Electroreceptors in Brachiopterygii and Dipnoi. Naturwissenschaften 60: 106.

Schultze, H.-P. (1986). Dipnoans as sarcopterygians. Journal of Morphology Supplement 1: 39-74.

Schultze, H.P. and Chorn, J. (1998). The Permo-Carboniferous genus *Sagenodus* and the beginning of modern lungfish. Contributions to Zoology 67: 9-70.

Shafland, P.L. and Lewis, W.M. (1984). Terminology associated with introduced organisms. Fisheries 9(4): 17-18.

Skelton, P.H. (2001). A Complete Guide to Freshwater Fishes of Southern Africa. Struik Publishers, Cape Town, South Africa.

Smith, H.W. (1930). Metabolism of the lungfish, *Protopterus aethiopicus*. Journal of Biological Chemistry 88: 97-130.

Smith, H.W. (1931). Observations on the African Lungfish, *Protopterus aethiopicus*, and on the evolution from water to land environments. Ecology 12: 164-181.

Ssentongo, G.W. (1995). Report on the present fisheries situation of Lake Baringo. FAO Fisheries Department Report. FAO, Rome, Italy.

Stephens, D.W. and Krebs, J.R. (1986). Foraging Theory. Princton University Press, New Jersey, USA.

Stewart, K. (2001). The freshwater fish of Neogene Africa (Miocene–Pleistocene): systematics and biogeography. Fish and Fisheries 2(3): 177-230.

Tarras-Wahlberg, T., Harper, D. and Tarras-Wahlberg, N. (2003). A first limnological description of Lake Kichiritith, Kenya: a possible reference site for the freshwater lakes of the Gregory Rift Valley? South African Journal of Science 99: 1-3.

Thomson, K.S. (1969). The biology of lobe-finned fishes. Biological Reviews 44: 91-154.

Tokita, M., Okamoto, T. and Hikida, T. (2005). Evolutionary history of African lungfish: a hypothesis from molecular phylogeny. Molecular Phylogenetics and Evolution 35: 281-286.

Trewavas, E. (1954). On the presence in Africa, east of the Rift Valleys of two species of *Protopterus, P. annectens* and *P. amphibius*. Annual Royal Museum Congo No. 4(1): 83-100.

Wasawo, D.P.S. (1959). A dry season burrow of *Protopterus aethiopicus* Heckel. Revue de Zoologie et de Botanique Africaines 60: 65-70.

Watt, M., Evans, C.S. and Joss, J.P.M. (1999). Use of electoreception during foraging by Australian lungfish. Animal Behaviour 58(5): 1039-1045.

Wooton, R.J. (1990). Fish Ecology. Chapman and Hall, New York, USA.

CHAPTER 5

Biology of the South American Lungfish, *Lepidosiren paradoxa*

Vera Maria Fonseca de Almeida-Val[1,2,*], Sérgio Ricardo Nozawa[2], Nívia Pires Lopes[3], Paulo Henrique Rocha Aride[2,5], Lenise Socorro Mesquita-Saad[4], Maria de Nazaré Paula da Silva[1], Rubens Tomio Honda[2], Mônica S. Ferreira-Nozawa[2] and Adalberto Luis Val[1]

[1]Instituto Nacional de Pesquisas da Amazônia – INPA, Manaus, Brazil
[2]Centro Universitário Nilton Lins – UNINILTON LINS, Manaus, Brazil
[3]Centro Universitário do Norte – UNINORTE, Manaus, Brazil
[4]Universidade Federal do Amazonas – UFAM, Manaus, Brazil
[5]IFAM – Instituto Federal do Amazonas – Campus Presidente Figueiredo, Brazil

ABSTRACT

Lepidosiren paradoxa, locally known as pirambóia, belongs to the order Dipnoi, or true lungfishes. This group includes two other genera: a single Australian species of *Neoceratodus* (*N. forsteri*) and at least four African species of *Protopterus* (*Protopterus aethiopicus, P. amphibius, P. annectens,* and *P. dolloi*). Lungfishes are members of an ancient group of bony fishes, the Sarcopterygii or lobe-finned fishes, which were widespread during the Devonian period.

Since the expedition of Carter and Beadle in the year 1931, it has been known that the pirambóia dig burrows and aestivate during low water periods. The black body of this lungfish is very elongated, with two pairs of lobe fins, five gill arches; its length can go upto 1.5 m, it moves slowly and lives in swampy areas. This chapter describes the main characteristics of the South American lungfish, *Lepidosiren paradoxa*, and, when necessary, compares it to other lungfishes or teleosts, such as *Arapaima* and

**Corresponding author*: E-mail: veraval@inpa.gov.br

Symbranchus that have similar biological or ecological traits. The chapter is divided into sections where general conditions of living dipnoan genera are compared, and the main characteristics of South American lungfish distribution and habitat, aestivation, reproductive, and feeding behavior are reviewed. Additionally, some metabolic adaptations are described, reviewing the literature and discussing some new facts that were studied in our laboratories.

Keywords: *Lepidosiren*, Amazon, DNA content metabolic adjustments, behavior.

INTRODUCTION

Lepidosiren paradoxa (locally called pirambóia) is the single living representative of the order Dipnoi (lungfish) in South America. Along with its other living relatives around the world (genus *Protopterus* in Africa, and *Neoceratodus* in Australia) the South American lungfish, *Lepidosiren paradoxa*, is one of the most important living vertebrates for studies of the evolutionary transition from water to land. Thus, numerous studies have already shed light on vertebrate evolutionary processes and have shown that lungfishes constitute a step towards acquisition of the main characteristics of the first land vertebrates, the tetrapods (see Johanson and Ahlberg, this volume). Particularly, the South American and African lungfish, in contrast with Australian lungfish, developed several metabolic and physiological abilities that resemble amphibians (Janssens and Cohen, 1966; Hochachka and Somero 1973). However, given that *Lepidosiren* and *Protopterus* are more recently derived lungfish than *Neoceratodus,* these advanced adaptations shared with amphibians could also be viewed as examples of parallel evolution.

In this chapter, we review some of these characteristics, from morphology to metabolism, trying to put together the main adaptive traits that allow the South American lungfish its specialized form of life. The presence of a true lung and the ability to aestivate make this fish an important biological model for studies of several metabolic adaptations to conditions such as desiccation, oxygen depletion, temperature increase, toxic nitrogenous waste storage and excretion, and metabolic depression. Besides these important physiological issues, this group is also interesting from another perspective – that of genome size (largest vertebrate genome, Pederson, 1971), and the presence or absence of genes which are important in evolution of special metabolic characteristics such as vitamin C synthesis and nitric oxide production in its cells.

COMPARATIVE CONSIDERATIONS AMONG DIPNOAN GENERA

Lungfishes have an extensive fossil record in Australia, dating from the Devonian, about 380 million years ago (Long 1995). Fossil records come from almost every period of geological time and the remains of this group indicate that they lived in both freshwater and shallow sea environments in the Devonian and since the end of the Carboniferous period have inhabited only freshwater environments, some acquiring the ability to aestivate during the Permian. They were most speciose in the Devonian, followed by slow steady change during the latter part of the Paleozoic and Mesozoic. Based on fossil records of the Queensland (Australian) lungfish, Long (1995) has suggested that since the Mesozoic they have remained almost unchanged – see chapter by Clack (this volume) for detailed fossil history of lungfishes.

The lepidosirenid lungfishes are longer, slimmer fishes than *Neoceratodus* and have reduced pectoral and pelvic fins for sensory functioning. Both *Protopterus* and *Lepidosiren* can survive periods of dry season burrowing into the ground, an ability not shared by *Neoceratodus*.

All three genera of living lungfish may be considered as "living fossils". Living fossils are known by the retention of particular anatomical forms and ways of life. According to Stanley (1984) most living fossil groups are (i) "depauperate in species"; (ii) retain a large number of plesiomorphies; (iii) are ecologically eurytopic in many physiological and behavioral attributes; (iv) show great individual species longevity with (v) broad area and habitat distribution. Living lungfish have all these attributes. Currently, *Neoceratodus* has a very restricted habitat due to modern anthropomorphic actions in its original habitat.

Evolutionary rates have been claimed to be a consequence of mutation rates and, regarding living fossils, there has always been a trend to correlate the rates of genetic changes with the rates of anatomical and morphological change (Stanley 1984; Kumar and Hedges 1998). However, there is evidence in the literature that related groups with different rates of morphological evolution show similar rates of molecular changes (Avise *et al.* 1994).

DNA CONTENT

The so called 'C-value paradox' (named C-value enigma by T. Ryan Gregory) describes the profound variation in genome size among eukaryotes. Genome size (C-value) is positively related to cell size, and negatively related to cell division rates. Thus, developmental rates have been associated with genome size in numerous plant, vertebrate, and invertebrate groups. Gregory (2002) includes developmental complexity as also being related to genome size. Studies of the C-value paradox seek to find answers to (i) which mechanisms are responsible for genome size variation; (ii) what are the relationships between genome size and nuclear, cellular and organism characteristics; and (iii) which traits are responsible for the maintenance of non-coding DNA.

Considering the highly diverse fish group Actinopterygii, which has considerable diversity in genome size, Hinegardner (1968) stated that "the advanced species of fishes have less DNA than the primitive ones" and thus that "evolution and specialization in the teleosts have been accompanied by loss of DNA". Hinegardner's ideas have led to some discussion since they were proposed, but the answers are not yet resolved (Ohno 1970; Cavalier-Smith 1985).

Modern lungfishes have the largest genomes of all vertebrates. In a recent report, Vinogradov (2005) suggested that the largest genome size among animals is found in the South American lungfish, *Lepidosiren paradoxa* (80pg/haploid nucleus) and that the former data for marbled lungfish, *Protopterus* (>100pg) was greatly overestimated (Pedersen 1971). However, Vinogradov measured *Protopterus dolloi*, instead of *P. aethiopicus* (tetraploid), which has the largest animal genome reported (133pg/haploid nucleus).

Lungfishes possess highly variable or duplicated isozyme systems and loss of gene duplication (Mesquita-Saad *et al.* 2002), sharing the same patterns with amphibians and other tetrapods (Allendorf *et al.* 1983; Buth 1983; Whitt 1983). Thus, the higher C-value in lungfish may be explained by an increase in non-coding DNA.

Reviewing the literature we can characterize *Lepidosiren* genetic traits as follows: (i) medium to low diploid chromosome number; (ii) enormous size of individual chromosomes; and (iii) high contents of DNA (Oliveira *et al.* 1988).

It is not easy to identify what caused the large genome size in lungfish during the evolutionary process. Gregory (2002) suggested that genome size is inversely related to developmental complexity, i.e., when there is an overall reduction in developmental complexity, larger genome sizes result, as occurs among lungfish and salamanders. In amphibians (salamanders) and lungfish, genomes probably have become large in association with *paedomorphosis* (the maintenance of juvenile characters in the adult by the deletion of developmental steps) (Bemis 1984). In advanced teleost fishes, the proposed mechanism is exactly the reverse – small genomes of specialized fish are associated with the *terminal addition* of more developmental steps. However, this correlation with developmental complexity may be a result of genome size rather than a cause of it, since the loss of DNA must be accompanied by the gain of more complex gene regulatory systems.

Cavalier-Smith (1991) proposed that the enormous size of lungfish genomes is due to its unique DNA. Recent evidence for this has been given for *Neoceratodus* (Sirijovsky *et al.* 2005). A transposable element was described for this lungfish, *Nf*CR1. It is the most common duplicated element in this lungfish genome but it is no longer functional and demonstrates high levels of mutation, which together with many other such-like transposable elements, could account for most of the unique DNA making up the large genome of this species of lungfish.

DISTRIBUTION AND HABITAT OF *LEPIDOSIREN*

The South American lungfish was the first living lungfish described during the expedition of the Viennese naturalist Johann Natterer to the Amazon River, Brazil in 1836 (Hyrtl 1845, *apud* Graham 1997). Besides this, the species is not as well studied as the single Australian species *Neoceratodus forsteri* or the four African species that belong to the genus *Protopterus* (present chapter).

Lepidosiren paradoxa is found in the neotropics of South America, including Argentina, Bolivia, Brazil, Colombia, French Guiana, Paraguay, Peru, and Venezuela (Lowe-McConnell 1987; Planquette *et al.* 1996). It is preferentially found in the Amazon River basin, inhabiting stagnant or lentic water systems, such as swamps or lakes, which are associated with vegetation and low oxygen conditions. This species lives in a subtropical climate with temperatures ranging from 24 to 28°C. It is of particular interest that *Lepidosiren* has the most extensive distribution of all extant species of lungfishes, occurring as it does in the tributaries of the Amazon River and Parana-Paraguay River systems, also found in Guiana (Figure 1). For comparison, the distribution of *Protopterus* spp. and *Neoceratodus* are given in the chapters devoted to those two species (this volume).

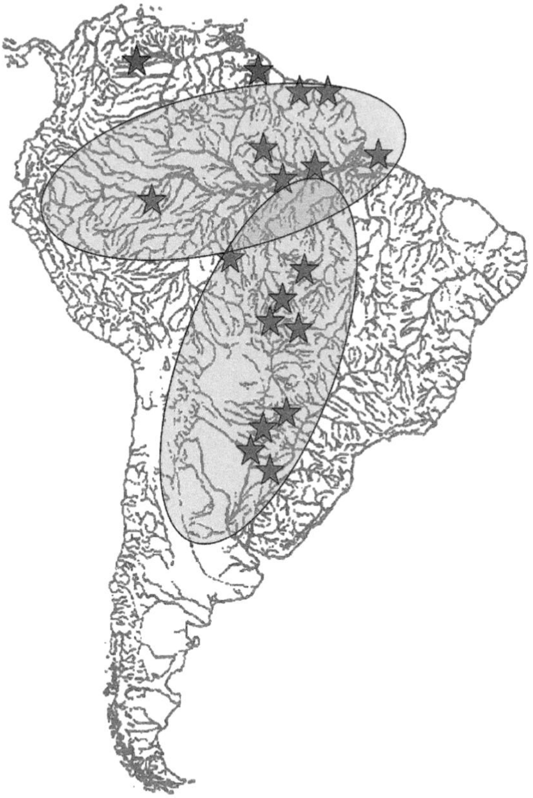

Fig. 1 Map showing distribution of *Lepidosiren paradoxa* (gray shaded area) and localities (gray stars) where *Lepidosiren* specimens were reported based on literature and personal collection.

Lepidosiren is a sluggish fish that stays close to the surface of the water, facilitating its air-breathing habit. This fish swims by sinuous body movements or by use of its pectoral and pelvic fins when moving close to the substrate (e.g. in the bottom of an aquarium), a habit that may be useful particularly to scavenge from the bottom of the waterways they inhabit. Different from the Australian lungfish, *Neoceratodus*, the South American species (and the African species) are obligate air-breathers. *Lepidosiren* has paired fins that are reduced to thin, flexible wisps which, while aiding minimally in locomotion, have tactile and chemosensory capabilities useful in orientation and prey capture in muddy water. In the Amazon region, this species becomes confined in a moist mud burrow during the dry season.

During high water season, it swims slowly and emerges to breath air, although not very actively. Some observations have been done in our laboratory comparing the swimming movements to reach air between juveniles and adults of this species (Figure 2). As we have observed, adults take a longer time between breaths and remain for longer periods at the bottom of the acquaria, showing a greater capacity to suppress activities compared to the juveniles. Previous observations described smaller intervals between breaths for this species (4-5 or 2-10 min) according to Johansen and Lenfant (1967). Our findings rely on measurements of five individuals of each size and controlled environmental conditions (dark/light regime; temperature; quiet room) and were recorded by a digital camera in front of the aquaria.

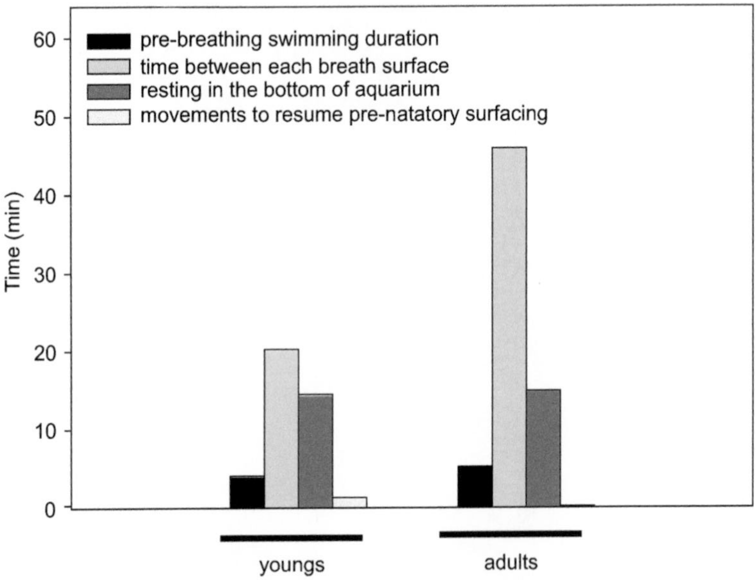

Fig. 2 Comparison of breathing swim activities between adult (n=5) and young (n=5) *Lepidosiren paradoxa* in acquaria. Adults hold their breaths longer than young that are more active.

REPRODUCTIVE BEHAVIOR

Lepidosiren has external fertilization and spawns seasonally during the wet season. Adult males guard and aerate the hatchlings and young temporarily. The larvae (which resemble tadpoles) respire through external gills that resemble feathers for one to two months after which they start to breathe air. It is interesting to note that the male is the one to guard eggs and developing larvae within nests, and during this period it develops vascular filaments on the paired fins (Kerr 1900; Krogh 1941; Sterba 1963). This role for the fin filaments in aeration of *Lepidosiren* eggs is suggested by the resemblance to gills. However, the function of such filaments in respiration is still to be confirmed; the older accepted thought was that these organs are auxiliary to respiration of the adult and enables it to guard the nest without leaving to breathe air; while another view suggests that the fin filaments permit the "emission" of aerially obtained oxygen into the nest water, causing the oxygenation of the water surrounding the eggs (Cunningham 1932; Cunningham and Reid 1933). Although Foxon (1933) defended the idea of auxiliary gills, the function of these filaments still requires confirmation (Graham 1997).

AESTIVATION

The South American (and the African) lungfish is distinguished from most of the other known air-breathing fishes because of its ability to burrow in the mud and aestivate (Johansen 1966; Johansen and Lenfant 1967). This species is profoundly tolerant of low oxygen (Johansen 1966; Lahiri *et al.* 1970; Johansen 1970).

During the dry season, *Lepidosiren* burrows in the mud up to 50 cm to avoid drying out (Berra 2001). Many of these swampy areas inhabited by *Lepidosiren* dry out on an annual basis, but the dryness varies in intensity according to the weather cycle of each year. When most of the water has dried up and only mud remains, *Lepidosiren* burrows, leaving two or three holes for breathing purposes. Although these animals are air-breathers, their ability to survive in this dry environment also involves adaptive mechanisms against desiccation. They start to decrease their metabolic rate as they enter aestivation.

Measurements of heart rates during aquatic and aestivation phases by Harder and co-workers show interesting facts. *Lepidosiren* decreases its heart rate during the first hour after burrowing, entering into bradycardia (heart rates below 46% of the control) and resumes its control levels after the second hour of mud immersion. During the following 24 hours, their heart rates remain as high as control (aquatic phase) levels. Interestingly, apnea periods decrease in the burrowed phase (Harder *et al.* 1999).This short term change in cardio-respiratory pattern to aestivation is also characterized by a significant increase in lung ventilation rate, which has also been observed in air-exposed individuals by Johansen and Lenfant (1967) and can partially contribute to the increased arterial carbon dioxide partial pressure in the animal. For comparison with *Protopterus*, DeLaney *et al.* (1974) have suggested that the aestivation period is characterized by different phases with changes in cardiovascular parameters. According to Harder *et al.* (1999) sequential physiological data on the course of aestivation in *Lepidosiren paradoxa* is not available, as they claim that what they measured during the 31 days of their experiment was the hypothetical initial phase of *Lepidosiren* aestivation. Inducing artificial aestivation leads to decreased oxygen uptake by 13-36% of the initial rate in this species (Perez-Gonzalez and Ginkraut 1971; Abe and Steffensen 1996). Thus, more observations are needed before we can conclude what are the actual physiological changes that occur during the entire period of aestivation in South American lungfish.

Lepidosiren does not form a mucous "seal" as does *Protopterus*. This latter genus is more resistant to desiccation. This state can go on for several months and awakening occurs only when water becomes available again (Figure 3) (reviewed by Val and Almeida-Val 1995; Abe and Steffensen 1996).

Aestivation has been documented also for Permian lungfishes after the finding of fossilized burrows (McAllister *et al.* 1988). The degree of aestivation varies among species, and has been particularly well studied in *Protopterus*. For both genera the excavation of the burrows is made through biting the soil and expelling mud through the gill openings. After finishing this procedure, the fish turns around and remains with its head facing the burrow opening, which allows it to obtain oxygen. Metabolic changes have been reported during aestivation periods, both to save energy and to endure the lack of moisture. During aestivation, lungfishes do not feed to sustain themselves, they metabolize fat initially and when these stores are depleted, they begin to metabolize their muscle mass. This latter activity is constrained by nitrogen excretion (Janssens and Cohen 1966). Changes in nitrogen metabolism have been described as one of the most important adaptations in the transition from water to air breathing animals (Hochachka and Somero 1973). According to Hochachka and Somero (2002), during their hypometabolic state (aestivation) amino acids are frequently used along with CO_2 and H_2O, catabolically to release a less toxic nitrogenous waste than ammonia (see further details in this chapter).

FOOD AND FEEDING BEHAVIOR

Lepidosiren paradoxa is primarily carnivorous. Their food items include bony fishes, shrimp, insects, clams, snails and some algae, weeds and terrestrial plants (stems), (reviewed by Berra 2001). Their larvae remain relatively inactive and attached to the nest until their yolk reserves have been depleted. After that, juveniles start foraging for food and eat insect larvae, crustaceans and snails since they are strictly aquatic. Adults primarily ingest their prey by sucking it into their mouths where they use their tooth plate and depressor mandibulae to crush food prior to swallowing (Bemis 1986a).

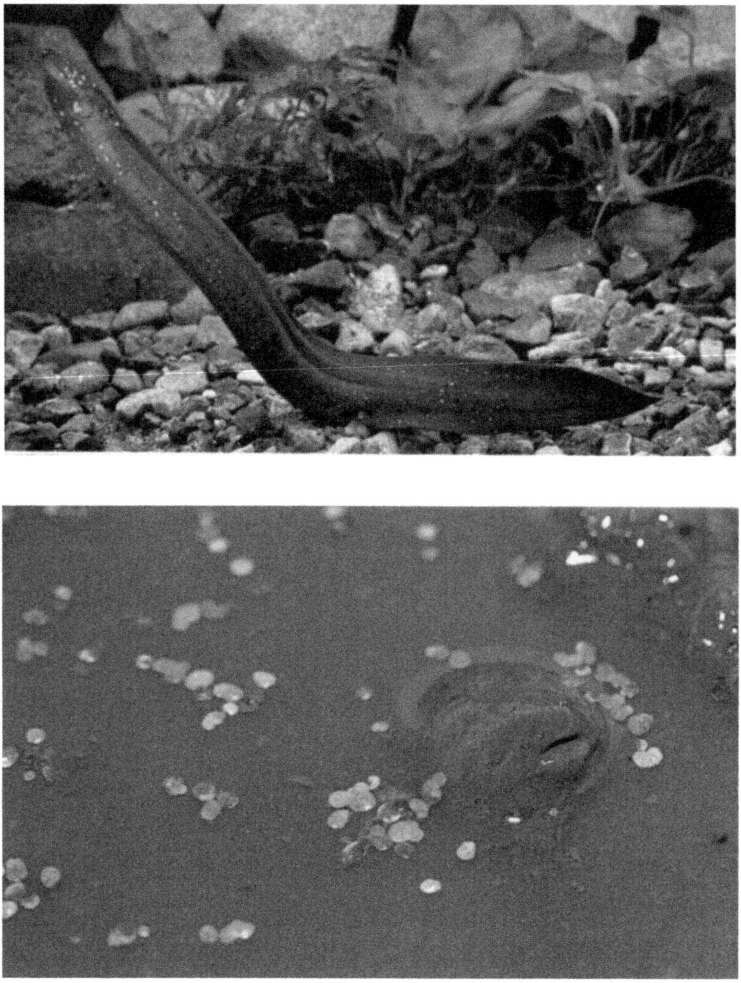

Fig. 3 *Lepidosiren paradoxa* "awaken" in the aquarium in the laboratory (top) and aestivating in the field (bottom). Photos by Lenise Mesquita-Saad during descending water period, 1997.

Color image of this figure appears in the color plate section at the end of the book.

There is no evidence of any natural predator of this species once they are adults and no features that may suggest some anti-predator characteristics. Lungfish share the characteristics of strong teeth and enlarged cranial ribs that serve as the site for the origin of the muscle that depresses the hyoid apparatus. This apparatus, together with the tongue, promotes the hydraulic transport and positioning of the prey inside the mouth (Bemis 1986b). There is no evidence of the use of their strong teeth for anti-predator defense, although there are some reports that describe painful bites with their strong jaws and sharp teeth if provoked in captivity. Some authors presume that larger carnivorous fishes and other vertebrates prey on lungfishes, particularly when they are young but there is no direct evidence for this as yet.

The morphology and function of the feeding apparatus of the South American lungfish, *Lepidosiren paradoxa*, has been studied and described by Bemis and Lauder (2005) as very specialized. They reported that this species' process of initial prey capture shares functional features with other primitive vertebrates, while feeding cycles, including alternate adduction and transport phases, is specialized.

METABOLIC ADAPTATIONS

Glycogen Storage and Metabolism

Studies of the ultrastructure of cardiac muscle, white muscle, and red muscle of South American lungfish have found that this species has an overwhelming dependence on glycogen as storage of carbon and as an energy source. Fish collected in Lake Janauacá at Rio Solimões and kept on board the *River Vessel Alpha Helix* were subjected to enzyme analysis and ultra-microscopy. The musculature of this lungfish consists of both white and red muscle. The heart, although presenting an overall structure similar to other fish species, is modified by its partial atrial septa and ventricular trabeculae, clearly foreshadowing the situation in terrestrial vertebrates (Johansen and Hanson 1968).

Lungfish white muscle differs from other fish species in the kind and amount of glycogen stored (Hochachka and Hulbert 1978; personal observation). The ultrastructure of this tissue is characterized by large amounts of glycogen, different from other fish species, although most of them do have their white muscle metabolism dependent on anaerobic glycolysis. The amount of stored glycogen granules is much lower in teleost fishes than found in lungfish (Pritchard *et al.* 1974) with the exception of tuna white muscle that also stores glycogen in a monoparticulate granular form (Guppy *et al.* 1977). The depots of glycogen found in lungfish white muscle by Hochachka and Hulbert (1978) are large-diameter alpha particles or glycogen rosettes. Glycogen rosettes appear to represent a more efficient way of packaging large amounts of glycogen, as in liver and other tissues. This unique situation (presence of this form of glycogen storage in white muscle) is presumably of advantage to the animal during episodes of hypoxia during aestivation. On the other hand, lungfish red muscle shows glycogen-membrane complexes, which are myofibrillar and peripheral in location. They usually occur in close association with mitochondria, which are far more abundant than in the white muscle. Occasionally, sections of lungfish red muscle show excessive peripheral concentrations of glycogen granules, which are so numerous and so uniformly dispersed that Hochachka and Hulbert (1978) referred to them as glycogen 'seas'. In all cases observed, glycogen 'seas' surround the immense nuclei that typify lungfish cells, Large glycogen reservoirs such as these glycogen 'seas' were considered clearly advantageous during hypoxic or anoxic episodes even in the red muscle, which is often considered to be a relatively aerobic, oxygen-dependent tissue. Furthermore, lungfish heart has a well developed sarcoplasmic reticulum (SR), unlike other fish hearts, where this structure is not so well developed. This unusually shaped SR is in close association with glycogen and more rarely with mitochondria. Thus, the ultrastructure of white and red muscles and the heart of *Lepidosiren paradoxa* were concluded to be adapted for diving, burrowing, and aestivating (Hochachka and Hulbert 1978).

Cardio-respiratory studies done by Costa *et al.* (2002) indicates that *Lepidosiren*, in spite of having a potentially functional and anatomically well-developed SR, does not present a high calcium-cycling capacity, which is not compatible with high heart frequencies as temperature increases. However, as temperature decreases, this organ (SR) has a fundamental role in calcium regulation. SR becomes essential to cardiac performance maintenance when temperature drops down to 15°C.

The activities of several selected enzymes used in energy metabolism of the South American lungfish heart showed a high anaerobic and a relatively low aerobic capacity in comparison with other fishes. Such ability is guaranteed by the extremely high activities of the LDH enzyme that is clearly used during prolonged submergence in the African lungfish (Dunn *et al.* 1983) and is most likely needed during periods of aestivation and burrowing (Almeida-Val and Hochachka 1995).

Reviewing the main biochemical metabolic characteristics of air-breathing fish, we have suggested that the lungfish heart is similar to the hearts of anoxia-resistant vertebrates in its substrate preferences (Almeida-Val and Hochachka 1995). Their enzyme profile is in accordance with preferential use of carbohydrates instead of lipid (as in the aquatic turtle, Almeida-Val *et al.* 1994). In the same review, we have also suggested that glycogen stores in lungfish heart and skeletal muscle are organized to meet energy needs during submergence and recovery (Almeida-Val and Hochachka 1995).

The metabolic adaptations found in African lungfish by Dunn and co-workers (1983) during submergence and subsequent recovery are summarized as (i) glycogen depletion, (ii) lactate accumulation; (iii) change in lactate/pyruvate ratio; (iv) change in adenylate concentrations, and (v) significant creatine phosphate depletion in epaxial muscles during submergence.

ENZYME PROFILE AND ADAPTATION

We have analyzed the enzyme system of South American lungfish during aestivation (Mesquita-Saad *et al.* 2002). Animals were collected (n=7) when aestivating in Lago do Canteiro on Careiro Island in the Amazon River, during the descending water period (August, 1997) and sacrificed with a sharp blow to their heads. Tissues were immediately dissected in the field and promptly stored in liquid nitrogen for transport to the laboratory at Manaus within 3 hours of collection. At the laboratory they were transferred to a low temperature freezer (–80°C). Maximum enzyme activities were determined and compared with those available in the literature. The comparison between the two conditions (aestivated fish *versus* non-aestivating fish) showed that the heart undergoes an absolute shutdown in metabolism during aestivation periods (Table 1). According to Hochachka (1980), besides the fact that lungfishes may retain relatively high glycolytic power in their hearts, they routinely utilize proteins and amino acids as metabolic fuels. The liver and kidney of these animals maintain a significant capacity for gluconeogenesis as well as amino acid metabolism. Further investigations on the variation of glycogen stores during aestivation and awakening periods need to be done before we may discover the metabolic preference of the South American lungfish and the utility of their huge amounts of stored glycogen in white and red muscles and heart.

Lungfish have large fat stores (Hochachka and Somero 1984) which, along with glycogen and protein, are utilized in energy metabolism. Graham (1997) summarized the main metabolic specializations of lungfish as follows: "although African and South American lungfish are capable of oxidizing a range of substrates, they, like the osteoglossids, have a low metabolic turnover relative to most fishes". Further studies in field and laboratory shall be carried out before we can be sure about the metabolic rate changes during the life cycle of the South American lungfish.

Table 1 Enzyme measurements (μmoles of substrate. min^{-1}.g wet $tissue^{-1}$) in heart of South American lungfish in two conditions (Awaken – Hulbert and Hochachka 1978; and Aestivating – Mesquita-Saad *et al.* 2002). The degree the enzymes are suppressed are indicated by the rate between awaken and aestivating fish. Pyruvate inhibition rates indicate predominance of LDH B subunits in the heart, suggesting strong control of acidic end product accumulation in heart, particularly during aestivation when the washout is decreased by lower heart rates).

Enzymes	*Heart awaken*	*Heart aestivating*	*Enzyme suppression rate*
LDH (1 mM)	1177	3.56	330
LDH (10 mM)	464	0.53	875
CS	11.5	1.29	8.9
MDH	307	2.71	113
Pyruvate Inhibition	2.5	6.7	
LDH/CS	102	2.8	
MDH/LDH	0.3	0.8	

RED CELL METABOLIC FEATURES

Lungfish possess, among other water-soluble phosphates such as ATP and GTP commonly present in fish erythrocytes, the compound Inositol diphosphate (Bartlett 1978b). This is a novel compound and no explanation has been found for the presence of this phosphate along with the more commonly found erythrocytic phosphates in fish. Inositol pentaphosphate has been found in another Amazonian air-breathing fish, *Arapaima gigas* (Bartlett 1978a), in which a strict correlation has been shown between animal size and the assumption of an air-breathing habit (Val *et al.* 1992, 2000). It is possible that inositol polyphosphates play a role in the modulation of the hemoglobin oxygen affinity in animals that acquired air-breathing respiration patterns while still living in water, where they are obliged to dive and hold their breath periodically. The tasks of oxygen uptake from the lung and oxygen release to the tissues are difficult when the animal must hold its breath while maintaining normal energy requirements.

Another interesting study utilizing lungfish red cells looked at their membrane permeability. According to Kim and Isaacks (1978) lungfish red cells are permeable to glucose and urea, facilitating the transport of both metabolites. Glucose is the main metabolic fuel for both anaerobic and aerobic glycolysis in most tissues and urea is a less toxic end product of nitrogen metabolism in lungfish (more commonly toxic ammonia in other fish) and plays a significant role in their adaptation to aestivation and in the osmoregulation of blood of several other vertebrate species.

NITROGEN METABOLISM

Most fish excrete nitrogenous waste as ammonia via their gills. Although ammonia (NH_3) can be considered highly soluble in tissues and plasma in its gaseous form, it strongly attracts protons and forms the ammonium ion (NH_4^+) in cellular milieu and may be kept diluted prior to excretion, or converted into less toxic compounds such as urea or uric acid (Hochachka and Somero 1973; Walsh and Henry 1991). If environmental water availability is limited, catabolically formed ammonia can be detoxified by conversion to urea (ureotelism) that may be voided via either the urine or gills (Mommsen and Walsh 1992). Ureotelism evolved prior to animals conquering the land, as a preadaptation like air-breathing.

In fish, urea is the predominant form of nitrogen excretion in marine elasmobranchs and the coelacanth (*Latimeria chalumnae*). Some air-breathers such as the African and the South American lungfish also form large amounts of urea (Smith 1930; Campbell 1973; Hochachka and Somero 1984; Mommsen and Walsh 1989). These latter species synthesize urea through the ornithine urea cycle (OUC) where ammonia is combined with bicarbonate to form urea (Hochachka and Somero 1984). The OUC is the major pathway for nitrogen excretion in terrestrial vertebrates and the presence of this pathway in the African and South American lungfishes, as well as in tetrapods, provides the rationale for the evolutionary link between the appearance of the OUC and the invasion of land by vertebrates. The OUC avoids problems caused by hypercapnia, desiccation, and nitrogen excretion imposed by air breathing and terrestriality. In water, the South American lungfish is ammoniotelic and has urea plasma levels equivalent to other freshwater fish species. During aestivation, however, the South American lungfish use the OUC to form and store urea (Hochachka and Somero 1973; reviewed by Graham 1997) and avoid toxicity caused by ammonia accumulation when there is no possibility of excretion to the surrounding environment. See also Chapter on Urinogenital System (this volume).

While summarizing, we can state that there is little information about metabolic changes during diving and aestivation for the South American lungfish. Most of the information is derived from the African lungfish (see Mlewa *et al.*, this volume). Although this is a valid comparison, the lack of knowledge about metabolic fates of the South American lungfish deserves more attention in the form of further investigations.

CONCLUDING REMARKS

Although many new developments have been made since we reviewed the biological features of *Lepidosiren paradoxa* (Val and Almeida-Val 1995) several questions remain to be answered regarding the impressive processes that allow these animals to remain burrowed for months of aestivation with low energy (ATP) turnover. Besides aestivation, it will be important to analyze the mechanisms involved in the degeneration of external gills, which occurs 45 days after hatching and has no noticeable effect on growth rate of the larvae. If apoptosis (programmed cell death) mechanisms are responsible for the "re-absorption" of these specialized "gills", how do these processes manage to rely on such low metabolic costs? We know that these animals are remarkably efficient in maintaining tissue oxygenation using gills during the early yolk absorption process. Also remarkable is the specialized capacity observed in adult males that, while taking care of their eggs stirring the surrounding water of their nests, develop a considerable number of respiratory filaments on their pelvic fins. Such transformation mechanisms are very specialized and should be investigated from a molecular point of view.

These and all other questions raised in this chapter may provide a good background for future evolutionary ecology studies. We have no doubt that the South American lungfish is a living proof that morphological, genetic, physiological, biochemical, and molecular processes have resulted from evolutionary changes based on the selection of many adaptive and specialized traits.

Acknowledgements

This work was funded by the National Research Council (CNPQ) of Brazil (PNOPG to VMFAV- processes # 400030/11-33) and a CNPq/FAPEAM (Amazon State Research Foundation) PRONEX grant to ALV. It is also part of the review documents to data bank for the National Institute ADAPTA funded by FAPEAM, CNPQ and MCT. ALV and VMFAV are the recipients of a Research fellowship from CNPQ.

References

Abe, A.S. and Steffensen, J.F. (1996). Respiração pulmonary e cutânea na Pirambóia, *Lepidosiren paradoxa*, durante a atividade e a estivação (Osteichthyes, Dipnoi). Revista Brasileira de Biologia 56: 485-489.

Allendorf, F.W., Learry, R.F. and Kundsen, K.L. (1983). Structural and regulatory variation of phosphoglucomutase in rainbow trout. In: M.C. Rattazzi, J.G. Scandalios and G.S. Whitt (eds). Isozymes: Current Topics in Biological and Medical Research. Alan R. Liss, New York, USA. pp. 123-142.

Almeida-Val, V.M.F. and Hochachka, P.W. (1995). Air-breathing fishes: metabolic biochemistry of the first diving vertebrates. In: P.W. Hochachka and T. Mommsen (eds). Biochemistry and Molecular Biology of Fishes, Environmental and Ecological Biochemistry. Elsevier Science, Amsterdam. The Netherlands. pp. 44-45.

Almeida-Val, V.M.F., Buck, L.T. and Hochachka, P.W. (1994). Substrate and temperature effects on turtle heart and liver mitochondria. American Journal of Physiology 266: R858-R862.

Avise, J.C. (1994). Molecular Markers, Natural History and Evolution. Chapman and Hall, New York, USA.

Bartlett, G.R. (1978a). Phosphates in red cells of two South American osteoglossids: *Arapaima gigas* and *Osteoglossum bicirrhosum*. Canadian Journal of Zoology 56(4): 878-881.

Bartlett, G.R. (1978b). Water-soluble phosphates of fish red cells. Canadian Journal of Zoology 56(4): 870-877.

Bemis, W.E. (1984). Paedomorphosis and the evolution of the Dipnoi. Paleobiology 10: 293-307.

Bemis, W.E. (1986a). Feeding systems of living dipnoi. Journal of Morphology supplement 1: 249-275.

Bemis, W.E. (1986b). Vertebrate Evolution: Evolutionary Biology of Primitive Fishes. Science 233(4759): 114-115.

Bemis, W.E. and Lauder, G.V. (2005). Morphology and function of the feeding apparatus of the lungfish, *Lepidosiren paradoxa* (Dipnoan). Journal of Morphology 187: 81-108.

Bemis, W.E., Fernandes, C.C., Castro, R.C., Zuanon, J., Py-Daniel, L., Alves-Gomes, J., Santos, G., Machado, F. and Malabarba, L. (2003). Notes on the Systematics, Distribution and Natural History of the South American Lungfishes in the genus *Lepidosiren* - Fitzinger, 1837 (Dipnoan: Lepidosirenidae). Online: www.bio.umass.edu/biology/bemis/lungfish_ms.pdf.

Berra, T.M. (2001). Freshwater Fish Distribution. Academic Press, San Diego, USA.

Brinkmann, H., Denl, A., Zitzier, J., Joss, J. and Meyer, A. (2004). Complete mitochondrial genome sequences of the South American and the Australian lungfishes: Testing of the phylogenetic performance of mitochondrial data sets for phylogenetic problems in tetrapod relationships. Journal of Molecular Evolution 59: 834-848.

Buth, D.G. (1983). Duplicate isozyme loci in fishes: origins, distribution, phyletic consequences and locus nomenclature. In: M.C. Rattazzi, J.G. Scandalios and G.S. Whitt (eds). Isozymes: Current Topics in Biological and Medical Research. Alan R. Liss, New York, USA. pp. 381-400.

Campbell, J.W. (1973). Nitrogen Excretion. In: C.L. Prosser (ed.). Comparative Animal Physiology. 3rd Edition. W.B. Saunders, Philadelphia, USA. pp. 279-316.

Carrol, R.L. (1997). Patterns and Processes of Vertebrate Evolution. Cambridge Paleobiology Series. Cambrigde Universty Press, Cambridge, UK.

Cavalier-Smith, T. (1985). Selfish DNA and the origin of introns. Nature 315: 283-284.

Cavalier-Smith, T. (1991). Coevolution of vertebrate genome, cell, and nuclear sizes. In: G. Ghiara, F. Angelini, E. Olmo and L. Varano (eds.). Symposia on the Evolution of Terrestrial Vertebrates. Selected Symposia and Monographs UZI 4, Mucchi. Modena, Italy. pp. 51-86.

Costa, M.J., Olle, C.D., Ratto, J.A., Anelli Jr., L.C., Kalinin, A.L. and Rantin, F.T. (2002). Effect of acute temperature transitions on chronotropic and inotropic responses of the South American lungfish. Journal of Thermal Biology 27: 39-45.

Cunningham, J.T. (1932). Experiments on the interchange of oxygen and carbon dioxide between the skin of *Lepidosiren* and the surrounding water, and the probable emission of oxygen by the male *Symbranchus*. Proceedings of the Zoological Society, London, 2: 876-887.

Cunningham, J.T. and Reid, D.M. (1933). Pelvic filaments of Lepidosiren. Nature 131: 913.

DeLaney, R.G., Lahiri, S. and Fishman, A.P. (1974). Aestivation of the African lungfish *Protopterus aethiopicus*: cardiovascular and respiratory functions. Journal of Experimental Physiology 61: 111-128.

Dunn, J.F., Hochachka, P.W., Davison, W. and Guppy, M. (1983). Metabolic adjustments to diving and recovery in the African lungfish. American Journal of Physiology 245: R651-R657.

Foxon, G.E.H. (1933). Pelvic filaments of *Lepidosiren*. Nature 131: 913-914.

Graham, J.B. (1997). Air-breathing Fishes: Evolution, Diversity, and Adaptation. Academic Press, San Diego, USA.

Gregory, T.R. (2002). Genome size and developmental complexity. Genetica 115: 131-146.

Guppy, M., Hulbert, W.C. and Hochachka, P.W. (1977). The tuna power plant and furnace. In: G. Sharp and A. Dizon (eds.). Physiological Ecology of Tuna. Academic Press, New York.

Harder, V., Souza, R.H.S., Severi, W., Rantin, F.T. and Bridges, C.R. (1999). The South American lungfish - adaptations to an extreme habitat. In: A.L. Val and V.M.F. Almeida-Val (eds.). Biology of Tropical Fishes. Editora do INPA, Manaus, AM. Brazil. pp. 87-98.

Hinegardner, R. (1968). Evolution of cellular DNA content in teleost fishes. American Naturalist 102: 517-523.

Hochachka, P.W. (1980). Living Without Oxygen. Harvard Univ. Press, Cambridge, Mass., USA.

Hochachka, P.W. and Somero, G.N. (1973). Strategies of Biochemical Adaptation. W.B. Saunders Company, Philadelphia, USA.

Hochachka, P.W. and Hulbert, W.C. (1978). Glycogen seas, glycogen bodies, and glycogen granules in heart and skeletal muscle of two air-breathing, burrowing fishes. Canadian Journal of Zoology 56: 774-786.

Hochachka, P.W. and Somero, G.N. (1984). Biochemical Adaptations. Princeton University Press, Princeton, USA.

Hochachka, P.W. and Somero, G.N. (2002). Biochemical Adaptation: Mechanism and Process in Physiological Evolution. Oxford University Press, New York.

Hyrtl, J. (1845). *Lepidosiren paradoxa*. Bohm. Gesell. Abh. 3: 605-668.

Janssens, P.A. and Cohen, P.P. (1966). Ornithine-urea cycle enzymes in the African lungfish *Protopterus aethiopicus*. Science 152: 358-359.

Johansen, K. (1966). Air breathing in the teleost *Symbranchus marmoratus*. Comparative Biochemistry and Physiology 18: 383-395.

Johansen, K. (1970). Air-breathing in fishes. In: W.S. Hoar and D.J. Randall (eds.). Fish Physiology. Volume 4. Academic Press, New York, NY, USA. pp. 361-411.

Johansen, K. and Lenfant, C. (1967). Respiratory functions in the South American lungfish, *Lepidosiren paradoxa*. Journal of Experimental Biology 46: 205-218.

Johansen, K. and Hanson, D. (1968). Functional anatomy of the hearts of lungfishes and amphibians. American Zoologist 8: 191-210.

Kerr, J.G. (1900). The external features in the development of *Lepidosiren paradoxa*, Fitz. Phil. Trans. R. Soc., London, 192B: 299-330.

Kim, H.D. and Isaacks, R.E. (1978). The membrane permeability of nonelectrolytes and carbohydrate metabolism of Amazon fish red cells. Canadian Journal of Zoology 56: 863-869.

Krogh, A. (1941). The Comparative Physiology of Respiratory Mechanisms. Univ. of Pennsylvania Press, Philadelphia, USA.

Kumar, S. and Hedges, B. (1998). A molecular timescale for vertebrate evolution. Nature 392: 917-920.

Lahiri, S., Szidon, J.P. and Fishman, A.P. (1970). Potential respiratory and circulatory adjustments to hypoxia in the African lungfish. Federation Proceedings 29: 1141-1148.

Long, J.A. (1995). The Rise of Fishes: 500 Million Years of Evolution. Johns Hopkins Univ. Press, Baltimore, USA.

Lowe-McConnell, R.H. (1987). Ecological Studies in Tropical Fish Communities. Cambridge University Press, Cambridge, UK.

McAllister, J.A., Dubiel, R.F., Blodgett, R.H. and Bown, T.M. (1988) Lungfish burrows in the Upper Triassic Chinle and Dolores formations, Colorado Plateau; comments on the recognition criteria of fossil lungfish burrows; discussion and reply. Journal of Sedimentary Research 58: 365-369.

Mesquita-Saad, L.S.B., Leitão, M.A.B., Paula-Silva, M.N., Chippari-Gomes, A.R. and Almeida-Val, V.M.F. (2002). Specialized metabolism and biochemical suppression during aestivation of the extant South American Lungfish – *Lepidosiren paradoxa*. Brazilian Journal of Biology 62(3): 495-501.

Mommsen, T.P. and Walsh, P.J. (1989). Evolution of urea synthesis in vertebrates: The piscine connection. Science 243: 72-75.

Mommsen, T.P. and Walsh, P.J. (1992). Biochemical and environmental perspectives on nitrogen metabolism in fishes. Experientia 48: 583-593.

Ohno, S. (1970). The enormous diversity in genome size of fish as a reflection of nature's extensive experiments with gene duplication. Transactions of American Fisheries Society 99: 120-130.

Oliveira, C., Almeida-Toledo, L.F., Foresti, F., Britski, H.A. and Toledo-Filho, S.A. (1988). Chromossome formulate of neotropical freshwater fishes. Revista Brasileira de Genetica 11: 577-624.

Pederson, R.A. (1971). DNA content, ribosomal gene multiplicity and cell size in fish. Journal of Experimental Zoology 177: 65-78.

Phelps, C. and Farmer, M. (1979). Equilibria and kinetics of oxygen and carbon monoxide binding to the hemoglobin of the South American lungfish, *Lepidosiren paradoxa*. Comparative Biochemistry and Physiology 62A: 139-143.

Perez-Gonzalez, M.D. and Grinkraut, C.N. (1971). Comportamento e metabolismo respiratório da *Lepidosiren paradoxa* durante a vida aquática e em estivação. Boletim de Zoologia e Biologia Marinha 28: 137-164.

Planquette, P., Keith, P. and Lebail, P.Y. (1996). Atlas des Poissons d'Eau Douce de Guyane, Tome 1. Museum National d'Histoire Naturelle. Paris, France.

Pritchard, A.W., Hunter, J.R. and Lasker, R. (1974). The relation between exercise and biochemical changes in red and white muscle and liver in the jack mackerel, *Trachurus symmetricus*. Fish Research Bulletin 69: 379-386.

Schwantes, M.L.B. and Schwantes, A.R. (1982). Adaptive features of ectothermic enzyme. I. Temperature effects on the malate dehydrogenase from a temperature fish *Leiostomus xanthurus*. Comparative Biochemistry and Physiology 72B: 49-58.

Sirijovsky, N., Woolnough, C., Rock, J. and Joss, J.M.P. (2005). *Nf*CR1, the first non-LTR retrotransposon characterised in the Australian lungfish genome, *Neoceratodus forsteri*, shows similarities to CR1-like elements. Journal of Experimental Zoology (Mol Dev Evol) 304B: 40-49.

Smith, H.M. (1930). Metabolism of the lungfish *Protopterus aethiopicus*. Journal of Biological Chemistry 88: 97-130.

Stanley, S.M. (1984). Simpon`s inverse: Bradytely and the phenomenon of living fossil. In: N. Eldredge and S.M. Staley (eds.) Living Fossil. Springer-Verlag, New York, USA. pp. 272-277.

Sterba, G. (1963). Freshwater Fishes of the World. Viking Press, New York, USA.

Val, A.L. (2000). Organic phosphates in the red blood cells of fish. Comparative Biochemistry and Physiology A: Comparative Physiology 125: 417-435.

Val, A.L. and Almeida-Val, V.M.F. (1995). Fishes of the Amazon and their Environments. Physiological and Biochemical Features. Springer Verlag, Heidelberg, Germany.

Val, A.L., Affonso, E.G. and Almeida-Val, V.M.F. (1992). Adaptive features of Amazon fishes: Blood characteristics of Curimata (Prochilodus nigricans, Osteichthyes). Physiological Zoology 65: 832-843.

Vinogradov, A.E. (2005). Genome size and chromatin condensation in vertebrates. Chromosoma 113: 362-369.

Walsh, P.J. and Henry, R.P. (1991). Carbon dioxide and ammonia metabolism and exchange. In: P.W. Hochachka and T.P. Mommsen (eds.). Biochemistry and Molecular Biology of Fishes, Volume 1. Elsevier, New York, USA. pp. 181-207.

Whitt, G.S. (1983). Isozymes as probe and participants in developmental and evolutionary genetics. In: M.C. Rattazzi, J.G. Scandalios and G.S. Whitt (eds.). Isozymes: Current Topics in Biological and Medical Research. Alan R. Liss., New York, USA. pp. 1-40.

CHAPTER 6

Early Head Development in the Australian Lungfish, *Neoceratodus forsteri*

Rolf Ericsson[1], Jean Joss[1] and Lennart Olsson[2,*]

[1]Department of Biological Sciences, Macquarie University, Sydney, New South Wales 2109, Australia

[2]Institut für Spezielle Zoologie und Evolutionsbiologie mit Phyletischem Museum, Friedrich-Schiller-Universität, Erbertstr, 1, D-07743 Jena, Germany

ABSTRACT

Early head development in the Australian lungfish *Neoceratodus forsteri* (Krefft 1870) has until recently received relatively little attention. In this chapter we summarize recently published studies, mostly from our own laboratories, and some unpublished data that document the early development of the head skeleton and the head musculature in this species. Like in other vertebrates, the cranial neural crest gives rise to a large part of the skull in the Australian lungfish. Unlike in related animals, cranial neural crest cells emerge late from the neural tube, but follow the general migratory pattern seen in other vertebrates, forming mandibular, hyoid and branchial streams. Labelling with the lipophilic fluorescent dye, DiI, allowed us to follow the fate of cranial neural crest cells until Kemp stage 43, when several cranial skeletal elements have started to differentiate. The cranial neural crest contributes cells to several parts of the head skeleton, including the trabecula cranii and derivatives of the mandibular arch (e.g. Meckel's cartilage, quadrate), the hyoid arch (e.g. the ceratohyal), and the branchial arches (ceratobranchials I–IV), as well as to the connective tissue surrounding the myofibers in cranial muscles. The myofibers of cranial muscles are likely to be derived from the head mesoderm, but no proper fate-mapping is availabe to prove this. Later they can be detected using antibodies that mark desmin, and are differentiating almost simultaneously in the mandibular and

[*]*Corresponding author:* E-mail: Lennart.Olsson@uni-jena.de

hyoid arches, but later in the branchial arches. Unlike the Mexican axolotl, which has an origin-to-insert pattern of differentiation, no obvious directionality can be observed in the differentiating cranial muscles of the Australian lungfish.

Keywords: head development, morphogenesis, cell migration, fate mapping

INTRODUCTION

Research into head development is an important subject in comparative vertebrate morphology, with a rich classical literature (see e.g. Gegenbaur 1888; de Beer 1937; Goodrich 1930 for reviews). In a modern 3-volume treatise on the vertebrate skull (Hanken and Hall 1993), one entire volume was devoted to development. Most of the recent studies have focused on the mechanisms that determine the early regionalization of the head and give the proper identity to the tissues in the head region (Grammatopoulos *et al.* 2000; Graham and Smith 2001; Pasqualetti *et al.* 2000; Rijli *et al.* 1993, 1998), and have used model species of birds (the quail/chicken system), rodents (mouse, rat), frogs (*Xenopus*) and salamanders (Mexican axolotl). Less is known about the later processes that regulate morphogenesis and are responsible for shaping the architecture of the developing head (but see, e.g. Depew *et al.* 2002; Depew and Simpson 2006), and about non-model species. The picture emerging from these studies is that, in contrast to the situation in the trunk, most skeletal elements in the head are derived from neural crest cells (Couly and Le Douarin 1990; Couly *et al.* 1992, 1993; Noden 1991). The neural crest is an embryonic cell population that gives rise to a large array of different cell and tissue types (Hall 1999; LeDouarin and Kalcheim 1999). Cranial neural crest cells emerge on top of the neural tube. They migrate in a ventral direction and form streams which feed into the pharyngeal arches. Neural crest cells form most of the head skeleton, but also connective tissue to which the mesoderm-derived muscles attach (Noden 1986a, b; Olsson *et al.* 2001; Ericsson *et al.* 2004). The muscles of one arch are attached to bones (or cartilages) from the same arch and innervated by cranial nerves emerging from the corresponding rhombomere (Köntges and Lumsden 1996; Olsson *et al.* 2001). Thus an early pattern of head organisation leaves an imprint even into the adult stages, although the adult vertebrate head does not show a clear segmentation (see Olsson *et al.* 2005 for review).

In the Australian lungfish *Neoceratodus forsteri* (Krefft1870), the development of the head, and early events in particular, are relatively poorly known (but see Fox 1965; Bartsch 1993, 1994; Bertmar 1966). Despite some recent work on neural (Kemp 2000) and neural crest (Ericsson *et al.* 2008; Falck *et al.* 2000; Kemp 1995, 2000) development, most of the literature is old (e.g. Greil 1908, 1913), using the methods then at hand, and material collected in the wild (Olsson *et al.* 2004). Sometimes they concern other species of lungfish (Agar 1906). Only the later stages of development and the larval anatomy have been the subject of thorough recent study (Bartsch 1993, 1994; Kemp 1999). In this chapter, we summarize our studies on cranial neural crest cells and conclude that they do emerge and migrate in the Australian lungfish in the same way as in other vertebrates, and that they form the skeletal derivatives expected from considerations of homology. We also describe the development of cranial muscle in the Australian lungfish, in part using our own unpublished work, and compare head muscle development in the Australian lungfish with our published work on the Mexican axolotl (Ericsson and Olsson 2004).

EMERGENCE AND MIGRATION OF CRANIAL NEURAL CREST CELLS

In comparison with amphibians, migration of cranial neural crest cells in the Australian lungfish starts very late, at stage 25 or 26. At stage 27 (Fig. 1), the most anterior neural crest cells, forming the mandibular stream (M), are separating into two differentially migrating streams of cells. Cells from one of the streams migrate rostrally between the epidermis and the neural tube, dorsal to the optic vesicle, while most cells migrate rostroventrally between the epidermis, the neural tube and the mesoderm (Figs. 1A, B, arrows in B show the two parts of the mandibular stream). Mandibular neural crest streams from the left and right side of the embryo eventually meet and merge on the ventral side of the embryo. The more posterior neural crest cells migrate out later and by stage 28 they have separated to form the hyoid (H) and the branchial (B) neural crest streams (Figs. 1C, D). Cells from both these latter streams migrate ventrolaterally between the epidermis and the neural tube around the otic placode (Figs. 1C, D, Ot). At stage 28, the mandibular neural crest cells begin to surround the optic vesicle (Fig. 1C, O), a process that is completed by stage 30. By stage 28, the hyoid neural crest cells (H) have migrated ventrally on the lateral side of the embryo between the epidermis and the mesoderm (Figs. 1C, D), anterior to the otic placode (Ot). They reach the ventral side of the embryo at stage 29. During the same stages (28-29), the branchial neural crest cells (B) migrate towards the otic placode (Figs. 1C, D).

Thus, scanning electron microscopy has shown that there are migrating cranial neural crest cells in lungfish embryos (Falck *et al.* 2000). In fact, they are very similar to the embryos of the Mexican axolotl (Falck *et al.* 2002). Cranial neural crest cells do emerge and migrate in the Australian lungfish in the same way as in other vertebrates, forming mandibular, hyoid and branchial streams (Fig. 1). The major difference is in the timing of the onset of cranial neural crest cell migration. It is delayed in the Australian lungfish in comparison with their living sister group the Lissamphibia (all recent amphibians). Furthermore, the delay in timing between the emergence of the hyoid and branchial crest streams is very long, indicating a steeper anterior-posterior gradient than in amphibians (Fig. 1).

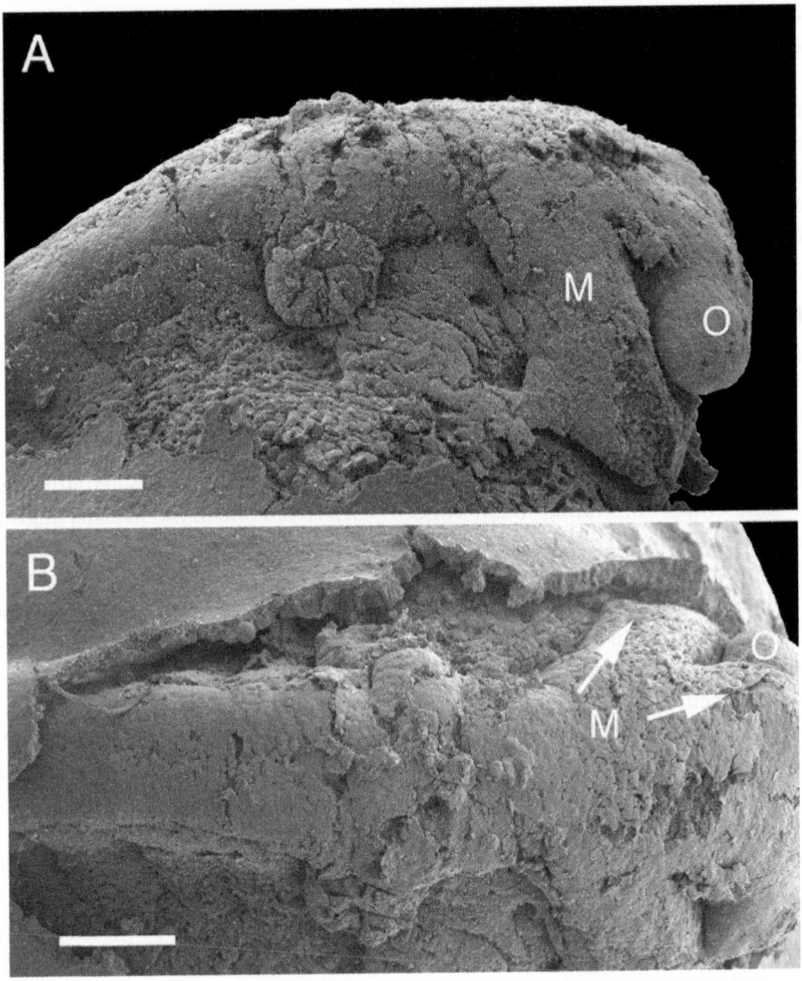

Fig. 1 Scanning electron micrographs of embryonic *Neoceratodus forsteri* showing the cranial neural crest from a lateral view in A, C and D, and from a dorsal view in B, anterior to the right. Overlying epidermis has been removed. Scale = 0.2 mm. Stages are according to Kemp (1982) A: A lateral view of a stage 27 embryo. There is a conspicuously migrating mandibular crest stream (M). B: A dorsal view of a stage 27 embryo. The mandibular (M) stream divides and migrates along two different routes. One migratory route goes rostrally between the epidermis and the neural tube dorsal to the optic vesicle (O) whereas the other takes a rostroventral route (arrows).

Fig. 1 Contd..

Fig. 1 Contd..

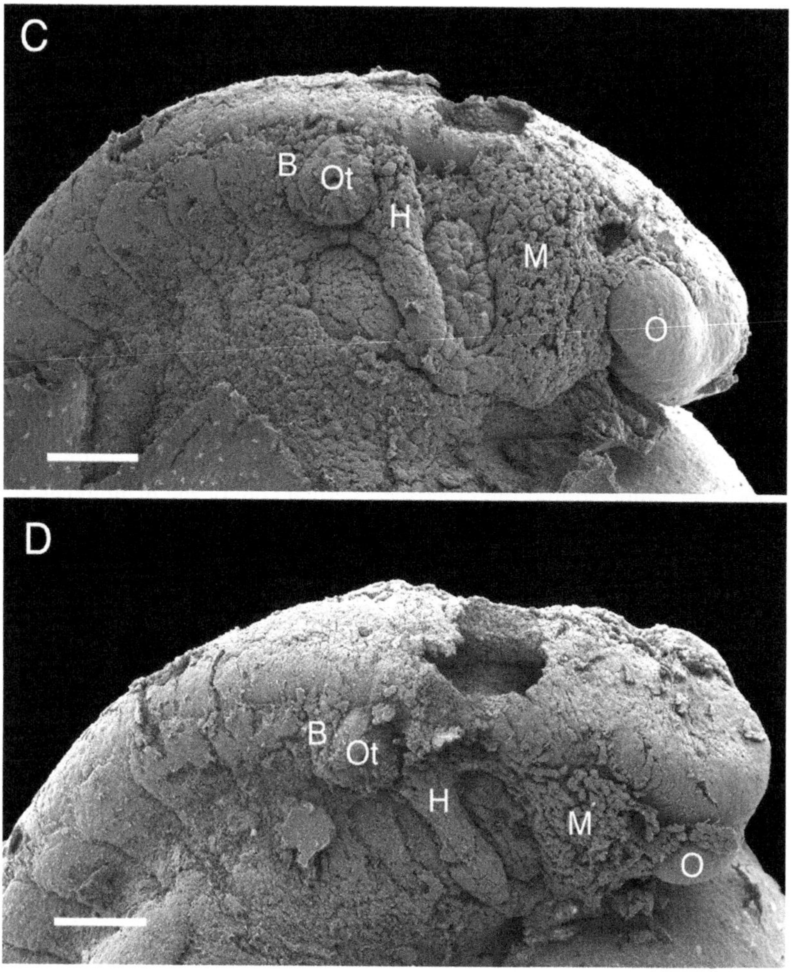

C: A stage 28 embryo showing the mandibular (M), hyoid (H) and branchial (B) neural crest streams, the developing optic vesicle (O) and the otic placode (Ot). D: The other side of the same animal (flipped 180 degrees so that anterior is to the right) showing the mandibular (M), hyoid (H), and branchial (B) neural crest streams.

THE FATE OF CRANIAL NEURAL CREST CELLS

A drawback with scanning electron microscopy is that at later stages, other cells (e.g. mesodermal cells) also start to migrate, and it becomes impossible to identify the cranial neural crest cells. For proper fate mapping, cells must be marked and followed through their migratory period. For this purpose we have used a fluorescent lipophilic dye, DiI, that was injected into living embryos. This method made it possible to produce a relatively long-term fate-map of cranial neural crest cells in the Australian lungfish. We could follow the fate until stage 43, when several cartilages and other tissues in the lungfish have started to differentiate.

In this study (Ericsson *et al.* 2008) the focus was on the later development and fate of cranial neural crest cells. Injections of DiI in the dorsal part of the rostral neural tube tagged a population of cells that had an early migratory pattern identical to what we have described earlier based on scanning electron microscopy. The pattern is also very similar to what happens in the Mexican axolotl (Ericsson and Olsson 2004; Falck *et al.* 2002). Later stages still show great similarities to Mexican axolotl larvae. Frontal sections of stage 40 larvae showed DiI present at the injection site in the neural tube and also in the ectoderm over this area (Figs. 2A, B). Because cells in these tissues do not migrate, they cannot be mistaken for neural crest cells. DiI was also found in the mesenchyme surrounding the eye (Fig. 2B) and in the pharyngeal arches (Figs. 2A, B). No labelled cells were observed in the otic capsule or somites (Fig. 2A). In more ventral sections of the same embryo, DiI-tagged cells were found in the epidermis (Fig. 3A, arrowheads) and, more importantly, in the cells forming a tube around the inner mesodermal core of the mandibular arch (Figs. 3A, B) as well as cells migrating rostrally, probably to form the trabecula cranii (Fig. 3B, lower left corner). As we move ventrally (Figs. 3C, D), the strong DiI staining in the mandibular arch (arrowheads) continues to the ventral tip of the mandibular arch (Fig. 3C, arrowheads). A close-up shows that the medial part was very strongly labelled (Fig. 3D) just like the medial part of the circle of cells surrounding the mesoderm core in Figure 3B. The Australian lungfish is very unusual in that cartilages, bones and other tissues differentiate late if at all (Kemp 1999), an example of heterochrony. At stage 43, the last stage of our study, many tissues were still relatively undifferentiated and only a few cartilages had started to differentiate in the skull. However, DiI is detected in several of these (Table 1). Elements assigned to the mandibular arch include Meckel's cartilage, the quadrate cartilage and the trabecula cranii (for terminology see de Beer 1937 and Kemp 1999). We detected DiI in Meckel's cartilage and in the trabecula cranii (Fig. 3A) and also in the quadrate cartilage (Fig. 5). Hyoid arch elements such as the ceratohyal also contained DiI-tagged cells (Fig. 4A). Further caudally, we found DiI in the developing ceratobranchials (Figs. 4A, B, Table 1). Cranial elements not marked with DiI after injection included the otic capsule and the parachordals. In the operculum, DiI was found in connective tissues (Fig. 5) and around the constrictor hyoideus muscles. Muscle fibres were not DiI marked.

In conclusion, our DiI fate-mapping data, although incomplete, leads us to suggest that the cranial neural crest cells have migratory pathways and fates virtually identical to those observed in related tetrapods with similar larvae, such as the Mexican axolotl. One limitation with the DiI method is that the fluorescence is of limited duration, and eventually disappears. In combination with the very slow differentiation of lungfish larvae, this causes problems for real long-term

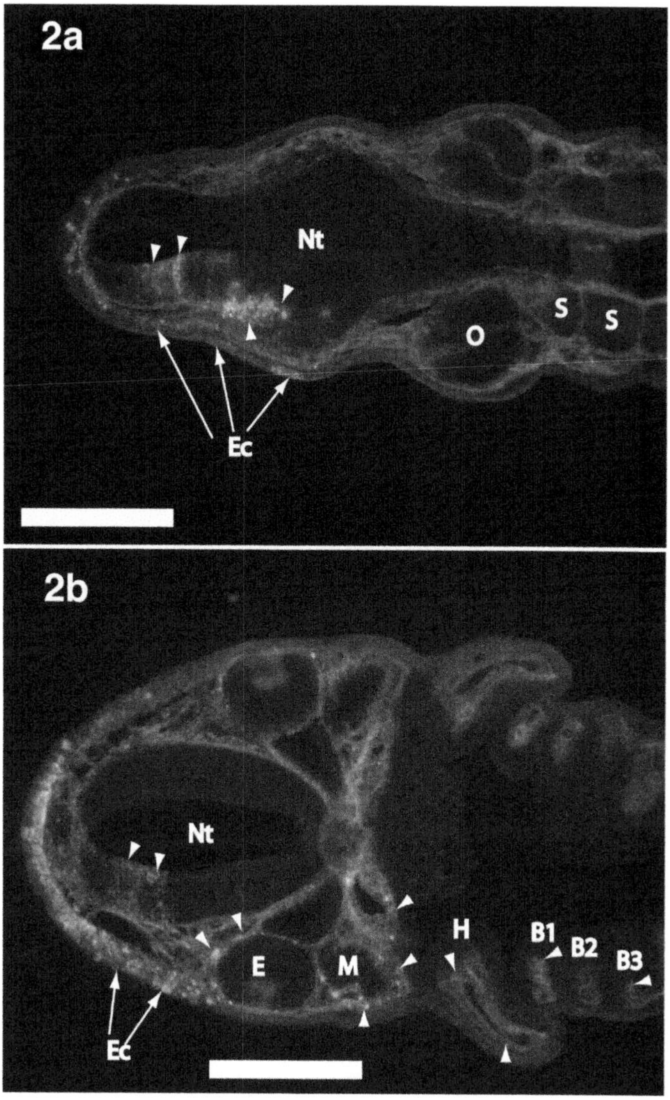

Fig. 2a Dorsal frontal vibratome section of a stage 40 *Neoceratodus forsteri* embryo. Red is DiI, green is fibronectin. Anterior is to the left. Arrowheads indicate DiI in the neural tube and in the mesenchyme. DiI is also seen in the ectoderm covering the neural tube. Ec, Ectoderm; Nt, Neural tube; O, otic vesicle; S, somite. Scale bar = 0.5 mm. (From Ericsson *et al.* 2008.) **2b:** Frontal vibratome section of the same embryo as in Figure 2a, slightly more ventral. Red is DiI, green is fibronectin. Arrowheads indicate DiI in the neural tube, in the mesenchyme surrounding the eye and in the pharyngeal arches. DiI is also found in the epidermis covering the neural tube. B1–B3, Branchial arches; E, Eye; Ec, Ectoderm; H, Hyoid arch; M, Mandibular arch; Nt, neural tube. Scale bar = 0.5 mm. (From Ericsson *et al.* 2008.)

Color image of this figure appears in the color plate section at the end of the book.

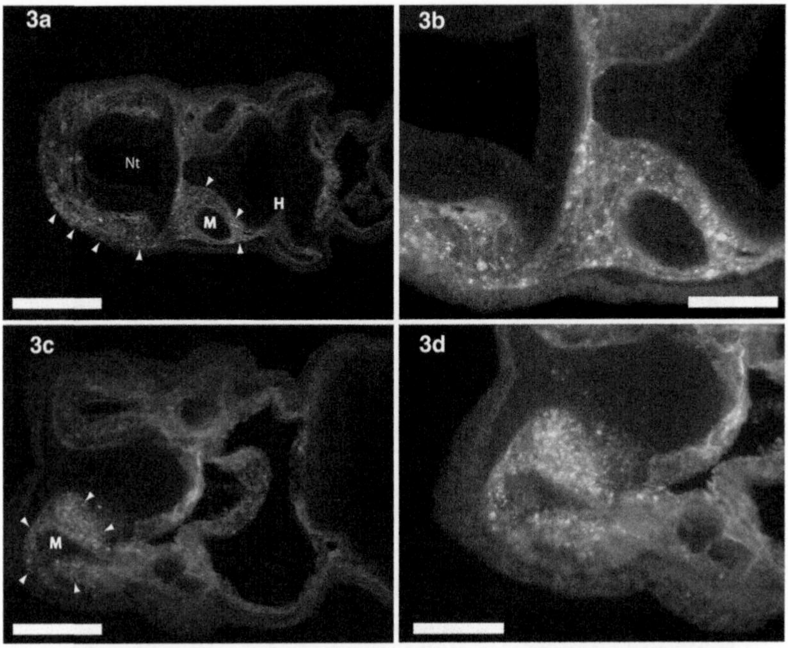

Fig. 3a Frontal vibratome section of the same embryo as in Figure 2a, slightly more ventral than Figure 2b. Red is Dil, green is fibronectin. Arrowheads indicate Dil in the epidermis covering the neural tube and in the mesenchyme of the mandibular arch. H, Hyoid arch; M, Mandibular arch; Nt, Neural tube. Scale bar = 0.5 mm. (From Ericsson *et al.* 2008.) **3b** Magnification of the central area in Figure 4a. Extensive Dil-staining can be observed in mandibular arch tissues surrounding the mesodermal core, especially in the medial part. Scale bar = 0.25 mm. (From Ericsson *et al.* 2008.) **3c** Frontal vibratome section of the same embryo as in Figure 2a, slightly more ventral than Figure 3a. Red is Dil, green is fibronectin. Arrowheads indicate Dil in the mesenchyme surrounding the ventral portion of the mandibular arch. Scale bar = 0.375 mm. (From Ericsson *et al.* 2008.) **3d** Magnification of the Dil-marked area in Figure 3c. Dil labels the medial part very strongly, as it does in Figures 2a and b. Scale bar = 0.5 mm. (From Ericsson *et al.* 2008.)

Color image of this figure appears in the color plate section at the end of the book.

fate mapping, such as identifying the developmental origin of late larval or adult structures. In amphibians, FITC-dextran has proven to be a reliable long-term marker, either in combination with transplantation from an injected donor to a "cold" host animal (Gross and Hanken 2004, 2005) or by direct focal injection (Piekarski and Olsson 2007). This technique can also be used in the Australian lungfish. This opens up the possibility of fate-mapping cells that take a very long time to reach their targets and differentiate in the neotenic Australian lungfish, such as fully differentiated bone and cartilage cells.

A logical next step is to investigate whether lungfish also share the mechanism of cranial muscle formation observed in amphibians and birds, where the muscle

Table 1 Fate mapping of head cartilages in *N. forsteri*

Lungfish cartilages	*Marked with DiI*
Meckel's cartilage	*Yes*
Trabecula cranii	*Yes*
Quadrate	*Yes*
Ceratohyal	*Yes*
Trabecular plate	*Yes*
Parachordals	*No*
Otic capsule	*No*
Ceratobranchials I-IV	Yes

fibres have an origin (mesoderm) different from that of the surrounding connective tissue cells, which are neural crest derived. The presence of neural crest cells has been shown to be crucial for normal muscle patterning in these species, and it is likely that the Australian lungfish shares this trait. First, however, we need to know more about the developmental anatomy of cranial muscles in the Australian lungfish. This is the topic to which we now turn.

CRANIAL MUSCLE DEVELOPMENT

There are several descriptions in the literature of cranial muscle development of *Neoceratodus forsteri*, some of the more notable ones being the work of Edgeworth (1935), Fox (1965) and Bartsch (1994). All of these studies have used histological sections, mostly from stages later than 41, when most muscles have already differentiated in the mandibular and hyoid arches. Labelling the cranial mesoderm with vital dyes has been attempted, but generally with poor results. Our knowledge of early events in the mesoderm, from the formation of the visceral arches to prior to the formation of muscle anlagen is, therefore, very limited.

The cranial muscles of the Australian lungfish develop from three distinct mesodermal cell populations. The extraocular muscles are derived from the premandibular cavity, which in its turn is possibly derived from prechordal mesoderm, the muscles attaching to the jaw and branchial apparatus are mostly derived from the paraxial head mesoderm, with a few exceptions derived from somites (Edgeworth 1935, Fox 1965). According to Edgeworth (1935), the cranial mesoderm is originally a continuous, hollow structure, from the premandibular cavity to the epicardium. Eventually it separates into mandibular/premandibular, hyoid and branchial mesoderm, although the ventral contact between the mandibular and hyoid parts is kept throughout development.

Two paired head cavities can be found in *Neoceratodus forsteri*, similar to those in sturgeon (Kuratani *et al.* 2000). The premandibular cavity, which lies medial to the developing eye, and the mandibular cavity, which lies just dorsal and caudal to the premandibular cavity. At stage 39, they have fused with each other and with

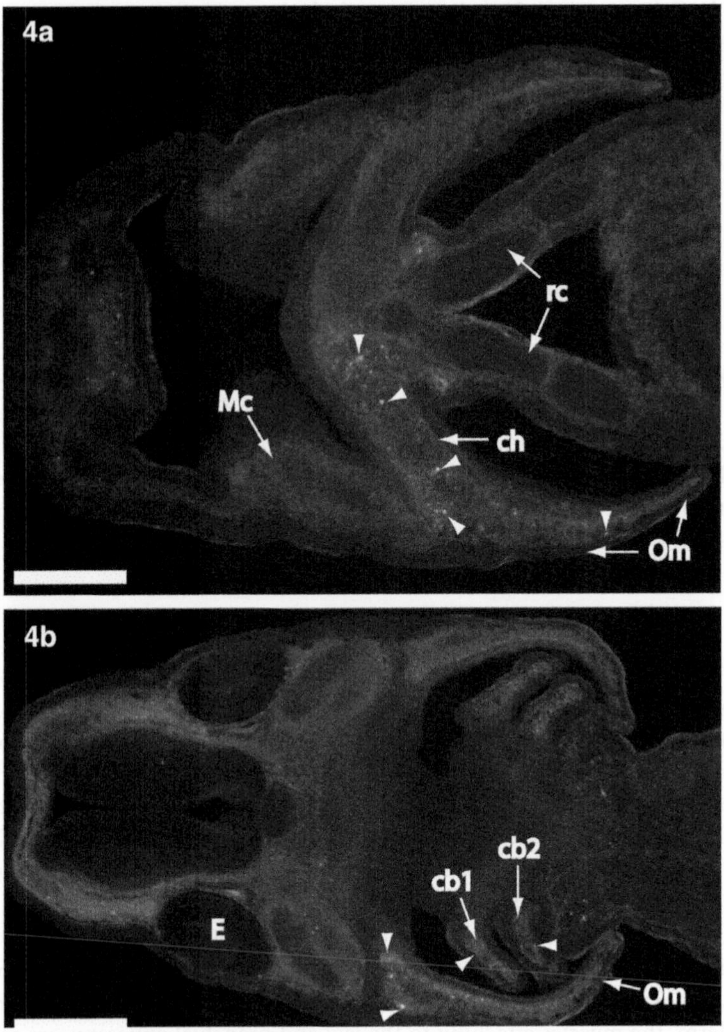

Fig. 4a Ventral frontal vibratome section of a stage 43 embryo. Red is DiI, green is fibronectin. Arrowheads indicate DiI in the cartilage of the ceratohyal and in the operculum. Cells in Meckel's cartilage are also labelled. ch, ceratohyal; Mc, Meckel's cartilage; Om, Operculum; rc, rectus cervicis. Scale bar = 0.5 mm. (From Ericsson *et al.* 2008.) **4b** Medial frontal vibratome section of a stage 43 embryo. Red is DiI, green is fibronectin. Arrowheads indicate DiI in the cartilage of ceratobranchials 1 and 2, and in the operculum. cb1-2, Ceratobranchials 1 and 2; E, Eye; Om, Operculum. Scale bar = 0.5 mm. (From Ericsson *et al.* 2008.)

Color image of this figure appears in the color plate section at the end of the book.

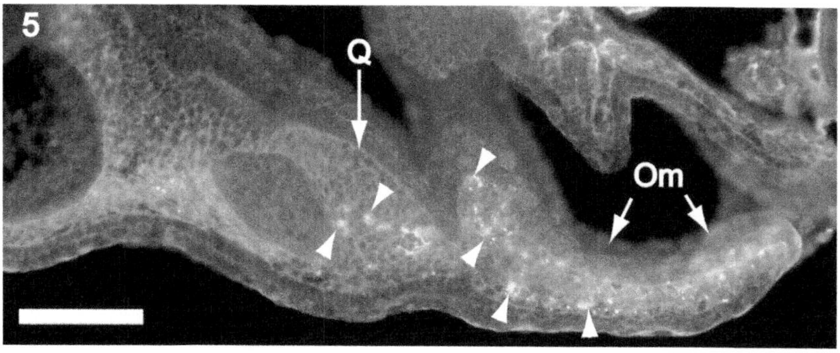

Fig. 5 Lateral frontal vibratome section of a stage 42 embryo, ventral to the eye. Red is DiI, green is fibronectin. Arrowheads indicate DiI in the pre-cartilage condensation of the quadrate and the connective tissue of the operculum. Om, Operculum; Q, Quadrate. Scale bar = 0.25 mm. (From Ericsson *et al.* 2008.)

Color image of this figure appears in the color plate section at the end of the book.

the mesoderm of the mandibular arch, which stretches from the eye to the ventral midline, extending and overlapping ventrally with the hyoid mesoderm (Fig. 6). At this stage, the mesoderm is still hollow, although the dorsal part is filling up with cells. Immunostaining with desmin antibodies show faint and sporadic staining at this stage (Fig. 7). At stage 40, the dorsal and ventral parts of the mesoderm are starting to be divided by the developing Meckel's cartilage. The separate anlagen of musculus levator mandibulae anterior and musculus levator mandibulae posterior can be distinguished at stage 41 (Fig. 8), just prior to the formation of muscle fibres in all the mandibular arch muscles. At stage 43, the levator mandibulae muscles are distinct and are attached at their inserts, on the dorsal surface of Meckel's cartilage (Fig. 9). The origins shift dorso-caudally through development, and in the adult these muscles take their origins from the chondrocranium and the covering bones (Edgeworth 1935). The anlage of the intermandibularis muscle is never more than a thin sheet of cells, as it develops between Meckel's cartilage and the ventral midline, just ventral to the developing interhyoideus muscle. At stage 43 it has the shape of a fairly sharp "V" (Fig. 9), but at later stages it develops into a relatively straight sheet of muscle fibres attaching to the ventral surfaces of the lower jaw and the insertion at the ventral midline (Fig. 11, Edgeworth 1935; Fox 1965).

The hyoid arch mesoderm develops as a sheet of cells from between the first and second pharyngeal pouch, stretching from the otic vesicle to the ventral midline (Fig. 6). At stage 39, desmin can be faintly detected by immunostaining (Fig. 7). The first muscle fibres are detected at stage 41 in the levator hyoideus muscle anlage (Fig. 8). Soon thereafter, muscle fibres can be detected in the interhyoideus and constrictor hyoideus muscles. As the operculum extends caudally to cover the branchial arches during development, the constrictor hyoideus follows (Fig. 9). It has its origin on the otic capsule and eventually fuses ventrally with the posterior

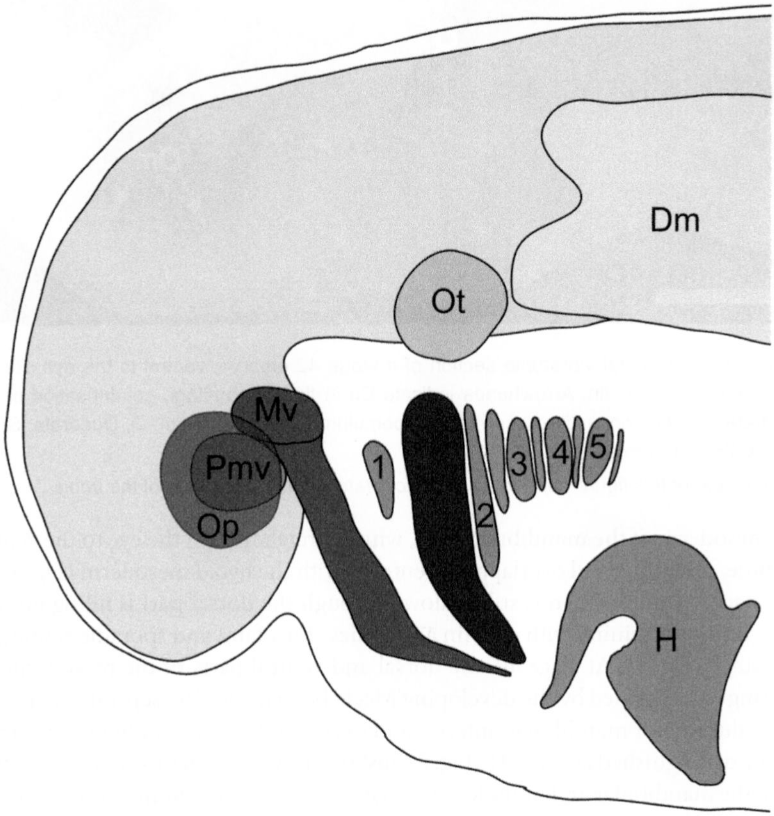

Fig. 6 Schematic view of the head of a stage 39 *Neoceratodus forsteri* embryo. Anterior is to the left. Brown is premandibular mesoderm, red is premandibular/mandibular arch mesoderm, blue is hyoid arch mesoderm, green is branchial arch mesoderm, and yellow is somitic mesoderm. At this stage, the premandibular and mandibular cavities have fused with the mandibular arch mesoderm. Dm, dorsal myotome; H, heart; Mv, mandibular vesicle; O, optic vesicle; Ot, otic vesicle; Pmv, premandibular vesicle; 1–5, pharyngeal pouches 1–5.

Color image of this figure appears in the color plate section at the end of the book.

part of the interhyoideus. The anterior part of the interhyoideus muscle is attached to the ventral surface of the ceratohyal cartilage and to the insertion at the ventral midline (Fig. 11). Unlike *Protopterus* and *Lepidosiren*, *Neoceratodus* does not have a depressor mandibulae muscle, connecting the posterior end of the lower jaw to the otic capsule. Instead, the levator hyoideus, considered to be homologous to the depressor mandibulae, has its origin on the otic capsule and inserts on the dorsal end of the ceratohyal (Figs. 10, 11). According to Fox (1965), there should also be a retractor mandibulae muscle, attaching to the postero-ventral surface of Meckel's cartilage and to the lateral surface of the ceratohyal. However, this muscle

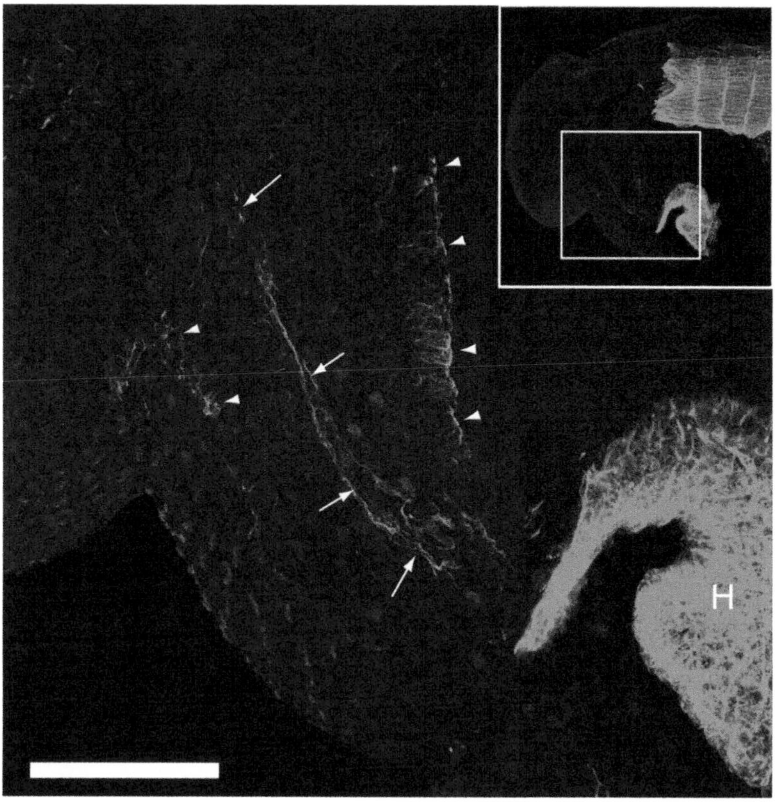

Fig. 7 Lateral view of stage 39 embryo of *Neoceratodus forsteri*, immunostained for desmin. Inset shows overview. Anterior is to the left. Arrowheads show desmin in mandibular and hyoid arch mesoderm. Arrows indicate afferent branchial artery I. H, heart. Scale bar is 250 μm.

Color image of this figure appears in the color plate section at the end of the book.

has not been described by earlier (Edgeworth 1926, 1935) or later (Bemis 1986; Bartsch 1994) studies. Our study has not been able to confirm the presence of this muscle.

The branchial arches develop later than the hyoid and the mandibular arches, and so do the branchial arch muscles. Four separate mesodermal cell populations can be found in the branchial region. Originally, these cell populations are connected with the pericardium, but they separate prior to the formation of muscle anlagen (Edgeworth 1935). At stage 41, the dorsal and ventral anlagen of the first branchial arch can be detected in histological sections, and at stage 42, the anlage of the ceratohyoideus internus is positioned between the hypohyal cartilage and the first ceratobranchial. The anlagen of the levatores arcus branchialis i-iii are first detected at stage 42 in histological sections, and can be detected with

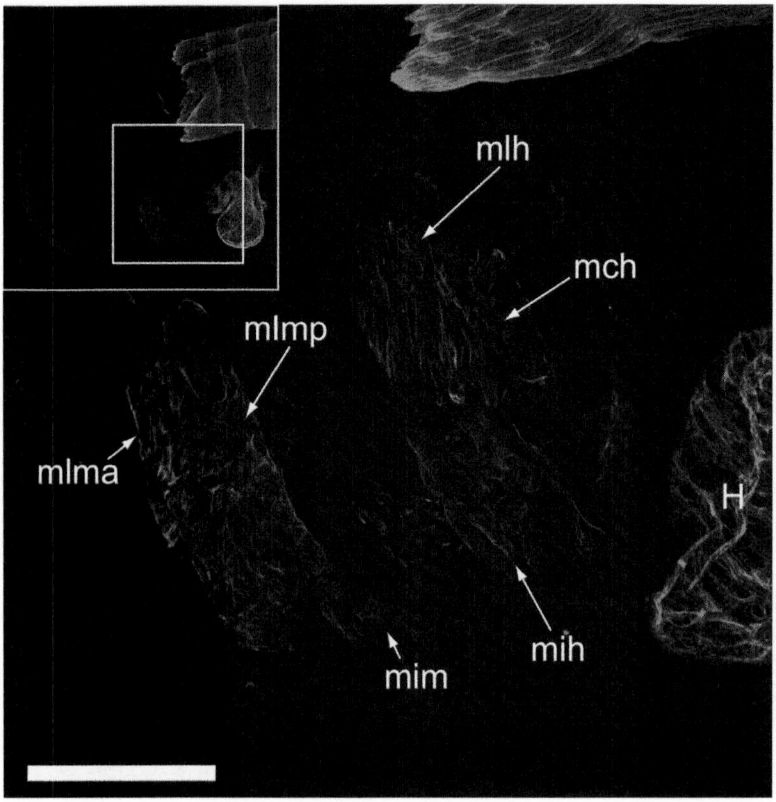

Fig. 8 Lateral view of stage 41 embryo of *Neoceratodus forsteri*, immunostained for desmin. Inset shows overview. Anterior is to the left. Desmin can be seen in the anlagen of all the mandibular and hyoid arch muscles. In mlh, muscle fibres have started forming. H, heart; mch, musculus constrictor hyoideus; mih, musculus interhyoideus; mim, musculus intermandibularis; mlh, musculus levator hyoideus; mlma, musculus levator mandibulae anterior; mlmp, musculus levator mandibulae posterior. Scale bar is 250 μm.

Color image of this figure appears in the color plate section at the end of the book.

desmin antibodies at stage 43 (Fig. 9). The first muscle fibres in the branchial arch muscles are found in the ceratohyoideus internus and transversus ventralis ii at stage 43. As development progresses, the more posterior branchial levator muscles develop, followed by the ventral subarcualis recti, constrictor branchiales and coracobranchiales (Fig. 10).

The somite-derived hypobranchial musculature in *Neoceratodus forsteri* is composed of two muscles, the geniocoracoideus and rectus cervicis. The anlage of the rectus cervicis can first be found at stage 40 in histological sections. Its two halves pass lateral to the heart, ventral to the branchial muscles but dorsal to the interhyoideus, and reach the developing hypohyal and ceratohyal. Muscle fibres have formed at stage 43 (Fig. 9). The posterior end partly attaches to the

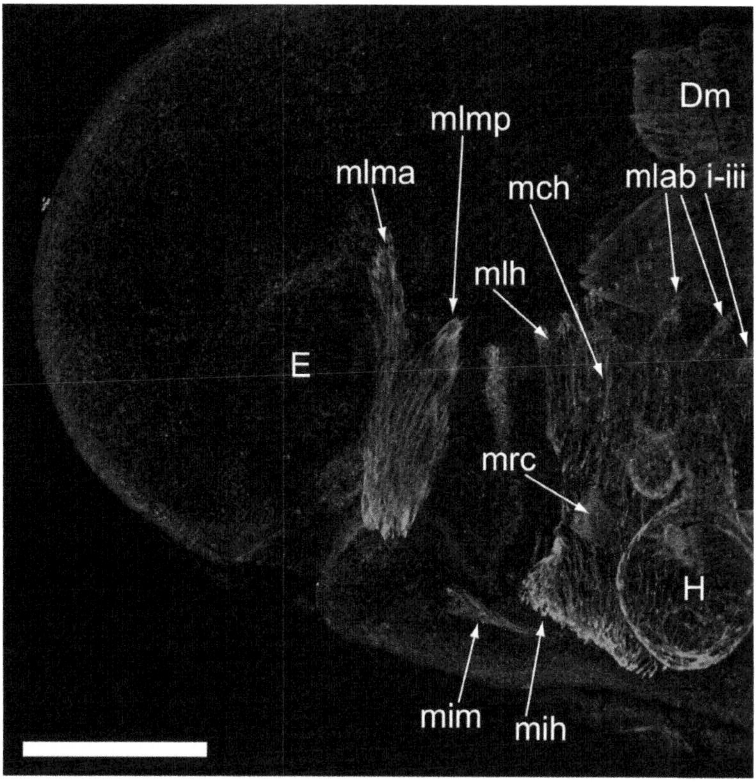

Fig. 9 Lateral view of a stage 43 larva of *Neoceratodus forsteri*, immunostained for desmin. Anterior is to the left. At this stage muscle fibres can be detected in all the mandibular and hyoid arch muscles, and in the hypobranchial rectus cervicis. Desmin is also detected in the anlagen of the levator arcus branchialis i–iii muscles. Dm, dorsal myotome; E, eye; H, heart; mch, musculus constrictor hyoideus; mih, musculus interhyoideus; mim, musculus intermandibularis; mlab i–iii, anlagen of musculus levator arcus branchialis 1–3; mlh, musculus levator hyoideus; mlma, musculus levator mandibulae anterior; mlmp, musculus levator mandibulae posterior; mrc, musculus rectus cervicis. Scale bar is 500 μm.

Color image of this figure appears in the color plate section at the end of the book.

pectoral girdle and partly fuses with the trunk muscles (Edgeworth 1935). The geniocoracoideus forms a single medial anlage extending from Meckel's cartilage, dorsal to the intermandibularis and interhyoideus muscles (Fig. 11), and fuses caudally with the rectus cervicis at the level of the first ceratobranchial at stage 43. Later in development it will attach to the pectoral girdle (Edgeworth 1935).

The extraocular muscles in *Neoceratodus forsteri* develop from the mesoderm surrounding the premandibular cavity. Edgeworth noted that these cells usually originate in the prechordal mesoderm, but for *Neoceratodus*, he considered them to be derived from the mandibular muscle plate. Considering that Edgeworth did

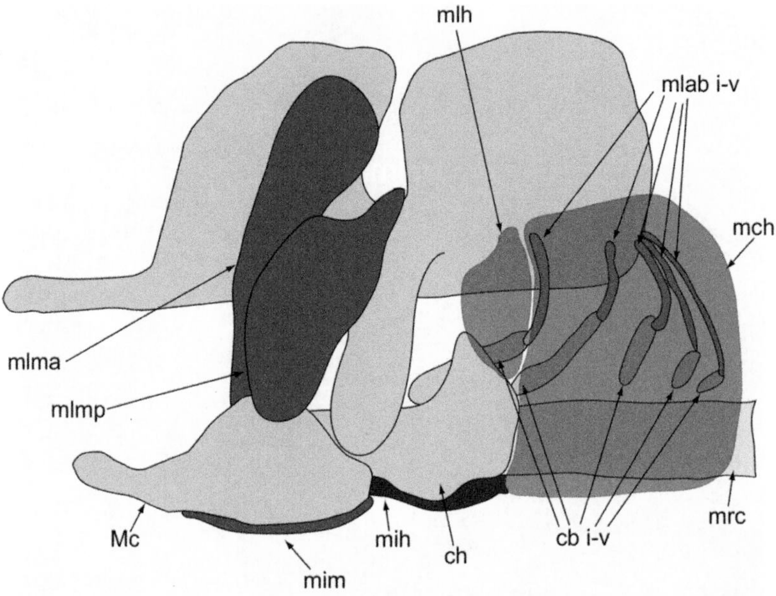

Fig. 10 Schematic, lateral view of the skull and cranial muscles of a stage 45+ *Neoceratodus forsteri* embryo. Anterior is to the left. The coracobranchiales and constrictor branchiales muscles have not yet developed at this stage. The constrictor hyoideus muscle is semitransparent to show the underlying structures. Red is mandibular arch muscles, blue is hyoid arch muscles, green is branchial arch muscles and yellow is hypobranchial muscles. cb i–v, ceratobranchial cartilage 1–5; ch, ceratohyal cartilage; Mc, Meckel's cartilage; mch, musculus constrictor hyoideus; mih, musculus interhyoideus; mim, musculus intermandibularis; mlab i–v, musculus levator arcus branchialis 1–5; mlh, musculus levator hyoideus; mlma, musculus levator mandibulae anterior; mlmp, musculus levator mandibulae posterior; mrc, musculus rectus cervicis. Modified from Fox (1965).

Color image of this figure appears in the color plate section at the end of the book.

not have access to embryos of sufficiently early stages, it is possible that further studies will reveal a prechordal origin for this cell population in *Neoceratodus* also. According to Bartsch (1994), there are six muscles controlling the eye: the recti superior, inferior, internus and externus, and the obliqui superior and inferior. The rectus internus is proximally fused with the rectus inferior, which makes them look like a single muscle. The first extraocular muscle to develop is the obliquus superior, the anlage of which can be detected in histological sections at stage 43. Soon thereafter, the obliquus inferior, rectus externus, inferior/internus and superior appear. Muscle fibres can be detected at stage 45.

When comparing the timing of development among the different muscles, it is clear that the dorsal mandibular and hyoid muscles are the first to appear, followed by the ventral muscles of the same arches. The branchial arch muscles develop later, in an anterior-to-posterior pattern. We have previously compared

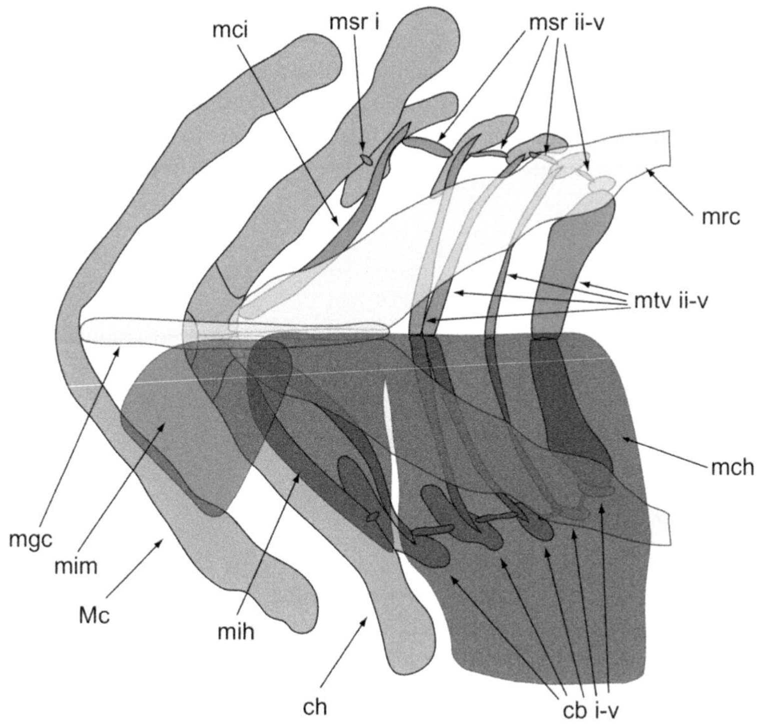

Fig. 11 Schematic ventral view of the lower jaw and associated muscles of a stage 45+ *Neoceratodus forsteri* embryo. Anterior is to the left. The coracobranchiales and constrictor branchiales muscles have not yet developed at this stage. The constrictor hyoideus, intermandibularis, geniocoracoideus and rectus cervicus muscles are semitransparent to show the underlying structures. Red is mandibular arch muscles, blue is hyoid arch muscles, green is branchial arch muscles and yellow is hypobranchial muscles. The mandibular and hyoid muscles are shown only on one side of the jaw. cb i–v, ceratobranchial cartilage 1–5; ch, ceratohyal cartilage; Mc, Meckel's cartilage; mci, musculus ceratohyoideus internus; mgc, musculus geniocoracoideus; mch, musculus constrictor hyoideus; mih, musculus interhyoideus; mim, musculus intermandibularis; mrc, musculus rectus cervicis; msr i–v, musculus subarcualis rectus 1–5; mtv ii–v, musculus transversus ventralis 2–5. Modified from Fox (1965).

Color image of this figure appears in the color plate section at the end of the book.

the sequence of cranial muscle development in vertebrates, and *Neoceratodus* comes out as an intermediate between amphibians and fish (Ericsson and Olsson 2004). In amphibians, cranial muscles can be detected simultaneously in all of the pharyngeal arches, while in zebrafish there is an anterior-to-posterior sequence of development. This comparison is unfortunately lacking information about the development of the other lungfishes, among many other taxa. A more detailed comparison of cranial muscle development between *Neoceratodus forsteri* and the Mexican axolotl, *Ambystoma mexicanum* reveals a potential difference in the pattern of differentiation. In *A. mexicanum*, the mandibular levator muscles, and in particular the depressor mandibulae muscle, differentiate in an origin-to-insert pattern. In *N. forsteri*, the current data indicates a more global onset of muscle markers such as desmin. Although the levator hyoideus may be homologous to depressor mandibulae, it's difficult to compare them due to its length, or rather lack thereof. Further studies of earlier stages of embryos will hopefully resolve this issue.

Acknowledgements

We acknowledge funding from Carl Tryggers stiftelse to Rolf Ericsson as well as a grant from the Deutsche Forschungsgemeinschaft (OL 134/2-4) to LO and from the Australian Research Council to JJ. We wish to thank Pierre Falck for scanning electron microscopy, and Katja Felbel for help with histology.

References

Agar, W. (1906). The development of the skull and visceral arches in *Lepidosiren* and *Protopterus*. Transactions of the Royal Society of Edinburgh 45: 49-64.

Bartsch, P. (1993). Development of the snout of the Australian lungfish *Neoceratodus forsteri* (Krefft, 1870), with special reference to cranial nerves. Acta Zoologica (Stockholm) 74: 15-29.

Bartsch, P. (1994). Development of the cranium of *Neoceratodus forsteri*, with a discussion of the suspensorium and the opercular apparatus in Dipnoi. Zoomorphology 114: 1-31.

Bemis, W. (1986). Feeding systems of living Dipnoi: Anatomy and function. Journal of Morphology Supplement 1: 249-275.

Bertmar, G. (1966). The development of skeleton, blood-vessels and nerves in the Dipnoan snout, with a discussion on the homology of the Dipnoan posterior nostrils. Acta Zoologica (Stockholm) 47: 82-150.

Couly, G. and Le Douarin, N. (1990). Head morphogenesis in embryonic avian chimeras: Evidence for a segmental pattern in the ectoderm corresponding to the neuromeres. Development 108: 543-558.

Couly, G.F., Coltey, P.M. and Le Douarin, N.M. (1992). The developmental fate of the cephalic mesoderm in quail-chick chimeras. Development 114: 1-15.

Couly, G.F., Coltey, P.M. and Le Douarin, N.M. (1993). The triple origin of the skull in higher vertebrates: a study in quail-chick chimeras. Development 117: 409-429.

de Beer, G.R. (1937). The Development of the Vertebrate Skull. Oxford University Press, Oxford, UK.

Depew, M.J. and Simpson, C.A. (2006). 21st century neontology and the comparative development of the vertebrate skull. Developmental Dynamics 235: 1256-1291.

Depew, M.J., Lufkin, T. and Rubenstein, J.L. (2002). Specification of jaw subdivisions by Dlx genes. Science 298: 381-385.

Edgeworth, F.H. (1926). On the hyomandibula of Selachii, Teleostomi and Ceratodus. Journal of Anatomy 60: 173-193.

Edgeworth, F.H. (1935). The Cranial Muscles in Vertebrates. Cambridge University Press, Cambridge, UK.

Ericsson, R. and Olsson, L. (2004). Patterns of spatial and temporal visceral arch muscle development in the Mexican axolotl *(Ambystoma mexicanum)*. Journal of Morphology 261: 131-140.

Ericsson, R., Cerny, R., Falck, P. and Olsson, L. (2004). The role of cranial neural crest cells in visceral arch muscle positioning and morphogenesis in the Mexican axolotl, *Ambystoma mexicanum*. Developmental Dynamics 231: 237-247.

Ericsson, R., Joss, J. and Olsson, L. (2008). Cranial neural crest cell migration and fate in the Australian lungfish *(Neoceratodus forsteri)*. Journal of Experimental Zoology. (Molecular and Developmental Evolution) 310: 345-354.

Falck, P., Hanken, J. and Olsson, L. (2002). Cranial neural crest emergence and migration in the Mexican axolotl *(Ambystoma mexicanum)*. Zoology 105: 195-202.

Falck, P., Joss, J. and Olsson, L. (2000). Cranial neural crest cell migration in the Australian lungfish, *Neoceratodus forsteri*. Evolution & Development 2: 179-185.

Fox, H. (1965). Early development of the head and pharynx of *Neoceratodus* with a consideration of its phylogeny. Proceedings of the Zoological Society of London 146: 470-554.

Gegenbaur, C. (1888). Die Metamerie des Kopfes und die Wirbeltheorie des Kopfskelettes. Morphologisches Jahrbuch 13: 1-144.

Goodrich, E.S. (1930). Studies on the Structure and Development of Vertebrates. Macmillan and Co., Ltd. London, UK.

Graham, A. and Smith, A. (2001). Patterning the pharyngeal arches. Bioessays 23: 54-61.

Grammatopoulos, G.A., Bell, E., Toole, L., Lumsden, A. and Tucker, A.S. (2000). Homeotic transformation of branchial arch identity after Hoxa2 overexpression. Development 127: 5355-5365.

Greil, A. (1908). Entwickelungsgeschichte des Kopfes und des Blutgefäßsystems von Ceratodus forsteri. I. Gesammtentwikkelung bis zum beginn der Blutzirkulation. Denkschriften der Medizinisch-Naturwissenschaftlichen Gesellschaft zu Jena 4: 661-934.

Greil, A. (1913). Entwickelungsgeschichte des Kopfes und des Blutgefäßsystems von Ceratodus forsteri. II. Die epigenetischen Erwerbungen während der Stadien 39-48. Denkschriften der Medizinisch-Naturwissenschaftlichen Gesellschaft zu Jena 9: 935-1492.

Gross, J.B. and Hanken, J. (2004). Use of fluorescent dextran conjugates as a long-term marker of osteogenic neural crest in frogs. Developmental Dynamics 230: 100-106.

Gross, J.B. and Hanken, J. (2005). Cranial neural crest contributes to the bony skull vault in adult *Xenopus laevis*: insights from cell labeling studies. Journal of Experimental Zoology. (Molecular and Developmental Evolution) 304: 169-176.

Hall, B.K. (1999). The Neural Crest in Development and Evolution. Springer-Verlag, New York, USA.

Hanken, J. and Hall, B.K. (eds). (1993). The Skull. Volumes I-III. University of Chicago Press, Chicago and London.

Kemp, A. (1982). The embryological development of the Queensland lungfish, *Neoceratodus forsteri* (Krefft). Memoirs of the Queensland Museum 20: 553-597.

Kemp, A. (1995). On the neural crest cells of the Australian lungfish. Bulletin du Museum National d'Histoire Naturelle Section C Sciences de la Terre Paleontologie Geologie Mineralogie 17: 343-357.

Kemp, A. (1999). Ontogeny of the skull of the Australian lungfish *Neoceratodus forsteri* (Osteichthyes: Dipnoi). Journal of Zoology (London) 248: 97-137.

Kemp, A. (2000). Early development of neural tissues and mesenchyme in the Australian lungfish *Neoceratodus forsteri* (Osteichthyes: Dipnoi). Journal of Zoology (London) 250: 347-372.

Köntges, G. and Lumsden, A. (1996). Rhombencephalic neural crest segmentation is preserved throughout craniofacial ontogeny. Development 122: 3229-3242.

Kuratani, S., Nobusada, Y., Saito, H. and Shigetani, S. (2000). Morphological characteristics of the developing cranial nerves and mesodermal head cavities in sturgeon embryos from early pharyngula to late larval stages. Zoological Science 17: 911-933.

LeDouarin, N.M. and Kalcheim, C. (1999). The Neural Crest. Cambridge University Press, Cambridge, UK.

Noden, D.M. (1986a). Origins and patterning of craniofacial mesenchymal tissues. Journal of Craniofacial Genetics and Developmental Biology Supplement 2: 15-31.

Noden, D.M. (1986b). Patterning of avian craniofacial muscles. Developmental Biology 116: 347-356.

Noden, D.M. (1991). Vertebrate craniofacial development: The relation between ontogenetic process and morphological outcome. Brain Behavior and Evolution 38: 190-225.

Olsson, L., Falck, P., Lopez, K., Cobb, J. and Hanken, J. (2001). Cranial neural crest cells contribute to connective tissue in cranial muscles in the anuran amphibian, *Bombina orientalis.* Developmental Biology 237: 354-367.

Olsson, L., Hossfeld, U., Bindl, R. and Joss, J. (2004). The development of the Australian lungfish *Neoceratodus forsteri* (Osteichthyes, Dipnoi, Neoceratodontidae): from Richard Semon's pioneering work to contemporary approaches. Rudolstädter naturhistorische Schriften 12: 51-128.

Olsson, L., Ericsson, R. and Cerny, R. (2005). Vertebrate head development: segmentation, novelties, and homology. Theory in Biosciences 124: 145-163.

Pasqualetti, M., Ori, M., Nardi, I. and Rijli, F.M. (2000). Ectopic Hoxa2 induction after neural crest migration results in homeosis of jaw elements in *Xenopus.* Development 127: 5367-5378.

Piekarski, N. and Olsson, L. (2007). A long-term fate map of the first somites in the Mexican axolotl (*Ambystoma mexicanum*). Evolution & Development 6: 566-578.

Rijli, F.M., Mark, M., Lakkaraju, S., Dierich, A., Dolle, P. and Chambon, P. (1993). A homeotic transformation is generated in the rostral branchial region of the head by disruption of Hoxa-2, which acts as a selector gene. Cell 75: 1333-1349.

Rijli, F.M., Gavalas, A. and Chambon, P. (1998). Segmentation and specification in the branchial region of the head: the role of the Hox selector genes. International Journal of Developmental Biology 42: 393-401.

CHAPTER 7

The Head Muscles of Dipnoans—A Review on the Homologies and Evolution of these Muscles within Vertebrates

Rui Diogo[1,*] and Virginia Abdala[2]

[1]Center for the Advanced Study of Hominid Paleobiology, Department of Anthropology, The George Washington University, 2110 G St. NW, Washington, DC 20052, USA

[2]Instituto de Herpetología, Fundación Miguel Lillo-CONICET, Facultad de Ciencias Naturales (UNT), Miguel Lillo, 251 4000 Tucumán, Argentina

ABSTRACT

This chapter provides an account on the homologies between the head muscles of dipnoans and the muscles of all the other major extant vertebrate groups, which is based on our own dissections of several vertebrate taxa and on an extensive review of the literature, both old and recent. Our observations and comparisons indicate that 13 mandibular, hyoid, branchial and hypobranchial muscles are present in adult dipnoans such as *Lepidosiren* (not including the branchial muscles sensu stricto), six are present in adults of at least some of the non-osteichthyan taxa listed in our tables (intermandibularis, A2, interhyoideus, protractor pectoralis, coracomandibularis and sternohyoideus), six are present in adults of at least some of the actinopterygian groups listed in these tables (intermandibularis, A2, A3', interhyoideus, protractor pectoralis and sternohyoideus), and eight are present in adults of at least some of the tetrapod taxa included in these Tables (A2, A2-PVM, interhyoideus, depressor mandibulae, protractor pectoralis, dilatator laryngis, constrictor laryngis and sternohyoideus). This shows that dipnoans may provide an important link to compare the anatomical structures of these three latter groups and

*Corresponding author: E-mail: ruidiogo@gwmail.gwu.edu

thus to discuss the homologies and evolution of these structures within the vertebrates as a whole. Importantly, dipnoans share some unique muscular synapomorphies with tetrapods, thus supporting the idea that these fishes are in fact the closest living relatives of the clade including extant amphibians and amniotes (e.g., presence of an adductor mandibulae A2-PVM; presence of a levator hyoideus in at least some developmental stages; absence of a recognizable adductor operculi in adults; absence of a recognizable adductor arcus palatini; and possibly presence of a depressor mandibulae in at least some developmental stages). This stresses that comprehensive comparative studies of muscles provide not only useful information for functional, evolutionary and ecomorphological studies, but also crucial data to disclose the relationships between major vertebrate groups.

Keywords: Dipnoi, cephalic muscles, evolution, homologies, vertebrates

INTRODUCTION

The head muscles of dipnoans have been studied by anatomists since the 19th century. Authors who have provided information about these muscles are Owen (1841), Humphry (1872a,b,c), Bischoff (1840), Albrecht (1876), Bridge (1898), Edgeworth (1911, 1923, 1926c, 1935), Luther (1913, 1914), Adams (1919), Lightoller (1939), Kesteven (1942-1945), Säve-Soderbergh (1944), Jarvik (1963, 1980), Fox (1965), McMahon (1969), Wiley (1979a,b), Bemis (1982, 1986), Jollie (1982), Bemis and Lauder (1986), Forey (1986), Miyake *et al.* (1992), Bartsch (1994), and Wilga *et al.* (2000). In this chapter, we provide a detailed list of the head muscles that are found in dipnoans, in Tables 1-12, as well as illustrations of most of these muscles, in Figures 1-4. The aim of this chapter is however not to restate all the information that has already been provided in the literature about attachments, innervation, configuration, ontogenetic development and function of each of these muscles. Instead, here we focus on a subject that in our opinion has been, unfortunately, somewhat neglected by anatomists, particularly over the most recent decades: the homologies between the dipnoan muscles and the muscles of other vertebrates.

Because of their phylogenetic position (Fig. 5), extant sarcopterygian fishes such as dipnoans are effectively a key group to clarify the homologies and evolution of muscles within vertebrates and namely to help comparing the muscles of other bony fishes, of tetrapods, and of non-osteichthyan vertebrates (e.g., Diogo 2007, 2008). Diogo (2007, 2008) and Diogo *et al.* (2008a,b) have recently discussed homologies between the head muscles of dipnoans and the muscles of other osteichthyans, but they did not compare these muscles with those of other vertebrates such as lampreys, elasmobranchs and holocephalans (see Fig. 5). In this chapter we will thus provide an account of the homologies between the head muscles of dipnoans and the muscles of all the other major extant vertebrate groups, which is based on our own dissections of several vertebrate taxa and on an extensive review of the literature, both old and recent. To our knowledge the type of information provided in Tables 1-12 about the detailed homologies of the mandibular, hyoid, branchial

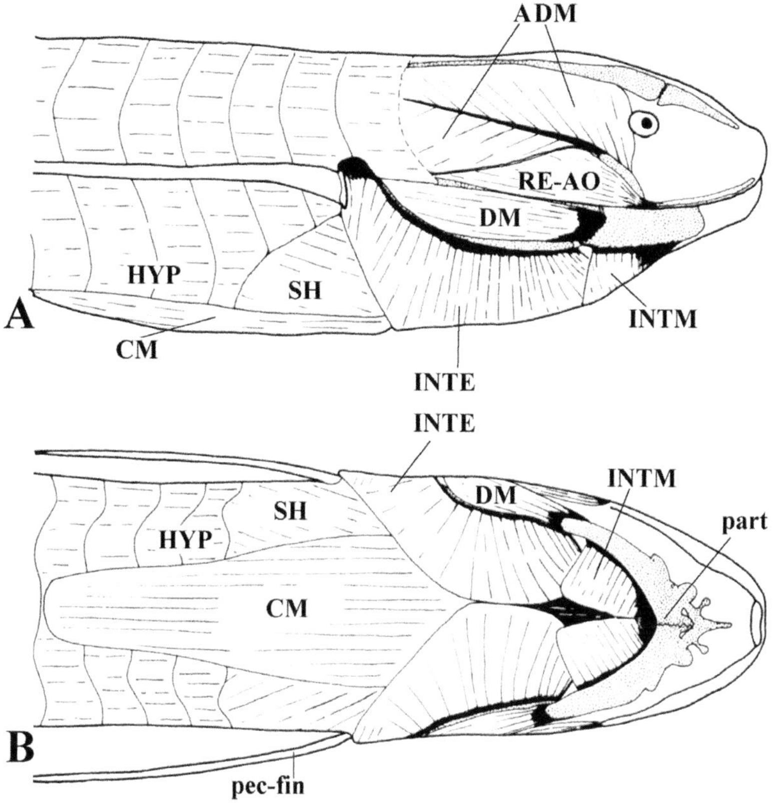

Fig. 1 *Lepidosiren paradoxa* (Dipnoi). Lateral (A) and ventral (B) views of the cephalic musculature (modified from Bemis and Lauder (1986); the nomenclature of the structures illustrated follows that used in the present work). ADM, adductor mandibulae complex; CM, coracomandibularis; DM, depressor mandibulae; HYP, hypoaxialis; INTE, interhyoideus; INTM, intermandibularis; part, prearticular; RE-AO, retractor anguli oris; SH, sternohyoideus.

and hypobranchial head muscles (see below) of representatives of all the major extant groups of vertebrates has never been integrated in a single book chapter. In addition to its value for anatomists and functional morphologists, we hope this comparative anatomical information will be useful to researchers working in developmental biology, genetics and/or evolutionary developmental biology for it should help them determine the wider homologies of these structures in dipnoans and non-dipnoan vertebrate model organisms.

MATERIALS AND METHODS

The myological nomenclature used in the present work follows that of Diogo (2004a, 2007, 2008) and Diogo *et al.* (2008a, b). Therefore, as did Edgeworth

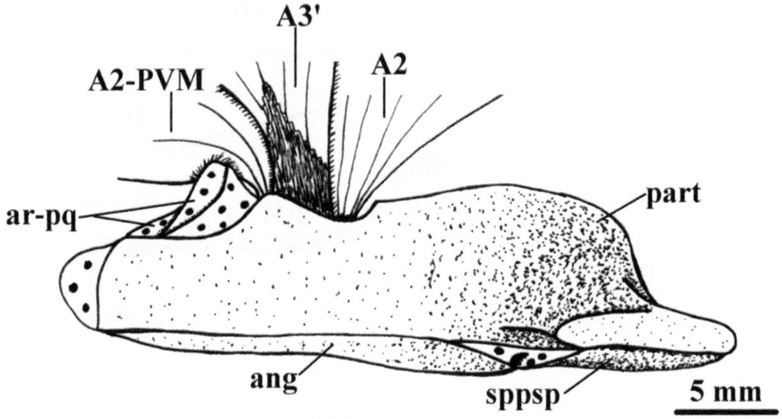

Fig. 2 *Neoceratodus forsteri* (Dipnoi). Mesial view of adductor mandibulae and mandible; the mandibular tooth-plates are not illustrated (modified from Diogo (2007); the nomenclature of the structures illustrated follows that used in the present work). A2, A2-PVM, A3', adductor mandibulae A2, A2-PVM and A3'; ang, angular; ar-pq, articulatory facet for palatoquadrate; part, prearticular; sppsp, splenio-postsplenial bone.

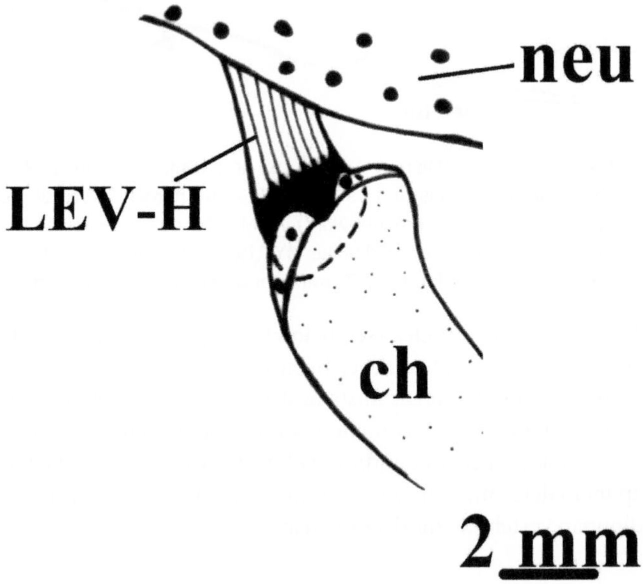

Fig. 3 *Lepidosiren paradoxa* (Dipnoi). Lateral view of levator hyoideus; the ventral portion of the ceratohyal was cut (modified from Diogo (2007); the nomenclature of the structures illustrated follows that used in the present work). LEV-H, levator hyoideus; neu, neurocranium.

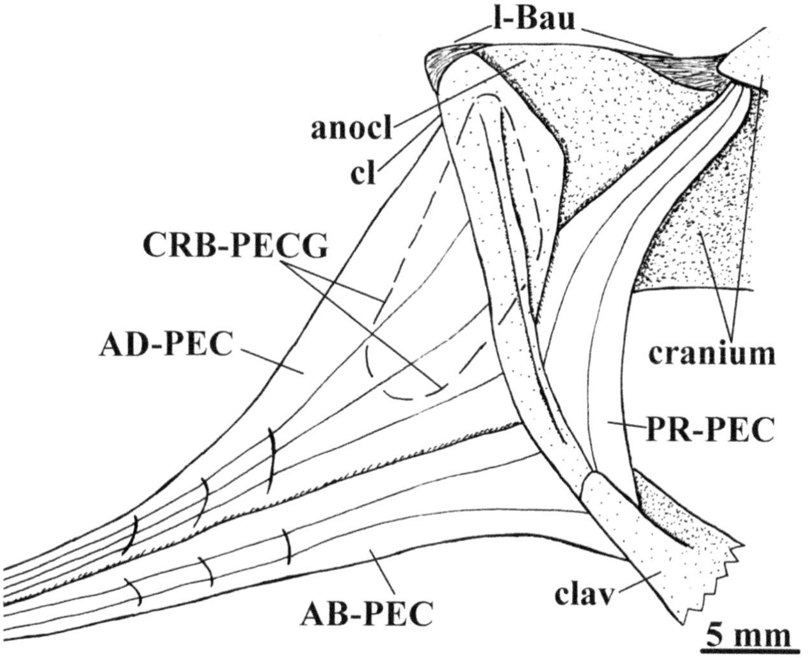

Fig. 4 *Neoceratodus forsteri* (Dipnoi). Lateral view of the protractor pectoralis; the clavicle was cut; the muscle connecting the cranial rib to the pectoral girdle, as well as the adductor and the abductor of the pectoral fin, are not discussed in the present work (modified from Diogo (2007); the nomenclature of the structures illustrated follows that used in the present work; for more details about these latter muscles and the homologies of the pectoral/upper limb muscles, see the recent works of Diogo (2007) and Diogo and Abdala (2007)). AB-PEC, abductor of the pectoral fin; AD-PEC, adductor of the pectoral fin; anocl, anocleithrum; cl, cleithrum; clav, clavicle; CRB-PECG, muscle connecting the cranial to the pectoral girdle; l-Bau, Baudelot's ligament; PR-PEC, protractor pectoralis.

(1902-1935), we recognize six main groups of head muscles: ocular, mandibular, hyoid, branchial, epibranchial, and hypobranchial. Edgeworth (1935) viewed the development of these muscles in the light of developmental pathways leading from presumptive premyogenic condensations to different states in each cranial arch (the condensations of the first and second arches corresponding respectively to Edgeworth's 'mandibular and hyoid muscle plates', and those of the more posterior, 'branchial' arches corresponding to his 'branchial muscle plates'). According to him, although exceptions may occur, the mandibular muscles are generally innervated by the Vth nerve, the hyoid muscles by the VIIth nerve and the branchial muscles by the IXth and Xth nerves. Also according to this author, the epibranchial and hypobranchial muscles are developed from the anterior myotomes of the body and thus are intrusive elements of the head; they retain a spinal innervation and

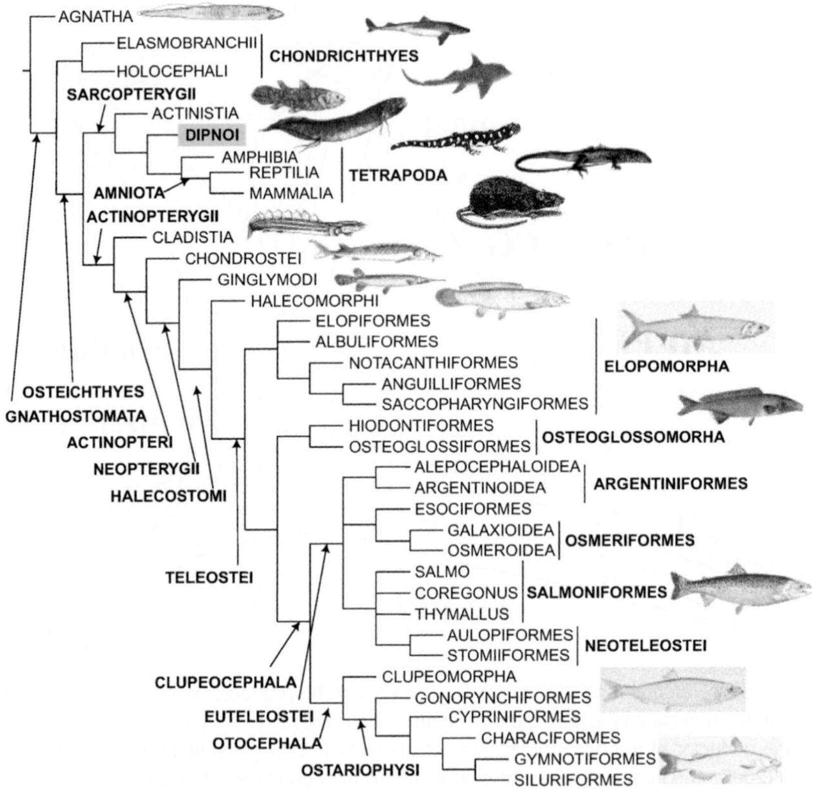

Fig. 5 Phylogenetic relationships among the major extant vertebrate groups included on Tables 1-12, mainly based on a recent cladistic analysis including 356 osteological and myological characters and 80 extant and fossil vertebrate terminal taxa (Diogo 2007). The relationships between the major non-osteichthyan follow Anderson's (2008) cladogram, which is based on Benton (2005). It should be noted that lampreys (Cephalaspidomorphi) have been traditionally considered to be the closest extant relatives of gnathostomes (the hagfishes, or Myxini, being thus the most plesiomorphic extant vertebrate group). However, in the present chapter we follow the now commonly accepted scenario in which living lampreys and hagfishes are included in a monophyletic clade, the Agnatha (jawless vertebrates), in opposition to the other major clade including extant vertebrates, the Gnathostomata (jawed vertebrates), because this scenario has been consistently supported by recent molecular cladistic studies (see, e.g., Dawkins 2004; Kuratani and Ota 2008).

may also be innervated by the XIth and XIIth nerves, but usually do not receive any branches from the Vth, VIIth, IXth and Xth nerves. In the present chapter we will not discuss the epibranchial muscles sensu Edgeworth, which are absent in extant osteichthyans and thus in dipnoans, nor the ocular muscles sensu this author, which are usually innervated by nerves III, IV and/or VI in vertebrates.

Table 1 Mandibular muscles of adults of representative non-osteichthyan extant vertebrate taxa. The nomenclature of the muscles shown in bold follows that of the present work, "ad. mand." meaning adductor mandibulae. In order to facilitate comparisons, in some cases certain names often used by other authors to designate a certain muscle/bundle are given in front of that muscle/bundle, in round brackets; additional comments are given in round brackets (for more details, see text).

Agnatha: *Lampetra japonica* (Japanese lamprey)	Elasmobranchii: *Squalus acanthias* (Spiny dogfish)	Holocephali: *Hydrolagus colliei* (Spotted ratfish)
— [a distinct, independent intermandibularis such as that found in chondrichthyans is seemingly not present in lampreys, but see velothyroideus below and also text]	**Intermandibularis**	**Intermandibularis**
— [a distinct, independent adductor mandibulae A2 such as that found in condrichthyans is seemingly not present in lampreys: see text]	**Ad. mand. A2** (part of adductor mandibulae complex sensu Miyake *et al.* 1992; adductor mandibulae sensu Anderson 2008)	Ad. mand. A2 (levator mandibulae anterior and posterior sensu Miyake *et al.* 1992; adductor mandibulae and posterior sensu Anderson 2008)
'Labial muscles' [see text]	**'Labial muscles'** [see text]	**'Labial muscles'** [see text]
Velothyroideus [does the velothyroideus sensu Holland *et al.* 1993 really corresponds to the levator arcus palatini of the LCA of gnathostomes? See text]	**Levator arcus palatini** (levator palatoquadrati sensu Miyake *et al.* 1992 and Anderson 2008)	—
Velohyoideus and velocranialis [if the velothyroideus is effectively derived from the constrictor dorsalis, are the other two muscles of the velum, i.e. the velohyoideus and velocranialis, also derived from the constrictor dorsalis, or do they possibly correspond to the intermandibularis and/or the adductor mandibulae of other vertebrates? See text]	—	—
—	**Spira-cularis, depressor palpebrae superioris, levator palpebrae nictitantis and retractor palpebrae superioris** [according to Miyake *et al.* 1992, all these muscles may be found in sharks, deriving ontogenetically from the constrictor dorsalis]	

Table 2 Mandibular muscles of adults of representative extant actinopterygian taxa (see caption of Table 1).

Probable plesiomorphic osteichthyan condition	Cladistia: *Polypterus bichir* (Bichir)	Chondrostei: *Psephurus gladius* (Chinese swordfish)	Ginglymodi: *Lepisosteus osseus* (Longnose gar)	Halecomorphi: *Amia calva* (Bowfin)	Teleostei – basal: *Elops saurus* (Ladyfish)	Teleostei-clupeocephalan: *Danio rerio* (Zebrafish)
Intermandibularis anterior and posterior plesiomorphically present in osteichthyans? See text	**Intermandibularis**	**Intermandibularis**	**Intermandibularis**	**Intermandibularis posterior**	**Intermandibularis posterior** [forming, together with interhyoideus, the protractor hyoideus]	**Intermandibularis posterior** [see cell on the left]
See cell above	—	—	—	**Intermandibularis anterior**	**Intermandibularis anterior**	**Intermandibularis anterior**
—	—	—	—	—	**Protractor hyoideus** [see cell above and text]	**Protractor hyoideus** [see cell above and text]
Ad. mand. A3'	**Ad. mand. A3'** (ad. mand. sensu Lauder 1980a)	—	**Ad. mand. A3'** (preorbitalis superficialis sensu Lauder 1980a)	**Ad. mand. A3'**	—	—
Ad. mand. A3"	**Ad. mand. A3"** (ad. mand. pterygoideus sensu Lauder 1980a)	—	**Ad. mand. A3"** (preorbitalis profundus sensu Lauder 1980a)	**Ad. mand. A3"**	—	—
Ad. mand. A2	**Ad. mand. A2** (ad. mand. posterolateral sensu Lauder 1980a)	**Ad. mand. A2** (ad. mand. sensu Carroll and Wainwright 2003)	**Ad. mand. A2** (ad. mand. posterolateral sensu Lauder 1980a)	**Ad. mand. A2**	**Ad. mand. A2**	**Ad. mand. A2**

Table 2 Contd..

Table 2 Contd..

'Labial muscles' plesiomorphically present in osteichthyans? See text	—	—	— [palatomandibularis minor and major of *Lepisosteus* correspond to/derive from the 'labial muscles' of non-osteichthyan vertebrates? See text]	— [levator maxillae superioris 3 and 4 of *Amia* correspond to/derive from the 'labial muscles' of non-osteichthyan vertebrates? See text]	—	—
—	—	—	**Palatomandibularis minor and major**	—	—	—
—	—	—	—	**Levator maxillae superioris 3 and 4**	—	—
—	—	—	—	—	—	**Ad. mand. A1-OST**
—	—	—	—	—	—	**Ad. mand. A0**
Ad. mand. Aω	**Ad. mand. Aω**	—	—	**Ad. mand. Aω**	**Ad. mand. Aω**	**Ad. mand. Aω**
Levator arcus palatini	**Levator arcus palatini**	**Protractor hyomandibulae** [seemingly originated from the portion of the hyoid muscle plate from which originate the adductor arcus palatini and dilatator operculi of other actinopterygians]	**Levator arcus palatini**	**Levator arcus palatini**	**Levator arcus palatini**	**Levator arcus palatini**
—	**Dilatator operculi**	—	**Dilatator operculi**	**Dilatator operculi**	**Dilatator operculi**	**Dilatator operculi**

Table 3 Mandibular muscles of adults of representative extant sarcopterygian taxa, including dipnoans (see caption of Table 1).

Probable plesiomorphic osteichthyan condition	Actinistia: *Latimeria chalumnae* (Coelacanth)	Dipnoi: *Lepidosiren paradoxa* (South American lungfish)	Amphibia: *Ambystoma ordinarium* (Michoacan stream salamander)	Reptilia: *Timon lepidus* (Ocellated lizard)	Mammalia: *Rattus norvegicus* (Norway *rat)*	Mammalia: *Homo sapiens* (Modern Human)
Intermandibularis anterior and posterior present in LCA of osteichthyans? See text	**Intermandibularis posterior**	**Intermandibularis (Fig. 1)**	**Intermandibularis posterior**	**Intermandibularis posterior**	**Mylohyoideus** [mylohyoideus and digastricus anterior of rats are derived from intermandibularis posterior]	**Mylohyoideus**
—	—	—	—	—	**Digastricus anterior** [see cell above]	**Digastricus anterior**
See cell above	**Intermandibularis anterior**	—	**Intermandibularis anterior**	**Intermandibularis anterior**	**Intermandibularis anterior** (transversus mandibularis sensu Greene 1935)	—
Ad. mand. A3'	**Ad. mand. A3'** (ad. mand. 'moyen' sensu Millot and Anthony 1958)	**Ad. mand. A3' (Fig. 2)** (ad. mand. anterior sensu Bemis and Lauder 1986)	**Pseudotemporalis** (pseudotemporalis posterior + pseudotemporalis anterior sensu Iordansky 1992) [seems to correspond to A3' + A3" of *Latimeria*]	**Pseudotemporalis** pseudotemporalis superficialis and profundus sensu Abdala and Moro 2003, and Holliday and Witmer 2007; adductor mandibulae A3' and A3" sensu Diogo 2007, 2008)	— [but see below]	— [but see below]

Table 3 Contd..

Table 3 Contd..

Ad. mand. A3"	**Ad. mand. A3"** (ad. man. 'profond' of e.g. Millot and Anthony [24])	—	—	—	— [but see below]	— [but see below]
—	—	—	—	**Pterygomandibularis** [seems to correspond to part of the pseudotemporalis of amphibians such as *Ambystoma*]	— [but see below]	— [but see below]
—	—	—	—	—	**Pterygoideus medialis** [seemingly corresponds to pseudotemporalis and/or pterygomandibularis of lizards]	**Pterygoideus medialis**
Ad. mand. A2	**Ad. mand. A2** (ad. mand. 'superficiel' sensu Millot and Anthony 1958)	**Ad. mand. A2 (Fig. 2)** (part of ad. mand. posterior sensu Bemis and Lauder 1986)	**Ad. mand. A2** (ad. mand. externus sensu Iordansky 1992)	**Ad. mand. A2** (ad. mand. externus sensu Abdala and Moro 2003)	**Masseter** [masseter, pterygoideus lateralis and temporalis of mammals seemingly derived from lateral, and eventually also medial, protions of ad. mand. of other tetrapods: Diogo *et al*. 2008b]	**Masseter**
—	—	—	—	—	**Temporalis** [see cell above]	**Temporalis**

Table 3 Contd..

Table 3 Contd..

Probable plesiomorphic osteichthyan condition	Actinistia: *Latimeria chalumnae* (Coelacanth)	Dipnoi: *Lepidosiren paradoxa* (South American lungfish)	Amphibia: *Ambystoma ordinarium* (Michoacan stream salamander)	Reptilia: *Timon lepidus* (Ocellated lizard)	Mammalia: *Rattus norvegicus* (Norway *rat)*	Mammalia: *Homo sapiens* (Modern Human)
—	—	—	—	—	**Pterygoideus lateralis** [see cell above]	**Pterygoideus lateralis**
—	—	**Ad. mand. A2-PVM (Fig. 2)** (part of ad. mand. posterior sensu Bemis and Lauder 1986)	**Ad. mand. A2-PVM** (ad. mand. posterior sensu Iordansky)	**Ad. mand. A2-PVM** (ad. mand. posterior sensu Abdala and Moro 2003)	**Tensor veli palatini** [tensor veli palatini and tensor tympani of mammals seemingly derived from ad. mand. A2-PVM of other tetrapods: see text]	**Tensor veli palatini**
—	—	—	—	—	**Tensor tympani** [see cell above]	**Tensor tympani**
—	—	**Retractor anguli oris (Fig. 1)** [seemingly derived from lateral portion of ad. mand.]	—	—	—	—
—	—	—	—	**Levator anguli oris mandibularis** [possibly derived/ modified from the retractor anguli oris, or at least from the portion of the mandibular muscle plate originating that muscle in other osteichthyan taxa]	—	—

Table 3 Contd..

Table 3 Contd..

Ad. mand. Aω	**Ad. mand. Aω** (intramandibular adductor sensu Lauder 1980a)	—	—	**Ad. mand. Aω**	—	—
'Labial muscles' plesiomorphically present in osteichthyans? See text	**'Labial muscle'** [Millot and Anthony 1958 and Anderson 2008 describe such a muscle in *Latimeria*, but Lauder 1980b contradicts this; if the muscle is present, does it correspond to/ derive from the 'labial muscles' of non-osteichthyan vertebrates? See text]	—	—	—	—	—
Levator arcus palatini	**Levator arcus palatini**	—	—	**Levator pterygoidei** [levator pterygoidei, protractor pterygoidei and levator bulbi are likely derived from constrictor dorsalis]	—	—
—	—	—	—	**Protractor pterygoidei** [see cell above]	—	—
—	—	—	**Levator bulbi** [see cell on the right]	**Levator bulbi** [see cell above]	—	—

Table 4 Hyoid muscles of adults of representative non-osteichthyan extant vertebrate taxa (see caption of Table 1).

Agnatha: *Lampetra japonica* (Japanese lamprey)	Elasmobranchii: *Squalus acanthias* (Spiny dogfish)	Holocephali: *Hydrolagus colliei* (Spotted ratfish)
— [a distinct, independent interhyoideus such as that found in gnathostomes does not seem to be present in lampreys: see text and also, e.g., Mallat 1996, 1997]	**Interhyoideus** (interhyoideus + superficial ventral constrictor of the second arch sensu Allis 1923, Lightoller 1939 and Anderson 2008; constrictor hyoideus ventralis sensu Miyake *et al.* 1992)	**Interhyoideus** (constrictor operculi ventralis sensu Anderson 2008) [see text]
— [a distinct, independent adductor operculi/constrictor hyoideus dorsalis such as that found in gnathostomes does not seem to be present in lampreys: see text and also, e.g., Mallat 1996, 1997]	**Constrictor hyoideus dorsalis** (dorsal hyoid constrictor sensu Anderson 2008) [although this muscle clearly seems to correspond to/be modified from to the adductor operculi of the LCA of gnathostomes, the name constrictor hyoideus dorsalis is probably more indicated for chondrichthyans, because elasmobranchs, for instance, do not have an opercle]	**Constrictor hyoideus dorsalis** (constrictor operculi dorsalis sensu Anderson 2008) [Miyake *et al.* 1992 did not describe an adductor operculi/constrictor hyoideus dorsalis in holocephalans)
— [a distinct, independent adductor arcus palatini such as that found in gnathostomes does not seem to be present in lampreys: see text and also, e.g., Mallat 1996, 1997]	**Adductor arcus palatini** (levator hyomandibulae sensu Miyake *et al.* 1992; levator hyoideus sensu Anderson 2008)	**Adductor arcus palatini** (levator hyomandibulae sensu Miyake *et al.* 1992; levator hyomandibularis sensu Anderson 2008)
—	**Additional hyoid muscles** [according to Miyake *et al.* 1992, elasmobranchs might have other hyoid muscles such as the 'levator rostri', 'depressor rostri' and 'depressor mandibulae'; this latter muscle should however have a different name, because it is very likely not homologous to the depressor mandibulae of dipnoans and tetrapods: see, e.g., Diogo 2007]	—

Table 5 Hyoid muscles of adults of representative extant actinopterygian taxa (see caption of Table 1).

Probable plesiomorphic osteichthyan condition	Cladistia: *Polypterus bichir* (Bichir)	Chondrostei: *Psephurus gladius* (Chinese swordfish)	Ginglymodi: *Lepisosteus osseus* (Longnose gar)	Halecomorphi: *Amia calva* (Bowfin)	Teleostei–basal: *Elops saurus* (Ladyfish)	Teleostei–clupeocephalan: *Danio rerio* (Zebrafish)
Interhyoideus	**Interhyoideus**	**Interhyoideus**	**Interhyoideus**	**Interhyoideus**	**Interhyoideus** [forming, together with intermandibularis posterior, the protractor hyoideus: see Table 2]	**Interhyoideus** [see cell on the left]
—	**Hyohyoideus**	**Hyohyoideus**	**Hyohyoideus**	**Hyohyoideus inferior**	**Hyohyoideus inferior**	**Hyohyoideus inferior**
—	—	—	—	**Hyohyoideus abductor** [often considered as part of a hyohyoideus superior]	**Hyohyoideus abductor** [see cell on the left]	**Hyohyoideus abductor** [see cell on the left]
—	—	—	—	**Hyohyoidei adductores** [often considered as part of a hyohyoideus superior]	**Hyohyoidei adductores** [see cell on the left]	**Hyohyoidei adductores** [see cell on the left]
Adductor operculi	**Adductor operculi**	**Adductor operculi** (opercularis sensu Carroll and Wainwright 2003)	**Adductor operculi**	**Adductor operculi**	**Adductor operculi**	**Adductor operculi**

Table 5 Contd..

Table 5 Contd..

Probable plesiomorphic osteichthyan condition	Cladistia: *Polypterus bichir* (Bichir)	Chondrostei: *Psephurus gladius* (Chinese swordfish)	Ginglymodi: *Lepisosteus osseus* (Longnose gar)	Halecomorphi: *Amia calva* (Bowfin)	Teleostei–basal: *Elops saurus* (Ladyfish)	Teleostei–clupeocephalan: *Danio rerio* (Zebrafish)
Adductor arcus palatini	**Adductor arcus palatini**	**Retractor hyomandibulae** [seemingly originated from the portion of the hyoid muscle plate from which originates the adductor arcus palatini of other actinopterygians]	**Adductor arcus palatini**	**Adductor arcus palatini**	**Adductor arcus palatini**	**Adductor arcus palatini**
—	—	—	—	—	—	**'Adductor hyomandibulae X'** [seemingly not homologous to the 'adductor hyomandibulae **Y**, of *Latimeria*: see text]
—	—	—	—	**Levator operculi** [seemingly not homologous to the 'levator operculi' of *Latimeria*: see text]	**Levator operculi** [see cell on the left]	**Levator operculi** [see cell on the left]

Table 6 Hyoid muscles of adults of representative extant sarcopterygian taxa, including the dipnoans (see caption of Table 1).

Probable plesiomorphic osteichthyan condition	Actinistia: *Latimeria chalumnae* (Coelacanth)	Dipnoi: *Lepidosiren paradoxa* (*South* American lungfish)	Amphibia: *Ambystoma ordinarium* (Michoacan stream salamander)	Reptilia: *Timon lepidus* (Ocellated lizard)	Mammalia: *Rattus norvegicus* (Norway rat)	Mammalia: *Homo sapiens* (Modern human)
Interhyoideus	**Interhyoideus** ('géniohyoïdien' + ''hyohyoïdien' sensu Millot and Anthony 1958)	**Interhyoideus (Fig. 1)**	**Interhyoideus** (interhyoideus anterior + interhyoideus posterior sensu Bauer 1992 and Ericsson and Olsson 2004]	**Interhyoideus** (constrictor colli of e.g. Herrel *et al.* 2005)	**Part of facial muscles** [facial muscles of mammals derive mostly from interhyo-ideus, but possibly also from cervicom-andibularis: e.g., Diogo *et al.*, 2008b]	**Part of facial muscles** [see cell on the left]
Adductor arcus palatini	**Adductor arcus palatini** (the adductor arcus palatini of *Latimeria* is not a mandibular muscle, as stated Anderson 2008, but a hyoid muscle: e.g. Miyake *et al.* 1992; Diogo 1997, 2008)	— [the portion of the hyoid muscle plate that gives rise to the levator hyoideus/ depressor mandi-bulae possibly corresponds to that giving rise to the adductor arcus palatini of other osteichthyans]	— [see cell on the left]	— [see cell on the left]	— [see cell on the left]	— [see cell on the left]

Table 6 Contd..

Table 6 Contd..

Probable plesiomorphic osteichthyan condition	Actinistia: *Latimeria chalumnae* (Coelacanth)	Dipnoi: *Lepidosiren paradoxa (South* American lungfish)	Amphibia: *Ambystoma ordinarium* (Michoacan stream salamander)	Reptilia: *Timon lepidus* (Ocellated lizard)	Mammalia: *Rattus norvegicus* (Norway rat)	Mammalia: *Homo sapiens* (Modern human)
—	—	**Levator hyoideus (Fig. 3)** [see cell above]	**Depressor mandibulae posterior** [the fibers corresponding to those of the levator hyoideus of dipnoans become also attached on the mandible, forming the depressor mandibulae posterior; the depressor mandibulae anterior thus seemingly corresponds to part of the depressor mandibulae of dipnoans)	— [but see cervicomandibularis]	**Stapedius** [see Diogo *et al.* 2008b]	**Stapedius** [see cell on the left]
—	—	**Depressor mandibulae (Fig. 1)**	**Depressor mandibulae anterior** [see cell above]	**Depressor mandibulae**	**Stylohyoideus** [seems to correspond to part of the depressor mandibulae of *Timon*: see Diogo *et al.* 2008b]	**Stylohyoideus** [see cell on the left]

Table 6 Contd..

Table 6 Contd..

—	—	—	—	—	**Digastricus posterior** [seems to correspond to part of the depressor mandibulae of *Timon*: see Diogo *et al.* 2008b]	**Digastricus posterior** [see cell on the left]
—	—	—	**Branchiohyoideus** (branchiohyoideus externus sensu Edgeworth and Ericsson and Olsson 2004) [seemingly corresponds to part of the depressor mandibulae of dipnoans]	— [as noted by Edgeworth 1935, the 'branchiohyoideus' of lizards seemingly corresponds to branchial muscle subarcualis rectus 1 of amphibians, and not to the hyoid muscle branchiohyoideus of the present work]	—	—
—	—	—	—	**Cervicomandibularis** [seemingly derived from levator hyoideus: e.g., Diogo *et al.* 2008b]	**Part of facial muscles** [see above]	**Part of facial muscles** [see cell on the left]

Table 6 Contd..

Table 6 Contd..

Probable plesiomorphic osteichthyan condition	Actinistia: *Latimeria chalumnae* (Coelacanth)	Dipnoi: *Lepidosiren paradoxa* (South American lungfish)	Amphibia: *Ambystoma ordinarium* (Michoacan stream salamander)	Reptilia: *Timon lepidus* (Ocellated lizard)	Mammalia: *Rattus norvegicus* (Norway rat)	Mammalia: *Homo sapiens* (Modern human)
—	**'Adductor hyomandibulae Y'** [seemingly not homologous to the 'adductor hyomandibulae X' of Table 5]	—	—	—	—	—
Adductor operculi	**Adductor operculi**	— [absent as a distinct, separate element in adults, but see text]	—	—	—	—
—	***Latimeria*'s 'levator operculi'** [seemingly not homologous to the levator operculi of Table 5]	—	—	—	—	—

Table 7 Branchial muscles of adults of representative non-osteichthyan extant vertebrate taxa (see caption of Table 1).

Agnatha: *Lampetra japonica* (Japanese lamprey)	Elasmobranchii: *Squalus acanthias* (Spiny dogfish)	Holocephali: *Hydrolagus colliei* (Spotted ratfish)
Branchial muscles sensu stricto [examples of branchial muscles that might be present in lampreys, according to Johanson 2003, are the interbranchiales, external branchial constrictors, internal dorsal and ventral diagonal constrictors, median muscle bands, and isolated muscle fibers associated with the interbranchial septum]	**Branchial muscles sensu stricto** [examples of branchial muscles that might be present in elasmobranchs, according to Edgeworth 1935, Miyake *et al.* 1992 and Anderson 2008, are the coracobranchiales, the superficial branchial constrictors, a dorsal branchial muscles that might be branchial muscle complex and an interbranchialis]	**Branchial muscles sensu stricto** [examples of present in elasmobranchs, according to Edgeworth 1935, Miyake *et al.* 1992, and Anderson 2008, are the coracobranchiales and a dorsal branchial muscle complex]
— [according to authors such as Kuratani *et al.* 2005 and Kusakabe and Kuratani 2005, the protractor pectoralis is seemingly absent as an independent muscle in living non-gnathostome animals, its presence thus probably constituting a synapomorphy of gnathostomes]	**Protractor pectoralis** (trapezius sensu Allis 1923; part or totality of cucullaris sensu Anderson 2008)	**Protractor pectoralis** (seems to correspond to the cucullaris superficial and probably also to the protractor pectoralis dorsalis sensu Anderson 2008; the cucullaris profundus sensu this author seems to be simply a branchial levator)

Table 8 Branchial muscles of adults of representative extant actinopterygian taxa (see caption of Table 1).

Probable plesiomorphic osteichthyan condition	Cladistia: *Polypterus bichir* (Bichir)	Chondrostei: *Psephurus gladius* (Chinese swordfish)	Ginglymodi: *Lepisosteus osseus* (Longnose gar)	Halecomorphi: *Amia calva* (Bowfin)	Teleostei–basal: *Elops saurus* (Ladyfish)	Teleostei–clupeocephalan: *Danio rerio* (Zebrafish)
Branchial muscles sensu stricto	**Branchial muscles sensu stricto** [according to Miyake *et al.* 1992, *Polypterus* might have a dorsal branchial muscle complex, transversi ventrales 2-4, a pharyngoclavicularis, a rectus communis and possibly an interarcualis ventralis 1]	**Branchial muscles sensu stricto** [according to Miyake *et al.* 1992, chondrosteans may have a dorsal branchial muscle complex, transversi ventrales 4-5, a pharyngoclavicularis, obliqui ventrales 2-3 and possibly an interarcualis ventralis 1]	**Branchial muscles sensu stricto** [according to ginglymodians may have a dorsal branchial muscle complex, a transversus ventralis 5, a pharyngoclavicularis, obliqui ventrales 1-4 and a possibly transversi ventralis 4-5]	**Branchial muscles sensu stricto** [according to Miyake *et al.* 1992, *Amia* might have a dorsal branchial muscle complex, obliqui ventrales 1-4, a transversus ventralis 5, a rectus ventralis 4, pharyngoclavicularis internus and externus, and possibly transversi ventralis 3-4]	**Branchial muscles sensu stricto** [according to Miyake *et al.* 1992, teleosts may have a dorsal branchial muscle complex, an interbranchialis adductoris, obliqui ventrales 1-4, transversi ventrales 1-4, recti ventrales 2-4, pharyngoclavicularis internus and externus, a rectus communis, and possibly an interarculalis ventralis 1 and/or a rectus ventralis 1]	**Branchial muscles sensu stricto** [see cell on the left]
Protractor pectoralis	**Protractor pectoralis** (cucullaris sensu Edgeworth 1935)	**Protractor pectoralis** (cucullaris sensu Edgeworth 1935)	—	**Protractor pectoralis** (cucullaris sensu Edgeworth 1935)	—	— [within the zebrafish specimens dissected by us, a distinct, independent protractor pectoralis was usually absent in adults, but was found in some old larvae]

Table 8 Contd..

Constrictor laryngis present in LCA of osteichthyans? See text and below	'Constrictor laryngis' and 'dilatator laryngis' of *Polypterus* are homologous to the constrictor laryngis and dilatator laryngis of sarcopterygians? See text and below	—	—	—	—	—
Dilatator laryngis present in LCA of osteichthyans? See text and below	See cell above	—	—	—	—	—
See cell on the right	**'Dilatator laryngis'** [according to Edgeworth 1935, this muscle is not homologous with the 'dilatator laryngis' of *Amia* and *Lepisosteus* nor with the dilatator laryngis of sarcopterygians]		—	—	—	—
See cell on the right	—	—	**'Dilatator laryngis'** [according to Edgeworth 1935, the 'dilatator laryngis' of *Lepisosteus* and *Amia* is not homologous with the 'dilatator laryngis' of *Polypterus* nor with the dilatator laryngis of sarcopterygians]	**'Dilatator laryngis'** [see cell on the left]	—	—

Table 8 Contd..

Table 8 Contd..

Probable plesiomorphic osteichthyan condition	Cladistia: *Polypterus bichir* (Bichir)	Chondrostei: *Psephurus gladius* (Chinese swordfish)	Ginglymodi: *Lepisosteus osseus* (Longnose gar)	Halecomorphi: *Amia calva* (Bowfin)	Teleostei–basal: *Elops saurus* (Ladyfish)	Teleostei–clupeocephalan: *Danio rerio* (Zebrafish)
See cell on the right	**'Constrictor laryngis'** [according to Edgeworth 1935, this muscle might well be homologous with the constrictor laryngis of sarcopterygians; if this is the case, the presence of this muscle would be plesiomorphic for osteichthyans]	— [seemingly absent, according to Edgeworth 1935]	— [seemingly absent, according to Edgeworth 1935]	—	—	—

Table 9 Branchial muscles of adults of representative extant sarcopterygian taxa, including the dipnoans (see caption of Table 1).

Probable plesiomorphic osteichthyan condition	Actinitia: *Latimeria chalumnae* (Coelacanth)	Dipnoi: *Lepidosiren paradoxa* (South American lungfish)	Amphibia: *Ambystoma ordinarium* (Michoacan stream salamander)	Reptilia: *Timon lepidus* (Ocellated lizard)	Mammalia *Rattus norvegicus* (Norwegian rat)	Mammalia *Homo sapiens* (Modern human)
Branchial muscles sensu stricto	**Branchial muscles sensu stricto** [according to Miyake *et al.* 1992, examples of branchial muscles that might be pre-sent in *Latimeria* are: a dorsalis branchial muscle complex, recti ventrales 2-4, transversi ventrales 2-4and possibly an intercualis ventralis 1, a rectus ventralis 1 and/or a constrictor superficialis ventralis]	**Branchial muscles sensu stricto** [according to Edgeworth 1935, examples of branchial muscles that might be present in *Lepidosiren* are the coracobranchiales, and possibly transversi ventralis 4 and 5]	**Branchial muscles sensu stricto** [according to Edgeworth 1935, examples of branchial muscles that might be present in *Ambystoma* are the levator arcus branchialis and the subarcualis rectus I]	— [absent as a group; adult lizards such as *Timon* lack all the branchial muscles sensu stricto except the hyobranchialis and ceratohyoideus, which are seemingly the result of a subdivision of the subarcualis rectus I: Diogo *et al.* 2008b]	— [absent as a group; the only branchial muscles sensu stricto that are present as independent structures in adult rodents such as the Norwegian rat are the ceratohyoideus and seemingly the stylopharyngeus: Diogo *et al.* 2008b]	— [absent as a group; the only branchial muscle sensu stricto that is present as an independent structure in adult modern humans is seemingly the stylopharyngeus: Diogo *et al.* 2008b]
—	— [see above]	— [see above]	— [see above]	**Hyobranchialis and 'ceratoideus'** [see cell above]	**Stylopharyngeus and ceratohyoideus** [see above]	**Stylopharyngeus** [see above; the ceratohyoideus is present in other primates]
Protractor pectoralis	—	**Protractor pectoralis** (Fig. 4) (cucullaris sensu Edgeworth 1935)	**Protractor pectoralis** (cucullaris sensu Edgeworth 1935)	**Trapezius** [the trapezius and sternocleidomastoideus derive from the protractor pectoralis]	**Acromiotrapezius** [the spinotrapezius and acromiotrapezius derive from the trapezius]	**Trapezius**

Table 9 Contd..

Table 9 Contd..

Probable plesiomorphic osteichthyan condition	Actinitia: *Latimeria chalumnae* (Coelacanth)	Dipnoi: *Lepidosiren paradoxa* (South American lungfish)	Amphibia: *Ambystoma ordinarium* (Michoacan stream salamander)	Reptilia: *Timon lepidus* (Ocellated lizard)	Mammalia *Rattus norvegicus* (Norwegian rat)	Mammalia *Homo sapiens* (Modern human)
—	—	—	—	—	**Spinotrapezius** [see cell above]	—
—	—	—	—	—	**Cleido-occipitalis** [probably derives from the sternoclei-domastöideus]	—
—	—	—	—	**Sternocleidoma-stoideus** [see cell above]	**Cleidomastoideus** [the cleidomastoideus and sternomastoideus derive from the sterno-cleidomastoideus]	**Sternocleidoma-stoideus**
—	—	—	—	—	**Sternomastoid-eus** [see cell above]	—
—	—	—	— [see on the right]	— [plesiomor-phically reptiles have no muscular pharynx; reptiles such as crocodi-lians do possess a secondary palate and a means to const-rict the pharynx, but this constric-tor is a derivative of an hyoid mus-cle, the interhyoi-deus: see Diogo *et al.* 2008b]	**Pharyngeal muscles** [the following pharyngeal muscles are present: const-rictor pharyngis medius, inferior and superior; cricothyroideus; pterygopharyn-geus; palatopha-ryngeus; levator veli palatini; and salpingopharyn-geus: see Diogo *et al.* 2008b]	**Pharyngeal muscles** [the following pharyngeal muscles present: constrictor pharyngis medius, inferior and superior; cricothyroideus; palatopharyn-geus; levator veli palatini; salping-opharyngeus; and musculus uvulae: see Diogo *et al.* 2008b]

Table 9 Contd..

Table 9 Contd..

Constrictor laryngis present in LCA os osteichthyans? See text	**Constrictor laryngis** present? See text	**Constrictor laryngis**	**Constrictor laryngis** [see text]	**Constrictor laryngis** [see text]	—	—
—	—	— [see text]	**Laryngei**	—	**Part of laryngeal muscles** [namely thyroarytenoideus, cricoarytenoideus lateralis, and arytenoideus]	**Part of laryngeal muscles** [namely thyroarytenoideus, cricoarytenoideus lateralis, vocalis, arytenoideus transversus and arytenoideus obliquus]
Dilatator laryngis present in LCA of osteichthyans? See text	Dilatator laryngis present? See text	**Dilatator laryngis** [see text]	**Dilatator laryngis** [see text]	**Dilatator laryngis** [see text]	**Part of laryngeal muscles** [namely cricoarytenoideus posterior]	**Part of laryngeal muscles** [namely cricoarytenoideus posterior]

Table 10 Hypobranchial muscles of adults of representative non-osteichthyan extant vertebrate taxa (see caption of Table 1).

Agnatha: *Lampetra japonica* (Japanese lamprey)	Elasmobranchii: *Squalus acanthias* (Spiny dogfish)	Holocephali: *Hydrolagus colliei* (Spotted ratfish)
Undifferentiated hypobranchial musculature (hypoglossal muscle or tongue muscle sensu Goodrich 1958)	**Coracomandibularis**	**Coracomandibularis** [Anderson 2008 and posterior in *Hydrolagus* but, as explained by him, this is more an artificial division than a real one]
—	**Sternohyoideus** [according to Miyake *et al.* 1992, in elasmobranchs the sternohyoideus is often subdivided into a coracohyoideus, a coracoarcualis, and/or a coracohyomandibularis]	**Sternohyoideus** (rectus cervicus sensu Miyake *et al.* 1992; coracohyoideus sensu Anderson 2008)

Table 11 Hypobranchial muscles of adults of representative extant actinopterygian taxa (see caption of Table 1).

Probable plesiomorphic osteichthyan condition	Cladistia: *Polypterus bichir* (Bichir)	Chondrostei: *Psephurus gladius* (Chinese swordfish)	Ginglymodi: *Lepisosteus osseus* (Longnose gar)	Halecomorphi: *Amia calva* (Bowfin)	Teleostei - basal: *Elops saurus* (Ladyfish)	Teleostei - clupeocephalan: *Danio rerio* (Zebrafish)
Coracomandibularis	**Branchiomandibularis** [modified from coracomandibularis]	**Branchiomandibularis** [see cell on the left]	—	**Branchiomandibularis** [*see cells on the left]	—	—
Sternohyoideus	**Sternohyoideus**	**Sternohyoideus**	**Sternohyoideus**	**Sternohyoideus**	**Sternohyoideus**	**Sternohyoideus**

Table 12 Hypobranchial muscles of adults of representative extant sarcopterygian taxa, including the dipnoans (see caption of Table 1).

Probable plesiomorphic osteichthyan condition	Actinistia: *Latimeria chalumnae* (Coelacanth)	Dipnoi: *Lepidosiren paradoxa* (South American lungfish)	Amphibia: *Ambystoma ordinarium* (Michoacan stream salamander)	Reptilia: *Timon lepidus* (Ocellated lizard)	Mammalia: *Rattus norvegicus* (Norway rat)	Mammalia: *Homo sapiens* (Modern human)
Coracomandibularis	**Coracomandibularis**	**Coracomandibularis** (Fig. 1) (geniothoracicus sensu Miyake *et al.* 1992)	**Geniohyoideus** [geniohyoideus does not correspond directly to coracomandibularis of bony fishes, because this latter also gave rise to tetrapod muscles such as the genioglossus and hyoglossus]	**Geniohyoideus** [geniohyoideus and/or at least part of mandibulohyoideus sensu Edgeworth 1935 and Herrel *et al.* 2005]	**Geniohyoideus**	**Geniohyoideus**
—	—	—	**Genioglossus** [according to e.g. Edgeworth 1935] the genioglossus of salamanders such as *Ambystoma* is derived from the coracomandibularis]	**Genioglossus** [according to Edgeworth 1935 the genioglossus of lizards such as *Timon* is derived from the coracomandibularis]	**Genioglossus**	**Genioglossus**

Table 12 Contd..

Table 12 Contd..

Probable plesiomorphic osteichthyan condition	Actinistia: *Latimeria chalumnae* (Coelacanth)	Dipnoi: *Lepidosiren paradoxa* (South American lungfish)	Amphibia: *Ambystoma ordinarium* (Michoacan stream salamander)	Reptilia: *Timon lepidus* (Ocellated lizard)	Mammalia: *Rattus norvegicus* (Norway rat)	Mammalia: *Homo sapiens* (Modern human)
—	—	—	**Hyoglossus** [the statements of Edgeworth 1935 concerning the origin of this muscle in salamanders such as *Ambystoma* are somewhat confuse: in page 196 he states that it originates from the sternohyoideus but in his page 211 he seems to indicate that, as in other amphibians as well as in amniotes, it derives from the coracomandibularis]	**Hyoglossus** [according to Edgeworth 1935 the hyoglossus of lizards such as *Timon* is derived from the coracom-andibularis]	**Hyoglossus**	**Hyoglossus**
—	—	—	—	**Intrinsic muscles of the tongue** [derive from the genioglos-sus and hyoglossus; examples of these muscles in amniotes are the longitudinalis superior and inferior, the transversus linguae and the verticalis linguae: e.g., Diogo *et al.* 2008b]	**Intrinsic muscles of the tongue** [see on the left]	**Intrinsic muscles of the tongue** [see on the left]

Table 12 Contd..

Table 12 Contd..

—	—	—	—	—	**Styloglossus** [seemingly derived from hyoglossus: see, e.g., Edgeworth 1935]	**Styloglossus**
—	—	—	—	—	—	**Palatoglossus** [seemingly derived from styloglossus: see, e.g., Edgeworth 1935]
—	—	—	**Interradialis** [according to Piatt 1938 in at least some adult *Ambystoma* there is a hypobranchial muscle interradialis, which derives ontogenetically from the genioglossus]	—	—	—
Sternohyoideus		**Sternohyoideus** (Fig. 1) (rectus cervicis sensu Bemis and Lauder 1986)	**Sternohyoideus** (rectus cervicis sensu Kardong 2002)	**Sternohyoideus** (rectus cervicis sensu Kardong 2002)	**Sternohyoideus**	**Sternohyoideus**

Table 12 Contd..

Table 12 Contd..

Probable plesiomorphic osteichthyan condition	Actinistia: *Latimeria chalumnae* (Coelacanth)	Dipnoi: *Lepidosiren paradoxa* (South American lungfish)	Amphibia: *Ambystoma ordinarium* (Michoacan stream salamander)	Reptilia: *Timon lepidus* (Ocellated lizard)	Mammalia: *Rattus norvegicus* (Norway rat)	Mammalia: *Homo sapiens* (Modern human)
—	—	—	—	—	**Sternothyroideus** [sternothyroideus and thyrohyoideus seemingly derived from sternohyoideus: see, e.g., Edgeworth 1935; Diogo *et al.* 2008b]	**Sternothyroideus**
—	—	—	—	—	**Thyrohyoideus** [see cell above]	**Thyrohyoideus**
—	—	—	**Omohyoideus** [seemingly derived from the sterno-hyoideus: see, e.g., Diogo *et al.* 2008b]	**Omohyoideus**	**Omohyoideus**	**Omohyoideus**

The phylogenetic framework for the comparisons provided in the present chapter is set out in Figure 5. In order to facilitate the comparisons between the mandibular, hyoid, branchial and hypobranchial muscles of dipnoans and those muscles of the other major extant vertebrate groups, we carefully chose to include in Tables 1-12: the lamprey *Lampetra japonica* (Agnatha: Cephalaspidomorphi), the shark *Squalus acanthias* (Elasmobranchii), the ratfish *Hydrolagus colliei* (Holocephali) [non-osteichthyan vertebrates; Tables 1, 4, 7, 10]; the bichir *Polypterus bichir* (Cladistia), the swordfish *Psephurus gladius* (Chondrostei), the gar *Lepisosteus osseus* (Ginglymodi), the bowfin *Amia calva* (Halecomorphi), the basal teleostean *Elops saurus* and the clupeocephalan teleostean *Danio rerio* (Teleostei) [actinopterygian osteichthyans; Tables 2, 5, 8, 11]; and the coelacanth *Latimeria chalumnae* (Actinistia), the lungfish *Lepidosiren paradoxa* (Dipnoi), the salamander *Ambystoma ordinarium* (Amphibia), the lizard *Timon lepidus* (Reptilia), and the mammals *Rattus norvegicus* and *Homo sapiens* (Mammalia) [sarcopterygian osteichthyans; Tables 3, 6, 4, 12] (see Fig. 5). As a base for the data presented in these tables, we have dissected numerous specimens of vertebrate taxa as diverse as dipnoans, anurans, caecilians, lizards, turtles, birds, monotremes, rodents, tree shrews, flying lemurs, primates, polypteriforms, chondrosteans, lepisosteiforms, amiiforms, and teleosts (a detailed list of these specimens is provided in Diogo *et al.* 2008b). We did not dissect chondrichthyans (including holocephalans and elasmobranchs) nor agnathans (including hagfishes and lampreys) (see Fig. 5). Therefore, contrary to the data that we provide for osteichthyans, which are in a great part based on our own dissections and compiled from Diogo (2007, 2008) and Diogo *et al.* (2008a,b), the data concerning the muscles of non-osteichthyan vertebrates are essentially based on a review of the literature. As in all our works, we made an effort to take into account as much information as possible, from classic anatomical descriptions such as those provided by Gegenbaur (1872), Cole (1896), Marion (1905), Allis (1917, 1923, 1931), Alcock (1898), Edgeworth (1902, 1911, 1923, 1926a,b,c, 1928, 1935), Luther (1938) and Roberts (1950), to more recent reviews such as Miyake *et al.* (1992), Mallat (1996, 1997), Johanson (2003) and Anderson (2008). Importantly, apart from the data described in these anatomical studies, our hypotheses about the homologies and evolution of each muscle also take into account the developmental and molecular data obtained in evo-devo works published over the last two decades (Holland *et al.* 1993, in press; Kuratani *et al.* 2002, 2004; Graham 2003; Manzanares and Nieto 2003; Santagati and Rijli 2003; Trainor *et al.* 2003; Cerny *et al.* 2004; Kuratani 2004, 2005a,b, in press; Takio *et al.* 2004; Helms *et al.* 2005; Kusakabe and Kuratani 2005; Northcutt 2005; Olsson *et al.* 2005; Shigetani *et al.* 2005; Knight *et al.* 2008; Kuratani and Ota 2008; Kuratani and Schilling 2008). These hypotheses are thus based on a comprehensive analysis of all the lines of evidence available (e.g., innervation, relationships with other muscular structures, relationships with hard tissues, configuration/orientation of the fibers, development, function, phylogeny, expression domains for homeobox genes, presence/absence/configuration in fossils, etc.; for more details about the choice and use of each of the these lines of evidence, see Diogo 2007, 2008; Diogo *et al.* 2008a,b). However, because we did not directly dissect chondrichthyan and agnathan specimens, and also because osteichthyans, agnathans and chondrichthyans have been evolving separately for hundreds of millions of years and each of these lineages has given rise to remarkably peculiar and unique phenotypes, it should be kept in mind that the hypotheses of homology proposed in Tables 1, 4, 7 and 10 are not as solid as those proposed in the other tables.

HOMOLOGIES BETWEEN THE HEAD MUSCLES OF DIPNOANS AND THE MUSCLES OF OTHER VERTEBRATES

It is obviously not possible to provide in a single chapter such as this an extensive discussion about all the lines of evidence supporting each of the hypotheses of homology for each of the mandibular, hyoid, branchial and hypobranchial muscles found in all the major extant vertebrate groups. Therefore, in the sections that follow we will essentially provide a brief summary of the information presented in Tables 1-12 and pay special attention to certain topics that remain particularly controversial among comparative anatomists. For more details about these the specific configuration, attachments, innervation, function and ontogenetic development of each of these head muscles, the readers should thus refer to the original descriptions given in the works cited in this chapter and in the studies cited in those works.

Mandibular Muscles (Tables 1-3)

According to Edgeworth (1935), in numerous vertebrates the embryonic mandibular muscle plate gives rise dorsally to the premyogenic condensation constrictor dorsalis and medially to the premyogenic condensation adductor mandibulae. To this can be added, ventrally, the intermandibularis (Tables 1-3). Molecular developmental studies have supported the existence of the constrictor dorsalis in the cranial region of teleosts, which gives rise to the levator arcus palatini and dilatator operculi of these fishes (Hatta *et al.* 1990, 1991; Table 2). Expression of engrailed genes marks muscle cells associated with the dorsal region of the first arch. The levator arcus palatini was very likely the only constrictor dorsalis muscle found in the last common ancestor (LCA) of gnathostomes, as well as in the LCA of osteichthyans (Tables 1-3). The spiracularis and "spiracularis" found in batoids and the spiracularis, depressor palpebrae superioris, levator palpebrae nictitantis and the retractor palpebrae superioris found in sharks (sensu Miyake *et al.* 1992; Table 1) were probably acquired only in elasmobranchs, because: 1) none of these latter muscles seems to be found in non-gnathostome extant vertebrates (e.g.,

Edgeworth 1935; Holland *et al.* 1993; Mallat 1996, 1997); 2) all the constrictor dorsalis muscles are absent in the closest extant relatives of elasmobranchs, the holocephalans (e.g., Miyake *et al.* 1992; Anderson 2008); 3) basal osteichthyans such as *Latimeria* have a single constrictor dorsalis muscle, the levator arcus palatini (e.g., Milllot and Anthony 1958; Diogo 2007, 2008); 4) contrary to some descriptions, our dissections indicated that *Polypterus* does not have a muscle spiracularis or a muscle 'spiracularis' such as that found in elasmobranchs, sensu Miyake *et al.* 1992; see Diogo 2007, 2008). The constrictor dorsalis was independently lost in dipnoans and in mammals and significantly reduced in amphibians (Table 3). The constrictor dorsalis that gives rise to the levator arcus palatini in extant sarcopterygian fishes such as *Latimeria* is therefore homologous with the constrictor dorsalis that gives rise to the levator arcus palatini and dilatator operculi in actinopterygians, to the protractor pterygoidei, levator pterygoidei and levator bulbi in amniotes, and to the "spiracularis", spiracularis, depressor palpebrae superioris, levator palpebrae nictitantis and retractor palpebrae superioris in elasmobranchs (Tables 1-3).

Interestingly, according to Holland *et al.* (1993) the velothyroideus of lampreys may be derived from the constrictor dorsalis, because engrailed immunoreactivity has been detected in this muscle (as explained above, such immunoreactivity has also been detected in the constrictor dorsalis muscles of teleosts). Holland *et al.* (1993) state that the mandibular arch mesoderm of lampreys gives rise to one extrinsic eye muscle, three muscles of the velum (velothyroideus, velohyoideus and velocranialis), and seven muscles of the hood and lip (most, or all, of them probably corresponding to the 'labial' muscles sensu Anderson 2008; see Table 1). If one accepts the homology between the lamprey velothyroideus and the levator arcus palatini of the LCA of gnathostomes, could the two other muscles of the lamprey velum, i.e. the velohyoideus (dorsomedial) and velocranialis (anteromedial) sensu Holland *et al.* (1993), be homologous to the two other non-'labial' mandibular muscles of this LCA, i.e. the adductor mandibulae and the intermandibularis? Mallat (1996) seems to suggest that this is not the case, because according to him the adductor mandibulae complex sensu Anderson (2008) came from a 'medial band' of the interbranchialis of the first arch of basal vertebrates, and thus does not correspond to the velohyoideus and/or the velocranialis. It should also be stressed that the detection of engrailed immunoreactivity in certain muscles of a taxon A and in certain muscles of a taxon B does not necessarily imply that the former muscles are directly homologous to the latter. For instance, in the zebrafish engrailed immunoreactivity is only detected in muscles that are derived from the constrictor dorsalis, i.e. the levator arcus palatini + dilatator operculi, while in the mouse it is detected in mandibular muscles that are very likely derived from the 'adductor mandibulae' portion of that plate (i.e. masseter, temporalis, pterygoideus medialis and/or pterygoideus lateralis; see Table 3) (Knight *et al.* 2008). Therefore, detection of immunoreactivity in the lamprey velothyroideus and in the zebrafish levator arcus palatini and dilatator operculi does not necessarily imply that the latter muscles are directly homologous to, or derived from, the former muscle.

Authors such as Kuratani and Ota (2008) have even suggested that lampreys and probably hagfishes lack 'somitomeres', i.e. that agnathans probably do not have 'mandibular', 'hyoid' and 'branchial' muscular plates (sensu Edgeworth 1935) such as those present in living gnathostomes. However, authors such as Holland *et al.* (1993), Mallat (1996) and Knight *et al.* (2008) consider that at least some of these plates (e.g., the mandibular) are present in extant agnathans, thus implying that they were present in the LCA of vertebrates.

Within gnathostomes, the ventral portion of the mandibular muscle plate usually gives rise to the intermandibularis (Tables 2, 3). In dipnoans (Fig. 1), as well as in adult extant members of the Actinistia, Chondrostei and Ginglymodi, the intermandibularis is mainly undivided (Tables 2, 3). Adults of *Amia*, *Latimeria*, and numerous amphibian, amniote and teleostean taxa exhibit an intermandibularis anterior and an intermandibularis posterior (Tables 2, 3). In most teleosts the intermandibularis posterior and the hyoid muscle interhyoideus (see below) form the protractor hyoideus, which is thus derived from both the mandibular and hyoid muscle plates (Tables 2, 5). Although a protractor hyoideus is not found in a few teleosts such as *Albula* and *Mormyrus*, this muscle was seemingly present in the ancestors of extant teleosts (Diogo 2007, 2008). Based on the altered morphology of the protractor hyoideus in morpholino-mediated Hox PG2 (*hoxa2b* and *hoxa2a*) knock-down larvae, Hunter and Prince (2002) suggested that in the zebrafish "the basihyal (cartilage) may be important for the proper ontogenetic organization" of the intermandibularis posterior and the interhyoideus, and, thus, for the association of their fibers and the formation of the protractor hyoideus. Further studies are needed to check if this is so and if it is a general feature within the Teleostei. In some mammals, such as rats and tree-shrews, there is also an intermandibularis anterior, the intermandibularis posterior giving rise to the mylohyoideus and digastricus anterior (Table 3; Diogo *et al.* 2008b). The intermandibularis is often undivided in chondrichthyans, but authors such as Edgeworth (1935), Kesteven (1942-1945) and Miyake *et al.* (1992) have described subdivisions of this muscle in some of these fishes. Miyake *et al.* (1992) and Anderson (2008) suggest that the plesiomorphic condition for gnathostomes was that in which this muscle was mainly undivided. This was probably the case, and at least some of the intermandbularis divisions found in certain living gnathostomes were seemingly acquired independently (e.g. Diogo 2007, 2008b). But the fact that at least some elasmobranchs, some holocephalans and many osteichthyans have two intermandibularis divisions does not allow to completely discard the hypothesis that this muscle was plesiomorphically divided in the LCA of gnathostomes and in the LCA of osteichthyans (Tables 1-3).

The adductor mandibulae complex is found in members of all major gnathostome groups (Tables 1-3). The number of divisions of this muscle is highly variable within these groups. In elasmobranchs the adductor mandibulae A2 may be subdivided into superficial and deep sections, while in holocephalans it is often divided into anterior and posterior sections (e.g., Miyake *et al.* 1992; Anderson

2008). In osteichthyans, the A2 usually gives rise to various muscles, such as the A2, A2-PVM, A3', A3", A1, A1-OST, Aω, pseudotemporalis, levator anguli oris mandibularis, retractor anguli oris, pterygomandibularis, masseter, detrahens mandibulae, temporalis, pterygoideus lateralis, tensor tympani, tensor veli palatini and pterygoideus medialis (Tables 2-3). Dipnoans have an A2, an A2-PVM, a retractor anguli oris and an A3' (Figs. 1, 2; Table 3). The former three structures seem to correspond to the A2 of other bony fishes such as *Latimeria* (Diogo *et al.* 2008b). The masseter, temporalis and pterygoideus lateralis of mammals such as rats and humans correspond to the A2 of dipnoans. The tensor tympani and tensor veli palatini of mammals probably correspond to the A2-PVM of dipnoans, while the pterygoideus medialis of these tetrapods seems to be a remaining of the A3'/A3" of bony fishes (Diogo *et al.* 2008b).

As explained above, extant agnathans do not seem to have a distinct adductor mandibulae A2 such as that found in chondrichthyans (Table 1). Mallat (1996) suggests that the A2 of chondrichthyans corresponds to a 'medial band' of the interbranchialis of the first arch of agnathans such as lampreys, and thus not to the velohyoideus and/or the velocranialis. However, authors such as Luther (1938) and Lightoller (1939) defend that the A2 of condrichthyans probably derived from a lateral part of the interbranchialis of this arch (and not from a medial part of this muscle, as defended by Mallat 1996), or even from its branchial superficial constrictor, because in gnathostomes the A2 usually lies on the lateral, and not on the medial, surface of the arch. Another subject of controversy regarding the homologies among the mandibular muscles of vertebrates concerns the so-called 'labial muscles' sensu Anderson (2008). Authors such as Edgeworth (1935), Lightoller (1939) and Lauder (1980a,b) argued that these 'labial' muscles are very likely derived from the adductor mandibulae complex sensu Edgeworth (1935). Mallat (1996) calls these muscles 'oral' muscles, but he recognizes that at least some of them (e.g. his 'buccal constrictor') develop from the 'mandibular branchiomere' in lampreys. The 'labial' muscles sensu Anderson (2008) also seem to develop from the mandibular plate in elasmobranchs and osteichthyans (Edgeworth 1935). Therefore, whether these muscles are called 'labial' (Anderson 2008), 'oral' (Mallat 1996) or 'preorbital/suborbital mandible adductors' (Edgeworth 1935; Lauder 1980a), the fact is that they do seem to develop from the mandibular mesoderm, as does the adductor mandibulae complex. This idea was supported by the developmental work of Kuratani *et al.* (2004) (see below).

It is quite possible that there is actually no direct correspondence between any of the individual 'labial' muscles present in living lampreys (i.e., buccalis anterior, elevator labialis ventralis and buccal constrictor in ammocoetes, sensu Mallat 1996), living hagfishes (i.e., coronarious, basitentacularis, levator cartilaginis basalis, protractor cartilaginis basalis, craniobasialis and cornuosubnasal muscles sensu Mallat 1996), living holocephalans (i.e., levator prelabialis, levator anguli oris anterior, levator anguli oris posterior, labialis anterior and levator posterior sensu Mallat 1996), living elasmobranchs (i.e., preorbitalis/levator labii superioris sensu Mallat 1996 and Anderson 2008), and living osteichthyans (levator maxillae superioris 3 and 4 of *Amia* and palatomandibularis major and minor of *Lepisosteus* – sensu Diogo 2007 and Anderson 2008 – and possibly 'suborbital portion of adductor mandibulae' of acipenseriforms – sensu Lauder 1980a – and/or 'labial muscles' of *Latimeria* – sensu Millot and Anthony 1958 and Anderson 2008). That is, in our opinion one cannot discard the hypothesis that the individual 'labial' muscles (sensu Anderson 2008) present in extant lampreys, hagfishes, elasmobranchs, holocephalans and osteichthyans are the result of an independent differentiation (in the lineages that gave rise to these major vertebrate groups) of the mandibular mesoderm. In fact the developmental work of Kuratani *et al.* (2004) has not only indicated that at least some of the lamprey 'labial muscles' are the result of a secondary migration of part of the mandibular mesoderm, but also that these muscles are very likely not homologous with the 'labial' muscles of living gnathostomes: "experiments labeling the mandibular mesoderm of the early lamprey embryo, before the cheek process has differentiated into the upper lip anlage or the premandibular domain, indicate that a part of the mandibular mesoderm secondarily grows anteriorly and laterally and migrates into the upper lip domain; no such muscles are known in the gnathostomes, in which all the trigeminal-nerve-innervated muscles are restricted to derivatives of the upper and lower jaws". Further studies, ideally combining dissections and direct comparisons of various lampreys, hagfishes, elasmobranchs, holocephalans, osteichthyans, and non-vertebrate taxa such as amphioxus with molecular techniques such as those employed by Kuratani *et al.* (2004), will hopefully help to clarify the evolution and homologies of the so-called 'labial' muscles within these taxa.

Hyoid Muscles (Tables 4-6)

The hyoid musculature of the LCA of gnathostomes was probably divided into a ventral portion, the interhyoideus, and two dorsal portions, the adductor arcus palatini (often called 'levator hyomandibularis' in chondrichthyans, because in these fishes it usually attaches exclusively on the second arch, and not also, or exclusively, on the first arch, as is the case in many osteichthyans), and the adductor operculi/constrictor hyoideus dorsalis (the name constrictor hyoideus dorsalis seems more appropriate for chondrichthyans, because living elasmobranchs, for instance, do not have an opercle: see, e.g., Miyake *et al.* 1992) (Table 4). Anderson (2008) suggested that holocephalans do not have an interhyoideus. However, the holocephalan muscle that was named 'constrictor operculi ventralis' by that author clearly seems to correspond to the interhyoideus of other living gnathostomes and particularly to that of various elasmobranchs and of basal sarcopterygians such as dipnoans. In fact, the interhyoideus and the adductor operculi present in early ontogenetic stages of dipnoans are remarkably similar to the 'constrictor operculi ventralis' and the 'constrictor operculi dorsalis' of holocephalans (sensu

Anderson 2008), respectively: they lie in a superficial position, and together form a structure that resembles the continuous, superficial constrictor of the second arch of non-gnathostome taxa such as lampreys (see, e.g., Diogo 2007, 2008; Diogo *et al.* 2008b). In later ontogenetic stages of dipnoans the adductor operculi usually becomes completely undistinguished from the interhyoideus (Fig. 1; Table 6).

Apart from the adductor arcus palatini and the adductor operculi, other dorso-medial hyoid muscles are found in certain living osteichthyans (Tables 5-6). For example, most extant teleosts and the halecomorph *Amia* have a muscle levator operculi (Table 5). Millot and Anthony (1958) stated that *Latimeria* has a 'levator operculi' (Table 6). However, whether this muscle is homologous to the levator operculi of *Amia* and teleosts is doubtful for two main reasons. First, the muscles have distinct function: contrary to teleosts and to *Amia*, *Latimeria* does not have an interoperculo-mandibular ligament and, therefore, does not have an opercular mechanism mediating mandible depression (e.g., Lauder 1980a,c; Lauder and Liem 1983). Second, and more importantly, it is cladistically more parsimonious to consider that these muscles were independently acquired in actinistians and halecostomes (2 steps) than to have one acquisition (in the node leading to osteichthyans) and various independent losses (at least in non-actinistian sarcopterygians, in cladistians, in chondrosteans and in ginglymodians) (see Fig. 5; Diogo 2007). Similar reasoning applies to the 'adductor hyomandibulae' of *Latimeria* (Table 6) and to the adductor hyomandibulae muscles found in some teleosts (Table 5). The dipnoan 'levator operculi' illustrated by Kardong (2002) seemingly corresponds to the 'constrictor operculi' sensu Bemis and Lauder (1986), which may correspond to the adductor operculi of other bony fishes but usually forms, in extant adult dipnoans, a continuous sheet of fibers together with other cranial muscles such as the interhyoideus (see above).

Examples of other dorso-medial hyoid muscles present in osteichthyans are the levator hyoideus and the depressor mandibulae, which seemingly gave rise to the stylohyoideus, digastricus posterior, stapedius, and possibly part of the facial muscles of mammals (Table 6). The levator hyoideus (Fig. 3) is usually related with the elevation of the posterodorsal portion of the ceratohyal, whereas the depressor mandibulae (Fig. 1) is usually related with the opening of the mouth. The levator hyoideus is found in at least some developmental stages of extant dipnoans (Fig. 3; Table 6) and of numerous extant tetrapods (Edgeworth 1935). The depressor mandibulae of extant dipnoans such as *Protopterus* and *Lepidosiren* (Fig. 1; this muscle is missing in adult specimens of the genus *Neoceratodus*) seems to be homologous with part or the totality of the depressor mandibulae of tetrapods (e.g., Diogo *et al.* 2008b; Table 6). Interestingly, works such as Köntges and Lumsden (1996) have shown that in tetrapod taxa such as birds, the posterior region of the mandible to which the depressor mandibulae attaches is constituted by neural crest derivatives of the hyoid arch, and not of the mandibular arch. This is one of the several examples given by these authors to illustrate the highly constrained pattern of cranial skeletomuscular connectivity found in these tetrapods: each rhombomeric neural crest population remains coherent throughout ontogeny, forming both the connective tissues of specific muscles and their respective attachment sites onto the neuro- and viscerocranium. It would be interesting, therefore, to investigate if the depressor mandibulae of dipnoans such as *Protopterus* and *Lepidosiren* also attaches in a region of the mandible constituted by neural crest derivatives of the hyoid arch. If future investigation shows that the mandible of extant non-dipnoan bony fishes is exclusively formed by mandibular neural crest derivatives, this would indicate that the presence of a depressor mandibulae in tetrapods and dipnoans might be related with an evolutionary change in which hyoid neural crest derivatives have become incorporated in the formation of the lower jaw.

As explained above, the plesiomorphic condition for gnathostomes is seemingly that in which the ventral portion of the hyoid muscle plate gives rise to a single division, the interhyoideus (Table 4). This is precisely the condition found in extant dipnoans (Fig. 1; Table 6). In most extant actinopterygians part of the interhyoideus separates into a distinct muscle during development, the hyohyoideus. In most teleosts and *Amia* the hyohyoideus is in turn subdivided into hyohyoideus inferior, hyohyoideus abductor and hyohyoidei adductores (Table 5). An independent hyohyoideus is seemingly missing in extant sarcopterygii (Table 6). Although there are some sarcopterygians in which the portion of the hyoid muscle plate that gives rise to the interhyoideus and hyohyoideus in actinopterygians eventually becomes somewhat divided into bundles that resemble these two muscles, these bundles remain deeply mixed throughout all developmental stages. This is the case of the interhyoideus anterior and interhyoideus posterior of various salamanders (Table 6) and of the 'géniohyoïdien' and 'hyohyoïdien' described by Millot and Anthony (1958) in *Latimeria* (Diogo 2007, 2008). In mammals the interhyoideus gives rise to several facial muscles (Diogo *et al.* 2008b; Table 6).

Branchial Muscles (Tables 7-9)

The muscles listed in Tables 7-9 correspond to the branchial muscles *sensu lato* of Edgeworth (1935). They can be divided into three groups. The first comprises the 'true' branchial muscles, which are subdivided into A) the branchial muscles *sensu stricto* that are directly associated with the movements of the branchial arches and that in mammals are usually innervated by the glossopharyngeal nerve (CNIX); B) the protractor pectoralis and its derivatives, which are instead mainly associated with the pectoral girdle and are primarily innervated by the spinal accessory nerve (CNXI). The second group consists of the pharyngeal muscles, which are only present as independent structures in extant mammals. They are considered to be derived from arches 4-6, and they are usually innervated by the vagus nerve (CNX). Following Diogo *et al.* (2008b), the mammalian stylopharyngeus, which is considered to be derived from the third arch and is primarily innervated by the glossopharyngeal nerve, is grouped here with the 'true' branchial muscles, and not with the pharyngeal muscles. The third group is made up of the laryngeal muscles, which are considered to be derived from arches 4-6 and are usually innervated by the vagus nerve (CNX).

Most adult vertebrates retain various branchial muscles *sensu stricto* (Bischoff 1840; Owen 1841; Allis 1823; Edgeworth 1935; Millot and Anthony 1958; Wiley 1979a b; Jollie 1982; Bemis 1986; Miyake *et al.* 1992; Wilga *et al.* 2000; Kardong 2002; Carroll and Wainwright 2003; Diogo *et al.* 2008b) (Tables 7-9). However, the branchial muscles *sensu stricto* are not present as a group in extant reptiles and extant mammals (e.g., Diogo *et al.* 2008b; Table 9). For instance, many adult reptiles have only one branchial muscle *sensu stricto*, the subarcualis rectus I (which is often named 'branchiohyoideus', although it is not homologous with the hyoid muscle branchiohyoideus found in, for example, salamanders: Table 6). The two branchial muscles *sensu stricto* found in adult reptiles such as lizards, i.e. the hyobranchialis and 'ceratohyoideus', seem to be the result of a subdivision of the subarcualis rectus I (Table 9). Adult extant mammals lack all the branchial muscles *sensu stricto* except the subarcualis rectus I (which in most adult mammals gives rise to the ceratohyoideus and stylopharyngeus), the subarcualis rectus II (usually present only in adult marsupials) and the subarcualis rectus III (usually present only in adult monotremes) (e.g., Edgeworth 1935; Diogo *et al.* 2008b).

The mammalian acromiotrapezius, spinotrapezius, cleido-occipitalis, sternocleidomastoideus, cleidomastoideus and sternomastoideus correspond to the reptilian trapezius and sternocleidomastoideus and thus to the protractor pectoralis of dipnoans (Fig. 4), amphibians, and other vertebrates (Tables 7-9). The protractor pectoralis is not a branchial muscle *sensu stricto* because it is mainly involved in the movements of the pectoral girdle and not of the branchial arches (Edgeworth 1935; Diogo *et al.* 2008b). The results of recent developmental studies indicate that the protractor pectoralis of *Ambystoma* and the trapezius of chickens and mice may be at least partially derived from somites (Piekarski and Olsson 2007). It is thus not certain whether the protractor pectoralis was primarily derived from the paraxial mesoderm as suggested by Edgeworth (1935) and only later became ontogenetically also derived from the cranialmost somites, or if it was instead primarily derived from somites. Interestingly, the results of Piekarski and Olsson (2007) suggest that at least in certain cases even branchial muscles *sensu stricto*, such as the levatores arcuum branchialium, may also be partially derived from somites. If this is the case, then the fact that muscles such as the protractor pectoralis have a partial somitic origin does not necessarily mean that they cannot be considered to be part of the branchial musculature (Piekarski and Olsson (2007) claim that hyoid muscles such as the interhyoideus might be also partially derived ontogenetically from somites). Matsuoka *et al.* (2005) recognize that the trapezius of at least some tetrapods is partially derived ontogenetically from somites, but they argue that the sum of the data available (i.e., innervation, topology, development and phylogeny) provides more support for grouping it with the branchial muscles *sensu stricto* than for its inclusion in the postcranial axial muscles. In fact, it is important to stress that lineage tracing analyses in transgenic mice support the idea that the trapezius is essentially a branchial muscle: these analyses reveal that neural crest cells from a caudal pharyngeal arch travel with the trapezius myoblasts and form tendinous and skeletal cells within the spine of the scapula (Noden and Schneider 2006). According to Noden and Schneider (2006), "this excursion seemingly recapitulates movements established ancestrally, when parts of the pectoral girdle abutted caudal portions of the skull". The innervation of the trapezius by not only XI, but also, in many cases, C3 and 4 spinal cord segments adds weight to the argument that the muscle is derived from both the paraxial mesoderm and somites. Support for a branchial component also comes from the position of the accessory nucleus in the ventral horn of the spinal cord, which is in line with the more cranial branchiomotor nuclei (Wilson-Pauwels *et al.* 2002; Butler and Hodos 2005).

The mammalian laryngeal muscles thyroarytenoideus, vocalis, cricoarytenoideus lateralis and arytenoideus seemingly derive from the laryngei of non-mammalian tetrapods such as salamanders, which in turn derives from the constrictor laryngis of dipnoans such as *Lepidosiren*; the mammalian cricoarytenoideus posterior corresponds to the dilator laryngis of other tetrapods and of dipnoans (Diogo *et al.* 2008b; Table 9). Some non-sarcopterygian vertebrates such as *Polypterus*, *Lepisosteus* and *Amia* have a 'constrictor laryngis' and/or a 'dilatator laryngis', but it is not clear if these muscles do actually correspond to the constrictor laryngis and dilatator laryngis of sarcopterygians and thus if these latter muscles are plesiomorphically present in osteichthyans or not (Edgeworth 1935; Table 9). The few descriptions of the laryngeal region of *Latimeria* do not allow us to appropriately discern if these laryngeal muscles are, or are not, present in this taxon (Millot and Anthony 1958; Table 9). The laryngei of tetrapods does not seem to be plesiomorphically found in sarcopterygians, because it is seemingly absent in sarcopterygian fish as dipnoans; a detailed study of the laryngeal region of *Latimeria* is however needed in order to support, or to contradict, this hypothesis (Diogo *et al.* 2008b; Table 9).

Hypobranchial Muscles (Tables 10-12)

According to Mallat (1996) in the LCA of vertebrates the hypobranchial myotomes were probably superficial to the external branchial arches and pharyngeal musculature, because this is the condition found in extant lampreys and hagfishes. In hagfishes, the mainly undifferentiated hypobranchial musculature runs anteriorly to insert on the floor of the oral cavity, so it seems reasonable to propose that in this LCA this musculature pulled posteriorly on this floor to widen the oral cavity or open the mouth. In Mallat's (1996) view, the hypobranchial musculature must have shifted, in 'pre-gnathostomes', to a deeper location, in order to insert on the ventrolateral surfaces of the internal mandibular and hyoid arches. The LCA of gnathostomes, as well as the LCA of osteichthyans, probably had two hypobranchial muscles, the coracomandibularis and the sternohyoideus (Tables 10-12). Such a plesiomorphic condition is retained in adult actinistians as well as in adult dipnoans (Fig. 1; Table 12). As explained by Miyake *et al.* (1992), although the sternohyoideus remains undivided in some sharks, in numerous other elasmobranchs it is subdivided into a coracohyoideus, a coracoarcualis and occasionally (e.g., batoids) a coracohyomandibularis (Table 10). In extant cladistians, chondrosteans and halecomorphs the coracomandibularis is modified into a peculiar muscle branchiomandibularis connecting the branchial arches to the mandible (Table 11). A coracomandibularis/branchiomandibularis is missing in living ginglymodians and teleosts (Table 11). Therefore, contrary to what is sometimes stated in the literature, the geniohyoideus of tetrapods does not correspond to the protractor hyoidei of teleosts nor to any of its constituents (i.e. the intermandibularis posterior and the interhyoideus) (Tables 2, 5, 12). In extant tetrapods, there are various hypobranchial muscles that are not found in other extant osteichthyans. The geniohyoideus, genioglossus, hyoglossus, interradialis, styloglossus, palatoglossus and the instrinsic muscles of the tongue found in tetrapods correspond to the coracomandibularis of sarcopterygian fishes such as dipnoans; the tetrapod omohyoideus, sternothyroideus, sternohyoideus and thyrohyoideus are derived from the sternohyoideus of sarcopterygian fishes (Diogo *et al.* 2008b; Table 12).

GENERAL REMARKS: THE HEAD MUSCLES OF DIPNOANS AND THE COMPARATIVE ANATOMY OF VERTEBRATES

Of the 13 mandibular, hyoid, branchial and hypobranchial muscles present in adult dipnoans such as *Lepidosiren* (not including the branchial muscles *sensu stricto*; Tables 3, 6, 9, 12), six are present in adults of at least some of the non-osteichthyan taxa listed in Tables 1, 4, 7 and 10 (intermandibularis, A2, interhyoideus, protractor pectoralis, coracomandibularis and sternohyoideus), six are present in adults of at least some of the actinopterygian groups listed in Tables 2, 5, 8 and 11 (intermandibularis, A2, A3', interhyoideus, protractor pectoralis and sternohyoideus; the number could be raised to seven or eight if the 'dilatator laryngis' and/or the 'constrictor laryngis' of at least some actinopterygians would be considered to be homologous to the dilatator laryngis and/or constrictor laryngis of dipnoans, respectively), and eight are present in adults of at least some of the tetrapod taxa included in Tables 3, 6, 9 and 12 (A2, A2-PVM, interhyoideus, depressor mandibulae, protractor pectoralis, dilatator laryngis, constrictor laryngis and sternohyoideus). That is, phenotipically dipnoans share numerous muscular features with non-osteichthyan vertebrates, with actinopterygians and with tetrapods. This thus shows that dipnoans may effectively provide an important link to compare the head muscles of these three latter groups and thus to discuss the homologies and evolution of these muscles within the vertebrates as a whole.

Perhaps more important, dipnoans share some unique muscular synapomorphies with tetrapods, thus supporting the idea that these fishes are in fact the closest living relatives of the clade including extant amphibians and amniotes (see Diogo 2007). The extensive cladistic analysis of Diogo (2007) pointed out four unambiguous muscular synapomorphies supporting the close relationship between dipnoans and tetrapods: the presence of an adductor mandibulae A2-PVM (Fig. 2; Table 3); the presence of a levator hyoideus in at least some developmental stages (Fig. 3; Table 6); the absence of a recognizable adductor operculi in adults (Fig. 1; Table 6); and the absence of a recognizable adductor arcus palatini (Fig. 1; Table 6). Authors such as Rosen *et al.* (1981), Forey (1986) and Diogo *et al.* (2008a,b) have pointed out an additional muscular synapomorphy shared by dipnoans and tetrapods: the presence of a depressor mandibulae in at least some developmental stages (Fig. 1; Table 6; as explained above, this muscle is present in *Lepidosiren* and *Protopterus*, but not in *Neoceratodus*). Therefore, the study of dipnoans also provides crucial data to understand the origin and evolution of the tetrapod anatomical structures. The discovery of such distinct, clear muscular synapomorphies shared by dipnoans and tetrapods also stresses another important point: that comprehensive comparative studies of muscles provide not only useful information for functional, evolutionary and ecomorphological studies, but also crucial data to disclose the relationships between major vertebrate groups (see, Diogo and Chardon 2000; Diogo 2004a,b, 2007, 2008; Diogo and Abdala 2007; Diogo *et al.* 2008a,b).

Acknowledgements

We would like to thank T. Abreu, A. Zanata, F. Meunier, D. Adriaens, F. Wagemans, C. Oliveira, M. de Pinna, P. Skelton, F. Poyato-Ariza, T. Grande, H. Gebhardt, M. Ebach, A. Wyss, J. Waters, G. Cuny, L. Cavin, F. Santini, J. Briggs, L. Gahagan, M. Gayet, J. Alves-Gomes, G. Lecointre, L. Soares-Porto, P. Bockmann, B. Hall, F. Galis, T. Roberts, G. Arratia, L. Taverne, E. Trajano, C. Ferraris, M. Brito, R. Reis, R. Winterbottom, C. Borden, M. Chardon, P. Vandewalle, E. Parmentier, I. Doadrio, B. Wood, B. Richmond, P. Lucas, R. Knight, S. Devoto, S. Hughes and many other colleagues for their discussions on vertebrate anatomy, functional morphology, phylogeny and/or evolution. We are particularly grateful to Jørgen Mørup Jørgensen, Tobias Wang and Jean Joss for inviting us to participate in this project. R. Diogo received financial support from a 'Presidential Merit Fellowship' (George Washington University).

References

Abdala, V. and Moro, S. (2003). A cladistic analysis of ten lizard families (Reptilia: Squamata) based on cranial musculature. Russian Journal of Herpetology 10: 53-78.

Adams, L.A. (1919). A memoir on the phylogeny of the jaw muscles in recent and fossil vertebrates. Annals of the New York Academy of Sciences 28: 51-166.

Albrecht, P. (1876). Beitrag zur Morphologie des M. omo-hyoides und der ventralen inneren Interbranchialmusculatur in der Reihe der Wirbelthiere. Unpublished PhD thesis, Universität zu Kiel.Germany.

Alcock, R. (1898). The peripheral distribution of the cranial nerves of ammocoetes. Journal of Anatomy and Physiology 33: 131-154.

Allis, E.P. (1917). The prechordal portion of the chondrocranium of *Chimaera colliei*. Proceedings of the Zoological Society of London 1917: 105-143.

Allis, E.P. (1923). The cranial anatomy of *Chlamydoselachus anguineus*. Acta Zoologica 4: 162-219.

Allis, E.P. (1931). Concerning the mouth opening and certain features of the visceral endoskeleton of *Cephalaspis*. Journal of Anatomy 65: 509-527.

Anderson, P.S.L. (2008). Cranial muscle homology across modern gnathostomes. Biological Journal of the Linnean Society 94: 195-216.

Bartsch, P. (1992). On the constructional anatomy of the jaw suspension and the cranial base in the larva of *Neoceratodus forsteri* (Krefft, 1870). Zoologische Jharbucher. Abteilung fur Anatomie und Ontogenie der Tiere 122: 113-127.

Bartsch, P. (1994). Development of the cranium of *Neoceratodus forsteri*, with a discussion of the suspensorium and the opercular apparatus in Dipnoi. Zoomorphology 114: 1-31.

Bauer, W.J. (1992). A contribution to the morphology of the m. interhyoideus posterior (VII) of urodele Amphibiia. Zoologische Jharbucher. Abteilung fur Anatomie und Ontogenie der Tiere 122: 129-139.

Bemis, W.E. (1892). Studies on the evolutionary morphology of lepidosirenid lungfish (Pisces: Dipnoi). Unpublished PhD thesis, University of Berkeley, USA.

Bemis, W.E. (1986). Feeding mechanisms of living Dipnoi: anatomy and function. Journal of Morphology suppl. 1: 249-275.

Bemis, W.E. and Lauder, C.V. (1986). Morphology and function of the feeding apparatus of the lungfish, *Lepidosiren paradoxa* (Dipnoi). Journal of Morphology 187: 81-108.

Benton, M.J. (2005). Vertebrate Palaeontology. Blackwell Publishing, Oxford, UK.

Bischoff, T.L.W. (1840). Description anatomique du *Lepidosiren paradoxa*. Annales des Sciences Naturelles, Sér. 2, 14: 116-159.

Bridge, T.W. (1898). On the morphology of the skull in the Paraguayan *Lepidosiren* and in other dipnoids. Transactions of the Zoological Society of London 14: 325-376.

Butler, A.B. and Hodos, W. (2005). Comparative Vertebrate Neuroanatomy: Evolution and Adaptation. Wiley Interscience, Hoboken, USA.

Carroll, A.M. and Wainwright, P.C. (2003). Functional morphology of feeding in the sturgeon, *Scaphirhyncus albus*. Journal of Morphology 256: 270-284.

Cerny, R., Lwigale, P., Ericsson, R., Meulemans, D., Epperlein, H.H. and Bronner-Fraser, M. (2004). Developmental origins and evolution of jaws: new interpretation of "maxillary" and "mandibular". Developmental Biology 276: 225-36.

Cole, F.J. (1896). On the cranial nerves of *Chimaera monstrosa* (Linn. 1754); with a discussion of the lateral line system and the morphology of the chorda tympani. Transactions of the Royal Society of Edinburgh 38: 631-680.

Dawkins, R. (2004). The Ancestor's Tale – A Pilgrimage to the Dawn of Life. Houghton Mifflin, Boston, USA.

Diogo, R. (2004a). Morphological Evolution, Aptations, Homoplasies, Constraints, and Evolutionary Trends: Catfishes as a Case Study on General Phylogeny and Macroevolution. Science Publishers, Enfield, USA.

Diogo, R. (2004b). Muscles versus bones: catfishes as a case study for an analysis on the contribution of myological and osteological structures in phylogenetic reconstructions. Animal Biology 54: 373-391.

Diogo, R. (2007). On the Origin and Evolution of Higher-Clades: Osteology, Myology, Phylogeny and Macroevolution of Bony Fishes and the Rise of Tetrapods. Science Publishers, Enfield, USA.

Diogo, R. (2008). Comparative anatomy, homologies and evolution of the mandibular, hyoid and hypobranchial muscles in bony fish and tetrapods: a new insight. Animal Biology 58: 123-172.

Diogo, R. and Chardon, M. (2000). Homologies between different adductor mandibulae sections of teleostean fishes, with a special regard to catfishes (Teleostei: Siluriformes). Journal of Morphology 243: 193-208.

Diogo, R. and Abdala, V. (2007). Comparative anatomy, homologies and evolution of the pectoral muscles of bony fish and tetrapods: a new insight. Journal of Morphology 268: 504-517.

Diogo, R., Hinits, Y. and Hughes, S.M. (2008a). Development of mandibular, hyoid and hypobranchial muscles in the zebrafish: homologies and evolution of these muscles within bony fishes and tetrapods. BMC Developmental Biology (in press).

Diogo, R., Abdala, V., Lonergan, N. and Wood, B. (2008b). From fish to moderm humans – comparative anatomy, homologies and evolution of the head and neck musculature. Journal of Anatomy 213: 391-424.

Edgeworth, F.H. (1902). The development of the head muscles in *Scyllium canicula*. Journal of Anatomy and Physiology 37: 73-88.

Edgeworth, F.H. (1911). On the morphology of the cranial muscles in some vertebrates. Quarterly Journal of Microscopical Sciences New Series 56: 167-316.

Edgeworth, F.H. (1923). On the development of the hypobranchial, branchial and laryngeal muscles of *Ceratodus*, with a note on the development of ther quadrate and epihyal. Quarterly Journal of Microscopical Sciences New Series 67: 325-368.

Edgeworth, F.H. (1926a). On the hyomandibula of Selachii, Teleostomi and *Ceratodus*. Journal of Anatomy and Physiology 60: 173-193.

Edgeworth, F.H. (1926b). On the development of the coracobranchialis and cucullaris in *Scyllium canicula*. Journal of Anatomy and Physiology 60: 298-308.

Edgeworth, F.H. (1926c). On the development of the cranial muscles in *Protopterus* and *Lepidosiren*. Transactions of the Royal Society of Edinburgh 54: 719-734.

Edgeworth, F.H. (1928). The development of some of the cranial muscles of ganoid fishes. Philosophical Transactions of the Royal Society London (Biology) 217: 39-89.

Edgeworth, F.H. (1935). The Cranial Muscles of Vertebrates. University Press, Cambridge, UK.

Ericsson, R. and Olsson, L. (2004). Patterns of spatial and temporal visceral arch muscle development in the Mexican axolotl (*Ambystoma mexicanum*). Journal of Morphology 261: 131-140.

Forey, P.L. (1986). Relationships of lungfishes. Journal of Morphology suppl. 1: 75-91.

Fox, H. (1965). Early development of the head and pharynx of *Neoceratodus* with a consideration of its phylogeny. Journal of Zoology (London) 146: 470-554.

Gegenbaur, C. (1872). Untersuchungen zur Vergleichenden Anatomie der Wirbelthiere, III, Das Kopfskelet der Selachier. Engelmann. Leipzig, Germany.

Goodrich, E.S. (1958). Studies on the Structure and Development of Vertebrates. Dover Publisher, New York, USA.

Graham, A. (2003). Development of the pharyngeal arches. American Journal of Medical Genetics 119A: 251-256.

Greene, E.C. (1935). Anatomy of the Rat. Hafner Publishing Co., New York, USA.

Hatta, K., Schilling, T.F., Bremiller, R. and Kimmel, C.B. (1990). Specification of jaw muscle identity in zebrafish: correlation with engrailed-homeoprotein expression. Science 250: 802-805.

Hatta, K., Bremiller, R., Westerfield, M. and Kimmel, C.B. (1991). Diversity of expression of engrailed-like antigens in zebrafish. Development 112: 821-832.

Helms, J.A., Cordero, D. and Tapadia, M.D. (2005). New insights into craniofacial morphogenesis. Development 132: 851-61.

Herrel, A., Canbek, M., Özelmas, Ü., Uyanoglu, M. and Karakaya, M. (2005). Comparative functional analysis of the hyolingual anatomy in lacertid lizards. The Anatomical Record Part A 284: 561-573.

Holland, L.Z., Holland, N.D. and Gilland, E. *Amphioxus* and the evolution of head segmentation. Integrative and Comparative Biology. (In press).

Holland, N.D., Holland, L.Z., Honma, Y. and Fudjii, T. (1993). Engrailed expression during development of a lamprey, *Lampetra japonica*: a possible clue to homologies between agnathan and gnathostome muscles of the mandibular arch. Development, Growth and Differentiation 35: 153-160.

Holliday, C.M. and Witmer, L.M. (2007). Archosaur adductor chamber evolution: integration of musculoskeletal and topological criteria in jaw muscle homology. Journal of Morphology 268: 457-484.

Humphry, G.M. (1872a). Observations in Myology - The Myology of Cryptobranch, *Lepidosiren*, Dog-fish, *Ceratodus*, and *Pseudopus Palasii*, with the Nerves of Cryptobranch and *Lepidorsiren* and the Disposition of Muscles in Vertebrate Animals. Macmillan & Co., Cambridge, UK.

Humphry, G.M. (1872b). The muscles of *Ceratodus*. Journal of Anatomy and Physiology 6: 279-287.

Humphry, G.M. (1872c). The muscles of *Lepidosiren annectens*, with the cranial nerves. Journal of Anatomy and Physiology 6: 253-270.

Hunter, M.P. and Prince, V.E. (2002). Zebrafish Hox Paralogue group 2 genes function redundantly as selector genes to pattern the second pharyngeal arch. Developmental Biology 247: 367-389.

Iordansky, N.N. (1992). Jaw muscles of the Urodela and Anura: some features of development, functions, and homology. Zoologische Jharbucher. Abteilung fur Anatomie und Ontogenie der Tiere 122: 225-232.

Jarvik, E. (1963). The composition of the intermandibular division of the head in fishes and tetrapods and the diphyletic origin of the tetrapod tongue. Kungliga Svenska Vetenskapsakademien Handlingar 9: 1-74.

Jarvik, E. (1980). Basic Structure and Evolution of Vertebrates. Academic Press, London, UK.

Johanson, Z. (2003). Placoderm branchial and hypobranchial muscles and origins in jawed vertebrates. Journal of Vertebrate Paleontology 23: 735-749.

Jollie, M. (1982). Ventral branchial musculature and synapomorphies questioned. Zoological Journal of the Linnean Society 75: 35-47.

Kardong, K.V. (2002). Vertebrates: Comparative Anatomy, Function, Evolution, 3rd Edition. McGraw, Hill, New York, USA.

Kesteven, H.L. (1942-1945). The evolution of the skull and the cephalic muscles. Memoirs of the Australian Museum 8: 1-361.

Knight, R.D., Mebus, K. and Roehl, H.H. (2008). Mandibular arch muscle identity is regulated by a conserved molecular process during vertebrate development. Journal of Experimental Zoology (Molecular and Developmental Evolution) 310B: 355-369.

Köntges, G. and Lumsden, A. (1996). Rhombencephalic neural crest segmentation is preserved throughout craniofacial ontogeny. Development 122: 3229-3242.

Kuratani, S. (2004). Evolution of the vertebrate jaw: comparative embryology and molecular developmental biology reveal the factors behind evolutionary novelty. Journal of Anatomy 205: 335-47.

Kuratani, S. (2005a). Craniofacial development and the evolution of the vertebrates: the old problems on a new background. Zoological Science 22: 1-19.

Kuratani, S. (2005b). Cephalic neural crest cells and the evolution of craniofacial structures in vertebrates: morphological and embryological significance of the premandibular-mandibular boundary. Zoology 108: 13-25.

Kuratani, S. (2008). Evolutionary developmental studies of cyclostomes and the origin of the vertebrate neck. Development, Growth and Differentiation 50, Suppl. 1: 189-194.

Kuratani, S. and Ota, K.G. (2008). Primitive versus derived traits in the developmental program of the vertebrate head: views from cyclostome developmental studies. Journal of Experimental Zoology (Molecular and Developmental Evolution) 310B: 294-314.

Kuratani, S. and Schilling, T. (2008). Head segmentation in vertebrates. Integrative and Comparative Biology 48: 604-610.

Kuratani, S., Kuraku, S. and Murakami, Y. (2002). Lamprey as an evo-devo model: lessons from comparative embryology and molecular phylogenetics. Genesis 34: 175-83.

Kuratani, S., Murakami, Y., Nobusada, Y., Kusakabe, R. and Hirano, S. (2004). Developmental fate of the mandibular mesoderm in the lamprey, *Lethenteron japonicum*: comparative morphology and development of the gnathostome jaw with special reference to the nature of the trabecula cranii. Journal of Experimental Zoology (Molecular and Developmental Evolution) 302B: 458-68.

Kusakabe, R. and Kuratani, S. (2005). Evolution and developmental. patterning of the vertebrate skeletal muscles: perspectives from the lamprey. Developmental Dynamics 234: 824-834.

Lauder, G.V. (1980a). Evolution of the feeding mechanisms in primitive actinopterygian fishes: a functional anatomical analysis of *Polypterus*, *Lepisosteus*, and *Amia*. Journal of Morphology 163: 283-317.

Lauder, G.V. (1980b). On the evolution of the jaw adductor musculature in primitive gnathostome fishes. Breviora 460: 1-10.

Lauder, G.V. (1980c). The role of the hyoid apparatus in the feeding mechanism of the coelacanth *Latimeria chalumnae*. Copeia 1980: 1-9.

Lauder, G.V. and Liem, K.F. (1983). The evolution and interrelationships of the actinopterygian fishes. Bulletin of the Museum of Comparative Zoology 150: 95-197.

Lightoller, G.H.S. (1939). Probable homologues. A study of the comparative anatomy of the mandibular and hyoid arches and their musculature – Part I. Comparative myology. Transactions of the Zoological Society of London 24: 349-382.

Luther, A. (1913). Über die vom N trigeminus versorgte muskulatur des Ganoiden and Dipneusten. Acta Societatis Scientiarum Fennicae 41: 1-72.

Luther, A. (1914). Über die vom N trigeminus versorgte muskulatur der Amphibien, mit einem vergleichenden aublick über deu adductor mandibulae der Gnathostomen, und cinem beitrag zum verständnis der organisation der anurenlarven. Acta Societatis Scientiarum Fennicae 44: 1-151.

Luther, A. (1938). Die Visceralmuskulatur der Acranier, Cyclostomen, und Fische. A. Acranier, Cyclostomen, Selachier, Holocephalen, Ganoiden und Dipnoer. In: L. Bolk, E. Goppert, E. Kallius and W. Lubosch (eds.). Handbuch der vergleichenden Anatomie der Wirbeltiere, Volume 5. Urban & Schwarzenberg. Berlin, Germany. pp. 468–542.

Mallat, J. (1996). Ventilation and the origin of jawed vertebrates: a new mouth. Zoological Journal of the Linnean Society 117: 329-404.

Mallat, J. (1997). Shark pharyngeal muscles and early vertebrate evolution. Acta Zoologica 78: 279-294.

Manzanares, M. and Nieto, M.A. (2003). A celebration of the new head and an evaluation of the new mouth. Neuron 37: 895-898.

Marion, E. (1905). Mandibular and pharyngeal muscles of Acanthias and Raia. American Naturalist 39: 891–924.

Matsuoka, T., Ahlberg, P.E., Kessaris, N., Iannarelli, P., Dennehy, U., Richardson, W.D., McMahon, A.P. and Koentges, G. (2005). Neural crest origins of the neck and shoulder. Nature 436: 347-355.

McMahon, B.R. (1969). A functional analysis of the aquatic and aerial respiratory movements of an African lungfish, *Protopterus aethiopicus*, with reference to the evolution of lung ventilation movements in vertebrates. Journal of Experimental Biology 51: 407-430.

Millot, J. and Anthony, J. (1958). Anatomie de Latimeria chalumnae, I – squelette, muscles, et formation de soutiens. CNRS. Paris, France.

Miyake, T., McEachran, J.D. and Hall, B.K. (1992). Edgeworth's legacy of cranial muscle development with an analysis of muscles in the ventral gill arch region of batoid fishes (Chondrichthyes: Batoidea). Journal of Morphology 212: 213-256.

Noden, D.M. and Schneider, R.A. (2006). Neural crest cells and the community of plan for craniofacial development: historical debates and current perspectives. In: J. Saint-Jeannet (ed). Neural Crest Induction & Differentiation - Advances in Experimental Medicine and Biology. Volume 589, Landes Bioscience Georgetown, USA. pp. 1-31.

Northcutt, R.G. (2005). The new head hypothesis revisited. Journal of Experimental Zoology (Molecular and Developmental Evolution) 304B: 274-97

Olsson, L., Ericsson, R. and Cerny, R. (2005). Vertebrate head development: segmentation, novelties, and homology. Theory in Biosciences 124: 145-63.

Owen, R. (1841). Description of the *Lepidosiren annectens*. Transactions of the Linnean Society of London 18: 327-361.

Piatt, J. (1938). Morphogenesis of the cranial muscles of *Ambystoma punctatum*. Journal of Morphology 63: 531-587.

Piekarski, N. and Olsson, L. (2007). Muscular derivatives of the cranialmost somites revealed by long-term fate mapping in the Mexican axolotl (*Ambystoma mexicanum*). Evolution & Development 9: 566-578.

Roberts, T.D.M. (1950). The respiratory movements of the lamprey (*Lampetra fluviatilis*). Proceedings of the Royal Society of Edinburgh (Biology) 64: 235-252.

Rosen, D.E., Forey, P.L., Gardiner, B.G. and Patterson, C. (1981). Lungfishes, tetrapods, paleontology and plesiomorphy. Bulletin of the American Museum of Natural History 167: 159-276.

Santagati, F. and Rijli, F.M. (2003). Cranial neural crest and the building of the vertebrate head. Nature Reviews Neuroscience 4: 806-18.

Säve-Söderbergh, G. (1952). Notes on the trigeminal musculature in non-mammalian tetrapods. Nova Acta Regiae Scientiarum Upsaliensis 13: 821-839.

Shigetani, Y., Sugahara, F. and Kuratani, S. (2005). A new evolutionary scenario for the vertebrate jaw. Bioessays 27: 331-338.

Takio, Y., Pasqualetti, M., Kuraku, S., Hirano, S., Rijli, F.M. and Kuratani, S. (2004). Comment on: Lamprey Hox genes and the evolution of jaws. Nature 429: 1-2.

Trainor, P.A., Melton, K.R. and Manzanares, M. (2003). Origins and plasticity of neural crest cells and their roles in jaw and craniofacial evolution. International Journal of Development Biology 47: 541-453.

Wiley, E.O. (1979a). Ventral gill arch muscles and the interrelationships of gnathostomes, with a new classification of the Vertebrata. Journal of the Linnean Society (Zoology) 67: 149-179.

Wiley, E.O. (1979b). Ventral gill arch muscles and the phylogenetic interrelationships of *Latimeria*. Occasional papers of the California Academy of Science 134: 56-67.

Wilga, C.D., Wainwright, P.C. and Motta, P.J. (2000). Evolution of jaw depression mechanisms in aquatic vertebrates: insights from Chondrichthyes. Biological Journal of the Linnean Society 71: 165-185.

Wilson-Pauwels, L., Akesson, E.J., Stewart, P.A. and Spacey, S.D. (2002). Cranial Nerves in Health and Disease. B.C. Decker Inc., Hamilton, Ontario.

CHAPTER 8

The Dipnoan Dentition: A Unique Adaptation with a Longstanding Evolutionary Record

Moya Meredith Smith[1,2,*] **and Zerina Johanson**[3]

[1]King's College London, MRC Centre for Developmental Neurobiology, London SE1 1UL, UK

[2]King's College London, Dental Institute, London SE1 9RT, UK

[3]Department of Palaeontology, Natural History Museum,Cromwell Road, London, UK, SW7 5BD

ABSTRACT

Lungfish have a long evolutionary history, first appearing in the Early Devonian, with three genera extant. Lungfish dentitions were particularly diverse and have been a focus of study for many years. Although diverse, all dentitions can be derived from a toothplated dentition, where components of this dentition, in terms of tooth structures and the processes controlling development, have become dissociated and free to vary. Despite previous suggestions that lungfish dentitions are not homologous to dentitions in other sarcopterygian and actinopterygian taxa (osteichthyans), new research on *Neoceratodus forsteri* indicates several shared similarities in terms of genes involved in dental patterning, tooth origin and positioning on the jaw, and contribution of neural crest cells to tooth development. Future research should expand on these early results, and continue particularly to study genes from the 'core dental gene network', found in other fishes.

Keywords: dentitions, evolution, tooth pattern, *Necoceratodus*, *protopterus*, Devonian

**Corresponding author:* E-mail: moya.smith@kcl.ac.uk

INTRODUCTION

Here we will review the growth and construction of the lungfish dentition through evolutionary time of selected dipnoans both fossil and extant, but with a primary focus on the toothplated dentition. This represents the plesiomorphic lungfish dentition, and one from which all other lungfish dentitions can be derived, as all are based on some modification of the teeth characterising toothplated lungfish. The teeth of these in turn are homologous to other osteichthyans, although patterning on the dentate bones has been highly modified. The majority of lungfish species are supremely and uniquely adapted to crushing all food prey using paired, opposing tooth plates in upper and lower jaws (see Figure 1 for examples of these). This crushing action occurs several times between powerful jaws, equipped with these palatal and lingual tooth plates, before sucking in the partly regurgitated food mass to swallow. This behaviour is observed in the three genera now living in Australia, Africa and South American (Bemis and Lauder 1986). Although this is an extremely specialized dentition it can be shown to have formed early in the fossil record from a developmental dental pattern that has been conserved for 400 Myr since the early Devonian (Reisz and Smith 2001) and as the closest living relatives of Tetrapoda this raises some fundamental questions (Ahlberg *et al.* 2006; Hällstrom and Janke 2009).

Exactly how did the evolutionary process of creating such a unique form of dentition come about, early in the history of the group? This has been hard to reconcile with the stereotypical osteichthyan pattern. Both the arrangement of teeth and the fact that all teeth are retained throughout life contrasts with the osteichthyan dentition (including tetrapods), with teeth arranged in conventional marginal linear rows. In the osteichthyan tooth rows, loss of the old tooth usually occurs in a regulated way and is linked with timed addition of teeth in each alternate tooth position in the jaws. Undoubtedly the evolutionary change characterizing lungfish dentitions occurred through extensive modification of the developmental programme of the stereotypic osteichthyan dentition, with its marginal and palatal rows of replaceable unitary teeth. Nevertheless, similarities to the osteichthyan dentition can be demonstrated and some of these are conserved in the early pattern of the dentition (Smith *et al.* 2009).

A generalised model for the development of lungfish tooth plates was proposed, through a study of the pattern of tooth development and growth in the larval tooth plates of the African species *Protopterus aethiopicus* and by comparison with *Neoceratodus forsteri*, the extant Australian lungfish (Smith 1985). At that time this accounted for all published data (Kemp 1977, 1979) but many papers on *Neoceratodus* have since been published. However, recent insights into developmental regulation of the patterning process have been offered by *in situ* gene expression data combined with skeletal preparations, in a new analysis of the early larval stages of *Neoceratodus forsteri* (Smith *et al.* 2009). We also now know that despite earlier reports to the contrary (Kemp 1995b), very early in embryonic

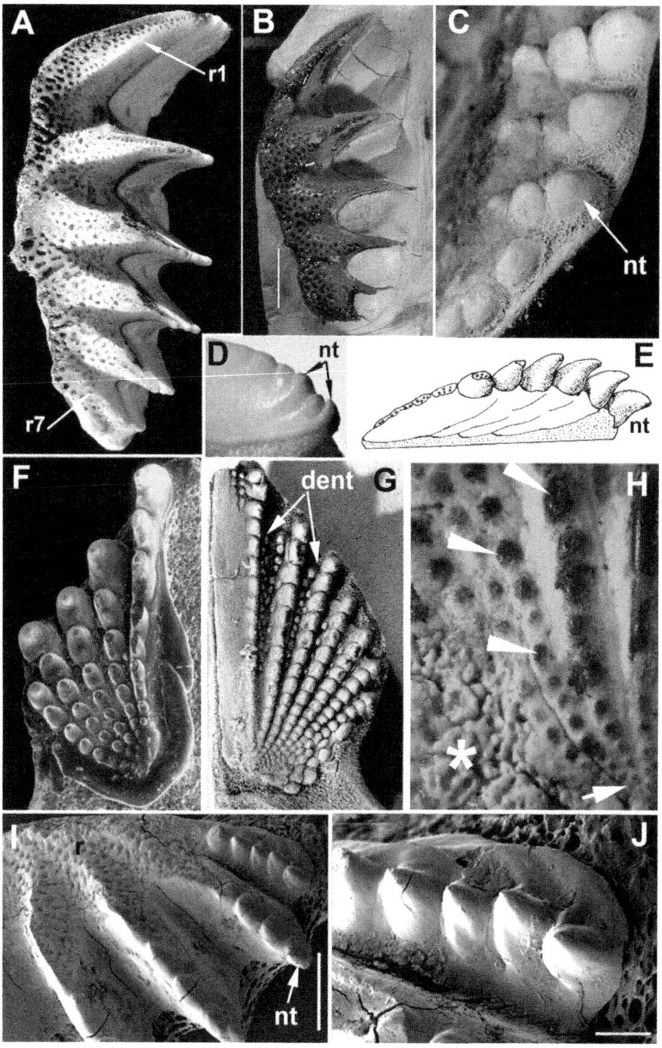

Fig. 1 The morphology of extant and fossil tooth plates to illustrate the separate teeth aligned into rows radiating antero-laterally from the medial aspect of each left and right plate, until they fuse to form either flat or ridged surfaces. A, B, D *Neoceratodus forsteri* right adult tooth plate with 7 ridges (r1, r7), worn punctate (pulp canal openings) surface of petrodentine with tiny teeth (nt) confined to the lateral margins and enamel bands covering them, and as seen in medial view of 1st ridge (D); all of this is covered by epithelial tissue in life (B). C, E are Devonian forms to show a series of whole teeth (nt) added in sequence to the lateral margin of each ridge; the 4 most posterior tooth rows of *Adololopas* in palatal surface view (C) and, medial view of 1st ridge in *Dipterus* with worn small teeth medially, new larger teeth laterally, covered by enamel bands (E), for direct comparison with that of *Neoceratodus*

Fig. 1 Contd..

development, neural crest cells participate in tooth development and this odontogenic fate of crest-derived mesenchyme (ectomesenchyme) is conserved at least from lungfish to tetrapods (Kundrat *et al.* 2008).

It is especially informative to compare detailed data on tooth development in extant forms with early growth stage information obtained from Devonian juvenile forms and late growth stages in the adult (Bemis and Lauder 1986; Reisz and Smith 2001; Smith and Krupina 2001; Smith *et al.* 2002). This type of data on developmental pattern can be obtained from amongst fossil dipnoans as the dentition is retained for life without shedding any of the separate teeth so that each tooth plate retains a history of its development (Denison 1974; Smith 1986) (see Figure 1F, G, H illustrating Devonian juvenile and adult plates of *Andreyevichthys epitomus* and *Gogodipterus paddyensis*). Adults have tooth plates on the palate and the lingual side of the lower jaws (prearticular), formed from anterior or lateral addition of teeth to radial rows diverging from the oldest teeth at the medial side of each (see Figure 1C, D, E for examples of Devonian and extant forms with new teeth added at the lateral margins). It has long been understood that individual teeth are consolidated into dental plates, some called tooth plates, without loss of teeth through shedding (statodont dentition: compare juvenile and adult form of the Devonian *Andreyevichthys* in Figure 1F, G). These retained teeth (Figure 1F, G, I, J), or flattened and worn tooth tissues, (as in *Neoceratodus* adult, Figure 1A, B) have extensive and continuous growth of their dentine below the biting surface. Not only does the dentine grow invasively into the bone (Smith 1985, 1986) but also it is formed of a special and unique extra hard type called petrodentine (Smith 1984). In this statodont dentition all teeth are retained and those formed at the earliest stages of construction of the tooth plates (as in the Devonian juvenile and adult plates, Figure 1F, G) were patterned in early development by regulatory systems (several aspects of these that are in common with other osteichthyan fishes are discussed below). These first teeth are often left *in situ* to allow one to interpret order of tooth initiation from their relative sizes, degree of wear and retained positions (Figure 1H). By comparison between the Late Devonian hatchling tooth plates of *Andreyevichthys epitomus* and the living form *Neoceratodus forsteri* (Smith and Krupina 2001; Smith *et al.* 2002), it was proposed that developmental

Fig. 1 Contd..

(D). F–H are Devonian fossils, which in juvenile *Andreyevichthys* (F) show unworn teeth in five radiating rows from a medial origin with increased size along each row, and the adult (G) where these tooth rows are fused and worn into 8+ ridges, with denticles (dent) formed between them in the furrows where epithelium would lie in life: posteromedial part of tooth plate of *Gogodipterus* (H) where the worn surface reveals the first formed, smallest teeth still in their original arrangement (arrow), but worn with cores of petrodentine (arrow heads), only random clusters of denticles form at the posterior margin (asterix), after minimal resorption of the tooth plate. I, J are scanning electron micrographs of *Neoceratodus* juvenile plate with separate large, unworn enamel covered teeth (nt) at the lateral ends of the posterior ridges (J), merging with the worn ridges (r) and punctate surface of the medial part of the plate (I).

constraints conserve this evolutionary pattern. That is, one where teeth are only added to one end of each radial row. This particular pattern has been conserved since the Early Devonian form *Diabolepis speratus*, recognized as the sister group of the 'undisputed' stem+crown dipnoans (Ahlberg *et al.* 2006, also see Johanson and Ahlberg, this volume). It has recently been proposed from the study of early tooth patterning in the lower jaw of *Neoceratodus* (Smith *et al.* 2009) that the first three teeth in the marginal row form in a pattern linked with that of stereotypic osteichthyans (these are illustrated in Figure 2C-H). This provides the basis for how this ancient dipnoan pattern was modified very early in its evolutionary history, at least in the lower jaw marginal pattern, and how this may have evolved is explored further in the sections below.

I – Building a Dentition Within Developmental Constraints to a Regulated Pattern

Growth by tooth addition can be interpreted from exclusively fossil groups, especially placoderms and acanthodians, according to the developmental sequential addition model (SAM theory) (Smith 2003) and these dentitions can be compared with those of extant chondrichthyan and osteichthyan dentitions that replace all teeth. From this analysis it has been proposed that for each specific pattern (chondrichthyan, or osteichthyan), the developing dentition has evolved independently and is unique for each higher-level clade of jawed vertebrates (Smith 2003; Smith and Coates 1998; Smith and Johanson 2003). However significantly, that of dipnoans has also emerged as distinctive from other osteichthyans and for the most part unique. The general model proposed for all osteichthyan dentitions (Smith 2003) is that each dentate region starts the process with a primordial tooth for each dentate bone on the jaws, followed by a strict pattern order on each bone for the first three teeth, or triad (see Figure 2E, F for larval *Neoceratodus*). Tooth initiation after that conforms to the stereotypic pattern for each toothed bone.

Kemp (1977) discussed the classic osteichthyan model to explain patterning of the dentition in the extant lungfish (*Neoceratodus forsteri*), where at embryonic stage 52, 4 radial rows of teeth have developed and a total of 15 teeth contribute to one upper tooth plate on the pterygoid bone. These first teeth fuse to each other and to the supporting bone forming the radial ridges of the composite tooth plate. An attempt was made to fit this pattern of tooth initiation into an order such as described for osteichthyan tooth families and tooth rows (Kemp 1977). We suggest that in the dentition of most lungfish, the rows form from one series of teeth without an alternate position tooth series and without families; thus without replacement teeth, as would be provided by tooth families for each odd and even position. Kemp (1977, 1979) provided a comprehensive account of early development and growth of the dentition in the Australian lungfish, and recognised that in each dental field there was a single or primordial tooth, before

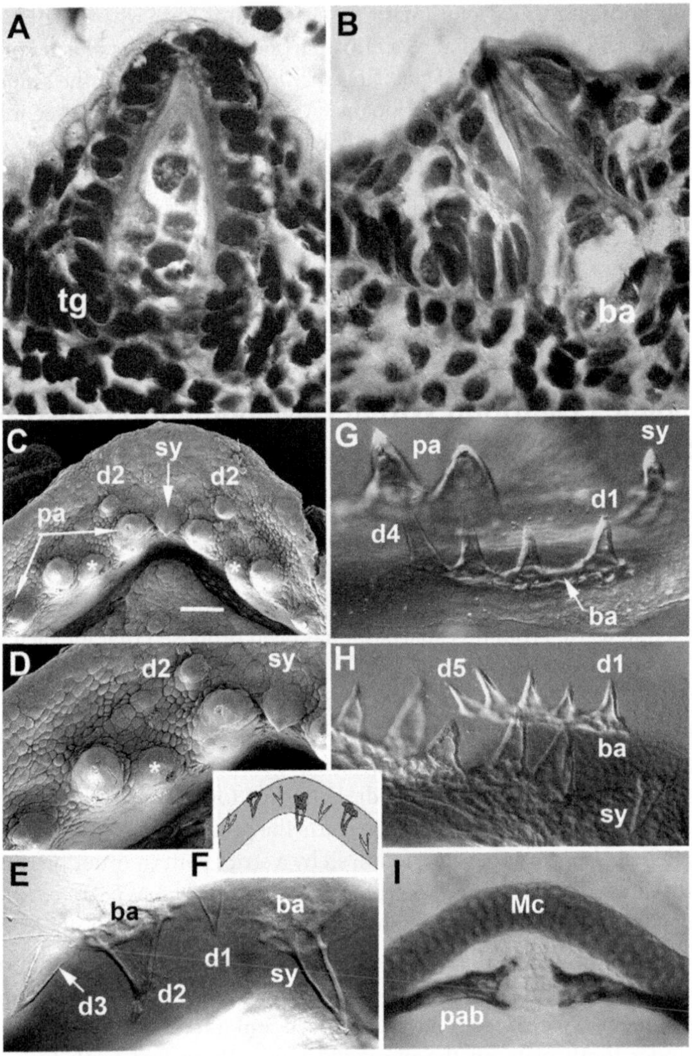

Fig. 2 Dentition of *Neoceratodus forsteri*, to illustrate the earliest stages of its formation in the lower jaw during larval development, tooth plate and outer marginal row both form from separate teeth. A, B are histological sections of stages 43 and 44 embryos, early teeth as a cone of dentine enclosing the pulp cells and adjacent early tooth germ cells (tg) in the oral epithelium containing yolk granules, with bone of attachment formed last (ba). C, D, scanning electron micrographs of oral surface of whole lower jaw at stage 43 and close up view of the left side, one symphyseal tooth (sy), two dentary teeth left and right at jaw position two (d2) and four prearticular teeth (pa); the first tooth to form in each region is the pioneer tooth (d2 and *). E–I, cleared Alcian blue stained skeletal preps of the whole, or one side of lower jaw, slide mounted for optical microscopy, show stages in developmental time of teeth added to

Fig. 2 Contd..

any others were added to make a radiating row pattern (Figure 2C, D, G, H). All row teeth are added after the first three teeth (Figure 2G, H), or triad proposed by Smith (2003) conforming to the SAM model. In the extant lungfish the vomerine tooth plates form as single rows as if they are restricted to forming just one of the radial rows on the palatal tooth plates, and also in the lower jaw field, first one, then three teeth form (Figure 2C, D, G, H) but these are aligned along the labial margin of Meckel's cartilage (the 'marginal dentition', Figure 2E). The first primordial teeth with a thin cone of dentine are formed very superficially in the oral epithelium, mostly from endoderm with many yolk granules in the cytoplasm, and the next tooth bud is formed within the dental epithelium of this tooth (Figure 2A). Later teeth have also formed the bone of attachment below the dentine (Figure 2B, G). Like the trout the first teeth are initiated from the odontogenic epithelium without a dental lamina forming (Fraser *et al.* 2004) and successive teeth do not form at all in the lungfish.

A historical review of the early papers on development of the snout of lungfish was given by Bertmar (1966) in which he also made exceptionally detailed reconstructions of the chondrocranium, teeth, blood vessels and nerves in *Neoceratodus forsteri*, *Protopterus annectens* and *Lepidosiren paradoxus*. He refers to the loss of the premaxilla and maxilla and their teeth, but the vomer just anterior to the pterygo-palatine has first one tooth then more are added as a single row just as all the tooth plates grow. In *Neoceratodus* he figures the developing blastema with teeth on the lower jaw, anterior to the prearticular tooth plate but lacking dermal bone, he called either the splenial, or dentary ('marginal dentition'). The development of marginal teeth on the lower jaw in *Neoceratodus*, has been discussed extensively by Kemp (1977, 1979, 1995a, 2002b) and more recently its significance debated by Smith *et al.* (2002, 2009). The addition of teeth to the dentary row, after the first three have formed is by proximal addition one at a time from four to seven (Figure 2D, F). These lower jaw teeth are also present in the Devonian fossil *Andreyevichthys epitomus* and like those of *Neoceratodus* are transitory, functional only in the hatchling and then lost by a process of resorption of the whole row of teeth (Smith and Krupina 2001; Smith *et al.* 2002). Smith *et al.* (2002) concluded

Fig. 2 Contd..

the marginal row situated above Meckel's cartilage (anterior toward top of page); E, stage 44, oral view shows order of time of development of first three teeth, oldest (d2, sy) have both added the bone of attachment below the dentine cone, very small dentine cone at jaw position one is youngest (d1) and larger at position three (d3) but neither has added bone to the dentine cone; F, drawing of both sides as shown in (E). G, stage 47 right side, anterior view in which fourth tooth (d4) has formed lateral to first 3 teeth, all joined together from d1 by their bone of attachment (ba, dentary bone development does not occur), symphyseal tooth (sy) and larger prearticular teeth (pa) on the oral side of the marginal tooth row. H, stage 50, oral view left side, five marginal row teeth (d1–d5) joined by their bone of attachment (ba). I, stage 47 midline view of lower jaw Meckel's cartilage (Mc) and prearticular bone (pab) after cyclopamine treatment at stage 42 for 8 days, during initiation of marginal teeth, none has appeared where normally five would develop.

that the programme of development, with teeth added only to the proximal end and then resorption starting from the distal end, is conserved for a substantial portion of lungfish evolutionary history.

In other osteichthyans, teeth are arranged in alternate rows at the margins, and a tooth family forms by sequential addition for each jaw position (Smith 2003), these replace the functional teeth at staggered times in alternate sites. In this sequential addition model (SAM) the developmental program is based on one primordial tooth for each dentate bone (Smith 2003), then two teeth are added either in adjacent positions, or alternate (Huysseune and Witten 2006) to an osteichthyan pattern. Although in the lungfish continuous tooth addition occurs in the growth phases at each end of the radiating rows, the replacement mode is missing (Kemp 1977; Smith 1985). Despite these differences the earliest pattern of tooth initiation of the marginal dentition in the lower jaw appears to be based on the osteichthyan one (Smith *et al.* 2009). However, after this, the pattern of tooth addition becomes one that is unique for dipnoans, whether the marginal dentition on the lower jaw cartilage, or on the palatal and lingual bones. Earlier, Smith (1985) suggested that the addition of teeth is patterned in a different way to that of stereotypic osteichthyans, for the statodont lungfish dentitions, and this is true for most of tooth development, except for the early ontogenetic stages. By comparison with the most phylogenetically basal dipnoan, *Diabolepis speratus*, also known from juveniles (Chang 1995), Smith and Krupina (2001) suggested that this pattern was derived within lungfish and conserved through evolution from the earliest known lungfish to the living forms. This developmental model is based on a radial pattern of successive tooth initiation in lungfish, where new teeth are only added to the antero-lateral end of each tooth row. Nevertheless, the restriction of new tooth sites to this lungfish pattern is probably regulated to some degree by an odontogenic gene network, as a periodic pattern generator that allows for diversity in dentitions (Fraser *et al.* 2008). This research showed that co-ordinated patterns of gene expression differ among species of lake Malawi cichlids and these prefigure the distribution of functional teeth in their spacing, size and number of rows.

It is axiomatic that the oral epithelium controls the pattern of a dentition for positioning of teeth and the timing of all their replacements at the initiation stage of their development (Lumsden 1988). Spatial and temporal co-localisation of molecular markers within this epithelium regulates the position of teeth prior to any morphological differences. These could show the earliest timing of odontogenic competence in a real time sequence. Implications for developmental models are profound and involve consideration of the germ layers ectoderm and endoderm to form the 'tooth competent' oral epithelium. Together with neural crest derived mesenchyme (ectomesenchyme), these epithelial cells impart pattern differences through regulation of gene expression of the primary signal to competent ectomesenchymal skeletal cells. Recent data from studies on cichlid dentitions (Fraser *et al.* 2008) have determined the combinatorial epithelial expression of *pitx2* and *shh* that govern both the location of tooth sites and of the additional tooth rows. These vary amongst the cichlid species where the gene expression prefigures the ordered distribution of the functional teeth. Both these genes (*shh* and *pitx2*) are expressed at the patterning stage of the dentition in the Australian lungfish (Smith *et al.* 2009) and *shh* is shown to be required for tooth initiation as use of cyclopamine, given to the *Neoceratodus* embryo at the earliest practical stage results in no teeth forming in the dentary field and none added to earlier formed prearticular teeth (Figure 2I). This reflects a shared developmental process at the level of the tooth module, explained as a core dental gene network (Fraser *et al.* 2009), but the unique way these modules are patterned in the lungfish dentition may depend on other genes, both epithelial and ectomesenchymal. Epithelial *wnt7b* and mesenchymal *eda* have been identified in cichlids as likely to regulate the spacing of the *shh*-positive unit teeth (Fraser *et al.* 2008), but data is currently lacking for *Neoceratodus*. Developmental modifications in the gene networks for iterative patterning could result in the distinctive dipnoan pattern and functional divergence from that of osteichthyans.

In addition, use of vital markers for neural crest cells shows for the first time that the mesenchymal population to make teeth does come from the neural folds in early embryogenesis. Experimental evidence that the neural crest participates in tooth development in any fish has so far been lacking, except for one study on *Neoceratodus* (Kundrat *et al.* 2008). Based on homologies of the tooth module in lungfish, tetrapods and ray-finned fish (actinopterygians, all three taxa grouped in the Osteichthyes), cranial neural crest does contribute to the dental ectomesenchyme in all these taxa (Kundrat *et al.* 2008).

Future work should obtain spatiotemporal gene expression data in all tooth fields of the embryo *Neoceratodus*, including for the three genes found to be expressed in trout odontogenesis for initiation of competence in oral epithelium of *shh* and *pitx2,* coincident with focal expression of *bmp4* in the reciprocal tissue dental mesenchyme. Other mesenchymal genes act in feed back mechanisms, such as *msx1* (Chen *et al.* 1996) and *bmp4,* which together propagate activation of tooth sites in the epithelium. Spatial and temporal patterning of the tooth induction sites has been found in mice to involve the transcription factor odd-skipped related-2 (*Osr2,* (Zhang *et al.* 2009). The *Osr2* mutant mice had a buccolingually expanded odontogenic field for tooth development, resulting in additional rows to the normal single row. Restriction of the dipnoan dentition to a single row on the lower jaw and to new teeth added only on the lateral or anterior side of the palatal tooth plates may involve some of these molecular mechanisms discovered for mammalian single rows of teeth.

II – Dentition Building Modules from the Developmental Repertoire are Used in Fossil Dentitions

Although our focus is on the toothplated dentition (Figure 1), it is important to note that a high diversity of lungfish dental morphologies existed during the Devonian and were present in a range of the earliest known lungfish from the Early Devonian, linking this diversity to the origins of the group. These morphologies include the toothplated dentition with rows of retained teeth and lateral addition of new teeth, in taxa such as *Diabolepis speratus* (Chang 1995; Chang and Yu 1984; Smith and Chang 1990) and *Tarachomylax oepiki* (Barwick *et al.* 1997). A dentition characterised by widespread denticles was found in the Early Devonian *Uranolophus wyomingensis* (Denison 1968) and a dentine-plated dentition, where the dental surfaces were covered in thick sheets of dentine, was seen in *Dipnorhynchus sussmilchi* (Thomson and Campbell 1971). Denison (1974) noted that experimentation with form, rather than stability, was a well-known characteristic in the early history of many groups, and would certainly be applicable to the lungfish. Denison divided lungfish dentitions into those with toothplates ('dental plates'), and those without, the latter including those characterised by the dentine sheets and by denticulated surfaces (Figure 3). He also noted that it was unlikely that these denticulated and dentine-plated dentitions could have been derived or evolved from the toothplated dentitions and suggested instead that the denticulated dentition was ancestral, with toothplated taxa arising more than once from this group. The primitive nature of denticles was based on observations that they occurred on the palate and gill arches of many fishes (Denison 1974; Miles 1977). Notably, Denison suggested that teeth arose from denticles that enlarged and became arranged into rows (i.e., not homologous to osteichthyan teeth; also (Campbell and Barwick 1991). The unique nature of the lungfish teeth was also stressed by White (1966), who described the independence of the lungfish dentitions from other sarcopterygian dentitions, stating that "these tooth-plates have no homologues among the Rhipidistia or the coelacanths...".

By comparison, Smith (1986) attempted to apply the model for the primitive gnathostome dentition as proposed by Reif (1982) to explore if it could explain the terminological ambiguity in these previous descriptions. That is, both oral denticles and teeth would be present at the origin of the gnathostomes and their relative distribution and presence or absence on the lungfish dental plates would account for two different developmental and morphological types, distinguished as 1) dentine plates (in which Smith included denticulated and dentine-plated dentitions) and 2) tooth plates. By accepting that denticles (non-spatially patterned) and teeth (ordered in space and time) could both exist as differently patterned developmental modules in construction of the dipnoan dental plate, it was possible to propose how the different morphological types could be derived from each other. Essentially, Smith (1986) accepted the definition of Reif (1982)

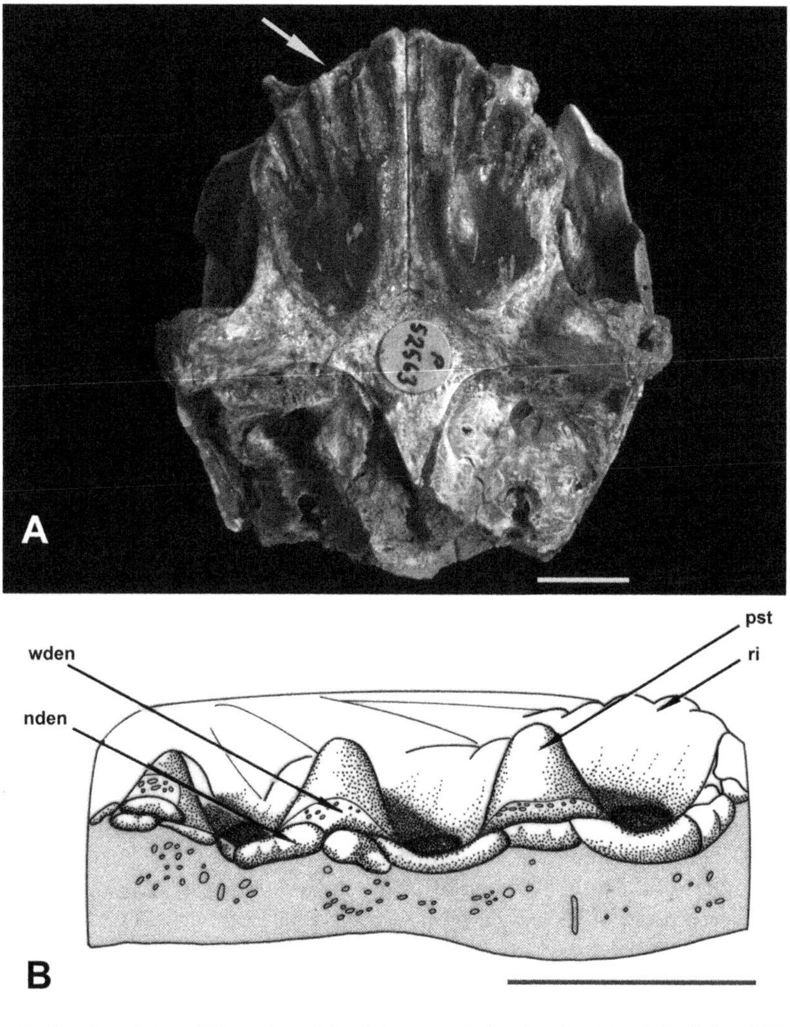

Fig. 3 Dentine plates of Devonian *Chirodipterus australis* showing lateral denticle addition. A, NHM P52563, white arrow indicates region shown in (B), five ridges and posteromedial plateau are regions of worn petrodentine (black). B, marginal view, showing addition of new dentine as denticles (nden) and 'pseudoteeth' (pst), inset from margin, forming ridges (ri), with worn denticles (wden) on their flanks. (From Smith and Campbell 1987.)

that different developmental pathways distinguish denticles, as those developed superficially on demand when space allowed and teeth, as those developed sequentially from a deep epithelial invagination (dental lamina), in advance of available space or need. Although the radial rows of sequentially added teeth are the dominant feature of the palatal tooth plates in the Early Devonian *Diabolepis speratus* (Chang 1995; Chang and Yu 1984; Smith and Chang 1990) and the Upper Devonian *Andreyevichthys epitomus* (Smith and Krupina 2001), denticles can appear between the rows of teeth in the adult dentition and seem part of the wear related process of addition (Figure 1G), or in response to resorption, as described further below.

Although the ideas of Smith (1986) were criticised (e.g., Campbell and Barwick 1991), it is imperative to note that lungfish are sarcopterygian and osteichthyan fishes, groups whose dentitions are characterised by individual teeth, albeit replacing ones. This includes taxa within the group Dipnomorpha, the sarcopterygian group to which lungfish are assigned (Ahlberg 1991). As noted above, *Diabolepis speratus* (Chang 1995; Chang and Yu 1984; Smith and Chang 1990) possesses a toothplated dentition and is resolved in most phylogenetic analyses as either, the most basal lungfish or, the sister group to the lungfishes. As discussed in the first section teeth in the lungfish toothplated dentitions are homologous in terms of structure, early developmental stages, neural crest contribution and gene expression, to other osteichthyan teeth including tetrapods; (Kundrat *et al.* 2008; Smith 1985; Smith and Krupina 2001; Smith *et al.* 2009). These observations strongly suggest that the plesiomorphic dentition of the Dipnoi is the toothplated form and that these teeth are homologous to osteichthyan teeth. Moreover, although there is a high diversity of lungfish dentitions in the Devonian, other lungfish dentition types are based on, or can be derived from, modification of these teeth and the tooth development program.

For example, Ahlberg *et al.* (2006) presented a simple diagram showing that the wide range of lungfish dentitions could be recovered by varying rates of only two processes: odontogenesis, or tooth formation, and tooth resorption, both characteristic of the typical osteichthyan dentition. Additive formation includes individual tooth development, with whole dentitions characterised by spacing, size and numbers of the unit teeth into rows. Resorption occurs in osteichthyan teeth during the process of individual tooth replacement. Although integrated, these processes should be viewed as potentially developmentally separate, and regulated by genes as discussed above. As well, the processes linked with the development of the individual tooth and its components (bone, dentine and enamel) should also be viewed as potentially separate, with respect to the lungfish dentition.

For example, an unusual type of dentition is found in the Devonian lungfish *Chirodipterus australis* (Figure 3), and is referred to as the dentine-plated dentition (Campbell and Barwick 1990; Smith 1986; Smith and Campbell 1987). Here, individual, discrete elements (denticles, blisters) are added to the periphery of the dentition (Figure 3B), but also internal to the lateral margin of the plate, tuberosities form that lack the enamel cover associated with osteichthyan teeth (Figure 3B). These were identified as 'pseudoteeth' (Ahlberg *et al.* 2006), and represent the ongoing production of individual dentine units in the form of a hard durable variety (petrodentine). These 'pseudoteeth' result from patterned deposition of petrodentine as if they were separate teeth, but within the bone of the plate and not added to the plate margin. This is where the tooth forming tissues are located in other osteichthyans, and in the toothplated lungfish (Campbell and Barwick 1998; Smith 1986).

As well, formation of individual lungfish dental elements and the patterning of these can be completely lost, along with loss of individual tooth components such as enamel, resulting in the production of a variety of dentine morphologies among Devonian lungfish. Indeed, dentine seems to be among the most consistently present yet variable tooth component in Devonian lungfish. For example, dentine blisters lack the patterning and positioning associated with teeth or 'pseudoteeth' (Figure 3B), and are seen in the early Devonian taxon *Speonesydrion iani* (Campbell and Barwick 1984; Smith 1986), as well as *Chirodipterus* (Smith and Campbell 1987), representing another contribution to dental diversity at this time. Broad flat areas (dentine bands) occur around the posterior margins of the dentition of Late Devonian *Adololopas moyasmithae* (Campbell and Barwick 1998) and sheets of dentine occur between the ridges in *Chirodipterus australis* (Smith and Campbell 1987). Flat dentine sheets characterise taxa such as the early Devonian *Dipnorhynchus* (several species) and *Sorbitorhynchus deleaskitus* (Campbell and Barwick 1990; Thomson and Campbell 1971; Wang *et al.* 1993). In *Dipnorhynchus*, dentine could also form large, irregularly shaped and positioned tubercles on the dental surface of the dentine-plated type (Campbell and Barwick 2000; Smith 1986).

Although the denticulated dentition in taxa such as *Uranolophus* and *Griphognathus* has been identified as forming a separate dental group within the lungfish (Campbell and Barwick 1990; Cloutier 1996), one interesting feature of the dipnoan dentition is that widespread occurrence of denticles appears in conjunction with wear or resorption (Figure 4). For example, the Devonian form *Holodipterus* was compared with tooth plated and denticle plated models and assigned by Denison (1974) to the group lacking dental plates. The study of histology and growth in *Holodipterus* by Campbell and Smith (1987) described massive areas of resorption that had occurred to remodel the dentine and bone of the plate surface and the taxon was identified as a denticulated form. However, (Smith 1986) called this dentition 'pseudoplates' and considered that they still retained radial rows of teeth at the lateral margins (Figure 4A, C, D), in other words, that resorption was a progressive process used to remodel the functional plate surface.

In lungfish, resorption processes have been decoupled from tooth replacement and are not timed to occur only with individual teeth but with the whole tooth row (see first section (Smith and Krupina 2001; Smith *et al.* 2002)). Resorption fronts can be seen in toothplated dentitions (Figure 4A, C, D) such as in the various species of *Holodipterus* (Campbell and Smith 1987; Pridmore *et al.* 1994), *Fleurantia denticulata* (Cloutier 1996), and *Melanognathus canadensis* (Schultze 2001).

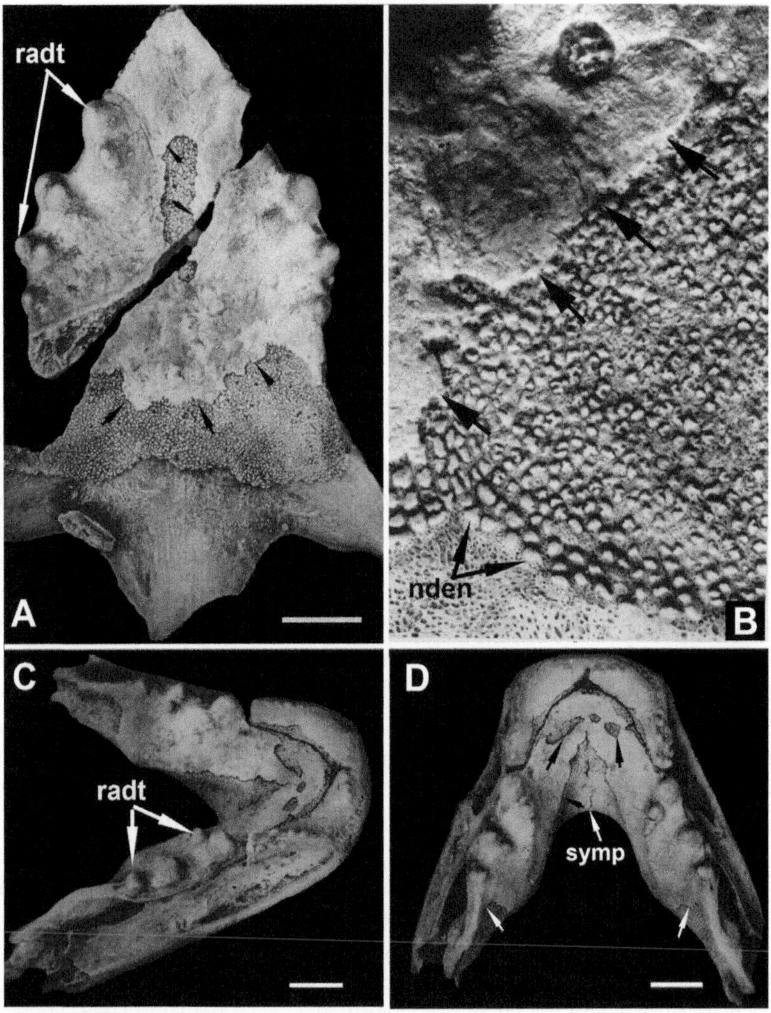

Fig. 4 Toothplates of *Holodipterus* with radial tooth rows, but extensive resorption and denticle covering. A, palatal (upper) dentition showing tooth rows (radt) and resorption occurring medially and posteriorly (black arrows, on the older worn tissue). B, closeup of resorption (black arrows) along an irregular margin, and new denticles added (nden). C, D, lower jaws in (C), lateral and (D), dorsal view, showing teeth in radial rows (radt) and resorption (black, white arrows), extensive at symphysis (symp) and posteriorly. (From Campbell and Smith 1987.)

Resorption here removes tooth rows along a broad and irregular resorption front, with rounded denticles being deposited on the newly exposed surfaces (Figure 4). Resorption of tooth rows also occurs in *Speonesydrion*, but the replacing denticles are of a vermiform morphology (Campbell and Barwick 1984). Resorption and deposition together consistute the bone remodelling process, where normally, resorbed bone is replaced by newly forming bone, in a tightly regulated association between the osteoblasts and osteoclasts, with each controlling the activity of the other at various points in the bone remodelling cycle (Henriksen *et al.* 2009; Martin *et al.* 2009; Witten and Huysseune 2009). Osteoblasts direct osteoclast activity during resorption, but osteoblasts responsible for depositing new bone, are subsequently directed by the osteoclasts (Martin *et al.* 2009). This tight association ensures that bone replacement will equal bone resorption. Remodelling can also occur in dermally derived structures such as fish scales (Persson *et al.* 1995; Witten and Huysseune 2009). Scale resorption has been linked to sexual maturity and spawning, with scale repair not occurring until after spawning. Interestingly, the presence of osteoclasts in resorption pits on these scales preceded the activity of the osteoblasts responsible for scale repair (Persson *et al.* 1995), indicating the activity of these cells may be linked as described for bone (Henriksen *et al.* 2009). In lungfish dentitions, in taxa such as *Speonesydrion, Holodipterus, Fleurantia* and *Melanognathus*, resorbing activity has been dissociated from the development of teeth at the lateral margins of the dental plate and instead resorption fronts are active medially (Figure 4A, B). These resorbing areas progress across the tooth plate, but we suggest they maintain the tight association with the matrix forming cells responsible for depositing new material, in much the same manner as described above. One important difference, however, is that since it is dentine of the tooth plate that has been predominantly resorbed, it is dentine that is redeposited, in the form of denticles onto the resorbed surface (Figure 4B). Although dentitions such as *Holodipterus* (Figure 4) and *Melanognathus* appear unusual in the massive amount of resorption, the processes are largely the same as in other vertebrates.

As noted above, recent work on cichlid fishes established a core gene network for dentitions, involving several genes that interact in the developing tooth module as a deep homology, but how the tooth module is regulated in space and time has not yet been fully resolved (Fraser *et al.* 2008, 2009). In osteichthyans, genes in this network are regulated and integrated to produce and maintain a functional dentition involving both tooth initiation and their replacement. Following Ahlberg *et al.* (2006) we suggest that parts of this network involved in production of the dentition were altered, resulting in the dissociation of many developmental processes. Although new work is highlighting similarities between the marginal dentition of *Neoceratodus* a toothplated lungfish and other osteichthyans (Smith *et al.* 2009), changes were involved that shifted most of the dentition-building process from the jaw margin to the palate and prearticular of the upper and lower jaws, respectively. These changes could have been related to the decoupling of the interconnected regulation of tooth formation and tooth loss originally present and presumed lost at the origin of the Dipnoi. Once free from this constraint the resorptive process was utilised for remodelling of the dentate surfaces, as either extensive or minimal, to keep pace with growth (Ahlberg *et al.* 2006). Dentine development was not restricted to a discrete tooth, but could produce 'pseudoteeth', large sheets over the dentition surface, or bands along the posterior plate margins. Later, during the Carboniferous period, the group settled to a pattern of dental development, represented by the toothplated dentition, but one decoupled from the stereotypic one for osteichthyans.

III – Petrodentine – Part of the Essential Dipnoan Developmental Repertoire

Without the ability to produce an exceptionally hard, highly mineralised form of dentine (petrodentine) from the pulp cells the dipnoan dentition would not be functionally effective (Bemis 1984; Bemis and Lauder 1986). Its distribution has been mentioned in all the extant and fossil dentitions discussed so far as a constant growth tissue, but produced beneath the functional surfaces of dentine plates, pseudoplates, tooth ridges and tooth rows. In a comprehensive account of microstructure, histogenesis and growth of petrodentine, Smith (1984) described its distribution in the Devonian forms as well as its development in the larval tooth plates of *Protopterus aethiopicus* (Smith 1985). This tissue is produced by specialised ectomesenchymal cells that differentiate from the dental pulp (petroblasts, Figure 5A, B, C, G) and is as hard and its matrix as completely mineralised as enamel (Figure 5F). Significantly, the latter tissue is deposited by superficial epithelial cells of the dental organ and is not able to be continually renewed.

The early literature of Lison, Denison and Ørvig, reviewed by Smith (1984), shows that the type of tissue first described by Lison (Lison 1941) in the extant lepidosirenids also occurred in the Devonian forms. This hard translucent tissue is localised to each of the separate teeth of the radial rows in *Dipterus valenciennesi* (Figure 5D) as seen in this horizontal section of the whole tooth plate, as also in the worn surfaces of the earliest teeth in *Gogodipterus paddyensis* (Figure 1H). Petrodentine is located in the centre of each tooth of the radial rows in the hatchling *Andreyevichthys epitomus* (Figure 1F) and is renewed by constant new formation within the pulp chamber beneath the surface of the fused teeth forming radial ridges in the adult (Figure 1G). The large enamel-like size of the hydroxyapatite crystals have been demonstrated at the formative surface in *Chirodipterus australis* in transmission electron microscopy of ion-beam thinned sections (Smith and Campbell 1987). Also, there are continuous and extensive growth regions of this tissue beneath the flat areas of the dentine plate (see Figure 3A) as previously illustrated (Smith 1984; Smith and Campbell 1987).

Sections taken vertically through the radial ridges of the extant lepidosirenid forms *Protopterus* and *Lepidosiren*, and the ceratodontid *Neoceratodus*, show the tissue is completely mineralised to the extent of enamel (Figure 5F), leaving only traces of organic matrix (Figure 5C, G). The function and hardness values of petrodentine have been documented by Bemis (1984), similarly its microstructure, formation and mineral composition by Ishiyama and Teraki (1990). All these investigations confirm its analogy with enamel and enameloid but they stress that its formation is by sole participation of the mesenchymal petroblasts, as illustrated here (Figure 5A, C, E, G). It represents a remarkable transformational adaptation to lack of replacement teeth and retention of all the original teeth conferring an ability to resist wear and form morphologies adapted to function, such as flat crushing surfaces or sharp cutting blades. These are both seen as crushing surfaces as in *Neoceratodus* (Figure 1A, B, I (Smith 1984)) and prominent cusps and sharpened blades in *Protopterus* and *Lepidosiren* (Bemis 1984; Ishiyama and Teraki 1990).

In early development of each tooth a row of characteristic cells (petroblasts) differentiated from odontoblasts forming the dentine, are seen in the pulp (Figure 5A, B). They secrete an organic-rich matrix that is reduced as it mineralizes through crystal growth of hydroxyapatite (Figure 5A, B). This type of tissue growth continues beneath each tooth as they become fused to each other and to the dental bones (Figure 5C-G). In the adult plates this growth is at a site deep to the tritural (functional) surface of the ridges in *Protopterus* (Figure 5E, G), and is the same beneath the flatter grinding regions of *Lepidosiren* (Smith 1984) and beneath the ridges and punctate worn surface of *Neoceratodus* (Figure 1A, B (Smith 1984)). The same tissue is referred to as 'central, columnar, or core material and interdenteonal dentine' in all publications by Kemp (1979, 1984, 1995a, 2002a, 2002b, 2002c).

In their phylogenetic analysis Ahlberg *et al.* (2006) used the character 'petrodentine cores' as one of the options in the data matrix for 'nature of large dentine elements'. It is clear that the production of this tissue early in the phylogeny of dipnoans was a major adaptive character for their diversity and success in surviving till today as two quite diverse forms.

IV – DISCUSSION – FUTURE DIRECTIONS

As lungfish have both living and fossil forms they are perfect animals for testing the hypothesis that a unique 'lungfish dental pattern' evolved from an osteichthyan one and has been conserved for 400 Myr. This is especially because all the ontogenetic stages are retained in the adult dentition and several examples of hatchling and juvenile stages are preserved in the Devonian fossil record (Reisz and Smith 2001; Smith and Chang 1990; Smith and Krupina 2001) so that these fossil species can disclose changing patterns through evolutionary time. How this pattern varied through evolutionary time was the subject of a new phylogenetic analysis by Ahlberg *et al.* (2006), using a novel approach based on detailed analysis of the dental characters. This approach involved analysis of two morphogenetic processes in each fossil species, additive for teeth or pseudo-teeth and subtractive through resorption of all tissues. In most osteichthyans these two processes (addition and

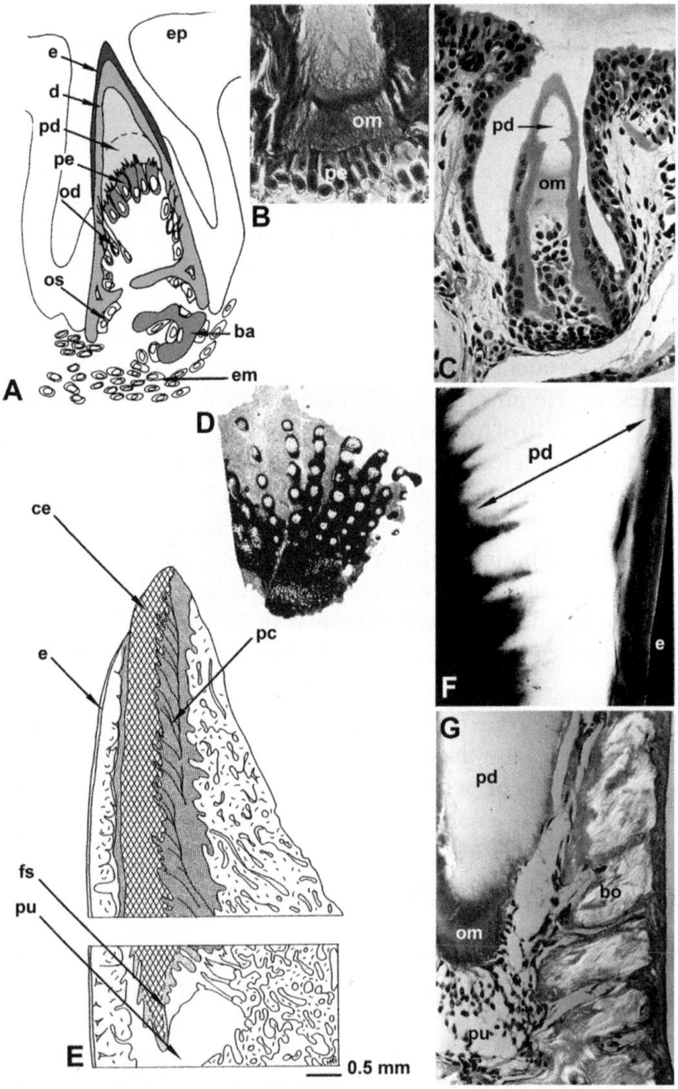

Fig. 5 Formation, growth and distribution of petrodentine in extant and fossil tooth plates of *Protopterus aethiopicus* juvenile and adult, with Devonian *Dipterus valenciennesi*. A, *Protopterus* embryo of 27.5 mm, drawing of a section through a tooth in which all the tissues have developed, enamel (e), dentine (d), petrodentine (pd), bone of attachment (ba), and the tooth has erupted through the epithelium (ep), cell types are petroblasts (pe), odontoblasts (od), osteoblasts (os), ectomesenchyme (em). B, C, 57 mm *Protopterus* juvenile, histological sections at high and low magnification to show the formative front of petrodentine deep in the tooth below the biting surface of the tooth ridge, petroblasts (pe) line up against organic rich early petrodentine (om), and highly mineralised petrodentine above with little matrix (pd).

Fig. 5 Contd..

loss) are tightly and stereotypically linked but in dipnoans they vary independently early in evolution and there is great disparity of dental form.

The remarkable difference in the developmental pattern of the dentition between that of a *Polypterus* type (Clemen *et al.* 1998), the most primitive living actinopterygian, and that of lungfish, must have been produced by significant differences in the regulatory genes. Now that we have some gene expression data for timing and position of the lungfish marginal dentition the different pattern in dipnoans can begin to be explained in relation to that of osteichthyans. Since this data is only so far for the marginal dentition, these could be a shared gene network for the marginal dentition and would not necessarily explain the different morphology on the palatal dentition of the prearticular and pterygoids. Fraser *et al.* (2004, 2008, 2009) proposed that pattern regulation of the crown-group gnathostome dentition occurred through differential spatio-temporal expression of several genes, which choreograph an evolutionary stable event in fish since the osteichthyan divergence. Moreover, they confidently predicted an origin of the 'ancient dental network' for the tooth module close to the origin of vertebrates (Fraser *et al.* 2009), with patterned dental units and their spacing genes occurring in jawless vertebrates within the pharyngeal cavity.

Attempts to track the evolutionary record of developmental changes of the marginal (dentary) teeth can only be attempted in one of the extant lungfish, *Neoceratodus forsteri*, as neither *Protopterus* nor *Lepidosiren* have retained marginal teeth during their evolution. With respect to two of these genes, *shh* and *pitx*, it can be assumed that this is shared with all toothed vertebrates as part of the core gene network identified in the cichlid dentition (Fraser *et al.* 2009). Smith *et al.* (2009) have suggested that investigation of *shh* expression pattern through the developmental stages can reveal exact details of relative timing of each tooth in the sequence along the tooth row. Moreover, this level of detail in spatio-temporal timing can show any asymmetry intrinsic to this process. Future research will expand these studies to the palatal and prearticular dentitions. Of particular interest will be an investigation of the potential for dissociating the addition of the tooth module from its removal by tooth resorption in the process of tooth replacement. The recent use of actinopterygians such as the cichlid as a laboratory

Fig. 5 Contd..

D, horizontal section through toothplate of *Dipterus valenciennesi* with translucent cores of petrodentine formed in each tooth. E, F, G, vertical sections through adult *Protopterus* tooth plate ridge tissue to show the extent of mineralization in petrodentine (pd arrowed line) in a microradiograph (F) deep to the enamel on the surface (e). E, drawing of all the tissues in two blocks separated by an extent of the same tissue, cutting edge (ce), forming surface (fs), pulp canals (pc) of trabecular dentine, pulp cavity (pu), enamel (e). G, histology of the forming tissue petrodentine (pd) with organic rich matrix (om), bone of attachment (bo), pulp cavity (pu). A, from Smith (1984), B, E, F from Smith (1985), C, from Smith *et al.* (2002).

animal and in the near future *Polypterus*, presents the possibility of targeted gene knockouts to test these hypotheses with respect to lungfish dental evolution.

Acknowledgements

We are greatly indebted to Jean Joss for the many visits made to her lab during which all these results were recorded, and for inviting us to participate in this volume. We also thank her many collaborators with whom we worked and who provided advice, especially Rolf Ericsson (Biological Sciences, Macquarie University). We thank Natasha Krupina, John Long and also many others who have made the fossil specimens available to us. For technical assistance we thank Debra Birch and Nicole Vella (Microscopy Unit, Biological Sciences, Macquarie University) and Verity Hodgkinson. Also Li Kershaw and Libby Eyre (Biological Sciences, Macquarie University) for histological work and animal care, respectively.

References

Ahlberg, P.E. (1991). A re-examination of sarcopterygian interrelationships, with special reference to the Porolepiformes. Zoological Journal of the Linnean Society 103: 241-287.

Ahlberg, P.E., Smith, M.M. and Johanson, Z. (2006). Developmental plasticity and disparity in early dipnoan (lungfish) dentitions. Evolution and Development 8: 331-349.

Barwick, R.E., Campbell, K.S.W. and Mark-Kurik, E. (1997). *Tarachomylax*: A new early Devonian dipnoan from Severnaya Zemlya, and its place in the evolution of the Dipnoi. Geobios 30: 45-73.

Bemis, W.E. (1984). Morphology and growth of lepidosirenid lungfish tooth plates (Pisces: Dipnoi). Journal of Morphology 179: 73-93.

Bemis, W.E. and Lauder, G.V. (1986). Morphology and function of the feeding apparatus of the lungfish, *Lepidosiren paradoxa* (Dipnoi). Journal of Morphology 187: 81-108.

Bertmar, G. (1966). The development of the skeleton, blood-vessels and nerves in the dipnoan snout, with a discussion on the homology of the dipnoan posterior nostrils. Acta Zoologica 47: 82-150.

Campbell, K.S.W. and Barwick, R.E. (1984). *Speonesydrion*, an Early Devonian dipnoan with primitive toothplates. Palaeo Ichthyologica 1-47.

Campbell, K.S.W. and Smith, M.M. (1987). The Devonian dipnoan *Holodipterus*: Dental form variation and remodelling growth mechanisms. Records of the Australian Museum 39: 131-167.

Campbell, K.S.W. and Barwick, R.E. (1990). Paleozoic dipnoan phylogeny; functional complexes and evolution without parsimony. Paleobiology 16: 143-169.

Campbell, K.S.W. and Barwick, R.E. (1991). Teeth and tooth plates in primitive lungfish and a new species of *Holodipterus*. In: M.-M. Chang, Y.-H. Liu and G.-R. Zhang (eds). Early Vertebrates and Related Problems of Evolutionary Biology. Science Press, Beijing, China.

Campbell, K.S.W. and Barwick, R.E. (1998). A new tooth-plated dipnoan from the Upper Devonian Gogo Formation and its relationships. Memoirs of the Queensland Museum 42: 403-437.

Campbell, K.S.W. and Barwick, R.E. (2000). The braincase, mandible and dental structures of the early Devonian lungfish, *Dipnorhynchus kurikae*, from Wee Jasper, New South Wales. Records of the Australian Museum 52: 103-128.

Chang, M.-M. (1995). *Diabolepis* and its bearing on the relationships between porolepiforms and dipnoans. Bulletin Museum National Histoire Naturelle, Paris 17: 235-268.

Chang, M.-M. and Yu, X. (1984). Structure and phylogenetic significance of *Diabolepis speratus* gen. et sp. nov., a new dipnoan-like form from the Lower Devonian of eastern Yunnan, China. Proceedings of the Linnean Society, New South Wales 107: 171-184.

Chen, Y., Bei, M., Woo, I., Satokata, I. and Maas, R. (1996). Msx1 controls inductive signaling in mammalian tooth morphogenesis. Development 122: 3035-3044.

Clemen, G., Bartsch, P. and Wacker, K. (1998). Dentition and dentigerous bones in juveniles and adults of *Polypterus senegalus* (Cladistia, Actinopterygii). Annals of Anatomy 180: 211-21.

Cloutier, R. (1996). Dipnoi (Akinetia: Sarcopterygii). In: H.-P. Schultze and R. Cloutier (eds). Devonian Fishes and Plants of Miguasha, Quebec, Canada. Verlag Dr. Friedrich Pfeil. Munich, Germany.

Denison, R.H. (1968). Early Devonian lungfishes from Wyoming, Utah and Idaho. Fieldiana: Geology 17: 353-413.

Denison, R. H. (1974). The structure and evolution of teeth in lungfishes. Fieldiana: Geology 33, 31-58.

Fraser, G.J., Graham, A. and Smith, M.M. (2004). Conserved deployment of genes during odontogenesis across osteichthyans. Proc R Soc Lond B 271: 2311-2317.

Fraser, G.J., Bloomquist, R.F. and Streelmann, J.T. (2008). A periodic pattern generator for dental diversity. BMC: Developmental Biology. (In press).

Fraser, G.J., Hulsey, C.D., Bloomquist, R.F., Uyesugi, K., Manley, N.R. and Streelman, J.T. (2009). An ancient gene network is co-opted for teeth on old and new jaws. PLoS Biology 7: e31.

Hällstrom, B.M. and Janke, A. (2009). Gnathostome phylogenomics utilizing lungfish EST sequences. Molecular Biology and Evolution 26: 463-471.

Henriksen, K., Neutzsky-Wulff, A.V., Bonewald, L.F. and Karsdal, M.A. (2009). Local communication on and within bone controls bone remodeling. Bone 44: 1026-1033.

Huysseune, A. and Witten, P.E. (2006). Developmental mechanisms underlying tooth patterning in continuously replacing osteichthyan dentitions. Journal of Experimental Zoology Part B: Molecular and Developmental Evolution 306: 204-215.

Ishiyama, M. and Teraki, Y. (1990). The fine structure and formation of hypermineralized petrodentine in the tooth plate of extant lungfish (*Lepidosiren paradoxa* and *Protopterus* sp). Archives of Histology and Cytology 53: 307-321.

Kemp, A. (1977). The pattern of toothplate formation in the Australian lungfish, *Neoceratodus forsteri* Krefft. Zoological Journal of the Linnean Society 66: 251-287.

Kemp, A. (1979). The histology of tooth formation in the Australian lungfish *Neoceratodus forsteri* Krefft. Zoological Journal of the Linnean Society 66: 251-287.

Kemp, A. (1984). A comparison of the developing dentition of *Neoceratodus forsteri* and *Callorhynchus milii.* Proceedings of the Linnean Society, New South Wales 107: 245-262.

Kemp, A. (1995a). Marginal tooth-bearing bones in the lower jaw of the recent Australian lungfish, *Neoceratodus forsteri* (Osteichthyes: Dipnoi). Journal of Morphology 225: 345-355.

Kemp, A. (1995b). On the neural crest cells of the Australian Lungfish. Bulletin du Museum National d'Histoire Naturelle Section C Sciences de la Terre Paleontologie Geologie Mineralogie 17.

Kemp, A. (2002a). Growth and hard tissue remodelling in the dentition of the Australian lungfish, *Neoceratodus forsteri* (Osteichthyes: Dipnoi). Journal of Zoology 257: 219-235.

Kemp, A. (2002b). The marginal dentition of the Australian lungfish, *Neoceratodus forsteri* (Osteichthyes: Dipnoi). Journal of Zoology 257: 325-331.

Kemp, A. (2002c). Unique dentition of lungfish. Microscopy Research and Technique 59: 435-448.

Kundrat, M., Joss, J. and Smith, M. (2008). Fate mapping in embryos of *Neoceratodus forsteri* reveals cranial neural crest participation in tooth development is conserved from lungfish to tetrapods. Evolution & Development 10: 531-536.

Lison, L. (1941). Recherches sur la structure et l'histogenese des dents des poissons dipneustes. Archives de Biologie, Paris 52: 279-320.

Lumsden, A.G. (1988). Spatial organization of the epithelium and the role of neural crest cells in the initiation of the mammalian tooth germ. Development 103 Suppl, 155-169.

Martin, T., Gooi, J.H. and Sims, N.A. (2009). Molecular mechanisms in coupling of bone formation to resorption. Critical Reviews in Eukaryotic Gene Expression 19: 73-88.

Miles, R.S. (1977). Dipnoan (lungfish) skulls and the relationships of the group: a study based on new species from the Devonian of Australia. Zoological Journal of the Linnean Society 61: 1-328.

Persson, P., Takagi, Y. and Björnsson, B.T. (1995). Tartrate resistant acid phosphatase as a marker for scale resorption in rainbow trout, *Oncorhynchus mykiss*: effects of estradiol-17 treatment and refeeding. Fish Physiology and Biochemistry 14: 329-339.

Pridmore, P.A., Campbell, K.S.W. and Barwick, R.E. (1994). Morphology and Phylogenetic Position of the Holodipteran Dipnoans of the Upper Devonian Gogo Formation of Northwestern Australia. Philosophical Transactions of the Royal Society London, B 344: 105-164.

Reif, W.E. (1982). Evolution of dermal skeleton and dentition in vertebrates: the odontode-regulation theory. Evolutionary Biology 15: 287-368.

Reisz, R.R. and Smith, M.M. (2001). Developmental biology. Lungfish dental pattern conserved for 360 Myr. Nature 411: 548.

Schultze, H.-P. (2001). *Melanognathus*, a primitive dipnoan from the Lower Devonian of the Canadian Arctic and the interrelationships of Devonian dipnoans. Journal of Vertebrate Paleontology 21: 781-794.

Smith, M.M. (1984). Petrodentine in extant and fossil dipnoan dentitions: microstructure, histogenesis and growth. Proceedings of the Linnean Society of New South Wales 107: 367-407.

Smith, M.M. (1985). The pattern of histogenesis and growth of tooth plates in larval stages of extant lungfish. Journal of Anatomy 140 (Pt 4): 627-643.

Smith, M.M. (1986). The dentition of Palaeozoic lungfishes: a consideration of the significance of teeth, denticles and tooth plates for dipnoan phylogeny. In: Teeth Revisited: Proceedings of the VIIth International Symposium on Dental Morphology, Volume 53, Paris, 1986. Memoires du Museum national d'histoire naturelle, Paris (serie C). pp. 179-194.

Smith, M.M. (2003). Vertebrate dentitions at the origin of jaws: when and how pattern evolved. Evolution and Development 5: 394-413.

Smith, M.M. and Campbell, K. S. W. (1987). Comparative morphology, histology and growth of the dental plates of the Devonian dipnoan *Chirodipterus*. Philosophical Transactions of the Royal Society London, Series B 317: 329-363.

Smith, M.M. and Chang, M.-M. (1990). The dentition of *Diabolepis speratus* Chang and Yu, with further considerations of its relationships and the primitive dipnoan dentition. Journal of Vertebrate Paleontology 10: 420-433.

Smith, M.M. and Coates, M.I. (1998). Evolutionary origin of the vertebrate dentition: phylogenetic patterns and developmental evolution. European Journal of Oral Sciences 106: 482-500.

Smith, M.M. and Krupina, N.I. (2001). Conserved developmental processes constrain evolution of lungfish dentitions. Journal of Anatomy 199: 161-168.

Smith, M. M. and Johanson, Z. (2003). Separate evolutionary origins of teeth from evidence in fossil jawed vertebrates. Science 299: 1235-1256.

Smith, M.M., Krupina, N.I. and Joss, J. (2002). Developmental constraints conserve evolutionary pattern in an osteichthyan dentition. Connective Tissue Research 43: 113-119.

Smith, M.M., Okabe, M. and Joss, J. (2009). Spatial and temporal pattern for the dentition in the Australian lungfish revealed with sonic hedgehog expression profile. Proceedings of the Royal Society B London 276: 623-631.

Thomson, K.S. and Campbell, K.S.W. (1971). The structure and relationships of the primitive Devonian lungfish – *Dipnorhynchus sussmilchi* (Etheridge). Bulletin of the Peabody Museum of Natural History 38: 1-109.

Wang, S., Drapala, V., Bawick, R.E. and Campbell, K.S.W. (1993). The dipnoan species *Sorbitorhynchus deleaskitus*, from the Lower Devonian of Guangxi, China. Philosophical Transactions of the Royal Society London B 340: 1-24.

White, E.I. (1966). Presidential address: A little on lung-fishes. Proceedings of the Linnean Society of London 177: 1-10.

Witten, P.E. and Huysseune, A. (2009). A comparative view on mechanisms and functions of skeletal remodelling in teleost fish, with special emphasis on osteoclasts and their function. Biological Reviews 84.

Zhang, Z., Lan, Y., Chai, Y. and Jiang, R. (2009). Antagonistic actions of Msx1 and Osr2 pattern mammalian teeth into a single row. Science 323: 1232-1234.

CHAPTER 9

The Integument of Lungfish: General Structure and Keratin Composition

Lorenzo Alibardi[1,*], Jean Joss[2] and Mattia Toni[3]

[1,3]Department of Evolutionary Experimental Biology, University of Bologna, Bologna, Italy

[2]School of Biological Science, Macquarie University, Sydney, New South Wales 2109, Australia

ABSTRACT

The aquatic environment poses less restrictions on mechanical and barrier performances in the integument of both marine and freshwater vertebrates, mainly fish and some amphibians, in comparison to the skin of terrestrial vertebrates (Whitear 1977; Zaccone *et al.* 2001; Alibardi 2006). As a result, apart from specific locations of the body in a few species of fishes and amphibians, the epidermis of aquatic vertebrates resembles the relatively poorly keratinised multi-layered epithelia found in the mucoses lining the respiratory or alimentary canals of terrestrial vertebrates (Whitear 1986a,b). In particular a corneal layer is missing over the general epidermis of fishes and perennibranchiate amphibians. This includes also sarcopterygian fish, both crossopterygians (*Latimeria*, saltwater) and dipnoans (freshwater).

This chapter is mainly concerned with the skin, especially the epidermis of the Australian lungfish, *Neoceratodus forsteri*, which is compared where studies are available, with the epidermis of other species of dipnoans (Kitzan and Sweeny 1968; Imaki and Chavin 1975a,b, 1984). In addition to a previously published study (Alibardi and Joss 2003), *N. forsteri* skin derived from two extra larval stages are described and the keratins have been partially characterised.

Keywords: lungfish, skin, epidermis, ultrastructure, keratins

*Corresponding author

Histology and Ultrastructure of Lungfish Epidermis during the Larval Stage

In the skin of a 1-2 months old hatchling lungfish, the epidermis comprises a flat periderm covering a flat to cuboidal layer of basal epidermis. The dermis beneath is composed of loose mesenchyme which is richly vascularized.

At more advanced larval stages (e.g. 2-6 months old), the dermis is still composed of loose connective tissue (Fig. 1 A) and the epidermis has 2 or 3 layers of cells (Fig. 1 A). The number of epidermal layers increases to 4-5 in juveniles about 2 years old and 16 cm in length, and 5-7 layers in older individuals with lengths from 21 to 26 cm (Alibardi and Joss 2003; see Figs. 1, 2). The skin of early larvae is differentiated into an intermediate epidermal layer above the germinal layer, external layers formed by polygonal cells and a cuticle (Fig. 1 A). The latter is resolved under the electron microscope as the apical part of the cytoplasm of keratinocytes where most of the mucus granules are localized and discharge their content on the epidermal surface to form the cuticle (Fig. 1 C).

Two main cell types are present, cells accumulating keratin filaments (keratinocytes) and mucous cells with differing degrees of granule accumulation. The apical cells contain keratin filaments that generally do not form dense tonofilaments as they do in amphibian and amniote keratinocytes. The microvilli that project from the outer surface of the apical cells are probably sections of micro-ridges (of unknown three-dimensional pattern) similar to the epidermis in actinopterigian fish (Bereither-Hahn *et al.* 1979; Whitear, 1986a). Hatchlings have a loose dermis with the uppermost layer in contact with the epidermis being formed from a layer of dermal melanosomes (Fig. 1 A). Sparse melanocytes are present in the epidermis and a few intracellular melanosomes can be seen in some keratinocytes.

In some areas of the trunk epidermis in a later larval stage *N. forsteri* (about 2-4 months post-hatching) epidermal cells are organized around an empty space that may be a big mucous cell remnant or a true space where mucus is secreted to reach the surface (Fig. 1 B). This collection of cells around a lumen is clearer in some areas than in others, e.g. right on Fig. 1 B, where it appears that an epidermal cell has evaginated into a glandular-like or sensory structure. These apparently multicellular, alveolar glands, resemble the forming mucous glands of metamorphosing amphibians (Delfino *et al.* 1982, 1985). These 'glands' were only observed in larval epidermis and not in adult skin. No studies to date have found glands in adult epidermis, probably due to the very heavy scales formed in the upper dermal layers and therefore it is likely that any amphibian-like glands in larval epidermis are lost at an early stage of development.

In the 5 months old larva, the epidermis is 20-40 m thick and rich in mucous cells at different stages of secretion (Fig. 2 A, B) or type of mucus. Some mucous cells are even seen extruded into the cuticular space, probably as degenerating

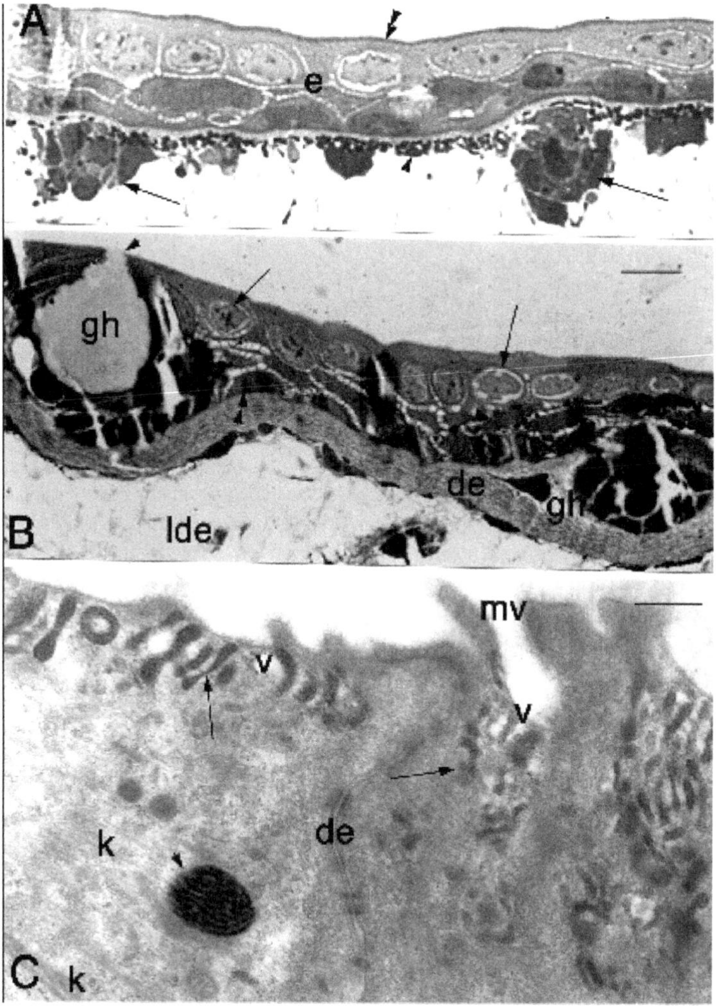

Fig. 1 A, Longitudinal section of epidermis of *N. forsteri* hatchling showing a superficial cuticle (double arrowhead) on the external layer of cuboidal pale cells of the epidermis (e). Intermediate and basal darker cells are present. In the dermis a lamina of pigment cells (arrowhead) underlies the epidermis while fibroblasts concentrate into discrete groups (arrows). Bar, 10 μm. B, other region of hatchling epidermis with 3-4 tiers of epidermal cells (double arrowhead indicates the basal layer). Pale cells (arrows) are seen beneath the outermost epidermal layer. Two possibly neuromasts are present (arrowheads). Dense dermal layer (dd) and deeper looser layer (d). Bar, 10 μm. C, detail of superficial epidermis of hatchling with microvilli (mv) and numerous mucus granules (arrows) and pale vesicles that discharge their content on to the surface (v). The arrowhead points to a melanosome among keratin filaments (k). de, desmosomes. Bar, 500 nm.

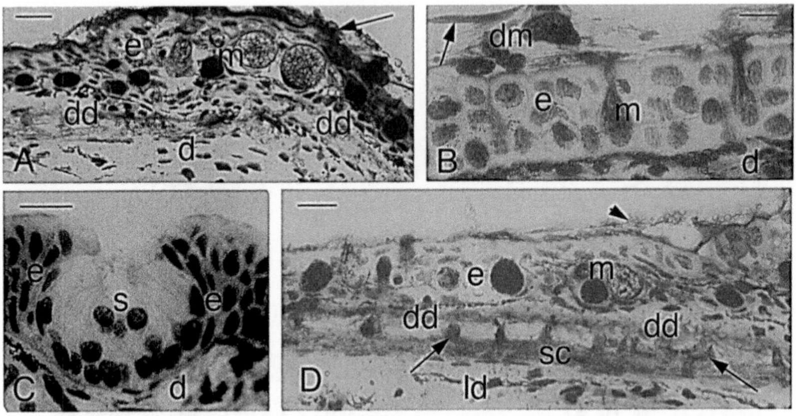

Fig. 2 Histology of the skin of a 5-month old juvenile. A, tail epidermis (e) with numerous mucous cells (m) with dense or pale content. The arrow indicates mucus covering the surface (arrow). d, dermis; dd, dense dermal layer; Bar, 20 µm. B, mucous cells (m) in tail epidermis (e) and degenerated mucus cells (dm) in the cuticular space (arrow). Bar, 10 µm. C, detail of an ampullar-organ (s) in trunk epidermis of sensory or glandular type, which slightly penetrate into the dermis (d). Bar, 20 µm. D, trunk epidermis (e) with numerous mucus cells (m) with underlaying bone tissue of scale (sc). The spinulae of the dermal bone (arrows) contact the overlying dense dermis (dd) and are confined within a loose dermis beneath. Bar, 10 µm.

Color image of this figure appears in the color plate section at the end of the book.

cells (see also ultrastructure). In both the tail and trunk there are some glandular-like organs, which in general appearance resemble a taste bud. In fact, their pale cytoplasm indicates that no secretory material is present (Fig. 2 C). As a consequence these ampullar-organs are interpreted as sensory organs in this specimen rather than glands as in the younger specimen (Fig. 2 C). Their true nature remains to be fully clarified.

Finally, the dermis shows a thin, denser layer and a deeper, loose layer. The dermal bone (scales) in a 5-month larva are already deposited beneath the thin dense layer, and form a series of pointed spines (Fig. 2 D).

Histology and Ultrastructure of Adult Epidermis

The skin of lungfishes, in particular of the adult Australian lungfish *N. forsteri*, is composed of a 100-150 µm thick multi-layered epithelium, which is not covered by a corneal layer while the dermis has the typical appearance of other fishes (Whitear 1986a, b).

In the epidermis of juvenile *N. forsteri* (16 cm, 19.5 and 26 cm long, over 2 years old) and adults of other species of dipnoans, the stratification of the epidermis increases to 5-7 cell tiers (layers). The external border of the outermost layer forms a compact superficial layer (Fig. 3 A). The main cell types in the epidermis are dark and pale keratinocytes, followed by mucous cells and other less frequent cell types (Fig. 3). The latter include epidermal melanocytes and granular cells most of which appear to be phagocytes. Detailed descriptions of these cells are reported in Kitzan and Sweeny (1968), Imaki and Chavin (1975a,b, 1984) and Alibardi and Joss (2003). Epidermal and mucous cells are believed to derive from a common progenitor stem cell, which, for mucus cells, follows a detour from the differentiation of keratin-accumulating cells to increase secretion from the Golgi apparatus in order to produce a large quantity of mucins.

Dark and pale keratinocytes (the latter present in the outermost 1-2 layers of the epidermis) contain sparse keratin filaments or small bundles of keratin filaments which do not form large tonofilaments as they do in keratinocytes of land vertebrates. These cells all contain typical cell organelles, but dark cells have more densely packed ribosomes, sparse rough endoplasmic reticulum and numerous desmosomes. Keratin filaments increase in the outermost layer of the epidermis. Pale keratinocytes often contain extended perinuclear endoplasmic reticulum and enlarged ergastoplasmic cisternae while there are less ribosomes than in darker cells. Some pale cells located on the surface are instead degenerating.

Mucous cells of variable but generally large diameter (20-30 × 50-80 μm), are present in the intermediate layer of the epidermis (Figs. 3A, 4 B). These cells have different shapes: pear-like, globular or elongate and contain PAS-positive granules of variable intensity (Kitzan and Sweeny 1968; Imaki and Chavin 1975a,b, 1984). Some cells discharge their secretion to the surface as a more or less continuous secretion but others tend to accumulate it and are completely packed with mature granules that are possibly secreted as the cell dies. Degenerating cells are occasionally observed within the epidermis, and some may represent the last stages of degenerating mucus cells (Fig. 4 C). Degenerating cells are surrounded by epidermal cells showing typical phagocytic phylopodia (Fig. 4 C) but often granulated macrophages (perhaps a further stage of differentiation from epidermal cells) are associated with these cells (Fig. 4 D). Finally, granules as dense as those present in macrophages are sometimes seen among the material of the cuticle on the external surface of the epidermis (Fig. 4 E). These degenerating cells may be derived from extruded mucous cells or epidermal phagocytes.

Ultrastructural studies have shown that two types of mucous cells are present: one contains non-homogenous mucin droplets (type A), and the other contains more homogenous droplets (type B). It is possible but not known whether these two types produce different components of the mucus. Also, it is not known yet whether mucus cells of lungfish respond to prolactin by increasing their secretion or are directly innervated like the mucus cells of teleosts. Similarly to the epidermis of other fish, however lungfish mucus has different functions, such as assisting swimming by reducing drag, controlling ion movements across the epidermis, giving mechanical protection by forming a superficial cuticle, and, in particular

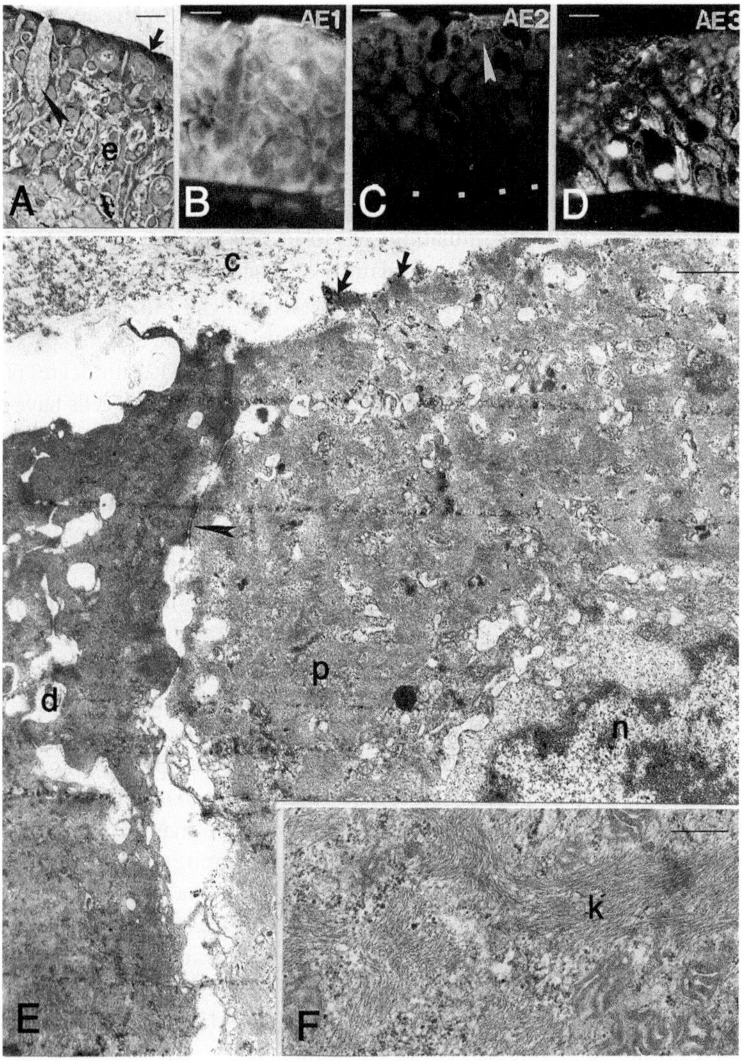

Fig. 3 A, Multi-layered adult epidermis with discharging mucus cells (arrowhead) and a thick cuticula (arrow) in *N. forsteri*. Bar. 10 μm. B, evenly immunofluorescent keratinocytes after immunostaining with AE1 antibody. Bar, 10 μm. C, after immunolabelling with the AE2 antibody, some fluorescence is seen in superficial cells (arrowhead). Dots underlie the basal layer. Bar, 10 μm. D, weaker immunolabelling of keratinocytes after application of the AE3 antibody. Bar, 10 μm. E, ultrastructural detail of the superficial cells of adult epidermis showing a diffuse network of keratin filaments (k) in both the pale (p) and dark (d) cells (n, nucleous). The cuticle (c) is made of amorphous material in contact with discharging mucus granules (arrows). Bar, 1 μm. F, detail of the cytoplasm of a superficial cell showing loose bundles of keratin filaments (k). Bar, 250 nm.

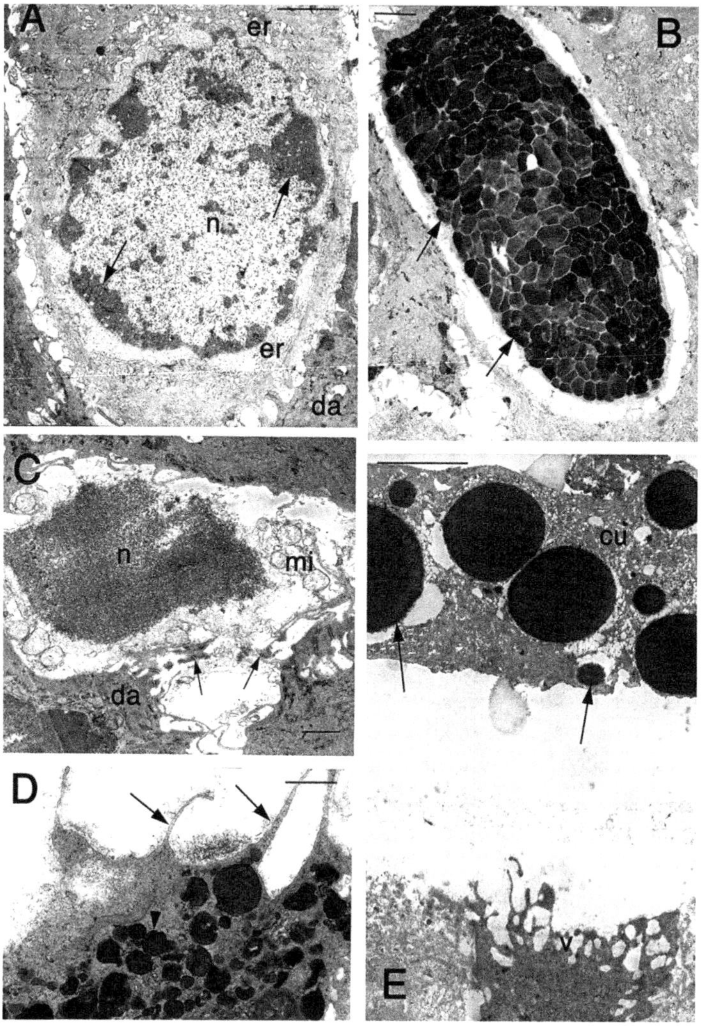

Fig. 4 A, General aspect of a pale keratinocyte (n, nucleous; arrows on heterochromatine) surrounded by the prevalent dark keratinocytes (da) in adult epidermis of *N. forsteri*. Scarce ergastoplasmic vesicles (er) are seen. Bar, 2 μm. B, condensed mucous granules (arrows) in a dark mucus cell among dark keratinocytes of adult epidermis. Bar, 2 μm. C, degenerating cell among dark keratinocytes (da). The condensed nuclear material (n) is surrounded by a vacuolated cytoplasm where organelles including mitochondria (mi) have mainly degenerated. Clumped keratin-like material (arrows) is seen. Bar 1 μm. D, granular (arrowhead) phagocytic cells with elongated filopodia (arrows), located among degenerating cells of the epidermis. Bar, 1 μm. E, dark granules (arrows) present among the amorphous material of the cuticle (cu) covering the external cells of the epidermis of adult. Numerous discharging vesicles containing mucus are seen. Bar, 2 μm.

in lungfishes protecting the body from desiccation out of the water (Zaccone *et al.* 2001). Another important role of mucus, emphasized in the more recent studies, is its anti-parasitic action (especially versus Gram+ and Gram–bacteria) as it contains proteases, lysozime, alkaline phosphatase and other antimicrobic molecules (Fast *et al.* 2002).

Melanophores are the only pigment cells so far described in lungfish and they comprise epidermal (melanocytes) and dermal melanophores. Epidermal melanocytes are generally located at different levels of the epidermis and are more frequent in *N. forsteri* than in the other species. They contain melanosomes 0.5-0.9 × 0.9-1.3 μm, which appear more oval-shaped than in the melanophores of the dermis. Few pre-melanosomes are seen in melanocytes, which contain concentric dense filaments with a regular periodicity pattern of 15-20 nm (Fig. 1 C). Although sparse, pre-melanosomes are more common in epidermal melanocytes than in dermal melanophores of *N. forsteri* and in melanocytes of *Protopterus* and *Lepidosiren* (Imaki and Chavin 1975a). This observation seems to be an indication of rapid production and utilization of melanosomes in the cytocrine activity of the Australian lungfish. In the other species there is less cytocrine activity. In fact, keratinocytes with melanosomes and pre-melanosomes are also found sparsely distributed in the epidermis of *N. forsteri* (Fig. 1 C).

Dermal melanosomes are often oriented along the basement membrane in larval and adult skin, and in some regions they form a more or less continuous pigmented layer. In other regions of the skin of adult individuals melanosomes are more irregular and are localized in deeper layers of the dermis. They may contact regions underneath the basement membrane and also fibroblast and osteogenic fibroblasts of the dermis (Fig. 5 A). In *Protopterus* and *Lepidosiren* it seems that dermal melanophores send thin elongations (pseudopods) into the epidermis that may be digested by epidermal phagocytes.

Cells engulfing melanosomes and other cell debris, seemingly phagocytes, are occasionally found among epidermal cells of *N. forsteri* but are more common in *Protopterus* and *Lepidosiren*. These cells have been implicated in the digestion of melanosomes and melanocytes in the epidermis. It seems that also keratinocytes can digest melanosomes after their incorporation into lysosomes.

Histology and Ultrastructure of Tadpole and Adult Dermis

The dermis of the tadpole in the trunk of *N. forsteri* is made of a vacuolated mesenchyme where fibroblasts tend to form in more or less regularly spaced aggregates of cells (Fig. 1 A). The basement membrane is very thin in these regions and is often connected to numerous, flat melanophores. In regions toward and in the tail, the dermis consists of a dense layer of 10-20 μm thickness of collagen fibrils located under the basement membrane, where there are numerous flat melanophores (Fig. 1 B). Beneath the dense stratum of collagen fibrils, a softer dermis is seen in which there are few cells with elongations. Among the extracellular dermal matrix, groups of fibroblasts are often seen to form aggregates. These initial aggregates of fibroblasts give rise in later stages to the dermal bones (scales).

The above transition is better appreciated under the electron microscope. In the skin of the trunk of hatchlings, cells are sparse in the dermis. In the tail a superficial layer of denser dermis is present, above the largely acellular dermis. In the dermis, beneath the dorsal fin of juvenile *N. forsteri* (2 yr or older), fibroblasts surround the forming scales (Alibardi and Joss, 2003). Ultrastructural examination of these areas in both larvae and juveniles reveals the presence of variably large (0.1-1.0 μm or more) and irregular granules composed of a substance of medium electron-density resembling elastin fibres, here indicated as osteoid (since it is mineralised in later stages). This material is surrounded by fibrillin-like filaments which resemble in aspect and dimension the collagen fibrils present in the dermis and associated with the basement membrane (Fig. 5 A, B). These fibrillin-like or collagen fibrils are apparently in contact with fibroblasts of the dermal aggregates (Fig. 5 A, B). The collagen or fibrillin-like fibrils are embedded within a matrix or ground substance of medium-low electron-density which is also in contact with the denser elastin-like, osteoid material.

The closeness of this osteoid substance with osteogenic fibroblasts strongly suggests that this material is derived from the synthetic apparatus of these cells. Osteoid material is never seen isolated or associated with epidermal cells. High magnification of the osteoid material shows that it is composed of a granular/fibrillar material, where the fibrils measure 10-20 nm. This elastin-like osteoid material appears to be generated from smaller granules that progressively merge with other granules to form larger and larger deposits. Some of the osteoid material probably makes the pointed structure of the dermal bone of 5-month old tadpoles, but an ultrastructural study on this tissue remains to be performed.

The examination of the external areas of forming or enlarging dermal bones in juvenile stages of *N. forsteri* reveal the first deposition of electron-dense mineralised material on the osteoid mass, surrounded by numerous collagen fibrils and, externally, by osteogenic fibroblasts (Fig. 6). In more central areas of dermal bones in the adult epidermis the opaque mineralization forms a fragile zone that is brittle and easily broken by sectioning (data not shown). This observation indicates that ossification in the skin of lungfishes follows the model of elasmoid scales (Sire 1993; Sire and Huysseune 2003), but a specific and complete study on this topic has yet to be undertaken.

Keratin Immunocytochemistry

Indirect immunofluorescence was carried out using general cytokeratin antibodies, such as AE1, AE2, and AE3 followed by FITC-conjugated secondary antibodies. Histochemistry for the -SH groups was shown by the Chevremon-Frederich

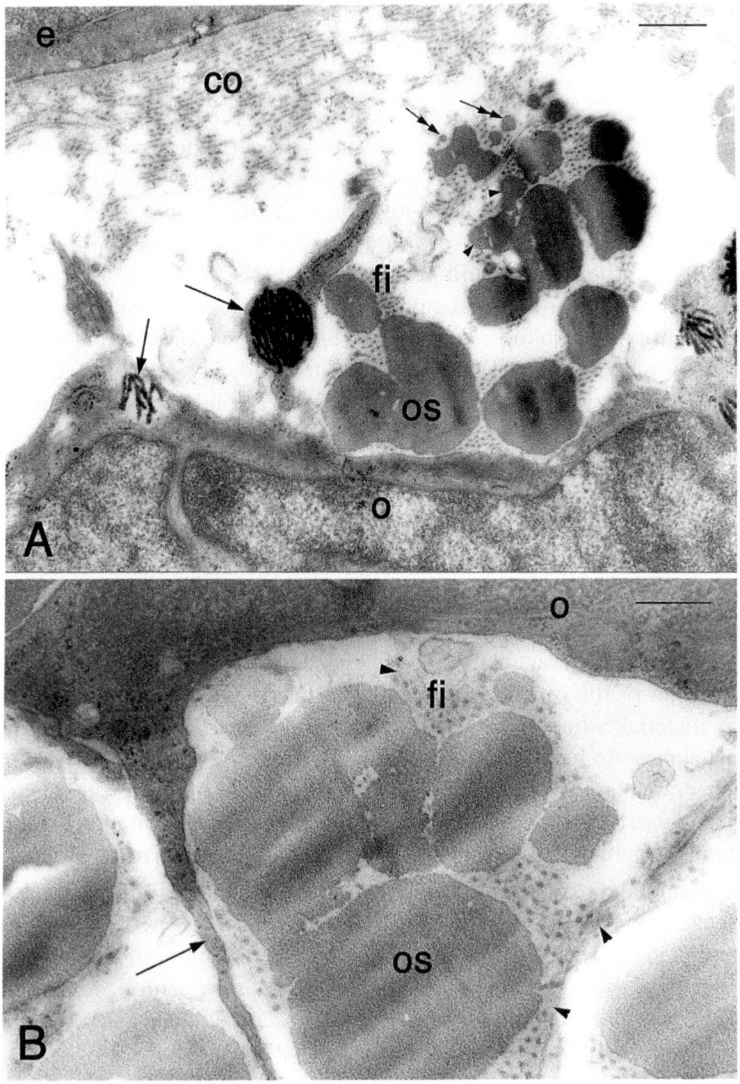

Fig. 5 A, Reticular/collagenous fibrils (co) beneath the epidermis (e) of a juvenile of *N. forsteri*. Fibrillin (fi) embedded in ground material (glycosaminoglycans) surround the elastin-like osteoid material (os) present on the surface of an osteogenic fibroblast (o). Arrows indicate melanosomes. Arrowhead indicate merging elastine-like material while the double arrows indicate very small elastine-like granules. Bar, 500 nm. B, detail of the granulation seen within the elastin-like material of the osteoid (os) in contact with a thin arm (arrow) of an osteogenic fibroblast (o). Fibrillin fibrils (fi) within amorphous substance (arrowheads) surround the elastine-like material. Bar, 250 nm.

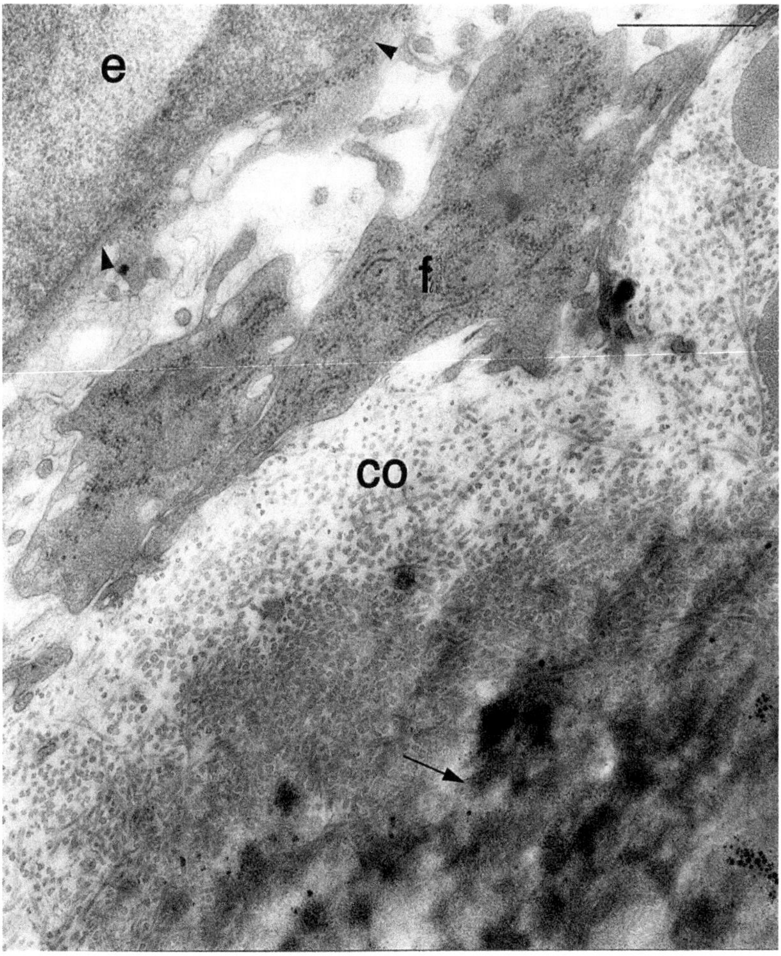

Fig. 6 Detail of the initial deposition of dense mineral material (arrow) over osteoid material in adult dermis of *N. forsteri*. Collagen fibrils (co) are produced in osteogenic fibroblast (fi) which also contact (arrowheads) the epidermis (e). Bar, 500 nm.

colorimetric reaction as previously described (Alibardi and Joss 2003). The presence of reactivity for –SH groups has been reported in the cuticle and in the more external layers of the epidermis of *N. forsteri* (Alibardi and Joss 2003). This observation has confirmed the above ultrastructural studies on the increase of keratin filaments in these layers (Figs. 1 C, 3 E, F). Sulfydryl groups indicate reactive keratin for the formation of intermolecular sulphydrilic groups in uppermost epidermal layers and in the cuticle. Also the presence of transglutaminase-immunoreactivity in the cuticular layer has suggested that the enzyme can participate in the hardening of the extracellular cuticle by forming cross-links with some secreted proteins (Alibardi and Toni 2004). The distribution of keratins in the epidermis of *N. forsterii* has shown that AE1, AE2, and AE3 antibodies (general antibodies for cytokeratins, see Sun *et al.* 1983; O'Guin *et al.* 1987) recognize epidermal cells but not mucus cells. AE1 in mammalian epidermis recognises acidic keratins of 40-52 kDa, AE2 the K1/K10 pair, and AE3, basic keratins of 50-66 kDa (Sun *et al.* 1983; O'Guin *et al.* 1987). AE1 intensely stains keratinocytes in all layers of the epidermis while the dermis remains completely unstained (Fig. 3 B). The AE2 antibody leaves unstained most of the epidermis but in some regions the superficial layers are clearly labelled and resemble the staining for –SH groups (Fig. 3 C). Also the AE3 antibody shows some immunoreactivity, although weaker and less uniform than the AE1 (Fig. 3 D).

No immunoreactivity for AE3 was noted for *P. aethiopicus* (Shaffeld *et al.* 2005) while AE2 was not tested in this species. Despite this difference, the outer and the cuticular layer of the two species show specific reactivity to the AE2 (*N. forsteri*) or to the 79.14 antibodies (*P. aethiopicus*, directed toward *X. laevis* cytokeratin 8). These two different antibodies are both directed at basic keratins, suggesting that basic keratins are present in the upper layers of the epidermis and in the discharged cell components of the cuticle. This observation indicates that, like amphibians, neutral or more basic keratins are present in lungfish epidermis (Alibardi and Joss 2003). This is further confirmed by the immunoblotting study carried out on both *N. forsteri* (Alibardi 2006) and *P. aethiopicus* (Shaffeld *et al.* 2005). An additional study was undertaken to further characterize *N. forsteri* keratins by using different anti-cytokeratin antibodies which reactivity is known (Moll *et al.* 1982; Coulombe and Omary 2002). Since this has not been published, the methods are appended at the end of the chapter.

Mono-dimensional gel separation of epidermal proteins found mainly proteins in the 40-62 kDa range, which is typical for cyto-(alpha)keratins (Fig. 7, lane A). Blotting of the extracted proteins on nitrocellulose, and probing with several antibodies found the following: (i) the Pan-cytokeratin antibody (a wide range cytokeratin antibody, directed to numerous types of cytokeratins, 1, 5, 6, 8, 10, 13, 18, Sigma catalogue C1810, clone C-11) immunostained bands at 40-50 kDa (Fig. 6, lane B); (ii) the AE1 antibody stained a broad range of proteins (40-70 kDa), which encompass the whole range of molecular sizes for cytokeratins (Fig. 7, lane C); (iii) the AE3 antibody labelled proteins of 40-50 kDa (Fig. 7, lane D). The use of more specific antibodies, directed against mammalian cytokeratins, found intense labelling in more specific bands than the general antibodies. The K6 antibody (for basic keratins) produced a band at 42 kDa but most labelled bands were at 50, 55 and 68-70 kDa (Fig. 7, lane E). The K8 antibody (for a basic keratin) showed labelled bands at 40-50 kDa and a thin band around 60-62 kDa (Fig. 7, lane F). The K10 antibody (against an acidic keratin) showed a band at 55-60 kDa (Fig. 7, lane G). The K13 antibody (for a specific acidic keratin) showed a main

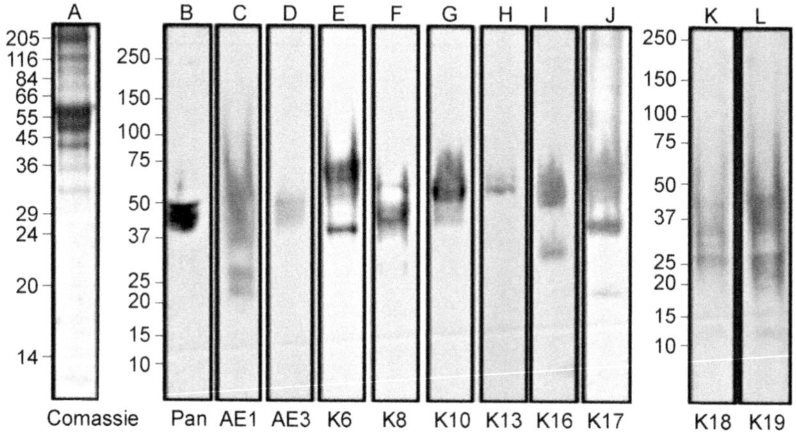

Fig. 7 One-dimensional electrophoretic pattern for proteins in whole epidermis of the Australian lungfish (*N. forsteri*). Coomassie blue (lane A), pan-cytokeratin (lane B), AE1 (lane C), AE3 (lane D), K6 (lane E), K8 (lane F), K10 (lane G), K13 (lane H), K16 (lane I), K17 (lane J), K18 (lane K), K19 (lane L). Lane A, 12% polyacrylamide gel; Lanes B-L, 4-12% gradient polyacrylamide gels.

band at 58-60 kDa (Fig. 7, lane H). The K16 antibody (also for a specific acidic keratin) produced two main bands at 35-38 kDa and at 50-60 kDa (Fig. 7, lane I). The K17 antibody immunolabelling identified three main bands at 40-42, 55 and 62-65 kDa (Fig. 7, lane J).

Antibodies K18 and K19 (for acidic keratins of the lowest molecular weight) labelled mainly bands at 27, 37-40 and 45-48 kDa (Fig. 7, lanes K and L).

The antibody results given above suggest that the epidermis of the Australian lungfish contains some cytokeratins with general epitopes recognised by several antibodies, although these cytokeratins do not correspond to specific mammalian cytokeratins as recognized by mammalian antibodies. However, the AE1 and the K16, 17, 18, and 19, showed labelled bands in the low range (25-45 kDa), as would be expected from these acidic (type II) keratins. More specific labelling however emerged from two-dimensional gel separation and immunoblotting.

After two-dimensional separation and silver staining, several spots were seen, mainly in the alpha-keratin range (45-62 kDa) and with pI at 4.8-6.8 (Fig. 8 A). A smaller number of spots were seen using Coomassie blue but those that were seen were also in the alpha-keratin range as above (Fig. 8 B). The Pan-cytokeratin antibody found immunoreactive keratins of 42-45 kDa with pI 5.0-5.2 and 6.5-7, and 50-52 kDa with pI 5.5 and 6.7-7.2 (Fig. 8 C). Therefore, the Pan-cytokeratin antibody, which can recognise basic keratins 1, 2, 5, 8, further confirms the presence of a small number of neutral and slightly basic keratins in the Australian lungfish skin. The few relatively high pI values found are unique fish keratins, which are

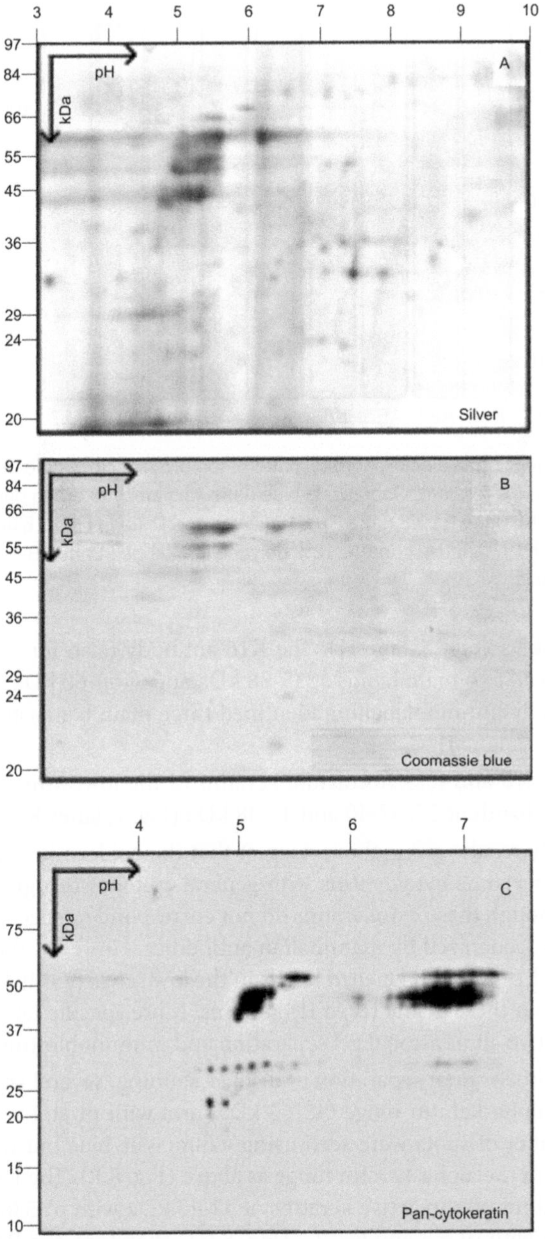

Fig. 8 Two-dimensional electrophoretic and pan-cytokeratin pattern for proteins in whole epidermis of the Australian lungfish (*N. forsteri*). A, silver stain; B, Coomassie blue; C, pan-cytokeratin. A-C, 10% polyacrylamide gels. Numbers in abscissa indicate pH, those in ordinals indicate molecular weight.

generally acidic, and have also been reported for *P. aethiopicus* by Schaffeld *et al.* (2005). In fact, in their Fig. 4 g, these authors show BD-gels with a slab of keratin recognised by Monoclonal antibody K 8.13 within a pI of 6.5-7.2,

Using antibodies as before to further identify these keratins, further results were obtained The AE1 antibody in some cases showed spots at 40-50 kDa with pI at 4.3, 5-5.8, and 6.0-6.2, and 6.8 (Fig. 9 A). Spots at 60-70 kDa had pI at 6.0-6.2. Another case showed spots at 50-70 kDa with some spots with a pI ranging from 5 to 6, and other spots with pI at 6.5-7.2 (Fig. 9 B). One more case showed keratins of 37-45 kDa and pI at 4.9-5.6, and others at 48-50 kDa and pI at 4-5, 5-6, and 6-6.8 (Fig. 8 C). Finally, using the AE3 antibody spots at 40-52 kDa with pI at 4.8-5.2 and 6,2-6.8 were seen (Fig. 10 A). Antother case showed spots at 50 kDa with pI 5.9 and spots at 66 kDa with pI at 5.5, 6, 6.5 (Fig. 10 B). One more case showed mainly spots at 42-45 kDa with pI at 5 and spots at 50-53 kDa with pI around 5.5 (Fig. 10 C).

The results shown for *N. forsteri* are consistent in most part with those found in the skin of *P. aethiopicus* for which 13 types of keratins have been sequenced, 10 of type I and 3 of type II (Schaffeld *et al.* 2005). Among these African lungfish keratins, the orthologues of human K8 (basic) and K18 (acidic) were found, in addition to keratins of 54 and 58 kDa (type II) with pI 5.5-6.5 (therefore acidic),which were immunolabelled with the K18 antibody. The same authors found other protein spots, identified as type I keratins, with molecular weights between 45-50 kDa and pIs of 5.0-5.2, i.e. relatively more acidic. Finally a large protein spot of type I keratin at 56 kDa with pI 5.9 was found in *P. aethiopicus*. The authors suggest that other major keratin spots still remain to be sequenced, especially (Pae-)K1, K2, K10, and K11, which would justify the suggestion that neutral and slightly basic keratins are also present in lungfish. A comparison of the amino acid sequences of these proteins may reveal interesting information on the evolution of the more basic keratins in tetrapods.

A difference between *Neoceratodus* and *Protopterus* appears to be the finding of a keratin with the highest molecular weight (66-70 kDa) so far reported for a fish in *N. forsteri*, which is a value more typical of tetrapods (Schaffeld *et al.* 2005; Alibardi 2006). In conclusion, although the amount of these neutral or slightly basic keratins is low in lungfishes, that the fact that they are present at all in the lineage of fish from which tetrapods originated allows them to be interpreted as a pre-adaptation of keratins toward the more basic types necessary for function in the skin of land vertebrates.

Cocoon Structure

The microscopic structure of the lungfish cocoon and of the aestivating epidermis have been described briefly (Greenwood 1984; Fishman *et al.* 1984; Sturla *et al.* 2002). From the few histological observations, it appears that in

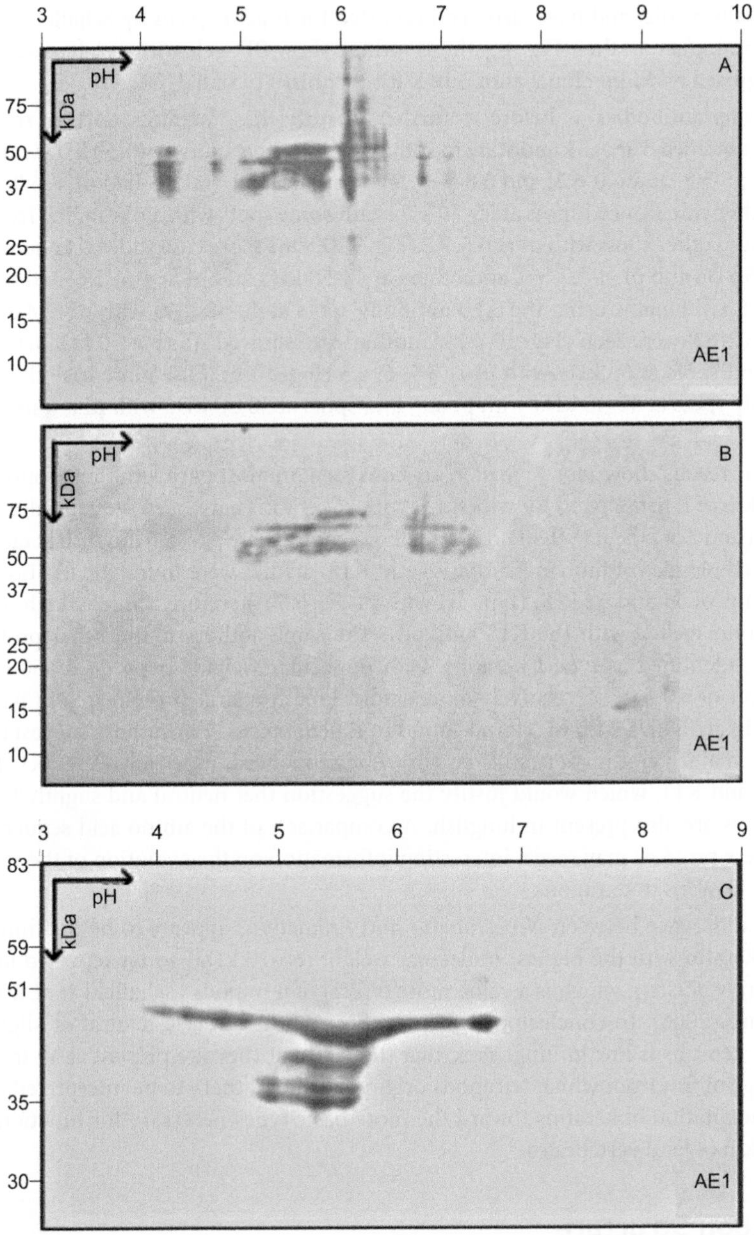

Fig. 9 Two-dimensional electrophoretic pattern using AE1 antibody for proteins in whole epidermis of the australian lungfish (*N. forsteri*). AE1 immunolabelled spots in three different samples are shown in Panels A, B, and C. A-C, 12% polyacrylamide gels. Numbers in abscissa indicate pH, those in ordinals indicate molecular weight.

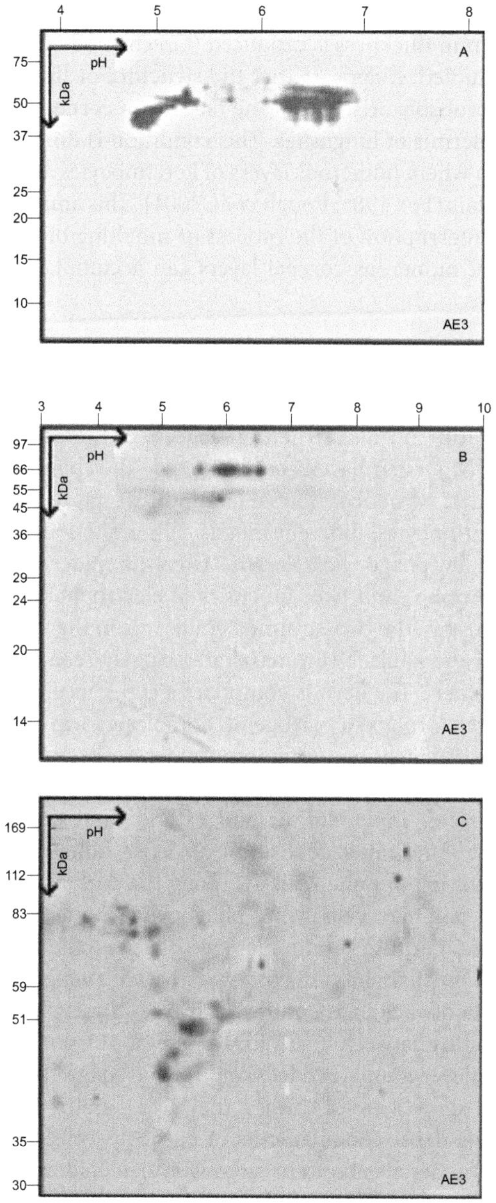

Fig. 10 Two-dimensional electrophoretic pattern using AE1 antibody for proteins in whole epidermis of the australian lungfish (*N. forsteri*). AE3 immunolabelled spots in three different samples are shown in Panels A, B, and C. A and B, 12% polyacrylamide gels. C, 10% polyacrylamide gels. Numbers in abscissa indicate pH, those in ordinals indicate molecular weight.

P. annectens above a 4-5 multilayered epidermis made of flat cells, a layer of dry mucus of 50-60 μm in thickness is produced (Greenwood 1984). Therefore in the dipnoans so far studied, it appears that the structure of the cocoon is acellular, which condition probably arises from the fact that a corneal layer is not-formed in the normal epidermis of lungfishes. This condition is different from that of an amphibian cocoon where numerous layers of keratinocytes are deposited over the aestivating epidermis (Fox 1986; Pough *et al.* 2001). This amphibian characteristic derives from the interruption of the process of moulting of the epidermis under aestivation, so that numerous corneal layers can accumulate one on top of the other.

Summary and Conclusions

The main histological and ultrastructural aspects of the epidermis and dermis, with emphasis on *N. forsteri* have been presented. The comparison with the skin of *P. annectens* and *L. paradoxa* shows a common structure of the epidermis and dermis. Aside keratinocytes, different mucus cells are found, a minor number of melanocytes and a few phagocytes. Keratins from the epidermis of *N. forsteri* have been analysed by mono- and two-dimensional electrophoresis and localised by immunocytochemistry. The two-stratified epidermis in the tadpole skin becomes 5-7 layers thick in the adult. Neuromasts are sparsely seen in advanced tadpole epidermis of *N. forsteri*. The dermis comprises a superficial, thin dense layer and a thicker loose layer. Groups of osteogenic fibroblasts form repeated units in the superficial dermis of hatchlings, and among these cells an initial deposition of collagen fibrils and later osteoid matrix occurs. In early juveniles these fibroblasts migrate into the softer inner dermis and extend the osteoid matrix, perhaps containing elastine. The initial deposition of dense mineral component occurs and form the beginning of bone scales. Among the dark and pale keratinocytes, mucus cells and granular cells with phagocytic characteristics are present. Keratinocytes contain diffuse keratin filaments that become more densely packed in external cells. These filaments are cross-reactive with general keratin antibodies but keratins are mainly acidic as confirmed by two-dimensional electrophoresis. Most lungfish keratins between 40-57 kDa posses a pI from 4.2 to 7.1. The latter value is the highest ever found for a fish keratin, and suggests that neutral keratins of sarcopterigians show a trend toward the basic keratins found in tetrapods. The latter make the dense tonofilaments of corneous cells forming the stratum corneum, which is generally absent in sarcopterygian epidermis.

The skin of the lungfish, while sharing general characteristics with other fish also shares a trend towards the more basic keratins typical of terrestrial vertebrates. Mucus production is very high, as is the case for the epidermis of other fish living in shallow fresh waters subject to periodical/occasional drought, where mucus protection is essential for survival. The epidermis comprises several layers but all contain nucleated live cells which should be capable of some gaseous exchange.

However, it is not highly vascularized as it is in most amphibians. The dermis produces large and heavy scales that would reduce the efficiency of exchanging gases between the blood and the external environment in these fish, which have both gills and lungs for this purpose. Thus as freshwater fish, the lungfish skin is predominantly performing the functions appropriate to this environment but also shares some characteristics of its keratins with those of tetrapods.

Methods for Typing Keratins in *N. forsteri*

Epidermal proteins were extracted by the method of Sybert *et al.* as previously reported (Alibardi and Toni 2004). Briefly, the skin was incubated in 5 mM EDTA in phosphate buffered saline (PBS) for 3-5 min at 50 C and 2-4 min in cold PBS. The epidermis was separated from the dermis by dissection using a stereomicroscope. The epidermis was homogenized in 8 M urea/50 mM Tris-HCl (pH 7.6)/0.1 M 2-mercaptoethanol/1 mM dithiothreithol/1 mM phenylmethylsulphonyl fluoride and let stirring for 5 hours at room temperature in nitrogen atmosphere to allow protein denaturation Then, the particulate matter was removed by centrifugation at 10,000 g for 10 min. The protein concentration was determined by the Lowry method. To calculate the protein concentration the O.D. was compared to a standard curve built by using BSA serial dilutions in the same extraction buffer used for protein homogenization.

For mono-dimensional electrophoresis experiments, proteins were denatured by boiling in the Sample Buffer for 5 minutes. Then, 50 mg of proteins were loaded in each lane and separated in SDS-polyacrylamide gels (SDS-PAGE) according to Laemmli. For bi-dimensional electrophoresis experiments, the Ettan IPGphor III IEF System (Amersham, U.K.) was used for the isoelectrofocusing (IEF). The protein sample (10 mg for silver staining or 150 mg for western blot) containing 2% CHAPS (Sigma, U.S.A.) and 1% carrier ampholyte mixture, pH 3,5-10 (Amersham, U.K.) was loaded on a 7 or 13 cm strips (pH 3-10, GeHealthcare, U.K.), depending on the experiment. Application of the strips and running procedure was carried out as described by the manufacturer. The following protocol was used. Rehydration was performed for 12 h at room temperature and was followed by the IEF, step by step, from 1/2 h 500 V, 1/2 h 1000 V, 1/2 h 5000 V or 8000 V gradually and for 1 h at 5000 V or 8000 V (5000 and 8000 V were applied to 7 and 13 cm strip, respectively). Strips were kept at 50 V until loaded on the second dimension. Before starting the second dimension, the strips were equilibrated for 10 min in 6 M urea, 30% glycerol, 50 mM Tris pH 6.8, and 2% DTT. Afterward, strips were briefly rinsed with double distilled water and equilibrated in 6 M urea, 30% glycerol, 50 mM Tris pH 6.8, and 2.5% iodoacetamide for an additional 10 minutes. The second dimension was carried out in a MiniProtean III electrophoresis apparatus (Biorad, U.S.A.) for 7 cm strips and in a 16 × 18 cm gel electrophoresis apparatus for 13 cm strips. 10%, 12% or 4-12% gradient (Criterion XT Precast Gel, BioRad, U.S.A.) SDS-polyacrylamide gels were used, depending on the experiment.

Successively, gels were stained in Blue Comassie, Silver (ProteoSilver Plus, Sigma, U.S.A.), or immunoblotted on nitrocellulose paper (Hybond C+ Extra, Amersham, U.K.), depending on the experiment. In electrophoresis experiments Wide Range (M.W. 6,500-205,000) (Sigma, USA), Prestained (Sigma, U.S.A.) or Precision Plus Protein Standard (M.W. 10,000-250,000; Biorad, U.S.A.) molecular weight marker were used. After western blot, membranes were stained with Ponceau red to verify the protein transfer and incubated with primary antibodies. The following primary antibodies were used: AE1, AE3, K13, K18, K19 and K8, diluted 1:50; K6, K16 and K17, diluted 1:1000; Pan-cytokeratin-ab (Sigma, U.S.A.), diluted 1:400; K10, diluted 1:250. AE1 and AE3 antibodies were a generous gift from Dr. T.T. Sun (NY University, U.S.A) and K6, K16 and K17 from Dr. P. Coulombe (Johns Hopkins University, U.S.A.; see Takahashi *et al.*, 1994; McGowan and Coulomb 1998a,b). The secondary antibodies HRP-conjugated (Sigma, USA) wad diluted 1:1000 in TBS-TWEEN + 5% non-fat milk powder. In controls, the primary antibody was omitted. Detection was done using the enhanced chemiluminescenceís procedure developed by Amersham (ECL, Amersham, U.K.).

Acknowledgments

The study was largely self-supported (L. Alibardi) and partially from a University of Bologna grant (60%, 2005-2006, L. Alibardi), and by the Australian Research Council (Joss, 40%).

References

Alibardi, L. (2001). Keratinization in the epidermis of amphibian and the lungfish: comparison with amniote keratinization. Tissue & Cell 33: 439-449

Alibardi, L. (2006). Structural and immunocytochemical characterization of keratinization in vertebrate epidermis and epidermal derivatives. International Review of Cytology 253: 177-259.

Alibardi, L. and Joss, J. (2003). Keratinization of the epidermis of the Australian lungfish *Neoceratodus forsteri* (Dipnoi). Journal of Morphology 256: 13-22.

Alibardi, L. and Toni, M. (2004). Immunolocalization of transglutaminase and cornification markers proteins in the epidermis of vertebrates suggests common processess of cornification across species. Journal of Experimental Zoology 302B: 526-549.

Berheither-Hahn, J., Osborn, M., Weber, K. and Voth, M. (1979). Filament organization and formation of microridges at the surface of the epidermis. Journal of Ultrastructural Research 69: 316-330.

Chavin, W. and Imaki, H. (1975). Ultrastructure of the integumental melanophores of the South American lungfish (*Lepidosiren paradoxa*) and of the African lungfish (*Protopterus* sp.). Cell & Tissue Research 158: 375-389.

Coulombe, P.A. and Omary, M.B. (2000). 'Hard' and 'soft' principles defining the structure, function and regulation of keratin intermediate filaments. Current Opinion in Cell Biology 14: 110-122.

Delfino, G., Brizzi, R. and Calloni, C. (1982). Development of cutaneous glands in *Salamandrina terdigitata* (Lacepede, 1788) (Amphibia: Urodela); findings by light and electron microscopy. Zeitchfrisch fur Mikroskopie und Anatomie Forschlung 96: 948-71.

Delfino, G., Brizzi, R. and Calloni, C. (1985). Dermo-epithelial interactions during the development of cutaneous gland anlagen in Amphibia: a light and electron microscope study on several species with some cytochemical findings. Zeitchfrisch fur Mikroskopie und Anatomie Forschlung 99: 225-253.

Fast, M.D., Sims, D.E., Burka, J.F., Mustafa, A. and Ross, N.W. (2002). Skin morphology and humoral non-specific defence parameters of mucus and plasma in rainbow trout, coho and Atlantic salmon. Comparative Biochemistry and Physiology 132A: 645-657.

Fox, H. (1986). Amphibian skin: dermis. In: X.Bereither-Hahn, G. Matoltsy and K. Sylvia-Richard (eds). Biology of the Integument. Springer-Verlag, Berlin-New York, pp. 78-110.

Greenwood, P.H. (1986). The natural history of African lungfishes. Journal of Morphology Supplement 1: 163-179.

Fishman, A.P., Pack, A.I., Delaney, R.G. and Galante, R.J. (1986). Estivation in *Protopterus*. Jounal of Morphology Supplement 1: 237-248.

Imaki, H. and Chavin, W. (1975a). Ultrastructure of the integumental melanophores of the South American lungfish (*Lepidosiren paradoxa*). Cell & Tissue Research 158: 375-389.

Imaki, H. and Chavin, W. (1975b). Ultrastructure of the integumental melanophores of the Australian lungfish, *Neoceratodus forsteri*. Cell & Tissue Research 158: 363-373.

Imaki, H. and Chevin, W. (1984). Ultrastructure of mucous cells in the sarcopterygian integument. Scanning Electron Microscopy 1984/1: 409-422.

Kitzan, S.M. and Sweeny, P.R. (1968). A light and electron microscope study of the structure of *Protopterus annectens* epidermis. I. Mucus production. Canadian Journal of Zoology 46: 767-779.

Moll, R., Franke, W.W., Schiller, D.L., Geiger, B. and Krepler, R. (1982). The catalogue of human cytokeratins: patterns of expression in normal epithelia, tumours and cultured cells. Cell 31: 11-24.

O'Guin, M.W., Galvin, S., Shermer, A. and Sun, T.T. (1987). Pattern of keratin expression define distinct pathways of epithelial development and differentiation. Current Topics in Developmental Biology 22: 97-125.

Pough, F.H., Andrews, R.M., Cadle, J.E., Crump, M.L., Savitzky, A.H. and Wells, K.D. (2001). Herpetology, 2nd Edition. Prentice Hall, Upper Saddle River, NJ, USA.

Sire, J.Y. (1993). Development and fine structure of the bony scutes in *Corydoras arcuatus* (Siluriformes, Callichthydae). Journal of Morphology 215: 225-244.

Sire, J.Y. and Huysseune, A. (2003). Formation of dermal skeletal tissues in fish: a comparative and evolutionary approach. Buletin of Biology 78: 219-249.

Schaffeld, M., Bremer, M., Hunzinger, C. and Markl, J. (2005). Evolution of tissue-specific keratins as deduced from novel cDNA sequences of the lungfish *Protopterus aethiopicus*. European Journal of Cell Biology 84: 363-377.

Sturla, M., Prato, P., Grattarola, C., Masini, M.A. and Uva, B.M. (2002). Effects of induced aestivation in *Protopterus annectens*: a histomorphological study. Journal of Experimental Zoology 292: 26-31.

Sun, T.T., Eichner, R., Nelson, W.G., Tseng, S.C.G., Weiss, R.A., Jarvinen, M. and Woodckoc-Mitchell, J. (1983). Keratin classess: molecular markers for different types of epithelial differentiation. Journal of Investigative Dermatology 81: 109-115.

Whitear, M. (1977). A functional comparison between the epidermis of fish and of amphibians. In: R.I.C. Spearman (ed). Comparative Biology of the Skin. Academic Press, London, UK. pp. 291-313.

Whitear, M. (1986a). The skin of fish including cyclostomes. Epidermis. In: A.G. Matoltsy, J. Bereither Hahn, K. Sylvia-Richards (eds). Biology of the Integument, Part B, Vertebrates. Springer-Verlag, Berlin and New York. pp. 8-38.

Whitear, M. (1986b). The skin of fish including cyclostomes. Dermis. In: A.G. Matoltsy, J. Bereither Hahn, K. Sylvia-Richards (eds). Biology of the Integument, Part B, Vertebrates. Springer-Verlag: Berlin and New York. pp. 39-55.

Zaccone, G., Kapoor, B.G., Fasulo, S., and Ainis, L. (2001). Structural, biochemical and functional aspects of the epidermis of fishes. Advances in Marine Biology 40: 255-348.

Zylberberg, L. and Meunier, F.J. (1996). Ultrastructure of the melanophores associated with the cellular elasmoid scales in *Leporinus friderici* (Teleostei: Ostariophisi, Anastomidae): their putative participation in scale matrix formation. Journal of Morphology 228: 155-164.

CHAPTER 10

Respiratory Function in Lungfish (Dipnoi) and a Comparison to Land Vertebrates

Mogens L. Glass*

Departamento de Fisiologia, Faculdade de Medicina de Ribeirão Preto, Universidade de São Paulo, Avenida Bandeirantes 3900, 14.049-900 Ribeirão Preto, SP, Brazil

ABSTRACT

The lungfish (*Dipnoi*) forms a probable sister group to the land vertebrates (Tetrapoda) and the two ramifications share many features of pulmonary function. The African lungfish, *Protopterus* (4 species) and the South American lungfish (*Lepidosiren paradoxa*) (1 species) have well developed lungs combined with rudimentary gills. Differently, the Australian lungfish (*Neoceratodus*) predominantly depends on gill ventilation, while its lung structure is more simple.

Like the land vertebrates, *Lepidosiren* and *Protopterus* possess central cerebral CO_2 and H^+ receptors that monitor acid-base status by adjustments of pulmonary ventilation, while less information is available for *Neoceratodus*. In land vertebrates, the central cerebral receptors account for 70 to 80% of the response to hypercarbia, while the peripheral blood screening CO_2/pH-receptors account for 20 to 30%. In *Lepidosiren*, the peripheral CO_2/H^+-receptors accounted for 20% of the hypercarbia-induced ventilatory responses, which suggests a very early origin of the central and peripheral CO_2/H^+ receptors. In tetrapods, the specific O_2-stimulus is O_2 partial pressure rather than O_2-content, which also applies to *Lepidosiren*. When exposed to higher temperatures, lungfish and amphibians become more dependent on the lung for CO_2-elimination. This increases intrapulmonary and arterial PCO_2 which, in turn, accounts for a negative $\Delta pHa/\Delta t$ as expected for ectothermic vertebrates. Recent publications emphasize a very similar respiratory control in lungfish and the amphibians, which raises a number of questions about a common ancestor.

*E-mail: mlglass@rfi.fmrp.usp.br

Keywords: lungfish, diffusing capacity, acid-base regulation, chemoreceptors, hypercarbia, hypoxia, aestivation, *Dipnoi*, temperature, *Lepidosiren*, *Neoceratodus*, *Protopterus*

INTRODUCTION

Sarcopterygians (lobe-finned fish) assume an evolutionary key position, because their descendants are the land vertebrates (Tetrapoda) and the lungfish (Dipnoi) that probably form a sister group (Meyer and Dolven 1992; Yokobori *et al.* 1994; Zardoya *et al.* 1998; Toyama *et al.* 2000; Brinkmann *et al.* 2004). A third descendant is *Latimeria,* the coelacanth (Actinistia) that also has a lung which, however, is filled with fat (Carroll 1988). For a long time, fossil evidence was missing for the ramifications tetrapods and lungfish, but recently an earlier unknown fossil (Styloichthys) was discovered, and its features fits those expected from a last common ancestor of tetrapods and lungfish, and the age for this animal was estimated to about 417 million years (Zhu and Xu 2002).

Extant lungfish include the rather heavily built Australian lungfish *(Neoceratodus forsteri*) that inhabits rivers in the Queensland region of Australia. It has a well developed and effective gill system combined with a single lung, which is a very unique feature. Only the right lung develops in *Neoceratodus* and blood is supplied by both pulmonary arteries (Perry 2007; Kind *et al.* 2002). Differently, Lepidosireniformes (*Lepidosiren* and *Protopterus*) are obligatory air-breathers with well-developed lungs and rudimentary gills, and are slender, weighing up to a few kilos. *Protopterus* is represented by four species in Africa, whereas the South American lungfish *Lepidosiren paradoxa* is the only species of its genus (Johansen and Lenfant 1967). Figure 1 shows *L. paradoxa* in the laboratory.

Lepidosireniform lungfish inhabit shallow slow moving waters that may dry out on a seasonal basis. *L. paradoxa* inhabits the Amazon and Paraná-Paraguai region, where shallow lakes can be partly covered by vegetation. This species feeds on mollusks and other invertebrates but may occasionally eat vegetables (Sawaya 1946). Like amphibians, lepidosirenid lungfish have "larval" forms equipped with external gills, and aerial respiration is initiated at a length of 2 to 3 cm in *P. aethiopicus* (Greenwood 1958).

This chapter is primarily dedicated to respiratory control in lungfish exposed to rather challenging ambient conditions, such as hypercarbia, hypoxia or the drying out of the habitat. The emphasis is acid-base regulation and O_2-homeostasis and will accentuate, that respiratory control in lungfish and tetrapods has many very specific features in common, which fortifies the concept of lungfish and tetrapods as a sister group. Temperature effects on respiratory function will be presented along with special conditions such as aestivation.

Fig. 1 Photo of the South American lungfish (*Lepidosiren paradoxa*) in the laboratory.

THE LUNG

Protopterus and *Lepidosiren* produce surfactants that resample those of amphibians, but the surfactant of *Neoceratodus* has a lipid composition similar to that of actinopterygian fish (Orgeig and Daniels 1995). In addition, there is evidence that *Neoceratodus* has preserved the same surfactant composition during 300 million years (Power *et al.* 1999).

The pulmonary gas diffusing capacity is defined by the equation (Bohr 1909),

$$D_LO_2 = VO_2 \cdot (P_LO_2 - PcO_2)^{-1},$$

where VO_2 is the O_2 flow across the barrier, separating lung gas and capillary blood. P_LO_2 is the PO_2 of the intrapulmonary gas, and PcO_2 represents the mean value for PO_2 in the pulmonary capillaries. The difference ($P_LO_2 - PcO_2$) is often abbreviated as ΔPO_2, and the equation can be simplified to,

$$D_LO_2 = VO_2 \cdot \Delta PO_2^{-1},$$

which defines D_LO_2 as an O_2 conductance, reminding of the law of Ohm: $G = I \cdot \Delta V$.

Using the approach above, it turned out that *L. paradoxa* has a D_LO_2 within the same range as the bullfrog *Rana catesbeiana* (Bassi *et al.* 2005; Glass *et al.* 1981a).

A parallel morphological and anatomical study proved that the lung of *L. paradoxa* is, by far, its dominant gas exchange organ (Morais *et al.* 2005). The morphological D_LO_2 is based on the Fick diffusing equation:

$$D_MO_2 = KO_2 \cdot A \cdot T^{-1} \cdot \Delta PO_2,$$

where A is the respiratory surface area of the lung; T = thickness of the membrane and KO_2 = the Krogh diffusion constant for gas-blood barrier of the lung. Morphological diffusing capacity always exceeds the values obtained by physiological methods (Scotto *et al.* 1987). The probable range for diffusing capacity of *L. paradoxa* can, however, be estimated as ($mlSTPD \cdot kg^{-1} \cdot min^{-1} \cdot mmHg^{-1}$): 0.04 (physiological value) < 0.10 (morphometric value). The values for diffusing capacity in *L. paradoxa* and *Rana catesbeiana* are low relative to those of some reptiles, including the tegu lizard *Tupinambis merianae* and the monitor lizard *Varanus exanthematicus*, but the difference is no more than 2-fold (Glass and Johansen 1982; Glass *et al.* 1981b). If the lungfish is compared to equal-sized mammals, then the alveolar lung has a D_LO_2, which is 16-fold higher (Takezawa *et al.* 1980). With the avian tubular system, the duck (*Carina moscata*) reaches a 40-fold higher D_LO_2, when compared to *L. paradoxa* (Shams and Scheid 1989). The relative values for D_LO_2, VO_2 and ΔO_2 for various groups of vertebrates are compared in Fig. 2.

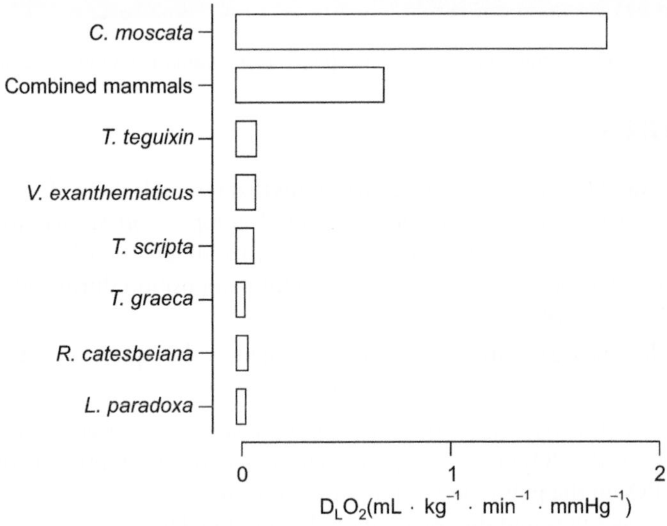

Fig 2 This figure illustrates the increase of pulmonary oxygen diffusing capacity (D_LO_2) of vertebrate lungs from lungfish to birds. The corresponding values for O_2-uptake (VO_2) are indicated within or next to the columns. Notice the large parallel increases of capacity and O_2 consumption. The animals are placed in increasing magnitude of the capacity (1) South American lungfish *L. paradoxa* (Bassi *et al.* 2005) (2) Bullfrog, *Chaunus schneideri* (earlier *Rana catesbeiana*) (Glass *et al.* 1981a) (3). Greek tortoise, *Testudo graeca* (Crawford *et al.* 1976). (4) Common slider *Trachemys scripta* (Earlier *Pseudemys scripta*) (Crawford *et al.* 1976). (5) Monitor lizard, *Varanus exanthematicus* (Glass *et al.* 1981b). (6). Tegu lizard, *Tupinambis teguixin* (Glass and Johansen, 1982). (7) Combined small mammals (Takezawa *et al.* 1980). (8) Pekin duck, *Carina moscata* (Shams and Scheid 1989).

D_LO_2 increases with higher temperatures, which is not immediately expected, because of the low Q_{10} of the Krogh diffusing constant (KO_2: $Q_{10} = 1.1$). Moreover, the Fick equation states that D_LO_2 depends on the lung structure expressed as gas exchange surface and the thickness of the gas-blood (see above). From these considerations, it is expected that D_LO_2 would depend little on temperature, but this is not necessarily the case when physiological measurements are considered. Instead, it turns out that the physiologically measured D_LO_2-values increase considerably at higher temperatures. As examples, D_LO_2 of *L. paradoxa* increased from 0.010 $mlSTPD \cdot kg^{-1} \cdot min^{-1} \cdot mmHg^{-1}$ (t= 25°C) to 0.044 $mlSTPD \cdot kg^{-1} \cdot min^{-1} \cdot mmHg^{-1}$ at 35°C (Bassi *et al.* 2005). Consistently, the D_LO_2 of the bullfrog (*Rana catesbeiana*) increased 2-fold between 20 and 30°C, while the capacity of the tegu lizard (*Tupinambis teguixin*) increased as much as 3-fold between 25 and 35°C (Glass and Johansen 1982). These large apparent increases of D_LO_2 can be due to reduced inhomogeneities and recruitment of more lung capillaries and/or distension of pulmonary capillaries, which would increase the contact surface between intrapulmonary gas and the pulmonary capillary blood (cf. Powell and Hempleman 1993). Looking at Fig. 2 it becomes clear that the large step for increases of D_LO_2 was associated with the transition from ectothermic to endothermic metabolism, which could only take place by increased surface area and reduced barrier thickness as seen in tubular or alveolar lungs.

REGULATION OF ACID-BASE STATUS AND OXYGEN LEVELS

The special features of respiratory control in lungfish become more obvious, after a short look at the holeost and teleost fish. Teleosts possess branchial O_2-receptors that monitor water and blood (Burleson and Milsom 1990, 1995a and b; Soncini and Glass 2000). As an important aspect, CO_2-receptors seem to be present in the rainbow trout (*Oncorhynchus mykiss*) (Perry and Reid 2002), whereas the existence of central chemoreceptors in teleost fish is an unsettled issue (Milsom 2002). Teleost fish such as the snakehead fish *Channa argus* may possess air-breathing organs and respond to gas phase hypoxia but failed to increase aerial ventilation in response to hypercarbia (Glass *et al.* 1986).

The holeost fish *Amia calva* is an air-breather, which motivated Hedrick *et al.* (1991) to superfuse mock CSF at positions that were likely to stimulate central receptors, but the stimulation had no effect on aerial ventilation. More promising, Wilson *et al.* (2000) worked on the *in vitro* brain stem preparation of the long-nosed gar pike (*Lepisosteus osseus*), and they reported an increased fictive air-breathing in response to low pH. This frequency response was, however, much higher than expected from measurement on the intact animal (Milsom 2002).

Teleosts regulate acid-base status through ion exchange (Heisler 1984), which liberates gill ventilation to control O_2-homeostasis (Dejours 1981; Milsom 2002).

Specialized cells in the gill epithelia carry out 90% of the acid-base relevant ion transfers, leaving 10% to the kidneys (Heisler 1984; Claiborne and Heisler 1986). This type of cell was not found within the rudimentary gill system of *L. paradoxa* (Moraes *et al.* 2005) and hypercarbia for 48 h left no sign of active increases of extra-cellular [HCO_3^-] levels. It is, however, possible that the intracellular compartment was modulated, while extra-cellular responses were absent (cf. Heisler *et al.* 1982).

Air-breathing organs may increase survival in hypoxic and hypercarbic water, which explains why pulmonary respiration evolved long before the transition from the aquatic environment to the terrestrial mode of life. Consistently, many components of respiratory control are common to lungfish and land vertebrates, which inspired Johansen *et al.* (1967) to state: "It is generally accepted that vertebrates acquired functional lungs before they possessed a locomotionary apparatus for invasion of the terrestrial environment". Like the land vertebrates, *L. paradoxa* and *P. annectens* possess central chemoreceptors (Sanchez *et al.* 2001; Gilmour *et al.* 2007) and it turns out that CO_2 and H^+ act as specific stimuli for the central receptors that monitor acid-base status (Amin-Naves *et al.* 2007a). Nattie (1999) suggested that "central chemoreceptors detect interstitial fluid pH and monitor the balance of arterial CO_2, cerebral blood flow, and central metabolism". A dual mode of stimulation (CO_2 and H^+) is present in mammals (Shams 1985), in toads (*Chaunus schneideri*; earlier *Bufo paracnemis*) (Smatresk and Smits 1991) and *L. paradoxa* (Amin-Naves *et al.* 2007a and b). In addition, the presence of central chemoreceptors in the South American lungfish were detected by superfusion of the 4th cerebral ventricle with mock CSF solutions, ranging from pH = 7.40 to 8.00. Stepwise reductions of pH were accompanied by corresponding increases of pulmonary ventilation (Sanchez *et al.* 2001a). Consistently, Gilmour *et al.* (2007) provided evidence for central chemoreceptors in the African lungfish *P. annectens*.

In addition, Amin-Naves *et al.* (2007b) assessed the relative roles of peripheral and central CO_2/H^+ receptors that underlie ventilatory responses, controlling acid-base status in *L. paradoxa*. Initially, the animal was kept in room air and well aerated water, after which hypercarbia was flushed to the gas phase and to the water (P_ICO_2=49 mmHg). This served to stimulate both central and peripheral CO_2/H^+ receptors. A large ventilatory response was reached after 3 h, after which hypercarbia was combined with central superfusion of the 4th ventricle with mock CSF at pH=7.45; PCO_2=21 mmHg, which returned CSF pH to normocarbic conditions. This reduced the ventilatory response, but the level of ventilation remained higher than the initial normocarbic control value. This remaining hyperventilation could then quantify the contribution of peripheral CO_2/H^+-receptors (see Fig. 3). The peripheral chemoreceptor drive turned out to account for 20% of the hypercarbia-induced response. In tetrapods, the peripheral chemoreceptors account for 20 to 40% of the hypercarbia-induced ventilatory

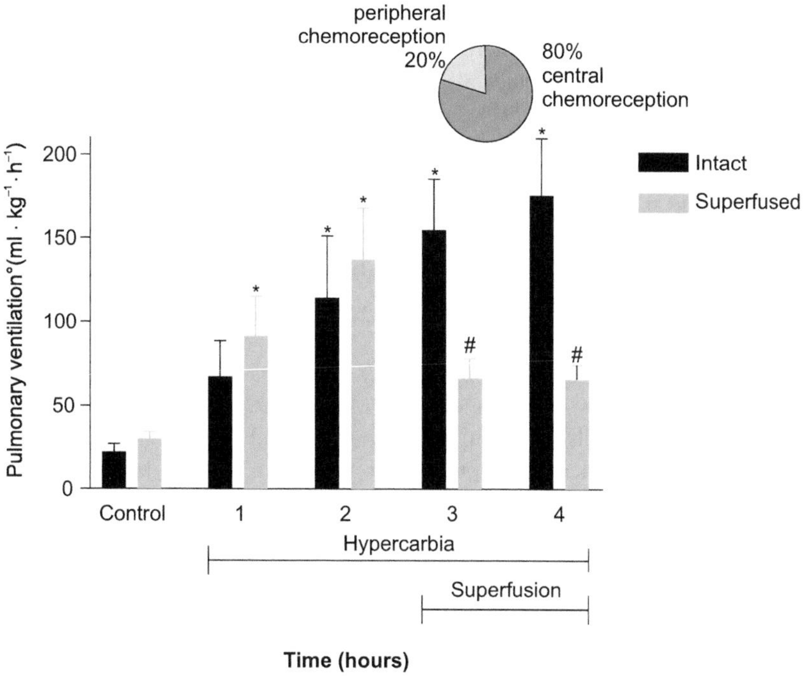

Fig. 3 This figure explains how to evaluate the relative roles of central and peripheral CO_2/H^+ drives to pulmonary ventilation as modified for *L. paradoxa* at 25°C. Intact animals were initially kept in normocarbic conditions (time zero), after which hypercarbia was applied for up to 5 h, during which the ventilatory response gradually increased. This procedure was repeated with the catheterized animals, which allowed to superfuse the cerebral ventricles for the last h with mock CSF with normocarbic mock CSF (pH=7.45; PCO_2=21 mmHg), so that only the peripheral CO_2/H^+-receptors would be stimulated. Thus, the remaining hyperventilation could be attributed to the peripheral CO_2/H^+-receptors. The result was that 20% of the ventilatory response could be attributed to the peripheral CO_2/H^+ receptors, whereas the central drive accounted for 80%. (Based on Amin-Naves *et al.* 2007a.)

responses, whereas the central CO_2/H^+-receptors contribute 60 to 80% (Amin-Naves *et al.* 2007b).

A specific question is: what is/or are the modality or modalities to acid-base related ventilatory responses? CO_2 can easily traverse the blood-brain barrier, whereas the blood-brain barrier is practically impermeable to H^+ and HCO_3^-. Once within the barrier, CO_2 reacts via the carbonic anhydrase reaction: $CO_2 + H_2O \Leftrightarrow H_2CO_3 \Leftrightarrow H^+ + [HCO_3^-]$, which could lead to the concept of H^+ as the only stimulus to the receptors (cf. Loeschcke *et al.* 1958; Lahiri and Forster 2003). Working on anaesthetized cats, Shams (1985) reported that both CO_2 and H^+ stimulate the receptor system. This finding was backed up by Harada *et al.* (1985), who worked on isolated brainstem preparations of newborn rats, and the authors found that an elevated PCO_2 increased respiratory output of the phrenic nerve, while pH was kept constant (see also Eldridge *et al.* (1985) and Nattie (1999)). Later, Smatresk and Smits (1991) reported that central chemoreceptors in toads (*Bufo marinus*) were stimulated not only by H^+ but also by CO_2. This dual mode of reception also applies to *L. paradoxa* (Amin-Naves *et al.* 2007b), since ventilation increased with elevated PCO_2 of the mock CSF, while pH was kept constant. Conversely, reduced pH increased ventilation, when PCO_2 was kept constant. The presence of these dual modes of stimulation both in lungfish, amphibians and mammals may suggest an origin from a common ancestor for lungfish and tetrapods (Fig. 4).

Many tetrapods possess intrapulmonary receptors that are stimulated by stretch and inhibited by increases of CO_2 (Milsom 1995, 2002), and these very specific receptors are also present in *L. paradoxa* and *Protopterus* (Delaney *et al.* 1983). As a hallmark of their presence, return from hypercarbia to room air elicits a transient hyperventilation that initially exceeds values during the CO_2 exposure. This response fades away to reach normocarbic values (Milsom 1995,

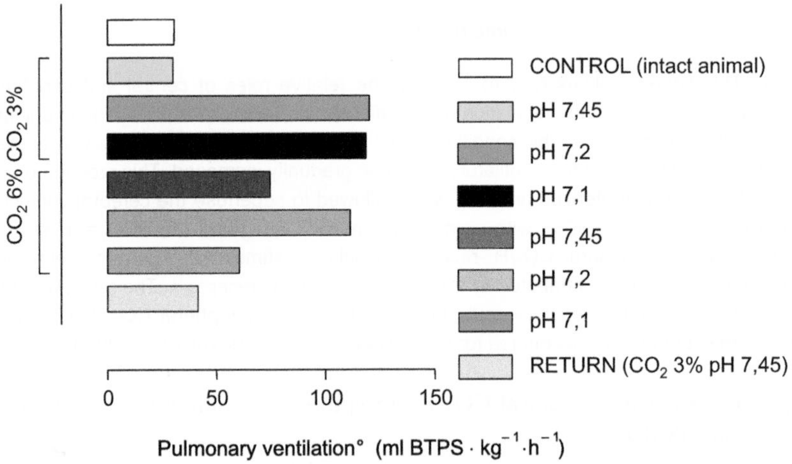

Fig. 4 This figure shows that both CO_2 and pH act as stimuli to the central ventilatory drive. Upper column is the control value for the intact animal. Below is the value for the catheterized animal at superfused with normocarbic mock CSF pH =7.45; PCO_2 ~ 21 mmHg). The third column shows the effects of reduction of pH from 7.45 to 7.20, which confirms that pH can stimulate, while PCO_2 kept at 21 mmHg. Conversely, the fifth column shows that superfusion with (pH=7.45; PCO_2 = 42 mmHg), if PCO_2 is increased from 21 to 42 mmHg, while mock CSF pH is kept constant at 7.45. A transition from pH 7.20 to 7.10 will, however, failed to increase ventilation, which could indicate that the value of pH was outside the physiological range. From Amin-Naves *et al.* (2007).

2002). *Lepidosiren* has this response (Sanchez and Glass 2001), which is consistent with the neural recordings by DeLaney *et al.* (1983). Like tetrapods, *P. aethiopicus* possess a Hering-Breuer reflex, which acts by buccal compression to inflate the lung (Pack *et al.* 1990, 1992).

As mentioned above, teleost fish possess O_2-receptors sensitive to reductions of $[O_2]$ (Soncini and Glass 2000). Differently, the normal O_2-stimulus in land vertebrates is the O_2 partial pressure and neither O_2-content nor HbO_2-saturation are involved (Comroe 1974). Studying *Protopterus* sp., Lahiri *et al.* (1970) injected hypoxic blood and cyanide into the afferent gill arteries, which caused a ventilatory response. Moreover, bilateral section of the first three gill arches reduced responses to hypoxic blood and practically abolished the response to cyanide. These data, however, cannot provide information about the specific O_2-stimulus.

The ventilatory response in the toad *Chaunus schneideri* was not affected by a reduction of the Hb-O_2 carrying capacity, which indicated that PaO_2 (the partial pressure of O_2 in arterial blood) is the regulated variable (Wang *et al.* 1994). Later, the CO method was used to reduce Hb-O_2 capacity, and the ventilatory responses once more confirmed PaO_2 as the specific O_2 stimulus (Branco and Glass 1995). These methods were also used to evaluate the specific O_2 stimulus in *L. paradoxa*. It turned out that aquatic hypoxia had no immediate effect on pulmonary ventilation, which indicated a primary role for internal rather than branchial receptors. Blood oxygen and hematocrit could be reduced by as much as 50%, while pulmonary ventilation was not stimulated at all, which points to PaO_2 as the specific O_2 stimulus of this lungfish.

The Australian lungfish, *Neoceratodus forsteri* Krefft looks like the Devonian fossilized ancestors such as *Dipterus* (Carroll 1988). In this context, it has been proposed that *N. forsteri* is an obligate neotene animal, which is a larval form that can reproduce. This idea gains interest, since it is backed up by deficiencies in thyroid function. Neotenic features in *Lepidosiren* and *Protopterus* are under discussion (Joss and Johanson 2007). *N. forsteri* inhabits river systems in south-east Queensland region and aestivation is absent. As a special feature, it has electroreceptors to locate prey (Watt *et al.* 1999). In a pioneering study, Johansen *et al.* (1967) evaluated respiratory function in *N. forsteri* and reported that more than 1 h could pass between air breaths (temp. 18°C). They also found that hypoxic water resulted in large increases of both branchial and pulmonary ventilation. More recently, Kind *et al.* (2002) reported a nearly 8-fold increase of air-breaths with reduction of O_2 from 120 mmHg to 40 mmHg. Air-breathing in *N. forsteri* is accompanied by a burst of branchial movements and a concomitant increase of pulmonary blood flow (Fritsche *et al.* 1993). These responses are consistent with the earlier data by Johansen *et al.* (1967). In *N. forsteri* the dependence on gill respiration leads to a low $PaCO_2$ (3.6 mmHg) and a high pH (7.64); temperature 18°C. Conversely, *L. paradoxa* and *Protopterus* sp. have high $PaCO_2$-values that fall within the range from 17 to 27 mmHg combined with low pH-values (7.45 to 7.55) (cf. Amin-Naves *et al.* in press). This is expected from Dejours (1981), who explains that aquatic breathing lowers $PaCO_2$, because the capacitance ratio ($ßwCO_2 \cdot ßwO_2^{-1}$) is low ($1{\cdot}30^{-1}$), while the capacitances for CO_2 and O_2 in air are the same.

AESTIVATION

Aestivation can be induced by unfavorable ambient conditions, which typically could be food shortage or drought (Abe 1995), and it is a dormancy that involves downregulation of metabolism. Aestivation differs from hibernation, because metabolism falls without any reduction of ambient temperature. In lungfish and amphibians, this condition is accompanied by huge reductions of O_2 uptake and downregulation of gas exchange, heart rate and cardiac work (Glass *et al.* 1997). The O_2 uptake of aestivating Urodele amphibians is often reduced to as little as 30% of the value for non-aestivating resting conditions (McClanahan *et al.* 1983; Flanigan *et al.* 1991). Metabolic reductions of the same magnitude have been reported for urodele amphibians (Etheridge 1990).

Aestivation in lungfish has attracted much research since the pioneering work by Smith (1935). In particular, cocooned *P. amphibius* can exist for 7 years, which seems to be the record (Lomholt 1993). In one animal, measurements were initially studied when in water, and pulmonary ventilation was 5 $mLBTPS{\cdot}kg^{-1}{\cdot}min^{-1}$, and 6 years of aestivation reduced ventilation to 0.7 $mLBTPS{\cdot}kg^{-1}{\cdot}min^{-1}$ (Lomholt 1993). *Protopterus* possesses very reduced gills and a well-developed lung (Fishman *et al.* 1986). When the lakes dry out, *P. aethipicus* secretes mucous that hardens over a few days to form a cocoon, protecting the animal. As an interesting point, Lomholt (1993) reported that as body mass was lost, but *P. amphibius* still filled the whole cocoon by an increased lung volume, which would reduce water losses and, in addition, eliminate non-pulmonary gas exchange. DeLaney *et al.* (1974) reported that mean blood pressure and heart rate of *P. aethiopicus* were downregulated to 50% of the value for the animal in water. Consistently, O_2-uptake became equally reduced as expected from the downregulated cardiac performance. Surprisingly, the animal is not completely unconscious during aestivation. Some noises cause transient bradycardia, which suggests that the animal was not in deep torpor (DeLaney *et al.* 1974). As an exception, *P. dolloi* does not aestivate in its normal habitat (Greenwood 1986). Nevertheless, specific laboratory conditions would be able to induce secretion of a cocoon that hardened over 4 to 5 days. In water, its O_2-uptake had been about 0.35 $mlSTPD{\cdot}kg^{-1}{\cdot}min^{-1}$(25°C) (Perry *et al.* 2007), which would be close to values for *P. aethiopicus* under the same conditions (~0.41 $mlSTPD{\cdot}kg^{-1}{\cdot}min^{-1}$, 21-25°C; Delaney 1974). Surprisingly, O_2-uptake was not downregulated. Just the opposite happened, O_2-uptake increased to 0.45 $mlSTPD{\cdot}kg^{-1}{\cdot}min^{-1}$, while $PaCO_2$ decreased from 18 mmHg to 14 mmHg in the cocoon. Perry *et al.* (2007) realized that this was not aestivation and replaced the condition of the animal to "terrestrialization". This condition had been maintained for more than one month, which excludes the possibility that a slow transition to aestivation could have developed.

This riddle might be resolved, by a better understanding of the mechanisms that control 1) cocoon formation and 2) metabolic levels during aestivation. A comparison between *P. dolloi* and *P. aethiopicus* could be insightful.

L. paradoxa also aestivates but without any formation of a cocoon (Harder *et al.* 1999), which can survive aestivation on a shorter seasonal basis. When the water dries out, it assumes a U-shaped position with the mouth close to the surface.

TEMPERATURE-DEPENDENT ACID-BASE REGULATION

In ectothermic vertebrates, the set-point for pH (pHa of arterial blood) decreases with rising temperature (Robin 1962; Rahn 1966; Reeves 1972; Heisler 1984). This discovery motivated Rahn (1966) to suggest that pHa declines in parallel to the pK of neutral water ($\Delta pK/\Delta t = -0.0176$ units·°C^{-1}; range 10 to 30°C). Alternatively, Reeves (1972) proposed that the degree of ionization of protein linked imidazole groups was the regulated variable, which predicts a $\Delta pK_{Im}\Delta t$ of −0.018 to −0.024 units·°C^{-1}, depending on ligands and steric arrangements (Edsall and Wyman 1958). Early studies were in agreement with these predicted $\Delta pHa/\Delta t$ values of the two models. In later studies, however, the arterial blood samples were obtained via implanted micro-catheters, which left the animal undisturbed. This resulted in values centered within a range of $\Delta pHa/\Delta t = -0.011$ to − 0.014 units/°C. Teleost fish fall within this range and the average for several turtle species was −0.010 units·°C^{-1} (Glass *et al.* 1985). Consistently, the $\Delta pHa/\Delta t$ for *L. paradoxa* is −0.014 units·°C^{-1} (Amin-Naves *et al.* 2004).

The use of adequate statistics became more common in the eighties and the models of Rahn and Reeves were not always supported by rigorous statistics (cf. Glass *et al.* 1985). With this background, it seemed more viable to ask how a negative $\Delta pH/\Delta t$ can be achieved. Amphibians and lepidosirenid lungfish have aquatic and aerial respiration. Cutaneous gas exchange is largely dependent on diffusion, although some capillary recruitment may increase the surface area for gas exchange (Feder and Burggren 1985). Cutaneous gas exchange is, however, mostly determined by the Krogh diffusion constants for CO_2 and O_2, which increase very little with temperature (Q_{10} of K_{CO_2}~1.1). This restricts CO_2-elimination by the skin and, as temperature increases, a larger fraction of total CO_2-production is eliminated via the lung (Jackson 1978; Mackenzie and Jackson 1978; Burggren and Moalli 1984; Jackson 1989; Wang *et al.* 1998; Amin-Naves *et al.* 2004). Thus, increased temperature favors CO_2-elimination by the lung rather than by the skin. The lung receives a larger fraction of total CO_2 produced, which increases intrapulmonary and arterial PCO_2, which, in *L. paradoxa*, leads to a $\Delta pHa/\Delta t$ of −0.014 units·°C^{-1}, while plasma $[HCO_3^-]$ remained constant over the tested range of temperatures.

This regulatory pattern has been verified for the toad *Chaunus schneideri*, (earlier *Bufo paracnemis*) in which the ΔpHa/Δt was – 0.014 units/°C (Wang *et al.*1998). How was ventilation adjusted in relation to temperature? Notably, both the lungfish and the amphibian kept intrapulmonary PO_2 constant over the studied range of temperature, which indicates a homeostatic regulation rather than a simple coincidence. Further, the PaO_2 of *L. paradoxa* is high and, moreover, constant in relation to temperature. This requires two conditions: 1) A very efficient separation of central vascular blood flow as reported by Szidon *et al.* (1969); 2) A constant intrapulmonary ratio of V_{eff}/VO_2, known as "air convection requirements" (Jackson 1978; Amin-Naves *et al.* 2004; Bassi *et al.* 2005). The high and largely temperature-independent PaO_2 in *L. paradoxa* differs substantially from the amphibian situation, which is characterized by large central vascular shunts that reduce saturations levels (Wood and Hicks 1985; Hedrick *et al.* 1999). A study by Jackson (1978) compared acid-base regulation in the turtle *Trachemys scripta elegans* (Earlier: *Pseudemys scripta elegans*) to that of a frog (*Rana catesbeiana*) and he made the comment that "so-called lower vertebrates promises not simplicity but rather complexity and confusion due to the diversity of adaptations represented", which seems a very realistic statement.

Acknowledgements

This research was supported by FAPESP (Fundação de Amparo á Pesquisa do Estado de São Paulo); Proc 98/06731-5, CNPq (Conselho Nacional de Desenvolvimento Científico o Tecnológico; Proc. 520769/93-7, FAEPA (Fundação de Apoio ao Ensino, Pesquisa e Assistência do Hospital das Clínicas da FMRP-USP).

References

Abe, A.S. (1995). Estivation in South American amphibians and reptiles. Brazilian Journal of Medical and Biological Research 28: 1241-1247.

Amin-Naves, J., Giusti, H. and Glass, M.L. (2004). Effects of acute temperature changes on aerial and aquatic gas exchange, pulmonary ventilation and blood gas status in the South American lungfish, *Lepidosiren paradoxa*. Comparative Biochemistry and Physiology A 138: 133-139.

Amin-Naves, J., Giusti, H., Hoffman, A. and Glass, M.L. (2007a). Components to the acid-base related ventilatory drives in the South American lungfish *Lepidosiren paradoxa*. Respiratory Physiology & Neurobiology 155(1): 35-40.

Amin-Naves, J., Giusti, H., Hoffman, A. and Glass, M.L. (2007b). Central ventilatory control in the South American lungfish, *Lepidosiren paradoxa*: contributions of pH and CO_2. Journal of Comparative Physiology B 177: 529-534.

Amin-Naves, J., Sanchez, A.P., Bassi, M., Giusti, H., Rantin, F.T. and Glass, M.L. (In press). Blood Gases of the South American Lungfish *Lepidosiren paradoxa*: A Comparison to Other Air-breathing Fish and to Amphibians. In: M.N. Fernandes, F.T. Rantin,

M.L. Glass and B.G. Kapoor (eds). Fish Respiration and Environment. Science Publishers Inc., Enfield, New Hampshire, USA, pp. 243-252.

Bassi, M., Klein, W., Fernandes, M.N., Perry, S.F. and Glass, M.L. (2005). Pulmonary oxygen diffusing capacity of the South American lungfish *Lepidosiren paradoxa*: Physiological values by the Bohr integration method. Physiological and Biochemical, Zoology, 78(4): 560-569.

Bohr, C. (1909). Über die spezifische Tätigkeit der Lungen bei der respiratorischen Gasaufnahme und ihr Verhalten zu der durch die Alveolenwand stattfindende Gasdiffusion. Skandinavisches Arciv für. Physiologie: 22: 221-280.

Branco, L.G.S. and Glass, M.L. (1995). Ventilatory responses to carboxihaemoglobinaemia and hypoxic hypoxia in *Bufo paracnemis*. Journal of Experimental Biology 198(6): 1417-1421.

Branco, L.G.S., Glass, M.L., Wang, T. and Hoffmann, A. (1993). Temperature and central chemoreceptor drive to ventilation in toad *(Bufo paracnemis)*. Respiration Physiology 93: 337-346.

Brinkmann, H., Denk, A., Zitzle, J., Joss, J.M.P. and Meyer, A. (2004). Complete Mitochondrial Genome Sequence of the South American and the Australian Lungfish: Testing of the Phylogenetic Performance of Mitochondrial Data Sets for Phylogenetic Problems in Tetrapod Relationships. Journal of Molecular Evolution 59: 834-848.

Burggren, W.W. and Moalli, R. (1984). Active regulation of cutaneous gas exchange in amphibians: experimental evidence and a revised model for skin respiration. Respiration Physiology 55: 379-392.

Burggren, W.W., Feder, M.E. and Pinder, A.W. (1983). Temperature and the balance between aerial and aquatic respiration in larvae of *Rana berlandieri* and *Rana catesbeiana*. Physiological Zoology 56(2): 263-273.

Burleson, M.L. and Milsom, W.K. (1990). Propanol inhibits O_2-sensitive chemoreceptor activity in trout gills. American Journal of Physiology 27: R1089-R1091.

Burleson, M.L. and Milsom, W.K. (1995a). Cardio-ventilatory control in rainbow trout: I. Pharmacology of branchial oxygen-sensitive chemoreceptors. Respiration Physiology 100: 231-238.

Burleson, M.L. and Milsom, W.K. (1995b). Cardio-ventilatory control in rainbow trout: II. Reflex effects of exogenous neurochemicals. Respiration Physiology 101: 289-299.

Carroll, R.L. (1988). Vertebrate Palaeontology and Evolution. 1st Edition. Freeman and Company, New York, USA.

Comroe, J. H. (1974). Physiology of Respiration — Year Book Medical Publishers, Inc., Chicago, USA.

Claiborne, J.B. Jr. and Heisler, N. (1986). Acid-base regulation and ion transfers in the carp (*Cyprinus carpio*): pH compensation during graded long and short-term environmental hypercapnia, and the effect of bicarbonate infusion. Journal of Experimental Biology 126: 41-61.

Crawford, E.C. Jr., Gatz, R.N., Magnussen, H., Perry, S.F. and Piiper, J. (1976). Lung volumes, pulmonary blood flow and carbon monoxide diffusing capacity of turtles. Journal of Comparative Physiology 107: 169-178.

Dejours, P. (1981). Principles of Comparative Respiratory Physiology. 2nd Edition, Elsevier/ North Holland Publishing Company, Amsterdam-New York, Oxford.

DeLaney, R.G., Lahiri, S. and Fishman, A.P. (1974) Aestivation of the African lungfish *Protopterus aethiopicus*: Cardiovascular and pulmonary function. Journal of Experimental Physiology 6: 111-128.

DeLaney, R.G., Lahiri, S., Hamilton, R. and Fishman, A.P. (1977). Acid-base balance and plasma composition in the aestivating lungfish (*Protopterus*). American Journal of Physiology 232(1): R10-R17.

Delaney, R.G., Laurent, P., Galante, R., Pack, A.I. and Fishman, A.P. (1983). Pulmonary mechanoreceptors in the dipnoi lungfish *Protopterus* and *Lepidosiren*. American Journal of Physiology 244: R418-R428.

Edsall, J.T. and Wyman, J. (1958). Biophysical Chemistry. Volume 1. Academic Press, New York, USA.

Eldridge, F.L., Kiley, J.P. and Millhorn, D.E. (1985). Respiratory response to medullary hydrogen ion changes: different effects of respiratory and metabolic acidosis. Journal of Physiology (London) 358: 285-297.

Etheridge, K. (1990). The energetics of aestivating sirenid salamanders (*Siren lacertina* and *Pseudobranchus striatus*). Herpetology 46: 407-414.

Feder, M.E. and Burggren, W.W. (1985). Cutaneous gas exchange in vertebrates: design, patterns, control and implications. Biological Reviews of the Cambridge Philosophical Society 60(1): 1-45.

Fishman, A.P., Pack, A.I., Delaney, R.G. and Galante, R.J. (1986). Estivation in *Protopterus*. Journal of Morphological Supplement 1: 237-248.

Flanigan, J.E.,Withers, P.C. and Guppy, M. (1991). *In vitro* metabolic depression of tissues from the aestivating frog *Neobatrachus pelobatoides*. Journal of Experimental Biology 161: 273-283.

Fritsche, R., Axelsson, M., Franklin, C.E., Grigg, G.C., Holmgren, S. and Nilsson, S. (1993). Respiratory and cardiovascular responses to hypoxia in the Australian lungfish. Respiration Physiology 94: 173-187.

Gilmour, K.M., Euverman, R.M., Esbaugh, A.J., Kenney, L., Chew, S.F., Ip, Y.K. and Perry, S.F. (2007). Mechanisms of acid-base regulation in the African lungfish *Protopterus annectens*. Journal of Experimental Biology 210: 1944-1959.

Glass, M.L., Burggren, W.W. and Johansen, K. (1981a). Pulmonary diffusing capacity of the bullfrog (*Rana catesbeiana*). Acta Physiologica Scandinavica 113: 485-490.

Glass, M.L., Johansen, K. and Abe, A.S. (1981b). Pulmonary diffusing capacity in reptiles (relations to temperature and O_2-uptake). Journal of Comparative Physiology 142B: 509-514.

Glass, M.L. and Johansen, K. (1982). Pulmonary oxygen capacity of lizard *Tupinambis teguixin*. Journal of Experimental Zoology 219: 385-388.

Glass, M.L., Boutilier, R.G. and Heisler, N. (1985). Effects of body temperature on respiration, blood gases and acid-base status in the turtle *Chysemys picta ellii*. Journal of Experimental Biology 114: 37-51.

Glass, M.L., Ishimatsu, A. and Johansen, K. (1986). Responses of aerial ventilation to hypoxia and hypercapnia in *Channa argus*, an air-breathing fish. Journal of Comparative Physiology B 156: 425-430.

Glass, M.L., Andersen, N.A., Kruhøffer, M., Williams, E.M. and Heisler, N. (1990). Combined effects of environmental PO_2 and temperature on ventilation and blood gases in the carp *Cyprinus carpio* L. Journal of Experimental Biology 148: 1-17.

Glass, M.L., Fernandes, M.S., Soncini, R., Glass, H. and Wasser, J.S. (1997). Effects of dry season dormancy on oxygen uptake, heart rate, and blood pressures in the toad *Bufo paracnemis*. Journal of Experimental Zoology 279: 330-336.

Greenwood, P.H. (1958). Reproduction in the East African lungfish *Protopterus aethiopicus* Heckel. Proceedings of the Zoological Society of London 130: 547-567.

Greenwood, P.H. (1986). The natural history of African lungfishes. Journal Morphological Suplement 1: 163-179.

Harder, V., Souza, R.H.S., Severi, W., Rantin, F.T. and Bridges, C.R. (1999). The South American lungfish adaptations to an extreme habitat. In: A.L. Val and V.M.F Almeida-Val (eds). Biology of Tropical Fishes. INPA. Manaus, Brazil, pp. 99-110,

Harada, Y., Wang, Y.Z. and Kuno, M. (1985). Central chemosensitivity to H^+ and CO_2 in respiratory center in vitro. Brain Research 333(2): 336-339.

Hedrick, M.S., Burleson, D.R., Jones, D.R. and Milsom, W.K. 1991. An examination of central chemosensitivity in an air-breathing fish (*Amia calva*). Journal of Experimental Biology 155: 165-174.

Hedrick, M.S., Palioca, W.B. and Hillmann, S.S. (1999). Effects of temperature and physical activity on blood flow shunts and intracardiac mixing in the toad *Bufo marinus.* Physiological Biochemistry and Zoology 72(5): 509-519.

Heisler, N. (1984). Acid-base regulation in fishes. In: W.S. Hoar and D.J. Randall (eds.). Fish Physiology. Volume XA. Academic Press, Orlando, New York. pp. 315-401.

Heisler, N., Forcht, G., Ultsch, G.R. and Anderson, J.F. (1982). Acid-base regulation to environmental hypercapnia in two aquatic salamanders, *Siren lacertina* and *Amphiuma means.* Respiration Physiology 49: 141-158.

Jackson, D.C. (1978). Respiratory control and CO_2 conductance: temperature effects in a turtle and a frog. Respiration Physiology 33: 103-114.

Jackson, D.C. (1989). Control of Breathing: Effects of Temperature. In: S.C. Wood (ed). Comparative Respiratory Physiology: Current Concepts. Marcel Dekker Inc., New York, USA. pp. 621-637.

Johansen, K. and Lenfant, C. (1967). Respiratory function in the South American lungfish, *Lepidosiren paradoxa.* Journal of Experimental Biology 46: 305-218.

Johansen, K. and Lenfant, C. (1968). Respiration in the African lungfish *Protopterus aethiopicus*: II. Control of breathing. Journal of Experimental Biology 49: 453-468.

Johansen, K., Lenfant, C. and Grigg, G.C. (1967). Respiratory control in the lungfish *Neoceratodus forsteri* (Krefft). Comparative Biochemistry and Physiology 20: 835-854.

Joss, J. and Johanson, Z. (2007). Is palaeospondylus gunni a fossil larval lungfish? Insights from *Neoceratodus forsteri.* Journal of Experimental Zoology Part B: Molecular and Developmental Evolution 308B: 163-171.

Kind, P.K., Grigg, G.C. and Booth, D.T. (2002). Physiological responses to prolonged aquatic hypoxia in the Queensland lungfish, *Neoceratodus forsteri.* Respiratory Physiology & Neurology 132: 179-190.

Kinkhead, R. and Milsom, W.K (1996). CO_2-sensitive olfactory and pulmonary receptor modulation of episodic breathing in bullfrogs. American Journal of Physiology 270: R134-R144.

Lahiri, S. and Forster, R.E. (2003). CO_2/H^+-sensing: peripheral and central chemoreception. International Journal of Biochemistry and Cell Biology 35(10): 1413-1435.

Lahiri, S., Szidon, J.P. and Fishman, A.P. (1970). Potential respiratory and circulatory adjustments to hypoxia in the African lungfish. Federation Proceedings 29(No. 2): 1141-1148.

Lenfant, C. K., Johansen, K. and Grigg, G.C. (1966). Respiratory properties of blood and pattern of gas exchange in the lungfish *Neoceratodus forsteri* (Krefft). Respiration Physiology 2: 1-21.

Lomholt, J.P (1993). Breathing in the aestivating African lungfish, *Protopterus amphibius*. Advances in Fish Research 1: 17-34.

Lundberg, J.G (1993). African fresh water fish clades and continental drift: problem with a paradigm. In: Goldblatt, J. (ed). Biological Relationships between Africa and South. America. Yale University Press, New Haven, CT, USA.

Loeschcke, H.H., Koepchen, H.P. and Gertz, K.H. (1958). Über den Einfluss von Wasserstoffionenkonzentration und CO_2-Druck im Liquor cerebrospinalis auf die Atmung. Pflügers Archiv 266: 569-585.

Mackenzie, J.A. and Jackson, D.C. (1978). The effect of temperature on cutaneous CO_2, loss and conductance in the bullfrog. Respiration Physiology 32(3): 313-323.

McClanahan, L., Ruibal, R. and Shoemaker, V.H. 1983. The rate of cocoon formation and its physiological correlates in a ceratophryd frog. Physiological Zoology 56: 430-435.

Meyer, A. and Dolven, S.I. (1992). Molecules, fossils, and the origin of tetrapods. Journal of Molecular Evolution 35: 102-113.

Milsom, W.K. (1995). The role of CO_2/pH chemoreceptors in ventilatory control. Brazilian Journal of Medical and Biological Research 28: 1147-1160.

Milsom, W.K. (2002). Phylogeny of CO_2/H^+ chemoreception in vertebrates. Respiratory Physiology & Neurobiology 131: 29-41.

Morais, M.F.P.G., Fernandes, M.N., Höller, S., Costa, O.P.F., Glass, M.L. and Perry, S.F. (2005). Morphometric comparison of the respiratory organs of the South American lungfish *Lepidosiren paradoxa* (Dipnoi). Physiological and Biochemical Zoology 78: 546-559.

Nattie, E. (1999). CO_2 brainstem chemoreceptors and breathing. Progress in Neurobiology 59: 299-331.

Orgeig, S. and Daniels, C.B. (1995). The evolutionary significance of pulmonary surfactant in lungfish (*Dipnoi*). American Journal of Respiratory Cell and Molecular Biology 13: 161-166.

Pack, A.J., Galante, R.J. and Fishman, A.P. (1990). Control of the interbreath interval in the African lungfish. American of Journal Physiology 259: R139-R146.

Pack, A.J., Galante, R.J. and Fishman, A.P. (1992). Role of lung inflation in control of air breath duration in African lungfish (*Protopterus annectens*). American Journal of Physiology 262: R879-884.

Perry, S.F. (2007). Swimbladder-Lung Homology in Basal Osteichthyes Revisited. In: M.N. Fernandes, F.T. Rantin, M.L. Glass and B.G. Kapoor (eds.). Fish Respiration and Environment. Science Publishers, Enfield, NH, USA. pp. 41-54.

Perry, S.F. and Reid, S.G. (2002). Cardiovascular adjustments during hypercarbia in rainbow trout *Oncorhynchus mykiss*. Journal of Experimental Biology 205 (pt21): 3357-3365.

Perry, S.F., Wilson, R.J.A., Straus, C., Harris, M.B. and Remmers, J. (2001). Which came first, the lung or the breath. Comparative Biochemistry and Physiology - Part A: Molecular and Integrative Physiology A 129: 37-47.

Perry, S.F., Euverman, R., Wang, T., Loong, A.M., Chew, S.F., Ip, Y.K. and Gilmour, K.M. (2007). Control of breathing in African lungfish (*Protopterus dolloi*): A comparison of aquatic and cocooned (terrestrialized) animals. Respiratory Physiology &. Neurobiology (In press).

Powell, F.L. and Hempleman, S.C. (1993). Diffusion limitation in comparative models of gas exchange. Respiration Physiology 91(1): 17-29.

Power, J.H., Doyle, I.R., Davidson, K. and Nicholas, T.E. (1999). Ultrastructural and protein analysis of surfactant in the Australian lungfish *Neoceratodus forsteri*: evidence for conservation of composition for 300 million years. Journal of Experimental Biology 202(Pt 18): 2543-2550.

Rahn, H. (1966). Aquatic gas exchange: theory. Respiration Physiology 1: 1-12.

Reeves, R.B. (1972). An imidazole alphastat hypothesis for vertebrate acid-base regulation: tissue carbon dioxide content and body temperature in bullfrogs. Respiration Physiology 14: 219-236.

Robin, E.D. (1962). Relationship between temperature and plasma pH and carbon dioxide tension in the turtle. Nature London 195: 249-251.

Sanchez, A.P. and Glass, M.L. (2001). Effects of environmental hypercapnia on pulmonary ventilation of the South American lungfish. Journal of Fish Biology 58: 1181-1189.

Sanchez, A.P., Hoffman, A., Rantin, F.T. and Glass, M.L. (2001a). The relationship between pH of the cerebro-spinal fluid and pulmonary ventilation of the South American lungfish, *Lepidosiren paradoxa*. Journal of Experimental Zoology 290: 421-425.

Sanchez, A.P., Soncini, R., Wang, T., Koldkjaer, P., Taylor, E.W. and Glass, M.L. (2001b). The differential cardio-respiratory responses to ambient hypoxia and systemic hypoxaemia in the South American lungfish *Lepidosiren paradoxa*. Comparative Biochemistry and Physiology A 130: 677-687.

Sanchez, A.P., Giusti, H., Bassi, M. and Glass, M.L. (2005). Acid-base regulation in the South American lungfish *Lepidosiren paradoxa*: Effects of prolonged hypercarbia on blood gases and pulmonary ventilation. Biochemical and Physiological Zoology 78: 908-915.

Sawaya, P. (1946). Sobre a biologia de alguns peixes de respiração aérea (*Lepidosiren paradoxa Fitzinger* e *Arapaima gigas Cuvier*). Boletim da Faculdade de Filosofia Ciências e Letras da. Universidade de. São Paulo 11: 255-286.

Scotto, P., Ichinose, L., Patané, L., Meyer, M. and Piiper, J. (1987). Alveolar-capillary diffusion of oxygen in dogs exercising in hypoxia. Respiration Physiology 68: 1-10.

Shams, H. (1985). Differential effects of CO_2 and H^+ as central stimuli of respiration in the cat. Journal Applied Physiology 58(2): 357-364.

Shams, H. and Scheid, P. (1989). Efficiency of parabronchial gas exchange in deep hypoxia: measurements in the resting duck. Respiration Physiology 77: 135-146.

Smatresk, N.J. and Cameron, J.N. (1982). Respiration and acid-base physiology of the spotted gar, a bimodal breather. II. Responses to temperature change and hypercapnia. Journal of Experimental Biology 96: 281-293.

Smatresk,, N.J. and Smits, A.W. (1991). Effects of central and peripheral chemoreceptor stimulation on ventilation in the marine toad, *Bufo paracnemis*. Respiration Physiology 83(2): 223-238.

Smith, H.M. (1935). The metabolism of a lungfish. I. General considerations of the fasting metabolism in an active fish. Journal of Cellular Comparative Physiology 6: 43-67.

Soncini, R. and Glass, M.L. (2000). Oxygen and acid-base related drives to gill ventilation in carp. Journal of Fish Biology. 56: 528-541.

Szidon, J.P., Lahiri, S., Lev, M. and Fishman, A.P. (1969). Heart rate of circulation of the African lungfish. Circulatory Research 25: 23-38.

Takezawa, J., Miller, F.J. and O`Neil, J.J. (1980). Single-breath diffusing capacity and lung volumes in small laboratory mammals. Journal of Applied Physiology 48(6): 1052-1059.

Toews, D.P. and Stiffler, D.F. (1990). Compensation of Progressive Hypercapnia in the Toad (*Bufo marinus*) and the bullfrog (*Rana catesbeiana*). Journal of Experimental Biology 148: 293-302.

Toyama, Y., Ichimiya, T., Kasama-Yoshida, H., Cao, Y., Hasegava, M., Kojima, H., Tamai, Y. and Kurihari, T. (2000). Phylogenertic relation of lungfish indicated by the amino acid sequence of myelin DM20. Molecular Brain Research 8: 256-259.

Wang, T., Branco, L.G.S. and Glass, M.L. (1994). Ventilatory responses to hypoxia in the toad (*Bufo paracnemis* Lutz) before and after reduction of HbO_2 concentration. Journal of Experimental Biology 186: 1-8.

Wang, T., Abe, A.S. and Glass, M.L. (1998). Temperature effects on lung and blood gases in the toad *Bufo paracnemis*: The consequences of bimodal gas exchange. Respiration Physiology 113: 231-238.

Watt, M., Evans, C.S. and Joss, J.M. (1999). Use of electroreception during foraging by the Australian lungfish. Animal Behavior 58: 1039-1045.

Wellner-Kienitz, M.-C. and Shams, H. (1998). CO_2-sensitive neurons in organotypic cultures of the fetal rat medulla. Respiration Physiology 111: 137-151.

Wilson, R.J., Harris, M.B., Remmers, J.E. and Perry, S.F. (2000). Evolution of air-breathing and central CO_2/H^+ respiratory chemosensitivity: new insights from an old fish? Journal of Experimental Biology 203: 3505-3512.

Wood, S.C. (1982). Effect of O_2 affinity on arterial PO_2 in animals with central vascular shunts. Journal of Applied Physiology 53(6): R1360-R1364.

Wood, S.C. and Hicks, J.W. (1985). Oxygen Homeostasis in Vertebrates with Cardiovascular Shunts. In: K. Johansen and W.W. Burggren (eds). Cardiovascular Shunts, Phylogenetic, Ontogenetic and Clinical Aspects. Munksgaard. Copenhagen, Denmark. pp. 354-372.

Yokabori, S., Hasegawa, M., Ueda, T., Okada, N., Nishikawa, K. and Watanabe, K. (1994). Relationship among Coelacanths, Lungfishes, and Tetrapods: A Phylogenetic Analysis Based on Mitochondrial Cytochrome Oxidase I Gene Sequences. Journal of Molecular Evolution 38: 602-609.

Zardoya, R., Cao, Y., Hasegava, M. and Meyer, A. (1998). Searching for the closest living relative(s) of Tetrapods through evolutionary analyses of mitochandrial and nuclear data. Journal of Molecular Biology and Evolution 15: 506-517.

Zhu, M. and Yu, X. (2002). A primitive fish close to the common ancestor of tetrapods and lungfish. Nature 418: 767-770.

CHAPTER 11

Respiratory Adaptations in Lungfish Blood and Hemoglobin

Roy E. Weber[1,*] and Frank Bo Jensen[2]

[1]Zoophysiology, Institute of Biological Sciences, University of Aarhus, DK 8000 Aarhus C, Denmark

[2]Institute of Biology, University of Southern Denmark, DK 5230 Odense-M, Denmark

ABSTRACT

The extant lungfish – that comprise the predominantly water-breathing *Neoceratodus forsteri* from Australia and the obligate air - breathing *Protopterus* and *Lepidosiren* from Africa and South America, respectively – face extraordinary variations in exogenous factors (O_2 and water availability and temperature) and endogenous constraints (bimodal breathing and estivation in moist or dried mud). These circumstances predictably result in inordinate variations in factors (blood levels of pH, O_2 and CO_2 tensions, urea and lactate levels and osmolality) that mandatorily affect O_2 and CO_2 binding by the circulating hemoglobin (Hb). This treatise focuses on the distinctive, compensatory adaptations in the gas-transporting functions of lungfish blood and Hb and the underlying molecular mechanisms that support aerobic metabolism in lungfish under harsh conditions.

Keywords: lungfish, hemoglobin, oxygen affinity, carbon dioxide, estivation

INTRODUCTION

In transporting respiratory gases, hemoglobin (Hb) in blood bridges wide and independent variations in tensions of these gases at the respiratory surfaces

**Corresponding author:* E-mail: roy.weber@biology.au.dk

and in the tissues. In meeting these requirements the gas binding properties of Hbs from ectotherm vertebrates display remarkable adaptability (Johansen and Lenfant 1971; Weber 1992). This applies particularly to lungfish, a sister group to tetrapod vertebrates, that have lungs and gills and are capable of bimodal breathing. Some lungfish species moreover survive drought by estivating in dried mud for extended periods [up to 7 years (Lomholt 1993)] risking subjection to strongly increased levels of excretory products (CO_2, protons, lactate, ammonium, urea) and body fluid osmolality (decreased water activity) – factors that are known to perturb the gas binding properties of Hb. Lungfish thus are ideal for studying the mechanistic basis underlying the ability of animals to survive combinations of harsh exogenous factors (oxygen and water availability, temperature, etc.) and endogenous constraints (activity, suspended animation, breathing mode, etc.).

The extant lungfishes that comprise *Neoceratodus forsteri* (family Ceratodontidae) found in Australia, and four species of *Protopterus* (*P. dolloi*, *P. aethiopicus*, *P. annectans* and *P. amphibius*) from Africa and *Lepidosiren forsteri* from South America (members of the family Epidosirenidae) differ in their breathing mode. Compared to the *Neoceratodus* that is a facultative air-breather, *Protopterus* and *Lepidosiren* are obligatory air-breathers with reduced gills. The roles the lungs compared to that of the gills and skin in gas exchange thus vary greatly in fish living in water with access to air. Whereas O_2 uptake and CO_2 elimination occur via gills and skin in *Neoceratodus*, O_2 uptake is predominantly pulmonary in *Protopterus* and almost entirely pulmonary in *Lepidosiren* – where the gills remain a major route for CO_2 elimination (Lenfant *et al.* 1970). Thus, in *Lepidosiren*, pulmonary gas exchange accounts for more than 95% of total O_2 uptake at elevated temperatures (Johansen and Lenfant 1967; Bassi *et al.* 2005) and 99% of the total morphological diffusion capacity lies in the lungs, compared to <1% in the skin and 0.001% in the gills (de Moraes *et al.* 2005).

This chapter focuses on respiratory properties of lungfish blood and Hbs, specifically the adaptations to exogenous and endogenous factors (environmental conditions, air/water breathing, estivation, etc.) and the implicated cellular and molecular mechanisms.

HB FUNCTION AND ADAPTABILITY

Tissue O_2 supply in vertebrates is integrally dependent on Hb function. As illustrated by the Fick equation, the O_2 transport (consumption) rate (VO_2) is a product of cardiac output and (Ca–Cv), the O_2 content difference between arterial and mixed-venous blood. It follows that for a given cardiac output, O_2 transport by blood is increased by any change in blood O_2 affinity (left- or right-shifting of the O_2 equilibrium curve) that increases the steepness of the O_2 equilibrium curve for a given arterio-venous PO_2 difference [i.e. the O_2 capacitance of the blood, $\beta_bO_2 = (Ca_{O_2}-Cv_{O_2})/(Pa_{O_2}-Pv_{O_2})$].

As in other vertebrates, lungfish Hbs are tetrameric molecules that undergo a conformational transition upon oxygenation: from a low-affinity, deoxygenated, tense (T) structure to a high-affinity, oxygenated relaxed (R) structure as the two α-β dimers slide relative to each other. This shift underlies cooperativity (higher affinity in partially oxygenated molecules, expressed in the sigmoid shape of the O_2 equilibrium curves) which increases O_2 loading in the gills and lungs and unloading in the tissues.

The O_2 affinity of blood (that varies inversely with P_{50}, the O_2 tension where the Hb is half-saturated) is a function of Hb's intrinsic O_2-binding affinity and cooperativity as well as its interactions with allosteric effectors (chiefly protons, polyanionic organic phosphates, and chloride ions) that normally stabilize the T-structure, and thus reduce Hb-O_2 affinity and enhance O_2 delivery in the tissues. The best-known example is the Bohr effect, ($\Delta \log P50/\Delta pH$), whereby the binding of H^+ to Hb as it passes tissue capillaries causes a transition to the low affinity T structure, which increases the release of O_2 in the respiring tissues (Bohr *et al.* 1904; Jensen 2004).

Typically short-term adaptations in blood-O_2 affinity (that may occur in individual animals) are mediated by changes in the intraerythrocytic levels of effectors. Compared to mammals where the predominating organic phosphate effector is 2,3-diphosphoglycerate (DPG), that in ectothermic vertebrates is predominantly ATP, although some fish species additionally have high red cell levels of guanosine triphosphate (GTP), while that in birds is inositol pentaphosphate (IPP). In contrast, long-term adaptations (such as those that were aquired in an evolutionary time scale and thus are evident when comparing different species) commonly involve changes in the structure of the Hb (and thus in its intrinsic properties). Typically these interspecific (between-species) adaptations comprise substitution of a few, key amino acid residues: those that are binding sites for effector ions or part of 'sliding surfaces' in the T/R transition. The residues at these sites [reviewed by (Weber and Fago 2004)] are thus of special interest in identifying molecular adaptations to exogenous and endogenous factors in lungfish Hbs.

BLOOD GAS TRANSPORT

A. Red Blood Cells

A striking peculiarity with lungfish is their extremely large red blood cells. The erythrocytes of lungfish are elliptical and have an oval nucleus, as generally observed in non-mammalian vertebrates (Nikinmaa 1990). The mean cellular volume of erythrocytes in *Protopterus aethiopicus* is approximately 6940 μm^3 (Koldkjaer *et al.* 2002; Jensen *et al.* 2003), which is 30-40 times larger than the typical teleost values of 185, 225, 171 and 172 μm^3 observed in carp, rainbow trout, eel and Atlantic cod, respectively (Jensen and Brahm 1995). The largest diameter of *Lepidosiren* erythrocytes is about 53 μm (Ribeiro *et al.* 2007), compared to approximately 14 μm in carp erythrocytes (Jensen and Brahm 1995). The large erythrocytes in lungfish is a consequence of their large genomes, as evident from the significant positive relationship between red cell size and genome size (Gregory 2001). The genome of *Protopterus aethiopicus* is the largest among all vertebrates (Gregory 2001). While the genome in *Lepidosiren* is almost as large, that in the third genus of living lungfish, *Neoceratodus*, is less than half as large, but still considerably larger than in most other vertebrates (Gregory 2002). The interspecific variation in genome size within the lungfishes is reflected in erythrocyte size. Thus, while the erythrocyte length is 53 μm in *Lepidosiren* (Ribeiro *et al.* 2007), it is about 32 μm in *Neoceratodus*, as evaluated from the photomicrographs given in Gregory (2001).

It has been speculated that large genomes and red cells may relate to life in hypoxic environments or aestivation, but this idea has been generally abandoned (Gregory 2001). The large size of erythrocytes is, however, bound to have consequences for the dimensions of capillaries. Relatively large capillary diameters are needed to allow passage of giant blood cells, and one may therefore envisage relatively long mean diffusion distances and low overall surface areas for oxygen and carbon dioxide exchange in lungs/gills and tissues, which could reduce the efficiency of gas exchange. This aspect has to our knowledge not been directly examined. It is, however, evident from the long evolutionary history of lungfish that their respiratory design is sufficient to cope with the respiratory challenges associated with their lifestyle.

B. O_2 Transport

1. O_2 carrying capacity

The O_2 carrying capacities of blood of non-estivating lungfish are typically close to 4 mM (2.8-4.4 mM – Table 1), which corresponds with the values found in bony and elasmobrach fishes and ectothermic vertebrates in general (Lenfant *et al.* 1966) but are only approximately half the values in mammals (~9 mmol l^{-1}). This difference reflects the lower metabolic rate and thus O_2 transport requirements in ectothermic vertebrates compared to endothermic vertebrates. Given the relative constancy of mean cellular Hb concentrations (Delaney *et al.* 1976), the variation in oxygen capacities is largely reflected in hematocrit values. Reported hematocrits in non-estivating lungfish are ~31% in *Neoceratodus*, 25-30% in *P. amphibius*, *P. aethiopicus* and *P. dolloi* (Lenfant *et al.* 1966; Johansen *et al.* 1976; Delaney *et al.* 1976; Wilkie *et al.* 2007), and 28% in *Lepidosiren* (da Silva *et al.* 2008) – compared to 45% in humans.

Does the O_2 carrying capacity of lungfish blood continuously adapt to exogenous and endogenous stresses? The answer to that question appears to vary with species and condition. Exposing *Neoceratodus* for 2-3 weeks to severe

Table 1 O_2 binding properties of lungfish blood and Hb

	Material; Acclimation/activity state	P_{50}	Conditions for P_{50} measurement			Bohr factor	O_2 capacity	Reference
		(mmHg)	°C	PCO_2 (mmHg)	pH		(mM)[#]	
Neoceratodus forsteri	Blood, normoxia (120 mmHg)	22	20	16	7.5	–0.48	3.52	(Kind *et al.* 2002)
Neoceratodus forsteri	Blood, hypoxia (60 mmHg)	19	20	16	7.7	–0.42	3.6	" "
Neoceratodus forsteri	Stripped hemolysate	5.3	25	0	7.5	–0.13		(Rasmussen *et al.* 2009)
Protopterus aethiopicus	Blood	32	25	40	7.4	–0.35	4.4*	(Swan and Hall 1966); *(Jensen *et al.* 2003)
Protopterus amphibius	Blood, active lungfish	32	26	22	7.21	–0.20	2.8	(Johansen *et al.* 1976)
Protopterus amphibius	Blood, estivating lungfish	7.2	26	29	7.62	–0.68	4.9-5.9	" "
Protopterus amphibius	Stripped hemolysate, active and estivating	1.3	26	0	7.40	–0.33		(Weber *et al.* 1977)
Protopterus annectans	Stripped hemolysate, active	5-6	25	0	7.4	–0.1		(Weber *et al.* 1977) and Fig.2.
Lepidosiren paradoxa	Blood	18.6	25	22	7.62	–0.66		(Bassi *et al.* 2005)
Lepidosiren paradoxa	Blood	26.9	25	44	7.38	–0.66		" "

#calculated from 'volume percent' values (multiplied by 10/22.4) or hemoglobin concentrations (multiplied by 4)

water hypoxia (PwO_2 = 40 mmHg) affects neither [Hb] nor hematocrit (Kind *et al.* 2002). In *Lepidosiren*, 40 days' estivation has no influence on hematocrit (da Silva *et al.* 2008). In *P. amphibius*, however, hematocrit, blood O_2 capacity and Hb concentrations increase by approximately 50% following prolonged (28-30 month) estivation in natural, dried mud (Johansen *et al.* 1976). Analogously, specimens of *P. aethiopicus* induced to estivate in muslin sacks for periods of up to 15 months show large (up to 82 %) increases in blood Hb concentration and hematocrit values due largely to hemoconcentration (Delaney *et al.* 1976).

2. Blood-O_2 affinity and its adaptive variation

Organic phosphate modulators

As with a majority teleost fishes that experience hypoxic conditions in nature, *P. amphibius* red cells contain high concentrations of GTP [GTP/ATP ratio = 1.6-2.1 in non-estivating specimens (Johansen *et al.* 1976)) (Fig. 1). GTP is also present in *P. aethiopicus* and *Lepidosiren* albeit at lower concentrations [GTP/ATP ratios ~0.6 (Bartlett 1978)]. Curiously the red cells of *P. aethiopicus* and *Lepidosiren* carry significant amounts of uridine triphoshate (UTP), with ATP:GTP:UTP ratios of approximately 53:30:17 and 55:33:12, respectively. Both species moreover have low levels of inositol diphosphate (IDP) (Bartlett 1978), which, however, is lacking in the Australian *N. forsteri* (Isaacks and Kim 1984). The presence of uridine phosphates and IDP in red cells appears to be a unique and novel characteristic of Dipnoi (Val 2000). As shown for carp Hb, where GTP that forms an additional hydrogen bond compared to ATP (Weber and Lykkeboe 1978; Gronenborn *et al.* 1984; Jensen *et al.* 1998a), GTP is a more potent modulator of O_2 affinity in *P. amphibius* and *Neoceratodus* Hbs (Weber *et al.* 1977; Rasmussen *et al.* 2009).

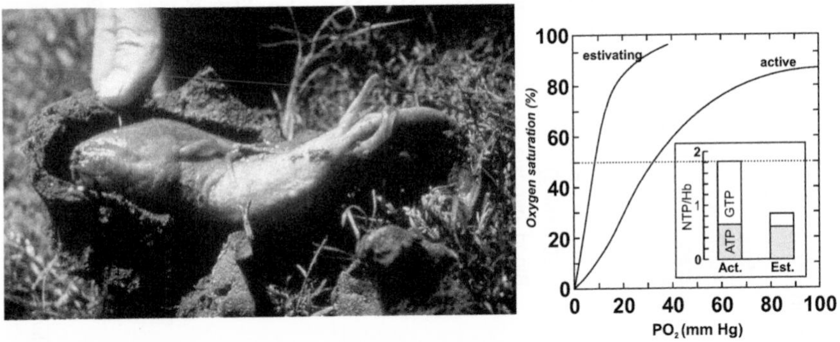

Fig. 1 *Protopterus amphibius* in its natural estivating cocoon (opened, with Kjell Johansen's thumb), O_2 equilibrium curves of whole blood from estivating and active (non-estivating, aquatic) specimens at 26°C, and (inset) blood NTP/Hb ratios, showing that the increased O_2 affinity under estivation correlates with a sharp decrease in the GTP/Hb (tetramer) concentration ratio. Modified after (Johansen *et al.* 1976).

Color image of this figure appears in the color plate section at the end of the book.

Although the effect of UTP on lungfish and other Hbs remains unknown, it likely is comparable to that of ATP based on the similar number of negative charges (Chanutin and Curnish 1967). IDP that structurally resembles IPP found in bird red cells carries little charge and therefore is unlikely to affect Hb-O_2 affinity (Isaacks *et al.* 1977).

The O_2 affinities reported for lungfish blood (cf. Table 1) reveal some general trends, despite the fact that they refer to different measuring conditions (pH, CO_2 tension, temperature, etc.) and acclimatory states. Thus, blood-O_2 affinities for active (aquatic, non-estivating) lungfish are relatively low. Also, as previously pointed out (Lenfant and Johansen 1968), O_2 affinities are lower in the obligate air-breathing *Protopterus* and *Lepidosiren* species than in the predominantly water-breathing *Neoceratodus*. This moreover aligns with the lower blood-O_2 affinities in bimodal than in water-breating teleosts fish (Johansen and Lenfant 1971). These correlations indicate that the left-shifted curves are adaptive to low O_2 tensions in water whereas right-shifted curves are better suited to the relatively constant high tensions in air.

Plotting arterial and venous oxygen tensions measured in *Lepidosiren* blood (81 and 43 mm Hg, respectively) into the blood-O_2 equilibrium curve measured under corresponding pH and CO_2 conditions (Bassi *et al.* 2005) indicates that only the upper part of the curves (O_2 saturations > 85%) are exploited in non-estivating animals at rest, leaving a large venous O_2 reserve that could be exploited under specific conditions (e.g. bouts of activity or temporary disruption in O_2 transport).

Aquatic hypoxia

Subjecting *Neoceratodus* to short-term (2-3 week) moderate aquatic hypoxia ($PwO_2 = 60$ mmHg) slightly increases blood O_2 affinity and decreases red cell ATP level (Kind *et al.* 2002) (Table 1). This mirrors the well-documented phosphate-induced affinity increase seen in teleosts [cf. (Wood and Johansen 1972; Weber *et al.* 1975; Weber and Lykkeboe 1978; Jensen and Weber 1982; Jensen and Weber 1985; Val 2000)] where the response is adaptive in enhancing O_2 loading in the gills (Wood and Johansen 1973a) (Nikinmaa 1992). However, unlike teleosts (e.g. trout) where the response may be graded (Tetens and Lykkeboe 1985), subjecting *Neoceratodus* to severe hypoxia (40 mmHg) does not further increase O_2 affinity, suggesting that additional left-shifting of the O_2 equilibrium curve is obviated by increased air-breathing (Kind *et al.* 2002).

Estivation

Estivation invariably involves drastic changes in the physico-chemical conditions in the circulating blood. Although the functional residual capacity (FRC) of the lungs in estivating fish may be large, the pulmonary ventilatory tidal volumes of fish enclosed in hard dried mud are small [cf. (Johansen *et al.* 1976)]. While reflecting the reduced metabolic rate during estivation, the low tidal volumes indicate increased dependence on diffusive processes (as opposed to convective processes) for renewal of O_2 at the respiratory surfaces. Additionally, the gaseous environment of the soil-entrapped cocoon may be hypoxic. Lung PO_2's will also vary with encasement, and likely differs for *Protopterus amphibius* and *P. aethiopicus* that estivate in cocoons, *P. dolloi* that estivates in open air in a layer of dried mucus (Chew *et al.* 2004) and *Lepidosiren* that estivates in simple mud burrows (da Silva *et al.* 2008). There is a dearth of information on ambient and pulmonary gas tensions in estivating lungfish.

Forty days' estivation in simple burrows of *Lepidosiren* - that does not form a cocoon - lowered PaO_2 (from 88 to 77 mm Hg), and increased $PaCO_2$ (from 22 to 38 mm Hg), plasma HCO_3^- (from 23 to 40 mM) and osmolality (from 232 to 267 mOsm·(kg water)$^{-1}$ (da Silva *et al.* 2008). During laboratory-induced estivation of *P. aethiopicus* in mud and muslin cloth bags suspended in moist surroundings for 0.5 to 9.5 months, arterial PCO_2 increases from 25-30 mm Hg to 45-70 mm Hg, pHa decreases from 7.55 -7.60 to 7.0-7.26, and PaO_2 increases from 25-40 to 50-58 mm Hg during the first 10 days and then returned to control. In the switch from water to moist air during the start of estivation, PaO_2 increased from 12 to 111 mm Hg within the span of one day (Delaney *et al.* 1974). In *P. aethiopicus* a marked respiratory acidosis occurring during the first two weeks of estivation is followed by a slow compensatory increase in plasma bicarbonate that partially restores arterial pH (Delaney *et al.* 1977). Estivation is moreover associated with drastic changes in the levels of other factors (blood osmolality, urea levels, etc.) that each may influence Hb-O_2 affinity and thus perturb O_2 transport. Additionally temperatures are high during summer draughts that induce estivation. Temperatures of 36 - 40°C characterize 'rock-hard' dried mud of White Nile valley localities where *Protopterus aethiopicus* estivates (Swan and Hall 1966).

Does Hb-O_2 affinity undergo adaptive changes during estivation? Compared to active *Protopterus amphibius* that had been kept in water for more than 2 years, specimens that had been estivating for 28-30 months showed phenomenal increases in blood O_2 affinity. This correlated neatly with a decrease in the red cell concentration of NTP that almost entirely can be ascribed to a massive (~78%) decrease in GTP concentration, whereas ATP concentration was little affected (Fig. 1, Table 1). The significance of GTP is further underscored by its greater effect on O_2 affinity of *P. amphibius* Hb than ATP at the same concentration (Weber *et al.* 1977). These responses are homologous to those seen when teleosts that carry both NTPs are subjected to water hypoxia (referred to above), but the functional significance may be radically different in the lungfish - where the O_2-affinity response moreover is much more pronounced. In contrast to active, hypoxic teleosts, where the increased O_2 affinity improves the arterial loading potential and secures O_2 transport, estivating lungfish face the vital need to conserve rather than to combust resources during prolonged estivation periods (Gorr *et al.* 1991).

In this light, rather than securing O_2 uptake and transport, the sharply increased affinity may serve to curtail O_2 unloading in the tissues (Weber 1982), thereby contributing to the massive down-regulation of organismal metabolism.

In contrast to the outspoken phosphate-induced increase in O_2 affinity of *P. amphibius* blood, the intrinsic Hb-O_2 affinity (measured in purified Hb) remains unchanged through >28 months' estivation in dry natural mud (Weber *et al.* 1977).

Thermal adaptation

Although the inverse relationship between temperature and blood O_2 affinity mandated by the exothermic nature of heme oxygenation normally is advantageous (increasing O_2 unloading when and where temperature, and thus metabolic rate, increases), it potentially hampers O_2 loading in warm habitats. The exothermicity of heme oxygenation is, however, lowered by endothermic dissociation of effectors. Commonly, as observed *in P. amphibius* (Weber *et al.* 1977), the overall heat of oxygenation is high, approaching the value of the heme groups [~–59 $kJ \cdot mol^{-1}$; (Atha and Ackers 1974)] at high pH, where the Bohr effect is low, but decreases at lower pH where the Bohr effect and binding of anionic effectors increase. In this light, the lower temperature sensitivity of O_2 affinity in *Neoceratodus* (Lenfant and Johansen 1967) aligns with the larger Bohr effect in *Neoceratodus* than in *Protopterus* blood (Table 1). The physiological implications of the pH dependence of the temperature effect in lungfish seem, however, to be unexplored.

L. paradoxa blood exhibits normal temperature effects (the apparent heat of oxygenation, $\Delta H' \sim -39$ and -35 $kJ \cdot mol^{-1}$ at 3% and 6% CO_2, respectively) (Bassi *et al.* 2005).

Other cellular regulatory mechanisms

In lungfish erythrocytes there is a passive "Donnan-like" distribution of protons across the membrane as observed in other vertebrates except lampreys (Jensen 2004). The distribution ratio of H^+, and thus intracellular pH, is controlled by the non-permeable negative charge carried by Hb and organic phosphates. It is therefore to be expected that the reduction in erythrocyte NTP during estivation increases O_2 affinity both directly (through lowered allosteric interaction with the Hb molecules) and indirectly by increasing red cell pH (Duhm 1971; Wood and Johansen 1973b). The relative contribution of these two mechanisms to the O_2 affinity increase has not been assessed in lungfish.

Another cellular mechanism that may increase O_2 affinity under hypoxia is β-adrenergic stimulation observed in red cells of many teleosts under hypoxia (Nikinmaa 1992). This stimulation activates the membrane bound Na^+(in)/H^+(out) exchanger (NHE) that raises intracellular pH and causes red cell swelling (Nikinmaa 1982). The response is greater at low than at high O_2 saturation

(Salama and Nikinmaa 1988) and increases O_2 affinity through alkalinization of the red cell cytoplasm and dilution the red cell Hb and NTP content. Red cells of *Protopterus aethiopicus* and *Lepidosiren paradoxa* possess β-adrenergic receptors and the number of receptors per unit red cell surface area in *P. aethiopicus* is about twice that in trout red cells (Koldkjaer *et al.* 2002). However, while the NHE in lungfish red cells can be activated by osmotic shrinkage, it is not activated by adrenergic stimulation (Koldkjaer *et al.* 2002). It appears that the β-adrenergic NHE is a special feature that evolved in higher teleosts relatively recently (Berenbrink *et al.* 2005).

An interesting question related to the highly variable O_2 tensions in lungfish blood is whether Hb oxygenation is involved in the control of glycolysis, as postulated for mammals. This relates to the fact that the major protein constituent of red cell membranes, Band 3 (that mediates Cl^-/HCO_3^- exchange) has a negatively-charged N-terminal cytoplasmic domain ('cdB3') that binds either deoxygenated Hb (at the same positively-charged site as organic phosphates) or glycolytic enzymes. It follows that low PO_2 leads to binding of deoxygenated Hb to cdB3 that releases glycolytic enzymes from shared binding sites and stimulates glycolysis (Low *et al.* 1993). However, unlike mammalian and bird Hbs tested (Weber *et al.* 2004), the O_2 affinity of *Protopterus* Hb is virtually insensitive to a peptide corresponding to trout cdB3 (Fig. 2) although this peptide does react with human Hb (Jensen *et al.* 1998b). Thus, there appears to be no (de)oxygenation-linked interaction between lungfish Hb and the membrane cdB3.

3. Hb's intrinsic O_2 binding properties

a. Functional heterogeneity

Interestingly, the major isoHbs of *P. amphibius* show a marked difference in O_2 affinity, irrespective of pH and the presence of ATP or GTP. Analogously, one of the three isoHbs from *Neoceratodus forsteri* shows a lower O_2 affinity than the other two (Rasmussen *et al.* 2009). In both species, O_2 affinities of the composite hemolysate are intermediate between those of the major components, indicating the lack of inter-isoHb interactions. The functional differentiation seen in these species likely extends the PO_2 range where the whole blood functions.

Do changes in Hb's structure and intrinsic O_2 binding characteristics resulting from altered expression of isoHbs contribute to the adaptive changes seen in whole blood? This is not likely. Stripped Hb from *P. amphibius* that estivated for 28-30 months show almost identical O_2 affinities and pH and ATP sensitivities as that from specimens that lived in water for 2 years. Moreover the Hb heterogeneity pattern (two major and one minor Hb components) was the same (Weber *et al.* 1977). In *P. aethiopicus* estivation is accompanied by some changes in relative concentrations of the 4 distinct components (Delaney *et al.* 1976). However no data is available on O_2 affinities of the components. *Lepisosiren* has only one component (Phelps *et al.* 1979).

b. Effector sensitivities

Studies on effector sensitivities of purified ('stripped') Hbs give insight into the implicated molecular adaptations.

The Hb of the only lungfish (*Lepidosiren paradoxa*) that has been sequenced has positively-charged β-chain residues at the 4 β-chain sites (Val-1, His-2, Lys-82 and Arg-143) that bind organic phosphates in vertebrate Hbs (Rodewald *et al.* 1984) (Table 2). The Bohr effects that are small in stripped Hbs are strongly increased by effectors (cf. Fig. 2). In *Lepidosiren* the large increase in the Bohr effect in presence of ATP correlates with a strong negative correlation between pH and the kinetic O_2 dissociation rate (Phelps *et al.* 1979). Interestingly the pH dependence of the rates of ligand (CO) dissociation indicates preferential binding of ATP to deoxygenated structure of *Lepidosiren* Hb even at high pH (>8).

A few studies have addressed the effects of excretory products and other factors that are known to change inordinately (CO_2, urea, lactate, plasma osmolality and water activity) on the oxygenation reaction of lungfish Hbs.

Urea and chloride. In *P. aethiopicus* plasma urea concentration rises to 203 mM after 13 months' estivation, where it accounts for 31% of plasma osmolality, compared to only 2% in the aquatic lungfish, while plasma osmolality increases 2.7-fold (to 650 mosmols) (Delaney *et al.* 1977). Urea is a potent destabilizer of protein structure and markedly raises the O_2 affinity of human Hb by promoting dissociation into subunits. In elasmobranch fishes that maintain high blood urea concentrations (approximately 0.5 M), methylamine osmolytes like

Table 2 Sites for binding of DPG, chloride ions (Cl) and protons (H) in human Hbs, and the amino acid residues at these sites in Hbs of *Lepidosiren* (Rodewald *et al.* 1984) and Andean frog *Telmatobius* (Weber *et al.* 2002).

	α-chain		*β-chain*				
Residue number in human Hb	1	131	1	2	82	143	146
Effector binding	(H) Cl[&]	(Cl)	D[1](Cl)[#]	D[2]	D[2] Cl[#]	D[2]	H
Human Hb	Val	Ser	Val	His	Lys	His	His
Lepidosiren paradoxus Hb	Met	Val	Val	His	Lys	Arg	His
Telmatobius peruvianus Hb	Ac –X¤	Ala	Val	His	Lys	Lys	His

[1]DPG binding to one β-chain; [2]DPG binding to both β-chains;

[&]Val-1α shares Cl⁻ binding with Ser-131α, and with Arg 141α of the opposite α-chain (O'Donnell *et al.* 1979);

#, Shared β-chain Cl⁻ binding site (Riggs 1988); ¤Unidentified acetylated residue.

trimethylamine oxide (TMAO) offsets urea's perturbating effects (Yancey 2001). However TMAO or its precursor TMA could not be detected in *P. dolloi* after 20 weeks' terrestrialization (Wilkie *et al.* 2007).

As with skate (elasmobranch), but in contrast to human Hb, the O_2 affinity of *P. amphibius* Hb is largely unaffected by urea, even at concentrations near 2M (Weber *et al.* 1977). However, unlike skate Hb, where Cl^- ions increase Hb-O_2 affinity reflecting increased electrostatic interactions between the subunits (Bonaventura *et al.* 1974), Cl^- ions markedly decrease O_2 affinities of *P. amphibius*, *P. annectans* and *Neoceratodus* Hbs [Fig. 2; (Weber *et al.* 1977; Rasmussen *et al.* 2009)] pointing to a novel molecular mechanism in lungfishes. Interestingly, polar residue α-131Ser that is a major Cl^- binding site in human Hb is replaced by non-polar Val in *Lepidosiren paradoxa* Hb (Rodewald *et al.* 1984) (Table 2). This indicates a lacking α-chain Cl^- binding site, as in Hb of high-altitude frog *Telmatobius* where Cl^- insensitivity (that accounts for the high O_2 affinity) correlates with non-polar Ala at this site (Weber *et al.* 2002) and calls for analysis of the Cl^- sensitivity of Hb-O_2 affinity in *Lepidosiren*.

CO_2 and lactate. High CO_2 tensions would increase carbamate formation, the reaction between CO_2 and the free amino groups of the N-termini of the α- and β-chains of Hb that occurs in air-breathing vertebrates but is physiologically-insignificant at the low PCO_2 values found in gill-breathing fish. CO_2 binding lowers Hb-O_2 affinity at high pH where the α-amino groups are not protonated (and thus decreases the Bohr factor). Given competition between CO_2 and organic phosphates for binding at the N-terminal residues of the β-chains (Table 2), carbamate formation may be reduced in the presence of organic phosphates. In *P. amphibius* Hb such interaction (and the greater affinity of Hb for GTP than ATP) is illustrated by the fact that the carbamate formation is completely obliterated by GTP but not by ATP (Weber and Johansen 1979).

Interestingly lactate that decreases the O_2 affinity and carbamation of human Hb (Nielsen and Weber 2007) does not accumulate in tissues of *P. dolloi* during estivating, indicating the absence of anaerobic glycolytic flux and catabolism of amino acid fuel stores during terrestrialization (Frick *et al.* 2008).

Water activity. As is well-established for human and teleost fish Hbs (Colombo *et al.* 1992; Hundahl *et al.* 2003), O_2 affinity also depends on water activity, based on the greater protein surface that is exposed to the solvent (and thus greater hydration) of Hb in the R (oxygenated) than in the T-state. The inordinately high osmolalities observed in terrestrialized lungfish begs investigation of possible differences in the sensitivity of Hb-O_2 affinity to water activity. Measurement of the relative effects of solvation [as previously detailed (Hundahl *et al.* 2003)] reveals that a reduction in water activity (induced by adding the inert solute betaine), only slightly decreases O_2 affinity of stripped *P. annectans* Hb, compared to a large effect on anodic HbA from the eel *Anguilla anguilla* (Fig. 3). Under a set of specific measuring conditions that were identical for the two Hbs, the relationship

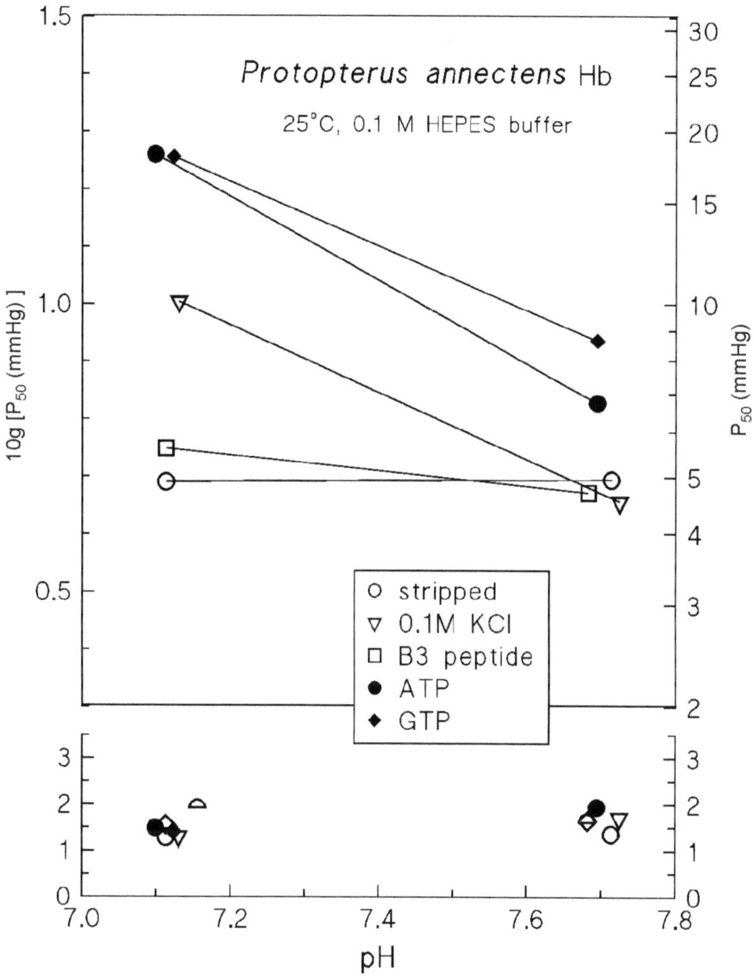

Fig. 2 O_2 tensions and cooperativity coefficients at 50% saturation (P_{50} and n_{50}) of 0.05 M solutions of *Protopterus annectens* Hb in the absence of effectors (stripped), and in the presence of 0.1 M KCl, a 10-mer peptide corresponding to the N-terminal domain of band 3 protein from trout red cells (B3 peptide at 5 molar excess over Hb tetrameric concentration) and of ATP and GTP at saturating concentrations (NTP/Hb > 10). Unpublished data of R.E. Weber, S. Pedersen, N. Serrano and A.R.W. Lassen (Århus) and Y.K. Ip (Singapore).

between water activity a_w [derived as $\ln a_w = -(Osm/M_w)$, where Osm in the solution osmolality and M_w is the molality of pure water (55.56 mol· kg^{-1})] and the shift in P_{50} indicate that an almost 7-fold lower number of water molecules bind to *P. annectans* compared to eel HbA (15 and 103, respectively) upon oxygenation.

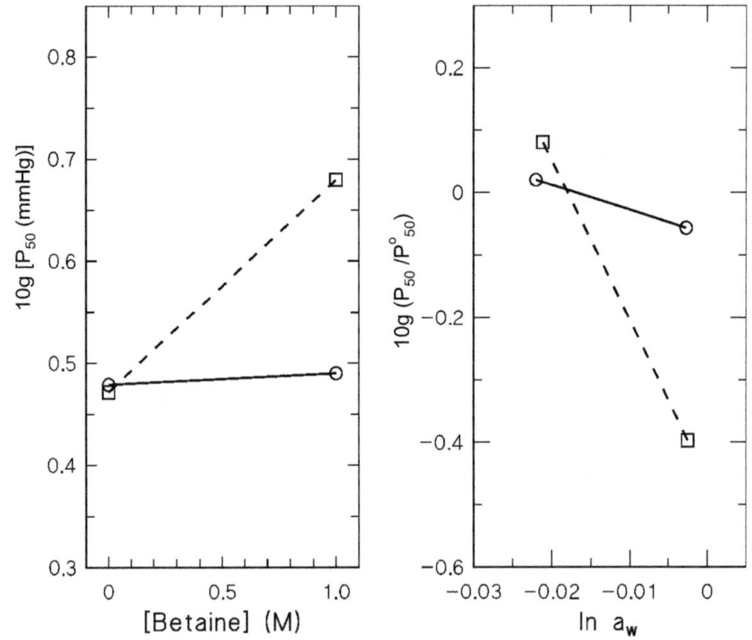

Fig. 3 Effects of 1 M betaine on P_{50} values (measured at pH 7.30 and 15°C) of stripped Hb of *Protopterus annectens* (o) and anodic HbA of eel *Anguilla anguilla* (□) (left panel), and the relative shifts in P_{50} related to water activity, *ln* a_w (right panel). (Unpublished data of R.E. Weber, C. Hundahl and Y.K. Ip.)

The reduction in the water effect in lungfish Hb may be adaptive in stabilizing O_2 affinity in the face of the large variations in blood osmolality during estivation.

C. CO_2 Transport

Blood transport of CO_2 in lungfish conforms to the general vertebrate scheme. Metabolically produced CO_2 diffuses into the erythrocytes, where carbonic anhydrase catalyses its hydration to H^+ and HCO_3^-. The produced H^+ is bound to Hb, whereas HCO_3^- is transported to plasma in exchange for Cl^- via the anion exchanger (known as AE1 or band 3). Most CO_2 is accordingly transported from the tissues to the respiratory surfaces (lungs and gills) as HCO_3^-, with $[HCO_3^-]$ being higher in plasma than in the red cells (Jensen *et al.* 2003). Some CO_2 may also be transported as carbamate, based on the reaction of the free N-terminal α-amino groups of lungfish Hb with CO_2 (see above). As the blood arrives at the respiratory surfaces, the reactions proceed in the opposite direction, and CO_2 is excreted with the expired medium. The CO_2 transporting capacity depends on the binding of H^+ to Hb, which is composite of the direct action of Hb as a buffer and the oxygenation-linked exchange of H^+ associated with the Bohr-Haldane effect. It has been noted that *Protopterus* blood shows no Haldane effect, because the total CO_2 content of oxygenated and deoxygenated blood are the same (Lenfant and Johansen 1968), which contrasts with the normally higher $[HCO_3^-]$ in deoxygenated blood that results from the additional binding of H^+ to Hb when it becomes deoxygenated (i.e. the fixed acid Haldane effect). However, since *Protopterus* blood has a Bohr effect (Lenfant and Johansen 1968; Johansen *et al.* 1976), and because the Bohr effect and the Haldane effect have the same molecular origin (Jensen 2004), a Haldane effect has to be present in *Protopterus* blood. It appears that anaerobic metabolism of *Protopterus* red cells during *in vitro* equilibration to the deoxygenated state proceeds at such a high rate that the lactic acid production titrates the extra HCO_3^- formed via the Haldane effect, leading to the apparent similar $[HCO_3^-]$ and total CO_2 in oxygenated and deoxygenated blood (Jensen *et al.* 2003).

The magnitude of the Bohr effect in *Protopterus* is moderate (Lenfant and Johansen 1968; Johansen *et al.* 1976) and the magnitude of the fixed acid Haldane effect in *Lepidosiren* Hb is low (Berenbrink *et al.* 2005). The specific buffer value of *Lepidosiren* Hb, on the other hand, is large, because the molecule has a large number of physiological buffer groups (histidines and α-amino groups) on its surface (Berenbrink *et al.* 2005). A high Hb-specific buffer value also characterizes the red cells in whole blood in *Protopterus* (Jensen *et al.* 2003). Low Haldane effects and high buffer values may be the ancestral condition for jawed vertebrates and have remained unchanged in lungfish (Berenbrink *et al.* 2005). These characteristics contrast with the large Bohr-Haldane effect and low buffer values (caused by an approximate halving of titratable histidines and α-amino groups) seen in teleosts (Jensen 1989), which evolved gradually within the lineage of ray-finned fishes after the evolutionary divergence of ray-finned fishes and lobe-finned fishes (Berenbrink *et al.* 2005).

HCO_3^-/Cl^- exchange across the red cell membrane is considered the rate-limiting step in blood CO_2 transport (Tufts and Perry 1998) and is deemed to be fast to reach completion during the short transit time of erythrocytes in the capillaries. The rate constant (k) for the unidirectional Cl^- transport via the anion exchanger is considerably lower in *Protopterus* than observed in other vertebrate species (Jensen *et al.* 2003). At first sight this could indicate a relatively slow HCO_3^-/Cl^- exchange in lungfish. However, when the large size of lungfish erythrocytes is taken into account by calculating the apparent chloride permeability P_{Cl} (the multiplication product of k with the ratio of cell water volume to membrane surface area), it turns out that P_{Cl} comes close to that observed in other vertebrates. A plot of ln P_{Cl} versus the inverse absolute temperature is left-shifted in tropical African lungfish compared to temperate species, but P_{Cl} is similar among animals when compared at their preferred temperatures (Jensen *et al.* 2003). HCO_3^-/Cl^- exchange across the red cell membrane is therefore not more of a bottle neck in the giant erythrocytes of lungfish than in other vertebrates, and a Q_{10} values of 2 for the anion exchange process (i.e. presumably close to that of CO_2 production) also suggests that the degree of rate-limitation does not vary considerably with temperature (Jensen *et al.* 2003).

It is generally recognized that the gills constitutes a major route for CO_2 excretion not only in the facultative air-breathing Australian lungfish *Neoceratodus* but also in the obligate air-breathing South American and African lungfishes (Graham 1997). Even though the latter have reduced gills with low diffusion conductance, it is believed that low water ventilation volumes are sufficient to sustain normal rates of CO_2 excretion due to the high CO_2 capacitance of water (Graham 1997). A recent study has modified this generalized picture by demonstrating that *Protopterus dolloi* excretes most of its CO_2 (76%) via the lungs (Perry *et al.* 2005). Thus, at least in this species, the lung is the major route for both O_2 uptake and CO_2 excretion, even though a significant part of CO_2 (24%) still exits via the gills/skin. This may reflect a species difference among African lungfish in the partitioning of CO_2 excretion between gills and lung, because the same authors note that they find the gill to be the predominant site of CO_2 excretion in *Protopterus annectens*, as reported in previous studies on this species (Perry *et al.* 2005).

Acknowledgments

The authors thank the Danish Natural Science Research Council and the Carlsberg Foundation for financial support.

References

Atha, D.H. and Ackers, G.K. (1974). Calorimetric determination of the heat of oxygenation of human hemoglobin as a function of pH and the extent of reaction. Biochemistry 13(11): 2376-2382.

Bartlett, G.R. (1978). Phosphates in the red cells of two lungfish: the South American, *Lepidosiren paradoxa* and the African, *Protopterus aethiopicus*. Canadian Journal of Zoology 56: 882-886.

Bassi, M., Klein, W., Fernandes, M.N., Perry, S.F. and Glass, M.L. (2005). Pulmonary oxygen diffusing capacity of the South American lungfish *Lepidosiren paradoxa*: physiological values by the Bohr method. Physiological and Biochemical Zoology 78: 560-569.

Berenbrink, M., Koldkjaer, P., Kepp, O. and Cossins, A.R. (2005). Evolution of oxygen secretion in fishes and the emergence of a complex physiological system. Science 307: 1752-1757.

Bohr, C., Hasselbalch, K. and Krogh, A. (1904). Über einen in biologischer Beziehung wichtigen Einfluss, den die Kohlensaurespannung des Blutes auf dessen Sauerstoffbindung ubt. Skandinavisches Archiv für Physiologie 16: 402-412.

Bonaventura, J., Bonaventura, C. and Sullivan, B. (1974). Urea tolerance as a molecular adaptation of elasmobranch hemoglobins. Science 186: 57-59.

Chanutin, A. and Curnish, R.R. (1967). Effect of organic and inorganic phosphates on the oxygen equilibrium of human erythrocytes. Archives of Biochemistry and Biophysics 121: 96-102.

Chew, S.F., Chan, N.K., Loong, A.M., Hiong, K.C., Tam, W.L. and Ip, Y.K. (2004). Nitrogen metabolism in the African lungfish (*Protopterus dolloi*) aestivating in a mucus cocoon on land. Journal of Experimental Biology 207: 777-786.

Colombo, M.F., Rau, D.C. and Parsegian, V.A. (1992). Protein solvation in allosteric regulation: a water effect on hemoglobin. Science 256: 655-659.

da Silva, G.D.F., Giusti, H., Sanchez, A.P., do Carmo, J.M. and Glass, M.L. (2008). Aestivation in the South American lungfish, *Lepidosiren paradoxa*: Effects on cardiovascular function, blood gases, osmolality and leptin levels. Respiratory Physiology & Neurobiology 164: 380-385.

de Moraes, M.F., Holler, S., da Costa, O.T., Glass, M.L., Fernandes, M.N. and Perry, S.F. (2005). Morphometric comparison of the respiratory organs in the South American lungfish *Lepidosiren paradoxa* (Dipnoi). Physiological and Biochemical Zoology 78: 546-559.

Delaney, R.G., Lahiri, S. and Fishman, A.P. (1974). Aestivation of the African lungfish *Protopterus aethiopicus*: cardiovascular and respiratory functions. Journal of Experimental Biology 61: 111-128.

Delaney, R.G., Shub, C. and Fishman, A.P. (1976). Hematologic observations on aquatic and estivating African lungfish, *Protopterus aethiopicus*. Copeia 1976: 423-434.

Delaney, R.G, Lahiri, S., Hamilton, R. and Fishman, P. (1977). Acid-base balance and plasma composition in the aestivating lungfish (*Protopterus*). American Journal of Physiology 232: R10-R17.

Duhm, J. (1971). Effects of 2,3-diphosphoglycerate and other organic phosphate compounds on oxygen affinity and intracellular pH of human erythrocytes. Pflügers Archiv 326: 341-356.

Frick, N.T., Bystriansky, J.S., Ip, Y.K., Chew, S.F. and Ballantyne, J.S. (2008). Carbohydrate and amino acid metabolism in fasting and aestivating African lungfish (*Protopterus dolloi*). Comparative Biochemistry and Physiology - Part A: Molecular & Integrative Physiology 151: 85-92.

Gorr, T., Kleinschmidt, T., Sgouros, J.G. and Kasang, L. (1991). A "living fossil" sequence: primary structure of the coelacanth (*Latimeria chalumnae*) hemoglobin evolutionary and functional aspects. Biological Chemistry Hoppe-Seyler 372: 599-612

Graham, J.B. (1997). Air-breathing Fishes. Evolution, Diversity, and Adaptation. Academic Press, San Diego, USA.

Gregory, T.R. (2001). The bigger the C-value, the larger the cell: genome size and red blood cell size in vertebrates. Blood Cells, Molecules, and Diseases 27: 830-843.

Gregory, T.R. (2002). Genome size and developmental complexity. Genetica 115: 131-146.

Gronenborn, A.M., Clore, G.M., Brunori, M., Giardina, B., Falcioni, G. and Perutz, M.F. (1984). Stereochemistry of ATP and GTP bound to fish haemoglobins. A transferred nuclear overhauser enhancement, ^{31}P-Nuclear Magnetic Resonance, oxygen equilibrium and molecular modelling study. Journal of Molecular Biology 178: 731-742.

Hundahl, C., Fago, A., Malte, H. and Weber, R.E. (2003). Allosteric effect of water in fish and human hemoglobins. Journal of Biological Chemistry 278: 42769-42773.

Isaacks, R.E. and Kim, H.D. (1984). Erythrocyte phosphate composition and osmotic fragility in the Australian lungfish, *Neoceratodus forsteri*, and osteoglossid, *Scleropages schneichardti*. Comparative Biochemistry and Physiology [A] 79A: 667-671.

Isaacks, R., Harkness, D., Sampsell R., Adler, J., Roth, S., Kim, C. and Goldman, P. (1977). Studies on avian erythrocyte metabolism. Inositol tetrakisphosphate: the major phosphate compound in the erythrocytes of the Ostrich (*Struthio camelus camelus*). European Journal of Biochemistry 77: 567-574.

Jensen, F.B. (1989). Hydrogen ion equilibria in fish haemoglobins. Journal of Experimental Biology 143: 225-234.

Jensen, F.B. (2004). Red blood cell pH, the Bohr effect, and other oxygenation-linked phenomena in blood O_2 and CO_2 transport. Acta Physiologica Scandinavia 182: 215-227.

Jensen, F.B. and Weber, R.E. (1982). Respiratory properties of tench blood and hemoglobin. Adaptation to hypoxic-hypercapnic water. Molecular Physiology 2: 235-250.

Jensen, F.B. and Weber, R.E. (1985). Kinetics of the acclimational response of tench to combined hypoxia and hypercapnia – I. Respiratory responses. Journal of Comparative Physiology B: Biochemical, Systemic, and Environmental Physiology 156: 197-203.

Jensen, F.B. and Brahm, J. (1995). Kinetics of chloride transport across fish red blood cell membranes. Journal of Experimental Biology 198: 2237-2244.

Jensen, F.B., Fago, A. and Weber, R.E. (1998a). Hemoglobin Structure and Function. In: S.F. Perry and B.L. Tufts (eds). In: Perry, S.F., Tufts, B. (eds.). Fish Respiration (Fish Physiology, Vol. 17). Academic Press, San Diego, USA. pp. 1-40.

Jensen, F.B., Jakobsen, M.H. and Weber, R.E. (1998b). Interaction between haemoglobin and synthetic peptides of the N-terminal cytoplasmic fragment of trout band 3 (AE1) protein. Journal of Experimental Biology 201: 2685-2690.

Jensen, F.B., Brahm, J., Koldkjaer, P., Wang, T., McKenzie, D.J. and Taylor, E.W. (2003). Anion exchange in the giant erythrocytes of African lungfish. Journal of Fish Biology 62: 1044-1052.

Johansen, K. and Lenfant, C. (1967). Respiratory function in the South American lungfish, *Lepidosiren paradoxa* (Fitz). Journal of Experimental Biology 46: 205-218.

Johansen, K. and Lenfant, C. (1971). A comparative approach to the adaptability of O_2-HB affinity. In: P. Astrup and M. Rørth (eds). Oxygen Affinity of Hemoglobin and Red Cell Acid-Base Status. Alfred Benzon Symposium IV. Munksgaard. Copenhagen, Denmark. pp. 750-780.

Johansen, K., Lykkeboe, G., Weber, R.E. and Maloiy, G.M.O. (1976). Respiratory properties of blood in awake and estivating lungfish, *Protopterus amphibius*. Respiration Physiology 27: 335-345.

Kind, P.K., Grigg, G.C. and Booth, D.T. (2002). Physiological responses to prolonged aquatic hypoxia in the Queensland lungfish *Neoceratodus forsteri*. Respiratory Physiology & Neurobiology 132: 179-190.

Koldkjaer, P., Taylor, E.W., Glass, M.L., Wang, T., Brahm, J., McKenzie, D.J. and Jensen, F.B. (2002). Adrenergic receptors, Na+/H+ exchange and volume regulation in lungfish erythrocytes. Journal of Comparative Physiology B: Biochemical, Systemic, and Environmental Physiology 172: 87-93.

Lenfant, C. and Johansen, K. (1967). Respiratory adaptations in selected amphibians. Respiration Physiology 2: 247-260.

Lenfant, C. and Johansen, K. (1968). Respiration in the African lungfish *Protopterus aethiopicus*. I. Respiratory properties of blood and normal patterns of breathing and gas exchange. Journal of Experimental Biology 49: 437-452.

Lenfant, C., Johansen, K. and Grigg, G.C. (1966). Respiratory properties of blood and pattern of gas exchange in the lungfish *Neoceratodus forsteri* (Krefft). Respiration Physiology 2: 1-21.

Lenfant, C., Johansen, K. and Hanson, D. (1970). Bimodal gas exchange and ventilation-perfusion relationship in lower vertebrates. Federation Proceedings 29: 1124-1129.

Lomholt, J.P. (1993). Breathing in the aestivating African lungfish, *Protopterus amphibius*. Advances in Fish Research 1: 17-34.

Low, P.S., Rathinavelu, P. and Harrison, M.L. (1993). Regulation of glycolysis via reversible enzyme binding to the membrane protein, band 3. Journal of Biological Chemistry 268: 14627-14631.

Nielsen, M.S. and Weber, R.E. (2007). Antagonistic interaction between oxygenation-linked lactate and CO_2 binding to human hemoglobin. Comparative Biochemistry and Physiology – Part A: Molecular & Integrative Physiology 146: 429-434.

Nikinmaa, M. (1982). Effects of adrenaline on red cell volume and concentration gradient of protons across the red cell membrane in the rainbow trout *Salmo gairdneri*. Molecular Physiology 2: 287-297.

Nikinmaa, M. (1990). Vertebrate Red Blood Cells. Adaptations of Function to Respiratory Requirements. Springer-Verlag, Berlin, Germany.

Nikinmaa, M. (1992). Membrane transport and control of hemoglobin-oxygen affinity in nucleated erythrocytes. Physiological Reviews 72: 301-321.

O'Donnell, S., Mandaro, R., Schuster, T.M. and Arnone, A. (1979). X-ray diffraction and solution studies of specifically carbamylated human hemoglobin A. Evidence for the location of a proton- and oxygen-linked chloride binding site at valine 1α. Journal of Biological Chemistry 254: 12204-12208.

Perry, S.F., Gilmour, K.M., Swenson, E.R., Vulesevic, B., Chew, S.F. and Ip, Y.K. (2005). An investigation of the role of carbonic anhydrase in aquatic and aerial gas transfer in the African lungfish *Protopterus dolloi*. Journal of Experimental Biology 208: 3805-3815.

Phelps, C., Farmer, M., Fyhn, H.J., Fyhn, U.E.H., Garlick, R.L., Noble, R.W. and Powers, D.A. (1979). Equilibria and kinetic of oxygen and carbon monoxide binding to the haemoglobin of the South American lungfish, *Lepidosiren paradoxa*. Comparative Biochemistry and Physiology Part A: Physiology 62: 139-143.

Rasmussen, J.R., Wells, R.M., Henty, K., Clark, T.D. and Brittain, T. (2009). Characterization of the hemoglobins of the Australian lungfish *Neoceratodus forsteri* (Krefft). Comparative Biochemistry and Physiology – Part A: Molecular & Integrative Physiology 152: 162-167.

Ribeiro, M.L., DaMatta, R.A., Diniz, J.A., de Souza, W., do Nascimento, J.L. and de Carvalho, T.M. (2007). Blood and inflammatory cells of the lungfish *Lepidosiren paradoxa*. Fish Shellfish Immunology 23: 178-187.

Riggs, A.F. (1988). The Bohr effect. Annual Review of Physiology 50: 181-204.

Rodewald, K., Stangl, A. and Braunitzer, G. (1984). Primary structure, biochemical and physiological aspects of hemoglobin from South American lungfish (*Lepidosiren paradoxus,* Dipnoi). Hoppe-Seylers Zeitschrift fur Physiologische Chemie 365: 639-649.

Salama, A. and Nikinmaa, M. (1988). The adrenergic responses of carp (*Cyprinus carpio*) red cells: effects of PO_2 and pH. Journal of Experimental Biology 136: 405-416.

Swan, H. and Hall, F.G. (1966). Oxygen-hemoglobin dissociation in *Protopterus aethiopicus*. American Journal of Physiology 210: 487-489.

Tetens, V. and Lykkeboe, G. (1985). Acute exposure of rainbow trout to mild and deep hypoxia: O_2 affinity and O_2 capacitance of arterial blood. Respiration Physiology 61: 221-235.

Tufts, B.L. and Perry, S.F. (1998). Carbon Dioxide Transport and Excretion. In: S.F. Perry and B.L. Tufts (eds.). Fish Respiration (Fish Physiology Vol. 17). Academic Press, San Diego, USA. pp. 229-281.

Val, A.L. (2000). Organic phosphates in the red blood cells of fish. Comparative Biochemistry and Physiology – Part A: Molecular & Integrative Physiology 125: 417-435.

Weber, R.E. (1982). Intraspecific adaptation of hemoglobin function in fish to oxygen availability. In: A.D.F. Addink and N. Spronk (eds). Exogenous and Endogenous Influences on Metabolic and Neural Control. Volume 1. Pergamon Press, Oxford, UK. pp. 87-102.

Weber, R.E. (1992). Molecular strategies in the adaptation in vertebrate hemoglobin function. In: S.C. Wood , R.E.Weber, A.R. Hargens and R.W. Millard (eds). Physiological Adaptations in Vertebrates; Respiration, Circulation, and Metabolism (Proc. Kjell Johansen Memorial Symp.). Marcel Dekker, New York, USA. pp. 257-277.

Weber, R.E. and Lykkeboe, G. (1978). Respiratory adaptations in carp blood. Influences of hypoxia, red cell organic phosphates, divalent cations and CO_2 on hemoglobin-oxygen affinity. Journal of Comparative Physiology 128B: 127-137.

Weber, R.E. and Johansen, K. (1979). Oxygenation-linked binding of carbon dioxide and allosteric phosphate cofactors by lungfish hemoglobin. In: R. Gilles (ed). Animals and Environmental Fitness. Pergamon Press, Oxford, UK. pp. 49-50.

Weber, R.E. and Fago, A. (2004). Functional adaptation and its molecular basis in vertebrate hemoglobins, neuroglobins and cytoglobins. Respiratory Physiology & Neurobiology 144: 141-159.

Weber, R.E., Lykkeboe, G. and Johansen, K. (1975). Biochemical aspects of the adaptation of hemoglobin-oxygen affinity of eels to hypoxia. Life Sciences 17: 1345-1350.

Weber, R.E., Johansen, K., Lykkeboe, G. and Maloiy, G.M.O. (1977). Oxygen-binding properties of hemoglobins from estivating and active African lungfish. Journal of Experimental Zoology 199: 85-96.

Weber, R.E., Ostojic, H., Fago, A., Dewilde, S., Van Hauwaert, M.L., Moens, L. and Monge, C. (2002). Novel mechanism for high-altitude adaptation in hemoglobin of the Andean frog *Telmatobius peruvianus*. American Journal of Physiology: Regulatory, Integrative and Comparative Physiology 283: R1052-R1060.

Weber, R.E., Voelter, W., Fago, A., Echner, H., Campanella, E. and Low, P.S. (2004). Modulation of red cell glycolysis: interactions between vertebrate hemoglobins and cytoplasmic domains of band 3 red cell membrane proteins. AJP-Regulatory, Integrative and Comparative Physiology 287: R454-R464.

Wilkie, M.P., Morgan, T.P., Galvez, F., Smith, R.W., Kajimura, M., Ip, Y.K. and Wood, C.M. (2007). The African lungfish (*Protopterus dolloi*): Ionoregulation and osmoregulation in a fish out of water. Physiological and Biochemical Zoology 80: 99-112.

Wood, S.C. and Johansen, K. (1972). Adaptation to hypoxia by increased HbO_2 affinity and decreased red cell ATP concentration. Nature New Biology 237: 278-279.

Wood, S.C. and Johansen, K. (1973a). Blood oxygen transport and acid-base balance in eels during hypoxia. American Journal of Physiology 225: 849-851.

Wood, S.C. and Johansen, K. (1973b). Organic phosphate metabolism in nucleated red cells: influence of hypoxia on eel HbO_2 affinity. Netherlands Journal of Sea Research 7: 328-338.

Yancey, P.H. (2001). Nitrogen compounds as osmolytes. In: P.A.Wright and P. Anderson (eds). Nitrogen Excretion. Academic Press, New York, USA. pp. 309-341.

CHAPTER 12

Lungfish Metabolism

James S. Ballantyne[1,*] **and Natasha T. Frick**[2]

[1]Department of Integrative Biology, University of Guelph, Guelph, Ontario, N1G 2W1, Canada

[2]Department of Biochemistry, University of Toronto, 1 King's College Circle, Medical Science Building, Toronto, Ontario, M5S 1A8, Canada

ABSTRACT

Lungfish provide a unique insight into the evolution of metabolism in the vertebrates. They are "living fossils anatomically" and this is also reflected in their metabolic organization. The complexity of enzyme isoforms and their tissue specific distribution are indicative of the early evolutionary position they occupy. The bile salts and forms of surfactant used in the lungs are also considered "primitive". The ability of some species to estivate provides an understanding of the mechanisms for down-regulating metabolism and the metabolic reorganization that must accompany such changes. During estivation, nitrogen metabolism is also modified to allow them to detoxify ammonia and convert it to urea reducing osmotic stress. The subcellular organization of the system for urea synthesis is different from that of mammals or elasmobranchs. Continued study of the metabolism of lungfish is certain to provide a more detailed understanding of the early stages of vertebrate evolution and the transition from an aquatic existence to a terrestrial lifestyle.

Keywords: metabolism, enzyme, lipid, estivation, urea

INTRODUCTION

Lungfish have long been considered of evolutionary interest. The phylogenetic relationship between lungfish and the tetrapods has been examined in a series

*Corresponding author: E-mail: jballant@uoguelph.ca

of studies with increasing evidence that lungfish are the closest ancestor of the tetrapods (Brinkmann *et al.* 2004; Meyer and Dolven 1992; Takezaki *et al.* 2004; Yokobori *et al.* 1994; Zardoya *et al.* 1998; Zardoya and Meyer 1996). Thus the six extant species of lungfish living in Africa (Protopteridae, *Protopterus* – 4 species), South America (Lepidosirenidae, *Lepidosiren* – one species) and Australia (Ceratodontidae, *Neoceratodus* – one species) provide a unique window into the evolution of metabolism of the vertebrates at that critical period when the transition from an aquatic to a terrestrial existence was made. Anatomical structures such as lungs and sturdy appendages were prerequisites for terrestrial survival but metabolic prerequisites were also required. Consequently, an important question is, if extant lungfish are living fossils anatomically, can we assume they also retain many of the primitive features of metabolism? As this review will show, there is considerable evidence that this assumption is valid.

This review outlines the current understanding of the metabolic organization of the extant lungfish. The 6 extant species of lungfish differ metabolically in several ways. The Australian species differs from the other species in its low reliance on aerial respiration and its inability to estivate. It is considered more primitive than the other species based on anatomical criteria. Thus it represents an early stage in the transition to a terrestrial existence. All 4 species of African lungfish have the ability to form cocoons to avoid desiccation, however only *P. annectens* has been routinely observed to do so in its natural environment (Greenwood 1986). This is likely due to differences in habitat type and the likelihood of experiencing complete seasonal drying.

The ability to breathe air with a primitive lung and to estivate for long periods, have interested physiologists for decades. The earliest recorded studies of the physiology of lungfish are those of Homer Smith (Smith 1930; Smith 1931; Smith 1935) who measured the rate of oxygen consumption in *Protopterus* in air and water, and demonstrated changes in the pattern of bimodal gas exchange occur during fasting and estivation and at different developmental stages. In the 1960's and 1970's Kjell Johansen contributed substantially to understanding the respiratory physiology with studies of the Australian (Johansen *et al.* 1967), South American (Johansen and Lenfant 1967) and African species (Johansen and Lenfant 1968). These studies demonstrated that in the primarily water breathing Australian lungfish, blood flow increased to the lung during hypoxia. In the late 1970's the studies of Fishman *et al.* on African lungfish provided more details on the regulation of respiration in active and estivating lungfish (Delaney and Fishman 1977; Laurent *et al.* 1978). They also examined cardiovascular physiology in active (Arbel *et al.* 1977) and estivating fish (Delaney *et al.* 1974). In Africa, G.M.O. Maloiy encouraged field and lab studies of African lungfish in Kenya (Dunn *et al.* 1981; Johansen *et al.* 1976a, b; Lomholt *et al.* 1975; Maina and Maloiy 1985; Weber *et al.* 1977). From the mid 1980's until recently, studies of the metabolism of lungfish were sporadic.

The resurgence in interest in the metabolism of lungfish is due to the initiatives of Profs. Y.K. Ip (National University of Singapore, Republic of Singapore) and S.F. Chew (Nanyang Technological University, Republic of Singapore). Their extensive research and collaborations (Chew *et al.* 2004; Frick *et al.* 2008a, b; Loong *et al.* 2005; Perry *et al.* 2005a, b; Wilkie *et al.* 2007; Wood *et al.* 2005a, b) have added much to the understanding of these important vertebrates. They have filled in many gaps in our knowledge and corrected errors in the literature that have persisted for decades. This review summarizes the current state of understanding of the metabolism of lungfish with the goal of not only describing the metabolism of the extant lungfish species but also placing the metabolic organization of lungfish in an evolutionary context.

DIGESTION

The basis for metabolism is the input of nutrients from food items. The extant lungfishes have been classified as omnivores (Bemis 1986; Greenwood 1986; Mlewa *et al.* 2009) based on ingested plant material. Much of this material may be unintentionally ingested since the foraging habits involve taking large amounts of substrate into the mouth. Species of *Protopterus* vary in the amount of plant material and other dietary components. Certainly some populations of *P. aethiopicus* are primarily piscivorous as adults (Mlewa and Green 2004) while others eat mainly mollusks (Corbet 1961). Cannibalism has also been reported (Curry-Lindahl 1956). Digestion has an associated specific dynamic action (SDA) that is comparable in duration (~ 24 hours) to that observed in other fish (Iftikar *et al.* 2008). The magnitude of the increase in respiration due to SDA is 9.5% of the caloric value of the meal ingested indicating an efficient conversion process (Iftikar *et al.* 2008).

Lungfish have a stomach and spiral valve intestine similar to that of other primitive fishes (Hassanpour and Joss 2009). The studies of Reeck *et al.* have characterized the digestive enzymes of African lungfish and have shown protein digestion is similar to that of other vertebrates. Lungfish have 2 amylases, 2 ribonucleases, 3-4 chymotrypsinogens, 2 procarboxpeptidases A, a procarboxypeptidase B and carboxypeptidase B (Reeck *et al.* 1970) and 2 proelastases (de Haen and Gertler 1974). Lungfish pancreatic procarboxypeptidase B and carboxypeptidase B have been purified and characterized (Reeck and Neurath 1972a). Carboxypeptidase B (MW 45,000) is chemically and enzymatically similar to the bovine, porcine, dogfish and rat enzyme (Reeck and Neurath 1972a). Three trypsinogens have been identified in *P. aethiopicus* (de Haen *et al.* 1977). The trypsinogen of lungfish is unique in that it lacks a tetraaspartyl sequence in the activation peptide found in all other species (Reeck and Neurath 1972b). In spite of this lungfish trypsinogen is activated by enterokinase as in mammals (de Haen *et al.* 1977). Lungfish trypsin is similar to mammalian trypsin in its stability at pH 3.0 (Reeck and Neurath 1972b).

Lipid digestion begins with the solubilization by bile salts. The bile of the Australian lungfish is unusual in that it contains no cholesterol (Moschetta *et al.* 2005). The major bile salts are C_{27} alcohols conjugated to sulfate and the main phospholipids in the bile are sphingomyelin (59%) and phosphatidylcholine (41%) (Moschetta *et al.* 2005). After lipid solubilization, lipases act on the lipids but to our knowledge there are no studies of the lipases of any species of lungfish.

OXIDATIVE METABOLISM

Primitive fishes generally have a lower aerobic capacity than the more advanced teleosts (Ewart and Driedzic 1987; Moyes *et al.* 1989; Sidell *et al.* 1987; Suarez *et al.* 1986). This appears to be true for lungfish as well.

Respiration Rates

The respiration rates of lungfish are low compared to those of most fish (Brett 1972; Hochachka and Guppy 1987; Seifert and Chapman 2006). Part of the observation that measured metabolic rates are low may be due to the fact that *Protopterus* spp. are nocturnal. Captive *P. annectens* display higher breathing rates at night than during the day (Johnels and Svensson 1954).

Several factors correlate with metabolic rate. Genome size correlates positively with cell size and inversely with metabolic rate. Lungfish genomes and erythrocytes are among the largest known for any vertebrate (Pedersen 1971) and this correlates with the metabolic rates that are among the lowest measured for any fish species (Brett 1972; Seifert and Chapman 2006). The relationship between genome size, cell size and metabolic rate has been analyzed in mammals and birds with the suggestion that the accumulation of noncoding DNA may influence cell size and thus metabolic rate (Kozlowski *et al.* 2003). It has been hypothesized that cell size began to increase in lungfish in the Carboniferous after the main diversification of the group occurred, resulting in the very large genomes of extant species (Thomson 1972). The large genome and large cell size of lungfish may be an important factor in their low metabolic rate and hence their ability to survive long periods in a dormant state.

Oxygen Delivery

In keeping with the low metabolic rate, the hematocrit and hemoglobin levels of lungfish (Lenfant *et al.* 1966; Lenfant and Johansen 1968) are lower than those reported for active teleosts (Grigg 1974). The shift from aquatic to aerial respiration correlates with decreased oxygen affinity of hemoglobin as an adaptation to take advantage of higher availability of oxygen in air (Johansen *et al.* 1976b). Lungfish have a unique mix of organic phosphates in erythrocytes that influence hemoglobin binding characteristics. These include ATP, GTP, IP_2 and UTP (Val 2000). GTP is quantitatively most important (Johansen *et al.* 1976b). The Bohr effect, an index of the pH/CO_2 sensitivity of oxygen binding to hemoglobin, is enhanced by ATP in *L. paradoxa* (Phelps *et al.* 1979). The Bohr effect is greater in *Neoceratodus* than in *Protopterus* and *Lepidosiren* (Lenfant and Johansen 1968). A reduced pH effect has been suggested to help maintain oxygen transport to tissues during hypoxia with its associated acidosis (Giardina *et al.* 2004). Since both *Protopterus* and *Lepidosiren* estivate they may be more likely to experience hypoxia in their habitat than *Neoceratodus*, that does not.

Protopterus shows no Haldane effect (Reciprocal of the Bohr effect) similar to the situation in elasmobranchs (Wood *et al.* 1994) while *Neoceratodus* and *Lepidosiren* do (Lenfant and Johansen 1968). The basis for this difference is unknown.

Mitochondrial Metabolism

As the foci of aerobic metabolism mitochondria are valuable indices of the importance of aerobic metabolism and substrate preferences of specific tissues. Mitochondrial abundance in red, intermediate and white muscle fibers of lungfish is lower than that of all other fish measured (Dunn *et al.* 1981). In addition to having fewer mitochondria, the existing mitochondria appear to have a reduced ability to oxidize substrates. In both liver and muscle tissues of *P. dolloi* succinate is oxidized at rates ~2 to 5-fold lower (Frick *et al.* 2008b) than those reported in other fish species (Chamberlin *et al.* 1991; Mishra and Shukla 1994; Moyes *et al.* 1986), but comparable to those reported in another primitive fish, *Amia calva* (Chamberlin *et al.* 1991). In addition to low mitochondrial oxidation rates, both tissue cytochrome c oxidase (CCO) and citrate synthase (CS) activities (enzymes considered good indicators of the oxidative capacity of a tissue) are generally low in lungfish. Specific comparisons of the activities of CS in African lungfish tissues (Dunn *et al.* 1983; Frick *et al.* 2008b) to values reported for other fish species, both primitive and advanced, (Battersby and Moyes 1998; Frick *et al.* 2007; LeBlanc *et al.* 1995; Pelletier *et al.* 1994; Singer *et al.* 1990; Singer and Ballantyne 1991; Speers-Roesch *et al.* 2006b; West *et al.* 1999) and in a number of species of amphibians, mammals, and birds (Sugden and Newsholme 1975) indicates the potential for oxidative metabolism in *Protopterus* is depressed. The low mitochondrial density in combination with the low oxidative capacity of isolated mitochondria correlate well with the overall low metabolic rate reported in lungfish (Hochachka and Guppy, 1987).

The role of mitochondria in balancing cytosolic redox has been examined in the heart of *L. paradoxa* where the malate-aspartate shuttle predominates similar to other air-breathing fishes and mammals unlike the situation in water breathing fish where both the malate-aspartate shuttle and the alpha-glycerophosphate shuttle are used (Hochachka *et al.* 1979).

CARBOHYDRATE METABOLISM

Glycolysis and Glycogenolysis

The amount of glycogen stored within a tissue indicates the tissue's anaerobic capacity (McDougal *et al.* 1968). In *Protopterus*, the greatest levels of glycogen are in the liver (~180 μmol glucosyl units·g wet tissue^{-1}) (Dunn *et al.* 1983; Frick *et al.* 2008a). Under conditions of oxygen shortage the main function of the liver is to supply glucose to extrahepatic tissues. Fish typically store small amounts of glycogen within the white muscle, however, lungfish have an abundance of glycogen stored between fibers and as large rosettes in myofibrillar regions (Hochachka *et al.* 1978b; Hochachka 1980). These glycogen rosettes are present in the heart, liver and red and white muscle tissues. Lungfish can use this energy substrate both aerobically and anaerobically (Hochachka and Hulbert 1978) but it is probably most advantageous to spare it for hypoxic episodes. This is discussed in more detail below. Brain glycogen levels in the lungfish are about 2-4 times higher than the brain glycogen stores of most vertebrates but similar to those of other anoxia tolerant species (Dunn *et al.* 1983; Kerem *et al.* 1973; McDougal *et al.* 1968).

In *P. dolloi* and *P. aethiopicus* the heart was found to have the greatest capacity for glycolysis and glycogenolysis with the activities of pyruvate kinase (PK), hexokinase (HK), phosphoglucoisomerase (PGI) and glycogen phosphorylase being highest in this tissue (Dunn *et al.* 1983; Frick *et al.* 2008a). The activity levels of HK in the tissues of *P. dolloi* were generally low compared to those reported in the other primitive fish including the bowfin (Singer and Ballantyne 1991), ratfish (Speers-Roesch *et al.* 2006b) and garfish (Frick *et al.* 2007), yet comparable to those of sturgeon (Singer *et al.* 1990), sea lamprey (LeBlanc *et al.* 1995) and several more advanced teleost fishes (Knox *et al.* 1980). The activity of PK in the liver of *P. dolloi* was high compared to the levels found in lamprey (LeBlanc *et al.* 1995), ratfish (Speers-Roesch *et al.* 2006b) and several teleosts (Knox *et al.* 1980, Pelletier *et al.* 1994; West *et al.* 1999) yet comparable to garfish (Frick *et al.* 2007). In other tissues, PK activities were similar to other fish species (Frick *et al.* 2007; Knox *et al.* 1980; LeBlanc *et al.* 1995; Pelletier *et al.* 1994; Speers-Roesch *et al.* 2006b). Glycogenolysis is stimulated by glucagon, adrenaline, arginine vasotocin and arginine vasopressin in *N. forsteri* indicating β-adrenergic control (Hanke and Janssens 1983).

Gluconeogenesis

African lungfish may have a substantial capacity for glucose and glycogen synthesis as the activity levels of fructose 1,6-bisphosphatase (FBPase), a key enzyme in the gluconeogenesis pathway (Suarez and Mommsen 1987) are high (up to 100-fold greater) compared to other fish species (e.g. *H. colliei. L. platyrhincus,* elasmobranchs, and actinopterygians (Battersby *et al.* 1996; Blier and Guderley 1988; Dickson *et al.* 1993; Moon and Mommsen 1987; Ritter *et al.* 1987; Singer *et al.* 1990; Speers-Roesch *et al.* 2006b; Treberg *et al.* 2003)). The activity of FBPase is highest in the liver (the major gluconeogenic tissue in fish and elasmobranchs (Moon and Mommsen 1987)) of both *P. dolloi* (Frick *et al.* 2008a) and *P. aethiopicus* (Dunn *et al.* 1983). Plasma glucose levels range from 0.2 mM in *P. aethiopicus* (Dunn *et al.* 1983) to 1.30 mM in *P. dolloi* (Frick *et al.* 2008a) and are in the range of values reported in teleost fish species but lower than those reported for marine elasmobranchs (Ballantyne 1997). Gluconeogenesis is not influenced by adrenaline in *N. forsteri* (Hanke and Janssens 1983) unlike the situation in teleost fishes where gluconeogenesis is activated by adrenaline (Iwama *et al.* 1999; Wright *et al.* 1989).

Lactate Metabolism

Blood lactate levels in *P. aethiopicus* are similar to other fish species (Dunn *et al.* 1983). In *P. dolloi* and *P. aethiopicus* the activity of lactate dehydrogenase (LDH) was greatest in the heart (Dunn *et al.* 1983; Frick *et al.* 2008a), higher than levels reported in tuna, salmon and Arapaima hearts (Hochachka *et al.* 1978a, c; Mommsen *et al.* 1980). *Protopterus* hearts display a high anaerobic potential, based on both the high levels of LDH and the large glycogen stores. Gar (Frick *et al.* 2007) and some air-breathing Amazonian fish (Almeida-Val and Hochachka 1995) also display high activities of LDH. The LDH/CS ratios are also high in lungfish (Dunn *et al.* 1983; Frick *et al.* 2008a, b), a common characteristic amongst air-breathing fish, which often inhabit environments with low oxygen availability and require a large anaerobic potential (Dunn *et al.* 1983).

NITROGEN METABOLISM

Amino Acid Metabolism

The early studies of Smith (Smith 1931) suggested *Protopterus* derives 50% of its energy requirements from protein even when lipid is available. Studies of *P. annectens* feeding on chironomid larvae indicate a nitrogen quotient of 0.27 typical of a very high reliance on protein as an aerobic fuel (Iftikar *et al.* 2008). Fish in general use amino acids to supply a substantial part of their energy requirements (Ballantyne 2001). Catabolism of amino acids occurs through transaminations and glutamate dehydrogenase (GDH). In *P. aethiopicus* GDH is kinetically similar to amphibian GDH (Janssens and Cohen 1968b). In the liver tissue of *P. dolloi* Frick *et al.* (Frick *et al.* 2008a) found that the activities of GDH, alanine aminotransferase (AlaAT), and aspartate aminotransferase (AspAT) are comparable to those of elasmobranchs (Ballantyne 1997), ratfish (Speers-Roesch *et al.* 2006b) and gar (Frick *et al.* 2007). Within other tissues of *P. dolloi* however, the capacity for amino acid metabolism appears to be reduced compared to other fish species (Ballantyne 1997; Frick *et al.* 2007; Speers-Roesch *et al.* 2006b;). The amino acid metabolism of lungfish erythrocytes seems to differ from that of other fish species in that glutamate is oxidized at higher rates than glucose (Mauro and Isaacks 1989). The physiological significance of this has not been determined but may help spare glucose for other tissues. Excess nitrogen is excreted either as ammonia or urea and the relative proportions vary as a function of feeding with a greater proportion of nitrogen being excreted as urea (>60%) after a meal (Lim *et al.* 2004).

Ammonia Metabolism

Ammonia is toxic to many aquatic species and is an even greater problem when fish move out of water when getting rid of ammonia with vast volumes of dilute urine is not an option. Exogenous ammonia only a problem for lungfish when in water. The reduced gills and highly impermeable skin reduce the influx of ammonia and allow them to survive high external ammonia (~100 mM) (Loong *et al.* 2007). During emersion metabolic ammonia buildup may be more of a problem especially during prolonged estivation. Air-breathing fish have many ways of dealing with the possibility of ammonia buildup. It has been suggested (Ip *et al.* 2004) there are six strategies tropical fish use to deal with ammonia toxicity. They are 1) Reduced amino acid catabolism and/or proteolysis, 2) Funneling amino acid nitrogen catabolism into alanine (partial amino acid catabolism), 3) Synthesis of glutamine, 4) Urea synthesis, 5) Volatilization of NH_3, and 6) Ammonia tolerance. Of these, lungfish primarily use glutamine and urea synthesis.

Urea Cycle

Early studies lead to the conclusion that the complete urea cycle is present in the liver of lungfish (Forster and Goldstein 1966; Janssens and Cohen 1966). All the enzymes of the urea cycle have been found in all extant species of lungfish (Table 1). The activities of urea cycle enzymes of *Lepidosiren* are intermediate between those of the African lungfish and those of the Australian lungfish (Funkhouser *et al.* 1972). On the other hand, studies of actual urea synthesis rates indicate the lowest capacity is in *Neoceratodus*, being 100-fold lower than that of African lungfish (Goldstein *et al.* 1967). This physiological limitation in Australian lungfish correlates with the inability of *Neoceratodus* to tolerate air exposure as effectively as its South American and African counterparts.

Until recently it was thought that the nitrogen donor for urea synthesis was ammonia since early studies found lungfish use carbamoyl phosphate synthetase (CPSase) I (the enzyme used in urea synthesis in amphibians and mammals

Table 1 Activities of enzymes involved in urea synthesis in the liver of lungfish species. GS=glutamine synthase; CPSase=carbamoyl phosphate synthetase III; OTC=ornithine.Transcarbamoylase; ASS+L=arginosuccinate synthetase + lyase. Units are μ mol min^{-1} g^{-1} wet mass unless indicated with a ‡, in which case units are μ moles mg $protein^{-1}$ $hour^{-1}$.

Enzyme	*P. dolloi*	*P. aethiopicus*	*P. annectens*	*L. paradoxa*	*N. forsteri*
GS	0.19±0.02[e]	0.47±0.12[a]	0.36±0.04[a]		
	0.42±0.04[f]	1.59±0.32[b]	1.03±0.19[b]		
CPSase III	0.41±0.06[e]	0.20±0.1[a]	0.061±0.01[a]		
	0.75±0.16[f]	0.34±0.12[b]	0.13±0.03[b]		
ASS+L	1.1[d]	0.36±0.07[a]	0.27±0.03[a]		0.22[d]
	0.37±0.05[e]	0.46±0.07[b]	0.448±0.079[b]		
	0.82±0.07[f]	1.8[g]‡			
		3.0[h]‡			
OTC	9.77±0.87[e]	25±5[a]	13±0.7[a]	10.4[c]	2.6[d]
	16.1±1.6[f]	33±10[b]	18±2[b]		
	41[g]‡				
	83[h]‡				
Arginase	580[d]	135±4[a]	113±13[a]	256	21[d]
	73.9±11.1[e]	176±25[b]	145±30[b]		
	96.8±6.1[f]	826[g]‡			
		796[h]‡			

[a]control (Loong *et al.* 2005); [b]6 days air exposed (Loong *et al.* 2005); [c](Funkhouser *et al.* 1972); [d](Goldstein *et al.* 1967); [e]control (Chew *et al.* 2004); [f]40 days air exposed (Chew *et al.* 2004); [g]control (Jaanssens and Cohen 1968); [h]estivated 78-129 days (Jaanssens and Cohen 1968).

(Anderson 2001)) for urea synthesis (Mommsen and Walsh 1989). Subsequent work indicated CPSase III is the enzyme used (Chew *et al.* 2003, 2004; Loong *et al.* 2005), and the immediate nitrogen source is therefore glutamine. Indeed, liver glutamine levels are very low in *P. aethiopicus* (< 0.001 mM) (Ip *et al.* 2005c) perhaps indicating the high utilization for urea synthesis.

The subcellular distribution of the enzymes of the urea cycle was determined in *P. annectens* and *P. aethiopicus* by Loong *et al.* (2005). CPSase III is localized exclusively within the mitochondria, and is found in terrestrial vertebrates, elasmobranchs and teleosts (Anderson 1991; Cao *et al.* 1991) (Figure 1). Arginase is largely cytosolic (76% in *P. annectens* and > 90% in *P. aethiopicus*), while glutamine synthase (GS) is localized in both the cytosol (75%) and the mitochondria (25%) (Loong *et al.* 2005). In elasmobranches and teleost fishes arginase and GS are located primarily in mitochondria, while in terrestrial vertebrates (mammals and amphibians) these enzymes are cytosolic (see review by Anderson 2001) (Figure 1). This suggests that the localization of urea cycle enzymes in lungfish resembles that of terrestrial vertebrates more so than elasmobranches and teleosts, reflecting the phylogenetic link between lungfish and the tetrapods.

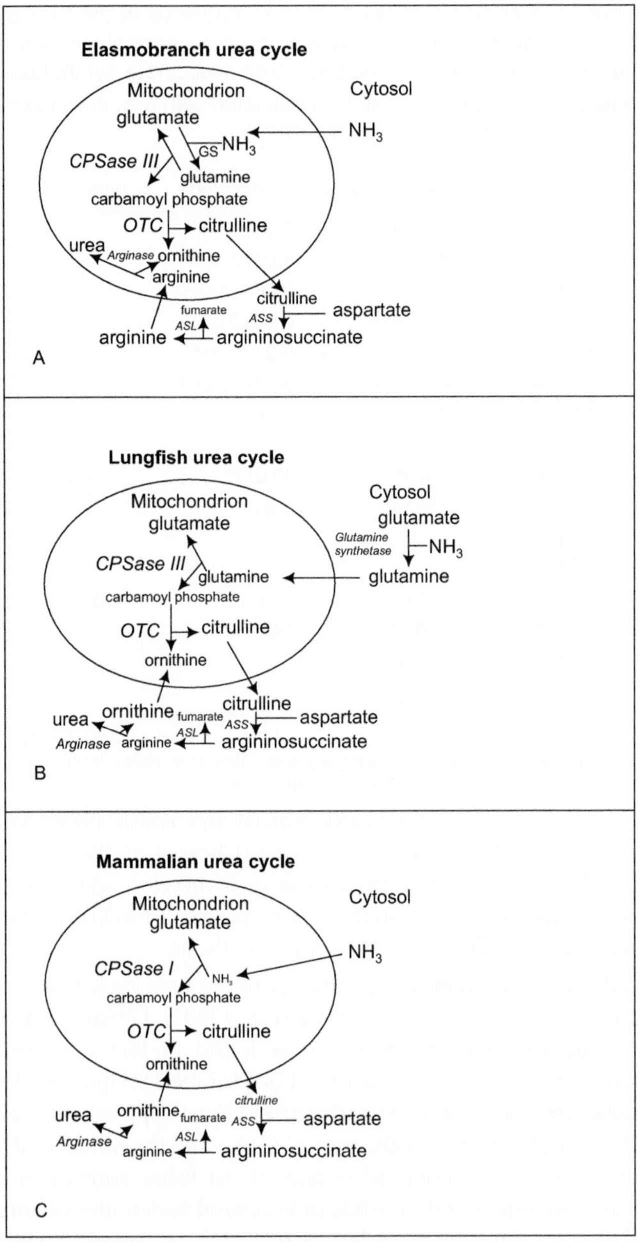

Fig. 1 Comparison of subcellular organization of enzyme reactions of the urea cycle in elasmobranchs (A), lungfish (B) and mammals (C). Compiled from the following references: A (Ballantyne 1997), B (Loong *et al.* 2005), C (Anderson 2001).

Uric Acid Metabolism

All the enzymes involved in uricolysis to form urea have been measured at relatively high levels in *P. aethiopicus* suggesting the presence of a "purine cycle in lungfish liver as a major source of urea during its aquatic existence" (Brown *et al.* 1966). Subsequently, it has been shown (Forster and Goldstein 1966) that the urea cycle contributes 100 times more urea than the uricolysis pathway.

LIPID AND KETONE BODY METABOLISM

Lipid Storage and Transport

Protopterus and *Lepidosiren* have large lipid stores along the posterior third of their body, and it has been postulated that these stores serve as important energy sources during periods of food deprivation (Hochachka and Hulbert, 1978). This lipid lies under the skin not in the muscle itself. Overall, the lipid content of muscle fibers is very low (Dunn *et al.* 1981) with a small amount of lipid present within red muscle, but not in white muscle fibers.

The mobilization and transport of lipids likely resembles that of other tetrapods. The plasma non-esterified fatty acid carrier, albumin, has been characterized in *Neoceratodus* and resembles tetrapod albumin more closely than the teleost protein (Metcalf *et al.* 2007). In lungfish, albumin has been shown to be the major fatty acid binding protein (FABP) in the plasma, but at levels (8 gm/L) much lower than found in mammals (30-50 gm/L) (Metcalf *et al.* 2007). Plasma nonesterified fatty acids have not been measured in any lungfish but would be expected to be low due to the low levels of albumin.

Intracellular transport of nonesterified fatty acids requires a fatty acid binding protein (FABP). A single form of FABP has been found in liver of *Lepidosiren* (DiPietro and Santome 2001). Other fish groups express more than one type of FABP in liver while tetrapods also only express one FABP in liver (DiPietro and Santome 2001). The FABP found in the liver of lungfish is more similar to the form found in birds, reptiles, teleost and elasmobranch fish than it is to mammalian FABPs (DiPietro and Santome 2001). The ligand binding characteristics of the lungfish protein are also similar to those of shark in that binding of lysophospholipids, retinoids, bilirubin, and bile salts are greater than those of fatty acids (DiPietro and Santome 2001). These characteristics may relate to the apparent low use of lipids as energy source.

Lipid Metabolism

The activity of 3-hydroxyacyl CoA dehydrogenase (HOAD), an enzyme indicative of lipid catabolism, is ~10-fold higher in the liver of *P. dolloi* (Frick *et al.* 2008b), compared to other primitive fish species (Frick *et al.* 2007; Speers-Roesch *et al.* 2006b). This is surprising given the low, levels of lipids within the plasma and tissues of lungfish. The activities of other enzymes involved in lipid metabolism are low, however, compared to other fish species. Carnitine palmitoyl CoA transferase (CPT) was not detectable in tissues other than in the liver and kidney (Frick *et al.* 2008b), where it was still comparatively, very low (Frick *et al.* 2007; Speers-Roesch *et al.* 2006b). The preferred fatty acyl-CoA substrate for carnitine acyl-CoA transferase can vary (Egginton 1996), and the apparent lack of CPT in heart may indicate acyl chains other than palmitate are preferred. The substrate preference of carnitine acyl-CoA transferase was examined in the liver tissue of *P. dolloi* (Frick *et al.* 2008b) and revealed a similar substrate preference for laurate, palmitate and palmitoleate. The activity of carnitine acyl-CoA transferase has not been measured in the heart using either laurate or palmitoleate as substrates. The activity of malic enzyme (ME) (an enzyme involved in the generation of NADPH required for fatty acid synthesis (Henderson and Tocher 1987) was generally low, (Frick *et al.* 2008b) compared to other fish species (Frick *et al.* 2007; Speers-Roesch *et al.* 2006b) suggesting the capacity for lipogenesis is limited in lungfish in keeping with their low metabolic rate.

Ketone Body Metabolism

Ketone bodies (e.g. β-hydroxybutyrate and acetoacetate) can be used as oxidative fuels to replace carbohydrates during periods of starvation and as lipid precursors (Newsholme and Leech 1983). Some primitive fishes such as elasmobranchs have a high reliance on ketone bodies under fed conditions (Ballantyne 1997). This does not seem to be the case for lungfish. Our knowledge of the ketone body metabolism of lungfish is limited to a study on the African lungfish, *P. dolloi*, (Frick *et al.* 2008b). Ketogenesis occurs primarily in the liver, thus β-hydroxybutyrate dehydrogenase (HBDH) in liver tissue would be involved in ketone body synthesis, suggesting this is low in *P. dolloi*. In extrahepatic tissues β-HBDH would operate in reverse, catabolizing ketone bodies once carbohydrate stores have been depleted. Interestingly, most studies on fish kidney tissue have reported much higher activities of D-β-HBDH compared to L-β-HBDH (LeBlanc and Ballantyne 2000; Speers-Roesch *et al.* 2006a, b), yet in lungfish the opposite was found. Succinyl coenzyme-A ketotransferase (SKT) was low or undetectable in the white muscle, kidney and liver tissues while β-HBDH was lowest in the lung and white muscle and highest in the liver. SKT was generally similar to the garfish, except for in the liver where activity was much lower in *P. dolloi*. In lungfish deprived of food for 60 days there were no changes in the activities of ketone body enzymes or ketone body levels in tissues (Frick *et al.* 2008b). Teleost fish show no enhanced capacity for ketone body metabolism during food deprivation (Segner 1997; Zammit and Newsholme 1979), and it is possible the same may be true for lungfish.

Surfactant Metabolism

Specialized lipids (surfactants) are needed to facilitate the expansion and contraction of the lungs during respiration. It has been suggested (Daniels and Orgeig 2003) that the surfactant mixtures of "primitive" vertebrates contain higher cholesterol levels.

The surfactants found in the lungs of lungfish have been characterized (Daniels and Orgeig 2003; Orgeig and Daniels 1995; Power *et al.* 1999). The cholesterol content of the lung of the more primitive *Neoceratodus* is higher than that of the *Lepidosiren* and *Protopterus* (Daniels and Orgeig 2003). The cholesterol/phospholipid (chol/pl) ratio (µg/µg) of *Neoceratodus* is high (~ 0.30) and more closely resembles that of teleost fish swimbladders (~ 0.27) and lungs of air-breathing fish (~0.20) while the chol/pl of *Protopterus* and *Lepidosiren* are low (~0.06) and more closely resembles that of amphibians, reptiles, birds and mammals (~0.06-0.075) (Daniels and Orgeig 2003).

LUNG AND GILL METABOLISM

In African and South American lungfish the lungs are paired and fused at the anterior portion. There is a corresponding paired pulmonary circulation (Graham 1997). In adults switching from gill to lung-based respiration involves a ductus arteriosus that shunts the circulation away from the gills when the ductus is closed. The ductus is dilated strongly by dopamine and more weakly by isoproterenol, prostaglandin E_2 and acetylcholine (Fishman *et al.* 1985) and constricted by noradrenaline (Fishman *et al.* 1985). Blood flow to the gills of *N. forsteri* is not influenced by nitric oxide (Jennings *et al.* 2008) similar to the situation in teleost fish (Olson 2002). Ultsch (1996) reported African and South American lungfish are obligatory air-breathers relying on their lungs for about 90% of their oxygen uptake and use the gills for 70-80% of their CO_2 excretion in normoxic water. The role of the gills is greatest in small fish compared to larger specimens of *P. amphibius* (Babiker 1979; Johansen *et al.* 1976a). The relative importance of aerial and aquatic respiration in *P. aethiopicus* also changes with water oxygen levels. Up to 40% of the respiration is via the gills in normoxic water while in hypoxic water 90-100% of respiration is via the lungs with a constant metabolic rate being maintained as the proportions of aerial versus aquatic respiration change (Seifert and Chapman 2006). Field studies suggest wild African lungfish do not rely exclusively on aerial respiration (Mlewa *et al.* 2007). In the laboratory, *P. dolloi* uses the lung for the majority of both O_2 uptake (91.0±2.9%) and CO_2 excretion (76.0±6.6%) (Perry *et al.* 2005a, b). The authors also showed aerial hypercapnia resulted in an increase in partial pressure of arterial blood CO_2 whereas aquatic hypercapnia had no effect, supporting the notion that the lung is the more important site of CO_2 excretion. The Australian lungfish on the other hand uses the gills for 100% of its oxygen uptake and CO_2 excretion in normoxic water (Ultsch 1996).

In fish that use the swimbladder for buoyancy rather than respiration the cells of the gas gland are specialized to produce lactic acid to acidify the blood and release oxygen via the Root effect (Pelster 1995). When the swimbladder is modified to act as an air-breathing organ, as in lungfish, the activity of LDH is reduced. The ratio of activities of LDH to CS of fish swimbladder is very high (530 to 1300) (Ewart and Driedzic 1990; Walsh and Milligan 1993) perhaps related to the role of lactate in "salting out" oxygen from hemoglobin to fill the gas bladder. Within the lung of *P. dolloi* the LDH/CS was calculated to be ~50 (Frick *et al.* 2008a, b), more comparable to the air-breathing organ of gar fish (*Lepisosteus platyrhincus*) (~24) (Frick *et al.* 2007), and the mammalian lung (~9 to 16) (Mirejovska *et al.* 1981; Murphy *et al.* 1980). This ratio does not change during aestivation when lungfish rely entirely on the lung for oxygen uptake (Frick *et al.* 2008a, b).

ESTIVATION METABOLISM

Of the three lungfish genera only the African and South American lungfishes are known to estivate during seasonal dry periods. This state may last as long as 6 years (Swan *et al.* 1968). Associated with prolonged estivation (> 17 months) in *Protopterus* is a shrinkage of the body including fins and skeletal elements (Conant 1976). Estivation involves reductions in heart rate and blood pressure (Fishman *et al.* 1986; Lahiri *et al.* 1970). Estivation also induces a reduction in respiration rate by up to 95%, depending on the duration of estivation (Delaney *et al.* 1974; Fishman *et al.* 1986; Hochachka and Guppy 1987; Lahiri *et al.* 1970; Meyer and Dolven 1992; Smith 1929, 1930; Swan *et al.* 1968).

Since lungfish do not feed during estivation some of the metabolic changes may simply be associated with starvation. A study of a single specimen of *Protopterus annectens* starved in water for three and one half years compared to a similar specimen estivating for the same period found similar changes in body weight (El Hakeem 1979). Recent studies indicate shorter (60 days) periods of food deprivation result in no changes in metabolic enzyme activities while key changes were observed in fish estivating for the same time (Frick *et al.* 2008a, b).

Recent work on the metabolism of African lungfish has caused some confusion as to the actual definition of estivation. Several studies have described the lungfish as being "terrestrialized" or "air-exposed" as opposed to "estivated" (Perry *et al.* 2008; Wilkie *et al.* 2007). The fish used in these studies lived in the absence of water, but were kept moist and did not secrete a complete cocoon (the cocoon material was present on the dorsal surface only) (A. Ip, pers. comm.). *P. dolloi* can be maintained out of water under conditions where the exterior is completely dry and a full cocoon (encasing the whole body) has been secreted (Frick *et al.* 2008a, b). The metabolic rate under these conditions has been measured in *P. dolloi* and is not different from that of aquatic fish (Perry *et al.* 2008) unlike the situation in other lungfish species where a metabolic depression occurs (Delaney *et al.* 1974; Fishman *et al.* 1986; Smith 1930). This species may not estivate in the same way as other lungfish. In spite of this, recent work indicates tissue specific metabolic depression in *P. dolloi* (Frick *et al.* 2008a, b).

The signals for the metabolic changes that allow lungfish to survive for very long periods have not been established. Slight increases in salt concentration have been shown to suppress amino acid catabolism in *P. dolloi* and has been suggested a possible signal for some of the metabolic changes associated with estivation. An "anti-metabolic" factor that reduced the metabolic rate of rats was extracted from the brains of estivating *P. aethiopicus* (Swan *et al.* 1968; Swan *et al.* 1969) but its identity has not been established. Intraperitoneal injection of urea lowers brain tryptophan levels similar to those observed during estivation perhaps indicating a role in regulating metabolism during estivation (Ip *et al.* 2005b). Further research in this area is needed.

Aerobic Metabolism

During estivation in *Protopterus* spp. the hematocrit, oxygen carrying capacity and hemoglobin concentrations increase by about 50% (Delaney *et al.* 1976; Johansen *et al.* 1976b). This is largely due to higher concentrations of erythrocytes in the blood perhaps due to dehydration (Johansen *et al.* 1976b). An increased hematocrit was not observed in *L. paradoxa* estivating for 20 days (da Silva *et al.* 2008). The oxygen affinity of lungfish hemoglobin also changes with estivation. The P_{50} value decreases from 33 mmHg to 7 mmHg (Johansen *et al.* 1976b). Most of this shift is due to decreases in GTP levels (Johansen *et al.* 1976b). The increased affinity may be in response to lower oxygen availability during estivation. PaO_2 is reduced from ~50 to 25 mmHg during aestivation (Delaney *et al.* 1974; Fishman *et al.* 1986) suggesting lungfish may experience hypoxic stress during this period. If there is indeed a reduction in oxygen availability during estivation it might be expected that there is an increased reliance on anaerobic metabolism. Based on the lack of accumulation of lactate in any tissue during estivation it has been suggested estivating fish are not in a hypoxic state (Frick *et al.* 2008a). In fact lactate levels decreased in the heart and muscle tissues of estivating *P. dolloi* (Frick *et al.* 2008a). Additionally, no increase in the activities of enzymes involved in glycolysis occur during estivation (Frick *et al.* 2008a). Based on these mixed findings it is difficult to say whether lungfish always experience hypoxia during estivation or not and further research is needed in this area.

Liver glycogen in estivating *Protopterus* spp. is similar to that of unfed fish (after 6 months) but lower than that of fed controls (Janssens 1964). Glycogen phosphorylase activity, however, increased in both starved and estivated liver (Janssens 1965). In the muscle, glycogen was greatly reduced in the starved group, but not in estivating fish suggesting conditions of fasting but not estivation result in glycogen breakdown (Janssens 1964). Glycogen phosphorylase activities in muscle were similar in fed, starved and estivated fish (Janssens 1965). These observations should be revisited since glycogen levels were expressed only as a percentage and the dehydration effect on muscle tissue was not considered. Lower liver glycogen levels have been found in *P. annectens* estivating for 2 years compared to free swimming fed fish (Babiker and El Hakeem 1979). In the same study, muscle glycogen in the estivated fish was higher than that of unfed and fed controls. Liver and kidney glycogen levels in estivated *P. dolloi* are higher than those of fasted fish, while in muscle and heart levels did not differ (Frick *et al.* 2008a). This indicates a very low rate of carbohydrate utilization perhaps coupled with an enhanced mechanism for glycogen resynthesis. Glucose 6-phosphatase activities in liver of *Protopterus* were lower in estivating animals than in starved or fed fish after 6 months indicating reduced supply of glucose from liver from other tissues (Janssens 1965).

In lungfish, metabolic depression in estivation is not associated with a substantial decrease in extra- or intracellular pH. Plasma pH only decreases by ~0.2 to 0.3 units in *Protopterus* (Delaney *et al.* 1974; Land and Bernier 1995) and negligibly in *L. paradoxa* (da Silva *et al.* 2008). Thus pH mediated down-regulation of enzymes for metabolic depression has not been demonstrated in lungfish.

While the maximal activities (V_{max}) of many metabolic enzymes are not affected by estivation, in certain tissues reduced activity of CCO may be an important way of down-regulating flux through all catabolic pathways (Frick *et al.* 2008a, b). CCO is the terminal enzyme in the electron transport chain and thus has the potential to reduce flux through pathways of aerobic amino acid, carbohydrate and lipid catabolism. The potential mechanisms used to regulate the activity of CCO during estivation were examined by Frick *et al.* (2010) in *P. dolloi*. This study suggested that the 67% reduction in CCO activity observed during estivation in isolated liver mitochondria occurs through both translational regulation and modifications of the phospholipid composition of the mitochondrial membrane. CCO subunit I protein expression was reduced by 46%, while no change in the mRNA expression levels of CCO subunits I, II, III, and IV were observed (Frick *et al.* 2010). CCO is a membrane bound enzyme that is influenced by the lipid composition of the surrounding membrane. In *P. dolloi* the phospholipid composition of the liver mitochondrial membrane changed during estivation, with a 2.3-fold reduction in the amount of cardiolipin. In addition, significant positive correlations were found between CCO activity and the amount of cardiolipin and phosphatidylethanolamine within the mitochondrial membrane (Frick *et al.* 2010), suggesting altering the membrane lipid composition may be part of the required physiological changes for the depression of CCO activity and for estivation to occur. Morphometric studies of the heart of *P. dolloi* indicate a reduction in mitochondrial number with estivation (Icardo *et al.* 2008). This correlates with a reduction in CCO activity in this tissue (Frick *et al.* 2008b). Activities of citrate synthase, another mitochondrial enzyme do not change in

Table 2 Tissue concentrations of urea in species of lungfish during estivation

Species	*Tissue*	*Duration (days)*	*Urea* (mM)
L. paradoxa[a]	Liver	149	133.2
	Muscle		50.9
	Kidney		53.0
L. paradoxa[a]	Liver	248	40.2
	Muscle		74.9
	Kidney		76.5
P. dolloi[b]	Muscle	40	55
	Liver		72
	Brain		61
	Gut		59
P. dolloi[c]	Plasma	30	51.0±10.3
P. aethiopicus[d]	All	78-129	28
P. annectens[e]	Muscle	6	8.23±1.94
	Liver		13.1±1.3
	Plasma		11.3±1.8
	Brain		9.91±1.54
P. aethiopicus[e]	Muscle	6	8.69±0.96
	Liver		9.96±1.31
	Plasma		9.97±0.72
	Brain		7.79±0.99
P. aethiopicus[f]	Muscle	12	33±4
	Liver		24±1
	Plasma		35±3
	Brain		22±2
P. aethiopicus[f]	Muscle	34	48±8
	Liver		42±5
	Plasma		46±6
	Brain		47±7
P. aethiopicus[f]	Muscle	46	53±11
	Liver		47±9
	Plasma		51±10
	Brain		52±13
P. aethiopicus[g]	Whole animal	380	111
P. aethiopicus[h]	Plasma	390	202.8

[a](Carlisky and Barrio 1972); [b](Chew *et al.* 2004); [c](Wood *et al.* 2005b); [d](Janssens and Cohen 1968a); [e](Loong *et al.* 2005); [f](Ip *et al.* 2005c); [g](Smith 1930); [h](Delaney *et al.* 1977)

liver during estivation (Frick *et al.* 2008b) or aerial exposure (Staples *et al.* 2008). Circulatory reductions to tissues like the kidney may be mediated by increases in expression and activity of endothelial nitric oxide synthases (Amelio *et al.* 2008). It has been speculated that this is accomplished by the known effects of nitric oxide in reducing the activity of cytochrome c oxidase and thus oxygen demand (Amelio *et al.* 2008). Further work in this area is needed to establish the importance of this and other mechanisms regulating oxidative metabolism.

Nitrogen Metabolism

In order to reduce the osmotic stress during estivation and to ameliorate the accumulation of toxic ammonia derived from protein degradation, lungfish produce and retain urea up to 203 mM (Delaney *et al.* 1977), depending on the species, duration of estivation and the tissue type (Table 2). It is well known that African lungfish accumulate urea within their tissues during estivation with the nitrogen coming from amino acids (Chew *et al.* 2003; Chew *et al.* 2004; Janssens 1964; Smith 1930). The conversion of ammonia to urea seems to be the most significant osmotic strategy used by lungfish to deal with ammonia accumulation during periods of terrestrial exposure.

The osmotic role of free amino acids seems to be minimal since total concentration of free amino acids in muscle of *P. dolloi* do not increase after 6 and 40 days of estivation although concentrations in the liver declined (Chew *et al.* 2004). An overall reduction in amino acid catabolism has been observed during terrestrial exposure in *Protopterus* (Loong *et al.* 2005). Suppression of amino acid catabolism to reduce ammonia production during terrestrial exposure has been reported in many fish species (Chew *et al.* 2005; Jow *et al.* 1999; Lim *et al.* 2001; Loong *et al.* 2005; Wilkie *et al.* 2007). Alanine does not accumulate in the tissues of estivating *P. dolloi* (Chew *et al.* 2005) and alanine aminotransferase activity does not change with estivation (Frick *et al.* 2008a), supporting the suggestion that lungfish do not use partial amino acid catabolism to reduce ammonia production during terrestrial exposure (Chew *et al.* 2005).

Unfed *P. aethiopicus* excretes 65% of the nitrogen as ammonia (Janssens and Cohen 1968a). The rate of urea synthesis was the same in unfed and estivating lungfish but ammonia production (excreted + accumulated) declined in the estivating fish as virtually all the ammonia was converted to urea and retained (Janssens and Cohen 1968a). In support of this plasma and tissue ammonia levels do not increase during terrestrial exposure in lungfish (Chew *et al.* 2004; Loong *et al.* 2005). Tissue free amino acid levels provide some evidence of the pathways involved in funneling ammonia nitrogen for urea synthesis. Lungfish deal with some ammonia production by converting ammonia to glutamine. Glutamate levels were significantly lower (1.7-fold) in the liver of estivating animals but glutamine levels were unchanged (Chew *et al.* 2004). Glutamine was shown to accumulate in the brain of terrestrially exposed lungfish (*P. dolloi*) (Chew *et al.* 2004; Ip *et al.* 2005c). Glutamine was also reported to increase in the muscle tissue of *P. annectens* exposed to air (Loong *et al.* 2005), but not in the muscle of *P. dolloi* (Chew *et al.* 2004).

GDH activity (measured in the aminating direction) is higher in the liver of estivating fish, suggesting GDH is an important conduit for ammonia detoxification, converting ammonia into glutamate (Frick *et al.* 2008a). Glutamine would be subsequently synthesized from this glutamate via GS. There is evidence in *P. annectens* that the kinetics of GDH are altered during estivation to reduce ammonia production (Loong *et al.* 2008a). A preferential activation by ADP of the aminating direction compared to the deaminating direction has been found in estivating fish and this is enhanced under hypoxic conditions (Loong *et al.* 2008a). It has been suggested differential expression of GDH isoforms may play a role in these changes as well (Loong *et al.* 2008a).

To enhance urea synthesis during estivation, the activities of several other enzymes are increased within the liver of *Protopterus*. These include GS and ornithine transcarbamoylase (OTC) after 6 days air exposure (Loong *et al.* 2005), and GS, CPSase III and argininosuccinate synthetase and lyases (ASS +L) following 12 and 46 days immersion (Ip *et al.* 2005c). Glutamine synthetase increases in liver but decreases in muscle of *P. dolloi* aerially exposed for 5 months (Staples *et al.* 2008).

The extent to which urea synthesis is enhanced during estivation is affected by the conditions under which this takes place. Estivation in mud rather than in air substantially reduces urea accumulation in tissues of *P. annectens* (Loong *et al.* 2008b). This is accomplished by a large depression of both urea synthesis and ammonia production perhaps due to a greater hypoxia indicated by lower blood pO_2 and tissue ATP, under these conditions (Loong *et al.* 2008b) although reduced desiccation stress may play a role. Estivation in mud would appear to be a more "natural" surrogate for studying this process but field studies should be undertaken to establish the physiological status of estivating fish in the wild.

It is not known if protein synthesis is suppressed during estivation but no changes in expression of myosin isoforms in muscle of *P. annectens* were found during estivation (Chanoine *et al.* 1994). Enhanced protein breakdown might be expected but protease (cathepsin) activity in liver and muscle is the same among unfed, fed and estivated *Protopterus* (Janssens 1964).

Lipid Metabolism

In spite of the potential importance of lipids as a fuel during estivation little is known of its quantitative importance. Total plasma lipids in lungfish estivating for 2 years were reduced to less than half of the levels found in control fish (Babiker and El Hakeem 1979). In estivating *P. dolloi* the activities of enzymes of lipid catabolism, 3-hydroxyacyl CoA dehydrogenase and carnitine palmitoyl transferase, were unchanged in the liver despite the ~42% reduction in CCO activity (Frick *et al.* 2008b), suggesting the potential for lipid metabolism is maintained.

There was no change in plasma leptin levels in *L. paradoxa* estivating for 20 days indicating no major shift in lipid utilization (da Silva *et al.* 2008).

Oxidative Stress

When an organism is awakened from a metabolically depressed state, a rapid elevation in oxygen consumption can occur, with rates ~2- and ~ 6-fold higher than control and dormant values, respectively, reported in several species of snails (Coles 1968; Hermes-Lima and Storey 1995; Herreid 1977). Lungfish increase their breathing frequency from approximately 8 to 40 breaths per hour at the onset of aestivation, and display an increase in arterial P_{O_2} from ~40 to ~110 mmHg (Delaney *et al.* 1974; Fishman *et al.* 1986). Although not proportional, positive correlations between reactive oxygen species (ROS) generation and both oxygen consumption (Adelman *et al.* 1988; Barja de Quiroga 1992; Davies *et al.* 1982; Finkel and Holbrook 2000; Spolter *et al.* 1965) and oxygen tension (Turrens *et al.* 1982) have been widely reported, suggesting the threat of damage from ROS may be augmented during estivation, and at awakening. Lungfish may experience periodic ischemia-reperfusion type insults at various stages of estivation, requiring the antioxidant defenses to respond accordingly. Frick *et al.* (unpublished data) examined the antioxidant defenses that exist in *P. dolloi* under control conditions and during different stages of estivation. *P. dolloi* increased their antioxidant capacity during the estivation process through elevated catalase (CAT) activity in the heart and brain tissues, possibly as a strategy to reduce the effects of ROS generation during arousal. However, in the liver tissue CAT activity was depressed during estivation, indicating tissue-specific differences exist. Superoxide dismutase (SOD) and glutathione reductase (GR) activities were generally depressed or unchanged during the different stages of estivation. The amount of lipid peroxidation products (thiobarbituric acid reactive substances (TBARS)) varied between tissues, with the kidney having the metabolic conditions to create the greatest scope for lipid peroxidative damage. No differences were found in the level of TBARS between control and estivating fish.

Little is known of the importance of antioxidant molecules during estivation. L-gulonolactone oxidase activity has been detected in *P. aethiopicus* (Touhata *et al.* 1995) and *L. paradoxa* (Fracalossi *et al.* 2001) indicating the capacity for synthesis of ascorbate. The ability to synthesize ascorbic acid in the kidney has been demonstrated in *Neoceratodus* (Dykhuizen *et al.* 1980) and in the liver and kidney of *Lepidosiren* (Mesquita-Saad *et al.* 1900). This is typical of more primitive fish groups (Cho *et al.* 2007).

HYPOXIA ADAPTATION

The Australian lungfish, *N. forsteri*, forced to undergo aquatic hypoxia (through nitrogen gas bubbling) was found to utilize three mechanisms to preserve oxygen delivery (Kind *et al.* 2002). Branchial ventilation, air-breathing frequency and haemoglobin oxygen affinity all increased in response to aquatic hypoxia. Several blood metabolites were measured in *N. forsteri* in the above study, including lactate and ATP, none of which was found to change (Kind *et al.* 2002). This is likely due to the ability of *N. forsteri* to maintain blood oxygen delivery through both behavioural and physiological modifications. *N. forsteri* exposed to moderate to severe hypoxia showed no change in hematocrit, hemoglobin, plasma lactate or erythrocyte ATP (Kind *et al.* 2002). In the same experiment whole blood oxygen affinity increased slightly in hypoxic fish. The weak response to hypoxia indicates air-breathing removes the need for a large hypoxic response (Kind *et al.* 2002). Metabolic rates of *P. aethiopicus* have been suggested to decrease by 80% under hypoxic conditions but direct measurements were not used to obtain this estimate (Lahiri *et al.* 1970). More recently, respiration rates have been demonstrated to decrease under aerial hypoxia by 85% (Perry *et al.* 2005b). As plasma oxygen levels decline, a bradycardia, reduced blood pH and an increase in lactate occur during submersion (inducing hypoxia) in *Protopterus* (Lahiri *et al.* 1970).

When hypoxic exposure cannot be avoided the metabolism of lungfish is altered. The metabolic consequences of this have been examined in lungfish (Dunn *et al.* 1983). In this study, lungfish (*P. aethiopicus*) were made hypoxic by forcing the fish to remain submerged for 12 hours ("diving"). Blood, heart, and brain glucose concentrations were also reported to increase significantly during "diving" and remain significantly elevated following a 12-hour recovery period (Dunn *et al.* 1983). Blood flow is redistributed to enable glucose to be transported to the heart and brain where lactate is removed (Dunn *et al.* 1983; Land and Bernier 1995).

The kinetics of LDH in the heart of *Lepidosiren* are adapted to hypoxia in that it serves more as a lactate oxidase for handling lactate washed out of other tissues after the hypoxic period ends (Hochachka and Hulbert 1978). Lactate levels in the heart, brain and blood showed a 2-3 fold increase in *P. aethiopicus* when prevented from breathing air for 12 hours, and then fall following a 12-hour recovery (Dunn *et al.* 1983). The 250 and 300% increase in lactate levels in the brain and heart respectively, are likely derived from both endogenous glycogen, as both brain and heart glycogen stores are depleted, and from the utilization of blood glucose, which can readily cross the blood-brain barrier.

During submergence, *P. aethiopicus* undergoes a marked reduction in liver glycogen and an increase in blood glucose levels. Liver glucose levels remain elevated above those of the blood to continuously favour the export of glucose to the extrahepatic tissues (Dunn *et al.* 1983). Lactate levels in laboratory estivating *P. aethiopicus* were similar to those of active animals after 7 months (Delaney *et al.* 1977). The extent of hypoxia during natural estivation needs to be determined and may differ among the estivating species.

EVOLUTIONARY CONSIDERATIONS

The study of the metabolism of lungfish has provided fascinating insights into the evolution of the vertebrates and in particular the transition from an aquatic existence to a terrestrial lifestyle. In order to retain "primitive" metabolic features after such a long evolutionary history, the rate of evolution must have been attenuated in this group. The questions of whether lungfish can be considered "living fossils" in a metabolic sense gains support from an analysis of the rates of evolution of the opioid/orphanin gene family (Lee *et al.* 2006) that found molecular evidence for slow rates of evolution to support the view that morphological evolution has been slow since the Permian. Analysis of the β-subunits of three pituitary glycopeptides hormones indicates a slow rate of evolution of luteinizing hormone (LH) and follicle-stimulating hormone (FSH) compared to that occurring after the emergence of the amphibians and in the teleosts and a slower rate of evolution of thyroid-stimulating hormone (TSH), than found in the teleosts (Querat *et al.* 2004). It has also been suggested that the hormonal regulation of carbohydrate metabolism (glycogenolysis) evolved from a simple β-adrenergic system such as is found in the lungfish into a more complex pattern with other receptor types in higher vertebrates (Hanke and Janssens 1983). Thus based on the evolution of these proteins, the extant lungfish can be considered "living fossils".

During the evolution of the vertebrates, 2 genome duplications are thought to have occurred. The first was before the Cambrian explosion and the second in the early Devonian. The lungfish genome reflects these events. A third duplication later in the Devonian after the separation of the actinopterygian lineage from the sarcopterygian resulted in actinopterygian fish having about twice as many genes as lungfish, amphibians, reptiles, birds and mammals (Meyer and Schartl 2007). It is possible to identify enzyme systems in extant lungfish where isozyme forms correlate with these events. Studies of trypsinogens of lungfish indicate the gene for trypsinogen was duplicated more than once (de Haen *et al.* 1977). Additionally, there are only 2 isoforms of creatine phosphokinase compared to three or more in more advanced groups and lungfish have a more generalized tissue expression (less tissue specificity of isoform expression) typical of primitive fishes (Fisher and Whitt 1978). Three LDH loci have been found in lungfish with the A4 isozyme dominating in all tissues unlike the situation in most other vertebrates where A4 is greatest in muscle and the B4 type in liver, heart and brain tissues (Basaglia 2002, 2003). The C locus for LDH has been detected in *P. annectens* and *N. forsteri* but not in *L. paradoxa* (Basaglia 2002a, 2002, 2003). Interestingly, the heart form of LDH is insensitive to pyruvate inhibition unlike that of more advanced fishes (Dunn *et al.* 1983). The heart form of LDH in lungfish is thus less optimally designed to funnel lactate carbon into the tricarboxylic acid cycle. It may represent a "primitive" LDH. Lungfish lack the E locus for LDH expressed in the eye of many teleost fish (Horowitz and Whitt 1972; Whitt and Horowitz 1970).

Alkaline phosphatase (ALP) apparently exists in lungfish with only one locus with the intestine-specific form being absent (Basaglia 2000a, 2002, 2003; Tarasoff *et al.* 1972). The intestine specific form is found in teleosts, amphibians, reptiles, birds and mammals.

Other enzyme systems display characteristics of either more advanced fishes or the tetrapods. For example, lungfish have 2 gene loci for glyceraldehyde 3-phosphate dehydrogenase (G3PDH) resembling Osteichthyes unlike the situation in Agnatha, amphibians, reptiles, birds and mammals where only one locus exists (Basaglia 2000a, 2002, 2003). Glucose phosphate isomerase (GPI) is coded by 2 gene loci similar to that of primitive teleosts and non-teleost fish but unlike the single locus found in tetrapods (Basaglia 2000a, b, 2002, 2003).

Studies on the expression of 10 enzymes in *Lepidosiren* and *Protopterus* found that 4 of the 10 (G3PDH, G6PDH, GPI and LDH) have more generalized tissue expression reflective of an ancestral pattern while 6 (alcohol dehydrogenase (ADH), ALP, fructose-bisphosphate aldolase (FBALD), glyceraldehyde-3-phosphate dehydrogenase (GAPDH), phosphoglucomutase (PGM) and SOD) have tissues specific expression typical of more advanced vertebrates (Basaglia 2000b, 2002, 2003). Similar findings have been reported for *N. ceratodus* (Basaglia 2000a).

Together, these findings support the idea that lungfish are "living fossils" and possess primitive features of metabolism in the complexity and tissue specificity of enzyme and hormonal regulatory systems. Lungfish thus provide valuable insights into the evolution of metabolism amongst the vertebrates.

SUMMARY

Lungfish display a variety of metabolic adaptations for estivation that provide the precursors for a terrestrial existence. These include the ability to breath air and the capacity for significant urea synthesis and accumulation to reduce osmotic stress and avoid toxic ammonia buildup. Studies of their metabolic biochemistry have thus provided valuable insights into the factors involved in the invasion of the land by the vertebrates. Further studies of lungfish metabolism are needed, particularly in the areas of the metabolism of the lung and the mechanisms involved in the coordination of the tissue-specific responses to estivation in order to expand our understanding of the metabolic adaptations required for a terrestrial existence.

References

Adelman, R., Saul, R.L. and Ames, B.N. (1988). Oxidative damage to DNA: relation to species metabolic rate and life span. Proceedings of the National Academy of the United States of America 85: 2706-2708.

Almeida-Val, V.M.F. and Hochachka, P.W. (1995). Air-breathing fishes: metabolic biochemistry of the first diving vertebrates. In: P.W. Hochachka and T.P. Mommsen (eds). Biochemistry and Molecular Biology of Fishes. Volume 5. Elsevier Science, Amsterdam, The Netherlands. pp. 45-55.

Amelio, D., Garofalo, F., Brunelli, E., Loong, A.M., Wong, W.P., Ip, Y.K., Tota, B. and Cerra, M.C. (2008). Differential NOS expression in freshwater and aestivating *Protopterus dolloi* (lungfish): heart vs kidney readjustments. Nitric Oxide 18: 1-10.

Anderson, P.M. (1991). Glutamine-dependent urea synthesis in elasmobranch fishes. Biochemistry and Cell Biology 69: 317-319.

Anderson, P.M. (2001). Urea and glutamine synthesis: environmental influences on nitrogen excretion. In: P.A. Wright and P.M. Anderson (eds). Fish Physiology. Volume 20. Nitrogen Excretion. Acacdemic Press, San Diego, USA. pp. 239-277.

Arbel, E.R., Liberthson, R., Langendorf, R., Pick, A., Lev, M. and Fishman, A.P. (1977). Electrophysiological and anatomical observations on the heart of the African lungfish. American Journal of Physiology 232: H24-H34.

Babiker, M.M. (1979). Respiratory behaviour, oxygen consumption and relative dependence on aerial respiration in the African lungfish (*Protopterus annectens*, Owen) and an air-breathing teleost (*Clarias lazera*, C.). Hydrobiologia 65: 177-187.

Babiker, M.M. and El Hakeem, O.H. (1979). Changes in blood characteristics and constituents associated with aestivation in African lungfish, *Protopterus annectens*. Zoologischer Anzeiger 202: 9-16.

Ballantyne, J.S. (1997). Jaws, the inside story. The metabolism of elasmobranch fishes. Comparative Biochemistry and Physiology 118B: 703-742.

Ballantyne, J.S. (2001). Amino acid metabolism. In: P.A. Wright and P.M. Anderson (eds). Fish Physiology. Volume 20. Nitrogen Excretion. Academic Press, San Diego, USA. pp. 77-107.

Barja de Quiroga, G. (1992). Brown fat thermogenesis and exercise: two examples of physiological oxidative stress. Free Radical Biology and Medicine 13: 325-340.

Basaglia, F. (2000a). Isozyme distrbution of eleven enzymes and their loci in Australian lungfish, *Neoceratodus forsteri* (Osteichthyes, Dipnoi). Italian Journal of Zoology 67: 51-56.

Basaglia, F. (2000b). Isozyme distribution of ten enzymes and their loci in South American lungfish, *Lepidosiren paradoxa* (Osteichthyes, Dipnoi). Comparative Biochemistry and Physiology 126B: 503-510.

Basaglia, F. (2002). Multilocus isozyme systems in African lungfish, *Protopterus annectens*: distribution, differential expression and variation in dipnoans. Comparative Biochemistry and Physiology 131B: 89-102.

Basaglia, F. (2003). Isozyme distribution of ten enzymes and their loci in South American lungfish, *Lepidosiren paradoxa* (Osteichthyes, Dipnoi). Comparative Biochemistry and Physiology 126B: 503-510.

Battersby, B.J. and Moyes, C.D. (1998). Influence of acclimation temperature on mitochondrial DNA, RNA, and enzymes in skeletal muscle. American Journal of Physiology 275: R905-R912.

Battersby, B.J., McFarlane, W.J. and Ballantyne, J.S. (1996). Short term effects of 3,5,3'-triiodothyronine on the intermediary metabolism of the dogfish shark *Squalus acanthias*: evidence from enzyme activities. Journal of Experimental Zoology 274: 157-162.

Bemis, W.E. (1986). Feeding systems of living Dipnoi: anatomy and function. Journal of Morphology Supplement 1: 249-275.

Blier, P. and Guderley, P. (1988). Metabolic responses to cold acclimation in the swimming musculatu re of lake whitefish, *Coregonus clupeaformis*. Journal of Experimental Zoology 246: 244-252.

Brett, J.R. (1972). The metabolic demand for oxygen in fish, particularly salmonids, and a comparison with other vertebrates. Respiratory Physiology 14: 151-170.

Brinkmann, H., Denk, A., Zitzler, J. and Meyer, A. (2004). Complete mitochondrial genome sequences of the South American and Australian lungfish: testing of the phylogenetic performance of mitochondrial data sets for phylogenetic problems in tetrapod relationships. Journal of Molecular Evolution 59: 834-848.

Brown, G.W., James, J., Henderson, R.J., Thomas, W.N., Robinson, R.O., Thompson, A.L., Brown, E. and Brown, S.G. (1966). Uricolytic enzymes in liver of the dipnoan *Protopterus aethiopicus*. Science 153: 1653-1654.

Cao, X., Kemp, J.R. and Anderson, P.M. (1991). Subcellular localization of two glutamine-dependent carbamoyl-phosphate synthetases and related enzymes in liver of *Micropterus salmoides* (largemouth bass) and properties of isolated liver mitochondria: comparative relationships with elasmobranchs. Journal of Experimental Zoology 258: 24-33.

Carlisky, N.J. and Barrio, A. (1972). Nitrogen metabolism of the South American lungfish *Lepidosiren paradoxa*. Comparative Biochemistry and Physiology 41B: 857-873.

Chamberlin, M.E., Glemet, H.C. and Ballantyne, J.S. (1991). Glutamine metabolism in a Holostean fish (*Amia calva*) and a teleost (*Salvelinus namaycush*). American Journal of Physiology 260: R159-R166.

Chanoine, C., El-Attari, A., Guyot-Lenfant, M., Ouedraogo, L. and Gallien, C.L. (1994). Myosin isoforms and their subunits in the lungfish *Protopterus annectens*: changes during development and the annual cycle. Journal of Experimental Zoology 269: 413-421.

Chew, S.F., Ong, T.F., Ho, L., Tam, W.L., Loong, A.M., Hiong, K.C., Wong, W.P. and Ip, Y.K. (2003). Urea synthesis in the African lungfish *Protopterus dolloi* – hepatic carbamoyl phosphate synthetase III and glutamine synthetase exposure are upregulated by 6 days of aerial exposure. Journal of Experimental Biology 206: 3615-3624.

Chew, S.F., Chan, N.K.Y., Loong, A.M., Hiong, K.C., Tam, W.L. and Ip, Y.K. (2004). Nitrogen metabolism in the African lungfish (*Protopterus dolloi*) aestivating in a mucus cocoon on land. Journal of Experimental Biology 207: 777-786.

Chew, S.F., Ho, L., Ong, T.F., Wong, W.P. and Ip, Y.K. (2005). The African lungfish, *Protopterus dolloi*, detoxifies ammonia to urea during environmental ammonia exposure. Physiological and Biochemical Zoology 78: 31-39.

Cho, Y.S., Douglas, S.E., Gallant, J.W., Kim, K.Y., Kim, D.S. and Nam, Y.K. (2007). Isolation and characterization of cDNA sequences of L-gulono-gamma-lactone oxidase, a key enzyme for biosytnthesis of ascorbic acid, from extant primitive fish groups. Comparative Biochemistry and Physiology 147B: 178-190.

Coles, G.C. (1968). The termination of aestivation in the large fresh-water snail *Pila ovata* (Ampulariidae)-I. Changes in oxygen uptake. Comparative Biochemistry and Physiology 25: 517-522.

Conant, E.B. (1976). Urea accumulation and other effects of estivation in the African lungfish, *Protopterus*. Virginia Journal of Science 27: 42.

Corbet, P.S. (1961). The food of non-cichlid fishes in the Lake Vistoria basin, with remarks on their evolution and adaptation to lacustrine conditions. Zoological Journal of the Linnean Society of London 37: 197-203.

Curry-Lindahl, K. (1956). On the ecology, feeding behaviour and territoriality of the African lungfish, *Protopterus aethiopicus* Heckel. Arkiv for Zoologi 9: 479-497.

da Silva, G.D.S.F., Giusti, H., Sanchez, A.P., do Carmo, J.M. and Glass, M.L. (2008). Aestivation in the South American lungfish, *Lepidosiren paradoxa*: effects on cardiovascular function, blood gases, osmolality and leptin levels. Respiratory Physiology and Neurobiology 164: 380-385.

Daniels, C.B. and Orgeig, S. (2003). Pulmonary surfactant: the key to the evolution of air breathing. News in Physiological Sciences 18: 151-157.

Davies, K.J.A., Quintanilha, A.T., Brooks, G.A. and Packer, L. (1982). Free radicals and tissue damage produced by exercise. Biochemical and Biophysical Research Communications 107: 1198-1205.

de Haen, C. and Gertler, A. (1974). Isolation and amino-terminal sequence analysis of two dissimilar pancreatic proelastases from the African lungfish, *Protopterus aethiopicus*. Biochemistry 13: 2673-2677.

de Haen, C., Walsh, K.A. and Neurath, H. (1977). Isolation and amino-terminal sequence analysis of a new pancreatic trypsinogen of the African lungfish *Protopterus aethiopicus*. Biochemistry 16: 4421-4425.

Delaney, R.G. and Fishman, A.P. (1977). Analysis of lung ventilation in the estivating lungfish *Protopterus aethiopicus*. American Journal of Physiology 233: R181-R187.

Delaney, R.G., Lahiri, S. and Fishman, A.P. (1974). Aestivation of the African lungfish *Protopterus aethiopicus*: cardiovascular and respiratory functions. Journal of Experimental Biology 61: 111-118.

Delaney, R.G., Shub, C. and Fishman, A.P. (1976). Hematologic observations on the aquatic and estivating African lungfish, *Protopterus aethiopicus*. Copeia 1976: 423-434.

Delaney, R.G., Lahiri, S., Hamilton, R. and Fishman, A.P. (1977). Acid-base balance and plasma composition in the aestivating lungfish (*Protopterus*). American Journal of Physiology 232: R10-R17.

Dickson, K.A., Gregorio, M.O., Gruber, S.J., Loefler, K.L., Tran, M. and Terrell C. (1993). Biochemical indices of aerobic and anaerobic capacity in muscle tissues of California elasmobranch fishes differing in typical activity level. Marine Biology 117: 185-193.

DiPietro, S.M. and Santome, J.A. (2001). Structural and biochemical characterization of the lungfish (*Lepidosiren paradoxa*) liver basic fatty acid binding protein. Archives of Biochemistry and Biophysics 388: 81-90.

Dunn, J.F., Davison, W., Maloiy, G.M.O., Hochachka, P.W. and Guppy, M. (1981). An ultrastructural and histochemical study of the axial musculature in the African lungfish. Cell Tissue Research 220: 599-609.

Dunn, J.F., Hochachka, P.W., Davison, W. and Guppy, M. (1983). Metabolic adjustments to diving and recovery in the African lungfish. American Journal of Physiology 245: R651-R657.

Dykhuizen, D.E., Harrison, K.M. and Richardson, B.J. (1980). Evolutionary implications of ascorbic acid production in the Australian lungfish. Experientia 36: 945-946.

Egginton, S. (1996). Effect of temperature on the optimal substrate for beta-oxidation. Journal of Fish Biology 49: 753-758.

El Hakeem, O.H. (1979). A lungfish that survives over three and half years of starvation under aquatic conditions. Zoologischer Anzeiger Jena 202: 17-19.

Ewart, H.S. and Driedzic, W.R. (1987). Enzymes of energy metabolism in salmonid hearts: spongy versus cortical myocardia. Canadian Journal of Zoology 65: 623-627.

Ewart, H.S. and Driedzic, W.R. (1990). Enzyme activity levels underestimate lactate production rates in cod (*Gadus morhua*) gas gland. Canadian Journal of Zoology 68: 193-197.

Finkel, T. and Holbrook, N.J. (2000). Oxidants, oxidative stress and the biology of ageing. Nature 408: 239-247.

Fisher, S. and Whitt, G.S. (1978). Evolution of isozyme loci and their differential tissue expression. Journal of Molecular Evolution 12: 25-55.

Fishman, A.P., Delaney, R.G. and Laurent, P. (1985). Circulatory adaptation to bimodal respiration in the dipnoan lungfish. Journal of Applied Physiology 59: 285-294.

Fishman, A.P., Pack, A.I., Delaney, R.G. and Galante, R.J. (1986). Estivation in *Protopterus*. Journal of Morphology Supplement 1: 237-248.

Forster, R.P. and Goldstein, L. (1966). Urea synthesis in the lungfish: relative importance of purine and ornithine cycle pathways. Science 153: 1650-1652.

Fracalossi, D.M., Allen, M.E., Yuyama, L.K. and Oftedal, O.T. (2001). Ascorbic acid biosynthesis in Amazonian fishes. Aquaculture 192: 321-332.

Frick, N.T., Bystriansky, J.S. and Ballantyne, J.S. (2007). The metabolic organization of a primitive air-breathing fish, the Florida gar (*Lepisosteus platyrhincus*). Journal of Experimental Zoology 307A: 7-17.

Frick, N.T., Bystriansky, J.S., Ip, Y.K., Chew, S.F. and Ballantyne, J.S. (2008a). Carbohydrates and amino acid metabolism following 60 days of fasting and aestivation in the African lungfish, *Protopterus dolloi*. Comparative Biochemistry and Physiology 151A: 85-92.

Frick, N.T., Bystriansky, J.S., Ip, Y.K., Chew, S.F. and Ballantyne, J.S. (2008b). Lipid, ketone body and oxidative metabolism in the African lungfish, *Protopterus dolloi* during periods of fasting and aestivation. Comparative Biochemistry and Physiology 151A: 93-101.

Frick, N.T., Bystriansky, J.S., Ip, Y.K., Chew, S.F. and Ballantyne, J.S. (2010). Cytochrome c oxidase is regulated by modulations in protein expression and mitochondrial membrane phospholipid composition in estivating African lungfish. American Journal of Physiology 298: R608-R616.

Funkhouser, D., Goldstein, L. and Forster, R.P. (1972). Urea biosynthesis in the South American lungfish, *Lepidosiren paradoxa*: relation to its ecology. Comparative Biochemistry and Physiology 41A: 439-443.

Giardina, B., Mosca, D. and De Rosa, M.C. (2004). The Bohr effect of haemoglobin in vertebrates: an example of molecular adaptation to different physiological requirements. Acta Physiologica Scandinavica 182: 229-244.

Goldstein, L., Janssens, P.A. and Forster, R.P. (1967). Lungfish *Neoceratodus forsteri*: activities of ornithine-urea cycle and enzymes. Science 157: 316-317.

Graham, J.B. (1997). Air-breathing Fishes. Evolution, Diversity, and Adaptation. Academic Press, San Diego, USA.

Greenwood, P.H. (1986). The natural history of African lungfishes. Journal of Morphology Supplement 1: 163-179.

Grigg, G.C. (1974). Respiratory function of blood in fishes. In: M. Florkin and B.T. Scheer (eds). Chemical Zoology. Volume VIII. Deuterostomes, Cyclostomes and Fishes. Academic Press, Inc., New York, USA. pp. 331-368.

Hanke, W. and Janssens, P.A. (1983). The role of hormones in regulation of carbohydrate metabolism in the Australian lungfish *Neoceratodus forsteri*. General and Comparative Endocrinology 51: 364-369.

Hassanpour, M. and Joss, J.M.P. (2009). The digestive system. In: J.M. Jorgensen, A. Kemp and T. Wang (eds). Biology of Lungfish. Science Publishers Inc., Enfield, New Hampshire, USA.

Henderson, R.J. and Tocher, D.R. (1987). The lipid composition and biochemistry of freshwater fish. Progress in Lipid Research 26: 281-347.

Hermes-Lima, M. and Storey, K.B. (1995). Relationship between anoxia exposure and antioxidant status in the frog *Rana pipiens*. American Journal of Physiology 271: R918-R925.

Herreid, C.F. (1977). Metabolism of land snails (*Otala lactea*) during dormancy, arousal and activity. Comparative Biochemistry and Physiology 56A: 211-215.

Hochachka, P.W. (1980). Living Without Oxygen: Closed and Open Systems in Hypoxia Tolerance. Harvard University Press, Cambridge, Mass. USA.

Hochachka, P.W. and Hulbert, W.C. (1978). Glycogen 'seas', glycogen bodies, and glycogen granules in heart and skeletal muscle of two air-breathing, burrowing fishes. Canadian Journal of Zoology 56: 774-786.

Hochachka, P.W. and Guppy, M. (1987). Metabolic Arrest and the Control of Biological Time. Harvard University Press, Cambrdige, Massachusetts, USA.

Hochachka, P.W., Guppy, M., Guderley, H., Storey, K.B. and Hulbert, W.C. (1978a). Metabolic biochemistry of water- vs. air-breathing osteoglossids: heart enzymes and ultrastructure. Canadian Journal of Zoology 56: 759-768.

Hochachka, P.W., Guppy, M., Guderley, H.E., Storey, K.B. and Hulbert, W.C. (1978b). Metabolic biochemistry of water- vs. air-breathing fishes: muscle enzymes and ultrastructure. Canadian Journal of Zoology 56: 736-750.

Hochachka, P.W., Hulbert, W.C. and Guppy, M. (1978c). I. The tuna power plant and furnace. In: G.D. Sharp and A.E. Dizon (eds). The Physiological Ecology of Tunas. Academic Press, Inc., New York, USA. pp. 153-174.

Hochachka, P.W., Storey, K.B., French, C.J. and Schneider, D.E. (1979). Hydrogen shuttles in air versus water breathing fishes. Comparative Biochemistry and Physiology 63B: 45-56.

Horowitz, J.J. and Whitt, G.S. (1972). Evolution of a nervous system specific lactate dehydrogenase isozyme in fish. Journal of Experimental Zoology 180: 13-32.

Icardo, J.M., Amelio, D., Garofalo, F., Colvee, E., Cerra, M.C., Wong, W.P., Tota, B. and Ip, Y.K. (2008). The structural characteristics of the heart ventricle of the African lungfish *Protopterus dolloi*: freshwater and aestivation. Journal of Anatomy 213: 106-119.

Iftikar, F.I., Patel, M., Ip, Y.K. and Wood, C.M. (2008). The influence of feeding on aerial and aquatic oxygen consumption, nitrogenous waste excretion, and metabolic fuel usage in the African lungfish *Protopterus annectens*. Canadian Journal of Zoology 86: 790-800.

Ip, Y.K., Chew, S.F. and Randall, D.J. (2004). Five tropical air-breathing fishes, six different strategies to defend against ammonia toxicity on land. Physiological and Biochemical Zoology 77: 768-782.

Ip, Y.K., Peh, B.K., Tam, W.L., Lee, S.L.M. and Chew, S.F. (2005a). Changes in salinity and ionic composition can act as environmental signals to induce a reduction in ammonia production in the African lungfish *Protopterus dolloi*. Journal of Experimental Zoology 303A: 456-463.

Ip, Y.K., Peh, B.K., Tam, W.L., Wong, W.P. and Chew, S.F. (2005b). Effects of intra-peritoneal injection with NH_4Cl, urea, or NH_4Cl + urea on nitrogen excretion and metabolism in the African lungfish *Protopterus dolloi*. Journal of Experimental Zoology 303A: 272-282.

Ip, Y.K., Yeo, P.J., Loong, A.M., Hiong, K.C., Wong, W.P. and Chew, S.F. (2005c). The interplay of increased urea synthesis and reduced ammonia production in the African lungfish *Protopterus aethiopicus* during 46 days of aestivation in a mucus cocoon. Journal of Experimental Zoology 303A: 1054-1065.

Iwama, G.K., Vijayan, M.M., Forsyth, R.B. and Ackerman, P.A. (1999). Heat shock proteins and physiological stress in fish. American Zoologist 39: 901-909.

Janssens, P.A. (1964). The metabolism of aestivating African lungfish. Comparative Biochemistry and Physiology 11: 105-117.

Janssens, P.A. (1965). Phosphorylase and glucose-6-phosphatase in the African lungfish. Comparative Biochemistry and Physiology 16: 317-319.

Janssens, P.A. and Cohen, P.P. (1966). Ornithine-urea cycle enzymes in the African lungfish, *Protopterus aethiopicus*. Science 152: 358-359.

Janssens, P.A. and Cohen, P.P. (1968a). Biosynthesis of urea in the estivating African lungfish and in *Xenopus laevis* under conditions of water-shortage. Comparative Biochemistry and Physiology 24: 887-898.

Janssens, P.A. and Cohen, P.P. (1968b). Nitrogen metabolism in the African lungfish. Comparative Biochemistry and Physiology 24: 879-886.

Jennings, B.L., Blake, R.E., Joss, J.M.P. and Donald, J.A. (2008). Vascular distribution of nitric oxide synthetase and vasodilation in the Australian lungfish, *Neoceratodus forsteri*. Comparative Biochemistry and Physiology 151A: 590-595.

Johansen, K. and Lenfant, C. (1967). Respiratory function in the South American lungfish, *Lepidosiren paradoxa* (Fitz). Journal of Experimental Biology 46: 205-218.

Johansen, K. and Lenfant, C. (1968). Respiration in the African lungfish *Protopterus aethiopicus*. II. Control of breathing. Journal of Experimental Biology 49: 453-468.

Johansen, K., Lenfant, C. and Grigg, G.C. (1967). Respiratory control in the lungfish, *Neoceratodus forsteri* (Krefft). Comparative Biochemistry and Physiology 20: 835-854.

Johansen, K., Lomholt, J.P. and Maloiy, G.M.O. (1976a). Importance of air and water breathing in relation to size of the African lungfish *Protopterus amphibius* Peters. Journal of Experimental Biology 65: 395-399.

Johansen, K., Lykkeboe, G., Weber, R.E. and Maloiy, G.M.O. (1976b). Respiratory properties of blood in awake and estivating lungfish, *Protopterus amphibius*. Respiration Physiology 27: 335-345.

Johnels, A.G. and Svensson, G.S.O. (1954). On the biology of *Protopterus annectens* (Owen). Arkiv for Zoologi 7: 131-158.

Jow, L.Y., Chew, S.F., Lim, C.B., Anderson, P.M. and Ip, Y.K. (1999). The marble goby *Oxyeleotris marmoratus* activates hepatic glutamine synthetase and detoxifies ammonia to glutamine during air exposure. Journal of Experimental Biology 202: 237-245.

Kerem, D., Hammond, D.D. and Elsner, R. (1973). Tissue glycogen levels in the weddell seal, *Leptonychotes weddelli*: a possible adaptation to asphyxial hypoxia. Comparative Biochemistry and Physiology 45A: 731-736.

Kind, P.K., Grigg, G.C. and Booth, D.T. (2002). Physiological responses to prolonged aquatic hypoxia in the Queensland lungfish *Neoceratodus forsteri*. Respiratory Physiology and Neurobiology 132: 179-190.

Knox, D., Walton, M.J. and Cowey, C.B. (1980). Distribution of enzymes of glycolysis and gluconeogenesis in fish tissues. Marine Biology 56: 7-10.

Kozlowski, J., Konarzewski, M. and Gawelczyk. A.T. (2003). Cell size as a link between noncoding DNA and metabolic rate scaling. Proceedings of the National Academy of Sciences of the United States of America 100: 14080-14085.

Lahiri, S., Szidon, J.P. and Fishman, A.P. (1970). Potential respiratory and circulatory adjustments to hypoxia in the African lungfish. Federation Proceedings 29: 1141-1148.

Land, S.C. and Bernier, N.J. (1995). Estivation: mechanism and control of metabolic suppression. In: P.W. Hochachka and T.P. Mommsen (eds). Biochemical and Molecular. Biology of Fishes. Volume 5. Elsevier Science, B.V., Amsterdam, The Netherlands. pp. 381-412.

Laurent, P., Delaney, R.G. and Fishman, A.P. (1978). The vasculature of the gills in the aquatic and aestivating lungfish (*Protopterus aethiopicus*). Journal of Morphology 156: 173-208.

LeBlanc, P.J. and Ballantyne, J.S. (2000). Novel aspects of the activities and subcellular distribution of enzymes of ketone body metabolism in liver and kidney of the goldfish, *Carassius auratus*. Journal of Experimental Zoology 286: 434-439.

LeBlanc, P.J., Gillis, T.E., Gerrits, M.F. and Ballantyne, J.S. (1995). Metabolic organization of liver and somatic muscle of landlocked sea lamprey, *Petromyzon marinus*, during the spawning migration. Canadian Journal of Zoology 73: 916-923.

Lee, J., Alrubaian, J. and Dores, R.M. (2006). Are lungfish living fossils? Observation on the evolution of the opioid/orphanin gene family. General and Comparative Endocrinology 148: 306-314.

Lenfant, C. and Johansen, K. (1968). Respiration in the African lungfish *Protopterus aethiopicus*. I. Respiratory properties of the blood and normal patterns of breathing and gas exchange. Journal of Experimental Biology 49: 437-452.

Lenfant, C., Johansen, K. and Grigg, G.C. (1966). Respiratory properties of blood and pattern of gas exchange in the lungfish *Neoceratodus forsteri* (Krefft). Respiration Physiology 2: 1-21.

Lim, C.B., Chew, S.F., Anderson, P.M. and Ip, Y.K. (2001). Reduction in the rates of protein and amino acid catabolism to slow down the accumulation of endgoenous ammonia: a strategy potentially adopted by mudskippers (*Periophthalmodon schlosseri* and *Boleophthalmus boddaerti*) during aerial exposure in constant darkness. Journal of Experimental Biology 204: 1605-1614.

Lim, C.K., Wong, W.P., Lee, S.M.L., Chew, S.F. and Ip, Y.K. (2004). The ammonotelic African lungfish, *Protopterus dolloi*, increases the rate of urea synthesis and becomes ureotelic after feeding. Journal of Comparative Physiology 174B: 555-564.

Lomholt, J.P., Johansen, K. and Maloiy, G.M.O. (1975). Is the aestivating lungfish the first vertebrate with suctional breathing? Nature 257: 787-788.

Loong, A.M., Hiong, K.C., Lee, S.M.L., Wong, W.P., Chew, S.F. and Ip, Y.K. (2005). Ornithine-urea cycle and urea synthesis in African lungfishes, *Protopterus aethiopicus* and *Protopterus annectens*, exposed to terrestrial conditions for 6 days. Journal of Experimental Zoology 303A: 354-365.

Loong, A.M., Tan, J.Y.L., Wong, W.P., Chew, S.F. and Ip, Y.K. (2007). Defense against environmental ammonia toxicity in the African lungfish, *Protopterus aethiopicus*: bimodal breathing, skin ammonia permeability and urea synthesis. Aquatic Toxicology 85: 76-86.

Loong, A.M., Ang, S.F., Wong, W.P., Portner, H.O., Bock, C., Wittig, R., Bridges, C.R., Chew, S.F. and Ip, Y.K. (2008a). Effects of hypoxia on the energy status and nitrogen metabolism of African lungfish during aestivation in a mucus cocoon. Journal of Comparative Physiology 178B: 853-865.

Loong, A.M., Pang, C.Y.M., Hiong, K.C., Wong, W.P., Chew, S.F. and Ip, Y.K. (2008b). Increased urea synthesis and/or suppressed ammonia production in the African lungfish, *Protopterus annectens*, during aestivation in air or mud. Journal of Comparative Physiology 178B: 351-363.

Maina, J.N. and Maloiy, G.M.O. (1985). The morphometry of the lung of the African lungfish (*Protopterus aethiopicus*): its structural-functional correlations. Proceedings of the Royal Society of London 224B: 399-420.

Mauro, N.A. and Isaacks, R.E. (1989). Relative oxidation of glutamate and glucose by vertebrate erythrocytes. Comparative Biochemistry and Physiology 94A: 95-97.

McDougal, D.B., Holowach, J., Hoew, M.C., Jones, E.M. and Thomas, C.A. (1968). The effects of anoxia upon energy sources and selected metabolic intermeidates in the brains of fish, frog and turtle. Journal of Neurochemistry 15: 577-588.

Mesquita-Saad, L. S. B., Val, A. L. and Almeida-Val, V. M. F. (1900). Evolutionary implications and kinetic properties of vitamin-C synthesizing enzyme (GLO) from South American lungfish (*Lepidosiren paradoxa*). (Unpublished work)

Metcalf, V.J., George, P.M. and Brennan, S.O. (2007). Lungfish albumin is more similar to tetrapod than to teleost albumins: purification and characterisation of albumin from the Australian lungfish, *Neoceratodus forsteri*. Comparative Biochemistry and Physiology 147B: 428-437.

Meyer, A. and Dolven, S.I. (1992). Molecules, fossils and the origin of tetrapods. Journal of Molecular Evolution 35: 102-113.

Meyer, A. and Schartl, M. (2007). Gene and genome duplications in vertebrates: the one-to-four (-to-eight in fish) rule and the evolution of novel gene functions. Current Opinion in Cell Biology 11: 699-704.

Mirejovska, E., Bass, A., Hurych, J. and Teisinger, J. (1981). Enzyme changes during experimental silicotic fibrosis. II Intermediary metabolism enzymes of the lungs. Environmental Research 25:434-440.

Mishra, R. and Shukla, S.P. (1994). Effects of endosulfan on bioenergetic properties of liver mitochondria from the freshwater catfish *Clarias batrachus*. Pesticide Biochemistry and Physiology 50: 240-246.

Mlewa, C.M. and Green, J.M. (2004). Biology of the marbled lungfish, *Protopterus aethiopicus* Heckel, in Lake Baringo, Kenya. African Journal of Ecology 42: 338-345.

Mlewa, C.M., Green, J.M. and Dumbrack, R. (2007). Are wild lungfish obligate air breathers? Some evidence from radioteletry. African Zoology 42: 131-134.

Mlewa, C.M., Green, J.M. and Dunbrack, R.L. (2009). The general natural history of the African lungfishes. In: J.M. Jorgensen, A. Kemp and T. Wang (eds). Biology of Lungfish. Science Publishers Inc., Enfield, New Hampshire, USA.

Mommsen, T.P. and Walsh, P.J. (1989). Evolution of urea synthesis in vertebrates: the piscine connection. Science 243: 72-75.

Mommsen, T.P., French, C.J. and Hochachka, P.W. (1980). Sites and patterns of protein and amino acid utilization during the spawning migration of salmon. Canadian Journal of Zoology 58: 1785-1799.

Moon, T.W. and Mommsen, T.P. (1987). Enzymes of intermediary metabolism in tissues of the little skate, *Raja erinacea*. Journal of Experimental Zoology 244: 9-15.

Moschetta, A., Xu, F., Hagey, L.R., van Berge-Henegouwen, G.P., van Erpecum, K.J., Brouwers, J.F., Cohen, J.C., Bierman, M., Hobbs, H.H., Steinbach, J.H. and Hofmann, A.F. (2005). A phylogenetic survey of biliary lipids in vertebrates. Journal of Lipid Research 46: 2221-2232.

Moyes, C.D., Moon, T.W. and Ballantyne, J.S. (1986). Oxidation of amino acids, Krebs cycle intermediates, lipid and ketone bodies by mitochondria from the liver of *Raja erinacea*. Journal of Experimental Zoology 237: 119-128.

Moyes, C.D., Buck, L.T., Hochachka, P.W. and Suarez, R.K. (1989). Oxidative properties of carp red and white muscle. Journal of Experimental Biology 143: 321-331.

Murphy, B., Zapol, W.M. and Hochachka, P.W. (1980). Metabolic activities of heart, lung and brain during diving and recovery in the Weddell seal. Journal of Applied Physiology 48: 596-605.

Newsholme, E.A. and Leech, A.R. (1983). Biochemistry for the Medical Sciences. Wiley, Toronto, Canada.

Olson, K.R. (2002). Gill circulation: regulation of perfusion distribution and metabolism of regulatory molecules. Journal of Experimental Zoology 293: 320-335.

Orgeig, S. and Daniels, C.B. (1995). The evolutionary significance of pulmonary surfactant in lungfish (Dipnoi). American Journal of Respiratory Cell and Molecular Biology 13: 161-166.

Pedersen, R.A. (1971). DNA content, ribosomal gene multiplicity, and cell size in fish. Journal of Experimental Zoology 177: 65-78.

Pelletier, D., Dutil, J., Blier, P. and Guderley, H. (1994). Relation between growth rate and metabolic organization of white muscle, liver and digestive tract in cod, *Gadus morhua*. Journal of Comparative Physiology 164B: 179-190.

Pelster, B. (1995). Metabolism of the swimbladder tissue. In: P.W. Hochachka and T.P. Mommsen (eds). Biochemistry and Molecular Biology of Fishes, Volume 4. Metabolic Biochemistry. Elsevier Science, Amsterdam, The Netherlands. pp. 101-118.

Perry, S.F., Gilmour, K.M., Swenson, E.R., Vulesevic, B., Chew, S.F. and Ip, Y.K. (2005a). An investigation of the role of carbonic anhydrase in aquatic and aerial gas transfer in the African lungfish *Protopterus dolloi*. Journal of Experimental Biology 208: 3805-3815.

Perry, S.F., Gilmour, K.M., Vulesevic, B., McNeil, B., Chew, S.F. and Ip, Y.K. (2005b). Circulating catecholamines and cardiorespiratory responses in hypoxic lungfish (*Protopterus dolloi*): a comparison of aquatic and aerial hypoxia. Physiological and Biochemical Zoology 78: 325-334.

Perry, S.F., Euverman, R., Wang, T., Loong, A.M., Chew, S.F., Ip, Y.K. and Gilmour, K.M. (2008). Control of breathing in African lungfish (*Protopterus dolloi*): a comparison of aquatic and cocooned (terrestrialized) animals. Respiratory Physiology and Neurobiology 160: 8-17.

Phelps, C., Farmer, M., Fyhn, H.J., Fyhn, U.E.H., Garlick, R.L., Noble, R.W. and Powers, D.A. (1979). Equilibria and kinetics of oxygen and carbon monoxide binding to the haemoglobin of the South American lungfish, *Lepidosiren paradoxa*. Comparative Biochemistry and Physiology 62A: 139-143.

Power, J.H.T., Doyle, I.R., Davidson, K. and Nicholas, T.E. (1999). Ultrastructural and protein analysis of surfactant in the Australian lungfish *Neoceratodus forsteri*: evidence for conservation of composition for 300 million years. Journal of Experimental Biology 202: 2543-2550.

Querat, B., Arai, Y., Henry, A., Akama, Y., Longhurst, T.J. and Joss, J.M.P. (2004). Pituitary glycoprotein hormone B subunits in the Australian lungfish and estimation of the relative evolution rate of these subunits within the vertebrates. Biology of Reproduction 70: 356-363.

Reeck, G.R. and Neurath, H. (1972a). Isolation and characterization of pancreatic procarboxypeptidase B and carboxypeptidase B of the African lungfish. Biochemistry 11: 3947-3955.

Reeck, G.R. and Neurath, H. (1972b). Pancreatic trypsinogen from the African lungfish. Biochemistry 11: 503-510.

Reeck, G.R., Winter, W.P. and Neurath, H. (1970). Pancreatic enzymes of the African lungfish *Protopterus aethiopicus*. Biochemistry 9: 1398-1403.

Ritter, N.M., Smith, D.D. and Campbell, J.W. (1987). Glutamine synthetase in liver and brain tissues of the Holocephalan, *Hydrolagus colliei*. Journal of Experimental Zoology 243: 181-188.

Segner, H. (1997). Ketone body metabolism in the carp *Cyprinus carpio*: biochemical and ^{1}H NMR spectroscopic analysis. Comparative Biochemistry and Physiology 116B: 257-262.

Seifert, A.W. and Chapman, L.J. (2006). Respiratory allocation and standard rate of metabolism in the African lungfish, *Protopterus aethiopicus*. Comparative Biochemistry and Physiology 143A: 142-148.

Sidell, B.D., Driedzic, W.R., Stowe, D.B. and Johnston, I.A. (1987). Biochemical correlations of power development and metabolic fuel preferenda in fish hearts. Physiological Zoology 60: 221-232.

Singer, T.D. and Ballantyne, J.S. (1991). Metabolic organization of a primitive fish, the bowfin (*Amia calva*). Canadian Journal of Fisheries and Aquatic Sciences 48: 611-618.

Singer, T.D., Mahadevappa, V.G. and Ballantyne, J.S. (1990). Aspects of the energy metabolism in the lake sturgeon, *Acipenser fulvescens*: with special emphasis on lipid and ketone body metabolism. Canadian Journal of Fisheries and Aquatic Sciences 47: 873-881.

Smith, H.W. (1929). The excretion of ammonia and urea by the gills of fish. Journal of Biological Chemistry 81: 727-742.

Smith, H.W. (1930). Metabolism of the lungfish, *Protopterus aethiopicus*. Journal of Biological Chemistry 88: 97-130.

Smith, H.W. (1931). Observations on the African lungfish, *Protopterus aethiopicus*, and on evolution from water to land environments. Ecology 12: 164-181.

Smith, H.W. (1935). The metabolism of the lungfish. I. General considerations of the fasting metabolism in active fish. Journal of Cellular and Comparative Physiology 6: 43-67.

Speers-Roesch, B., Ip, Y.K. and Ballantyne, J.S. (2006a). Metabolic organization of freshwater, euryhaline, and marine elasmobranchs: implications for the evolution of energy metabolism in sharks and rays. Journal of Experimental Biology 209: 2495-2508.

Speers-Roesch, B., Robinson, J.W. and Ballantyne, J.S. (2006b). Metabolic organization of the spotted ratfish, *Hydrolagus colliei* (Holocephali: Chimaeriformes): insight into the evolution energy metabolism in the chondrichthyan fishes. Journal of Experimental Zoology 306A: 631-644.

Spolter, P.D., Adelman, R.C. and Weinhouse, S. (1965). Distinctive properties of native and carboxypeptidase treated ald olases of rabbit muscle and liver. Journal of Biological Chemistry 240: 1327-1337.

Staples, J.F., Kajimura, M., Wood, C.M., Patel, M., Ip, Y.K. and McClelland, G.B. (2008). Enzymatic and mitochondrial responses to five months or aerial exposure in the slender lungfish *Protopterus dolloi*. Journal of Fish Biology 73: 608-622.

Suarez, R.K. and Mommsen, T.P. (1987). Gluconeogenesis in teleost fishes. Canadian Journal of Zoology 65: 1869-1882.

Suarez, R.K., Mallet, M.D., Daxboeck, C. and Hochachka, P.W. (1986). Enzymes of energy metabolism and gluconeogenesis in the Pacific blue marlin, *Makaira nigricans*. Canadian Journal of Zoology 64: 694-697.

Sugden, P.H. and Newsholme, E.A. (1975). Activities of citrate synthase, NAD+-linked and NADP+-linked isocitrate dehydrogenases, glutamate dehydrogenase, aspartate aminotransferase and alanine amino transferase in nervous tissues from vertebrates and invertebrates. Biochemical Journal 150: 105-111.

Swan, H., Jenkins, D. and Knox, K. (1968) Anti-metabolic extract from the brain of *Protopterus aethiopicus*. Nature 217: 671.

Swan, H., Jenkins, D. and Knox, K. (1969). Metabolic torpor in *Protopterus aethiopicus*: an antimetabolic agent from the brain. The American Naturalist 103: 247-258.

Takezaki, N., Figueroa, F., Zaleska-Rutczynska, Z., Takahata, N. and Klein, J. (2004). The phylogenetic relationship of tetrapod, coelacanth, and lungfish revealed by the sequences of forty-four nuclear genes. Molecular Biology and Evolution 21: 1512-1524.

Tarasoff, F.J., Bisaillom, A., Pierard, J. and Whitt, A.P. (1972). Locomotory patterns and external morphology of the river otter, sea otter and harp seal (Mammalia). Canadian Journal of Zoology 50: 915-929.

Thomson, K.S. (1972). An attempt to reconstruct evolutionary changes in the cellular DNA content of lungfish. Journal of Experimental Zoology 180: 363-372.

Touhata, K., Toyohara, H., Mitani, T., Kinoshita, M., Satou, M. and Sakaguchi, M. (1995). Distribution of L-gulono-1,4-lactone oxidase among fishes. Fisheries Science 61: 729-730.

Treberg, J.R., Martin, R.A. and Driedzic, W.R. (2003). Muscle enzyme activities in a deep-sea squaloid shark, *Centroscyllium fabricii*, compared with its shallow-living relative, *Squalus acanthias*. Journal of Experimental Zoology 300A: 133-139.

Turrens, J.F., Freeman, B.A., Levitt, G.J. and Crapo, J.D. (1982). The effects of hyperoxia on superoxide production by lung submitochondrial particles. Archives of Biochemistry and Biophysics 217: 401-410.

Ultsch, G.R. (1996). Gas exchange, hypercarbia and acid-base balance, paleoecology, and the evolutionary transition from water-breathing to air-breathing among vertebrates. Palaeogeography, Palaeoclimatology and Palaeoecology 123: 1-27.

Val, A.L. (2000). Organic phosphates in the red blood cells of fish. Comparative Biochemistry and Physiology 125A: 417-435.

Walsh, P.J. and Milligan, C.L. (1993). Role of buffering capacity and pentose phosphate pathway activity in the gas gland of the gulf toadfish *Opsanus beta*. Journal of Experimental Biology 176: 311-316.

Weber, R.E., Johansen, K., Lykkeboe, G. and Maloiy, G.M.O. (1977). Oxygen-binding properties of hemoglobins from estivating and active African lungfish. Journal of Experimental Zoology 199: 85-96.

West, J.L., Bailey, J.R., Almeida-Val, V.M.F., Val, A.L., Sidell, B.D. and Driedzic, W.R. (1999). Activity levels of enzymes of energy metabolism in heart and red muscle are higher in north-temperate-zone than in Amazonian teleosts. Canadian Journal of Zoology 77: 690-696.

Whitt, G.S. and Horowitz, J.J. (1970). Evolution of a retinal specific lactate dehydrogenase isozyme in teleosts. Experientia 26: 1302-1304.

Wilkie, M.P., Morgan, T.P., Galvez, F., Smith, R.W., Kajimura, M., Ip, Y.K. and Wood, C.M. (2007). The African lungfish (*Protopterus dolloi*): ionoregulation and osmoregulation in a fish–out of water. Physiological and Biochemical Zoology 80: 99-112.

Wood, C.M., Perry, S.F., Walsh, P.J. and Thomas, S. (1994). HCO_3^- dehydration by the blood of an elasmobranch in the absence of a Haldane effect. Respiration Physiology 98: 319-337.

Wood, C.M., Walsh, P.J., Chew, S.F. and Ip, Y.K. (2005a). Ammonia tolerance in the slender lungfish (*Protopterus dolloi*): the importance of environmental acidification. Canadian Journal of Zoology 83: 507-517.

Wood, C.M., Walsh, P.J., Chew, S.F. and Ip, Y.K. (2005b). Greatly elevated urea excretion after air exposure appears to be carrier mediated in the slender lungfish (*Protopterus dolloi*). Physiological and Biochemical Zoology 78: 893-907.

Wright, P.A., Perry, S.F. and Moon, T.W. (1989). Regulation of hepatic gluconeogenesis and glycogenolysis by catecholamines in rainbow trout during environmental hypoxia. Journal of Experimental Biology 147: 169-188.

Yokobori, S., Hasegawa, M., Ueda, T., Okada, N., Nishikawa, K. and Watanabe, K. (1994). Relationship among coelacanths, lungfishes, and tetrapods: a phylogenetic analysis based on mitochondrial cytochrome oxidase I gene sequences. Journal of Molecular Evolution 38: 602-609.

Zammit, V.A. and Newsholme, E.A. (1979). Activities of enzymes of fat and ketone body metabolism and effects of starvation on blood concentrations of glucose and fat fuels in teleost and elasmobranch fish. Biochemical Journal 184: 313-322.

Zardoya, R. and Meyer, A. (1996). The complete nucleotide sequence of the mitochondrial genome of the lungfish (*Protopterus dolloi*) supports its phylogenetic position as a close relative of land vertebrates. Genetics 142: 1249-1263.

Zardoya, R., Cao, Y., Hasegawa, M. and Meyer, A. (1998). Searching for the closest living relative(s) of tetrapods through evolutionary analyses of mitochondrial and nuclear data. Molecular Biology and Evolution 15: 506-517.

CHAPTER 13

The Lungfish Digestive System

Masoud Hassanpour* **and Jean Joss**

Department of Biological Sciences, Macquarie University, Sydney, NSW 2109, Australia

ABSTRACT

The gastrointestinal tract of lungfishes contains a spiral valve, as do the intestines of lampreys, elasmobranchs, sturgeons and coelacanths. A spiral valve intestine is generally considered an original way of achieving increased surface area for absorption of nutrients. The spiral valves are most commonly confined to the intestine distal to the duodenum or at least distal to the pyloris of the stomach. However, in *Neoceratodus*, the coiling commences immediately beyond the oesophagus at the point (glottis) where the pneumatic duct exits. This chapter discusses the spiral valve intestine of *Neoceratodus* and compares it to that of the lepidosirenid lungfishes. The intimate relationship between the spiral valve intestine and the two lobes of spleen and the pancreas is described. The chapter finishes with descriptions of the liver and bile salts, the haemopoietic nature of the spleen and the presence of lymphoid tissues in the lungfish gut. It also includes a note on the only digestive system-related trematode parasite to have been described in lungfish.

Keywords: *Neoceratodus*, spiral valve intestine, pyloric fold, lymphoid tissue, spleen and pancreas

INTRODUCTION

The digestive system of lungfishes is peculiar among bony fishes in that increased surface area for absorption of nutrients is provided by a spiral valve. There are several descriptions of lungfish spiral valve intestines in the older literature but there is only one relatively recent detailed anatomical description (Rafn and Wingstrand 1981) of *Neoceratodus forsteri*, which is arguably the most aberrant

**Corresponding author*: E-mail: masoudhg@gmail.com

of the three living lungfish genera. In this chapter we will discuss the study of Rafn and Wingstrand (1981), including where appropriate, our own more recent observations on the intestine of *Neoceratodus* (Hassanpour and Joss 2009). In our recent study more detailed anatomical and histological features of this structure are depicted. There has been almost nothing published on the physiology of lungfish digestion. The endocrine cells of lungfish intestine and pancreas are discussed in the chapter on the Endocrine System (this volume) and the unusual dentition of lungfish in the chapter on Teeth (this volume). In this chapter we will include lymphoid tissue associated with the lungfish intestine and spleen, and haemopoiesis associated with the spleen. We will also mention lungfish gut-associated parasites.

ANATOMY

Descriptions of the anatomy of the gut and associated organs of all three genera of extant lungfishes have been described in the old literature and reviewed in the only relatively recent published description by Rafn and Wingstrand (1981). While these latter authors had access to specimens of all three genera of living lungfish, their publication is primarily a detailed description of *Neoceratodus forsteri*, which is the only genus for which we have first hand knowledge and will, therefore form the basis of the following description.

Viewed from its outside, the intestine of *Neoceratodus* is as long as the body cavity and almost as wide in immature fish (Figure 1). In adults, the gonads are of considerable size and lie lateral and ventral to the gut, so that the gut occupies less, as a percentage, of the body cavity. Dorsal to the gut, anteriorly and medially lies

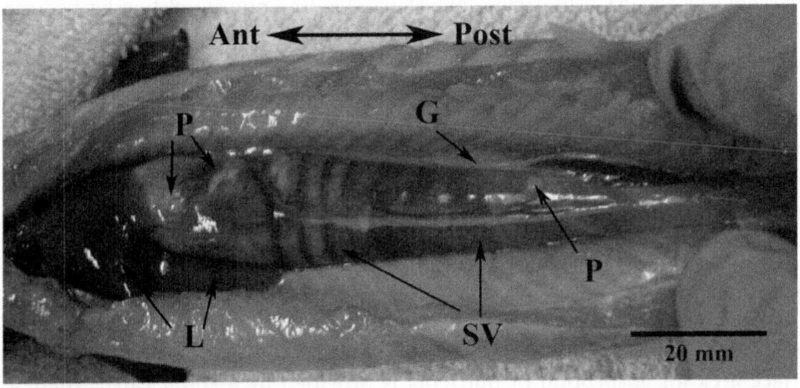

Fig. 1 Abdominal cavity of juvenile *N. forsteri*, anterior to the left. The intestine of *Neoceratodus* is as long as the body cavity and almost as wide in immature fish (Lungfish length: 23 cm, weight: 100 g. Intestinal length prior to fixation and cutting: 10 cm, weight: 3.5 g). SV: spiral valve, L: liver, G: immature gonads, and P: white nodules of parasite in gut wall, Ant: anterior direction, Post: posterior direction.

the single lung and posteriorly, the paired kidneys. Unlike most fish abdominal organs, those of the lungfish appear very 'tidy', being outwardly a wide tube running the length of the body cavity joining the oesophagus to the rectum. However, this apparent 'simplicity' is immediately banished as one seeks to understand the arrangement of the spiral valve, which in *Neoceratodus* commences immediately behind the very short oesophagus, at the entry (glottis) from the oesophagus into the pneumatic duct, leading to the lung.

The gastrointestinal tract is composed of the mouth and oral cavity, which lead into a short oesophagus and thence into the intestine with associated liver, spleen and pancreas and finally ends in a short rectum leading into a cloaca. These will be discussed in turn.

MOUTH AND ORAL CAVITY

Lungfish use powerful suction in order to capture their food. Unlike teleosts, which do the same, the method of achieving suction does not involve modification of the gape. Their mouths are terminal, non-extensible and their gape quite small. The suction is provided by the musculature of the hyoid and branchial arches and their unusual 'cranial rib', described in detail by Bemis (1986). After sucking food and usually some substrate into the expanded oral cavity, it is crushed by the robust tooth plates located centrally in the upper and lower jaws. The unique dentition of lungfish is described in detail in the chapter on Teeth and feeding behavior and action by Bemis (1986).

Lepidosiren and *Protopterus* are described as being largely carnivorous (see chapter on African Lungfish), whereas *Neoceratodus* consumes a lot of plant and detrital matter in addition to small animals, such as insect larvae, worms and molluscs. These latter are found under the substrate by electrolocation (Watt *et al.* 1999) and probably also olfaction. After locating the prey item, it is sucked into the oral cavity where it is crushed before being spat out to remove some of the substrate sucked in at the same time (generally sinks faster than food items), and then re-sucked into the mouth again. This may be repeated several times, adding saliva to the food items each time before swallowing via the oesophagus into the first spiral coil of the "stomach".

OESOPHAGUS

From the oral cavity, the food is transferred to the oesophagus, which in *Neoceratodus* is very short and poorly delimited. The internal wall of the oesophagus is wrinkled irregularly into longitudinal folds, which are lost as it merges with the gastric intestine (Rafn and Wingstrand 1981). These authors go on to describe the ciliated epithelium of the oesophagus and the very thick submucosa that is largely composed of adipose tissue. Leaving the short oesophagus on the right side is the pneumatic duct, which curves around to connect to the lung, lying dorsal to the intestine.

SPIRAL INTESTINE

The intestine of the lungfish is not straight; rather it comprised a complex spiral valve. In external view, the spiral valve is like a thick and relatively short pipe, inside which there is a screw that fits neatly to the inside of the pipe. Thus the pathway for intestinal contents is spiralled rather than straight (Figure 2).

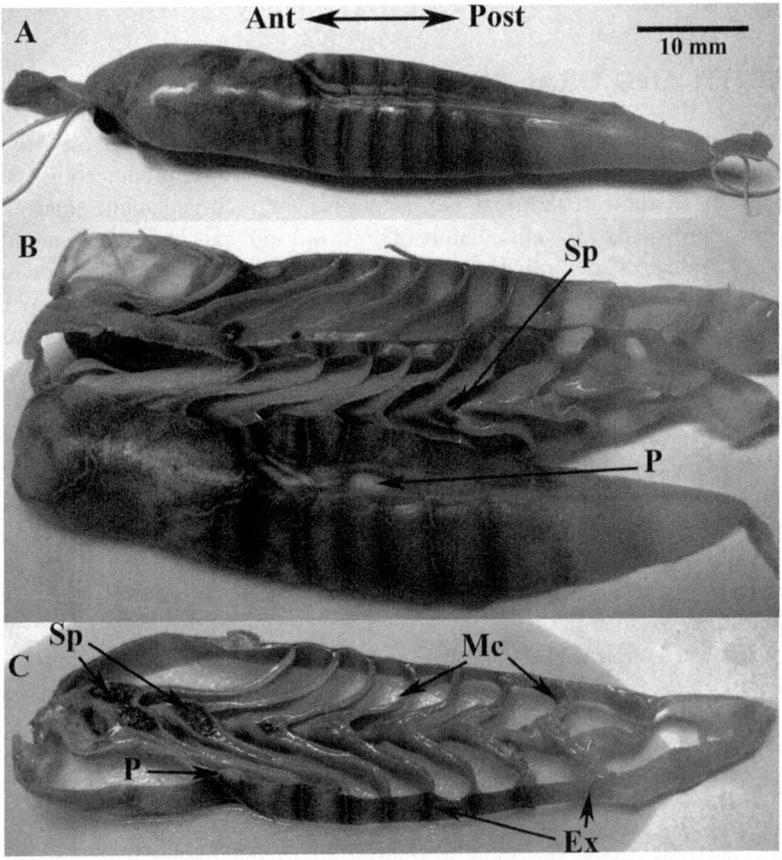

Fig. 2 A. The spindle-like spiral intestine of *N. forsteri*, distended with and fixed in Bouins fixative. Ant: anterior direction, Post: posterior direction. In external view, the spiral valve is like a thick and relatively short pipe, inside which there is a screw that fits neatly to the inside of the pipe. Nine spirals are visible. B, C Longitudinal sections of the spiral intestine. Ex: external wall, Mc: mucosal tissue, Sp: spleen, P: white nodules of parasite in gut wall.

Based on the Parker's classification (1885, cited by McAllister 1985) the spiral valve in lungfish is a type D spiral valve, in which the radius of the infolding mucosa is greater than the radius of the lumen and the spiraling cones' apexes are in an anterior direction (Figure 3). Based on our observations (Hassanpour and Joss 2009), the spiral valve in *N. forsteri* can be defined as a trapezium of mucosal tissue, in which the long side is twisted around the short side and the whole encased in an external covering to form a spindle-like structure (Figure 2A). This may be analogized to a tower with spiral stairs from the top to the bottom, which in this case is the short rectum. There are nine full coils, the first of which is high and occupies about one-third of the total length of the spiral valve intestine. The direction of winding is clockwise from anterior to posterior. The spirals in the anterior region of the intestine overlap each other, which Rafn and Wingstrand (1981) described as being cornet-like. As the spiral valve continues posteriorly it becomes less folded and resembles more the spiral valve intestine of elasmobranchs (Figure 3).

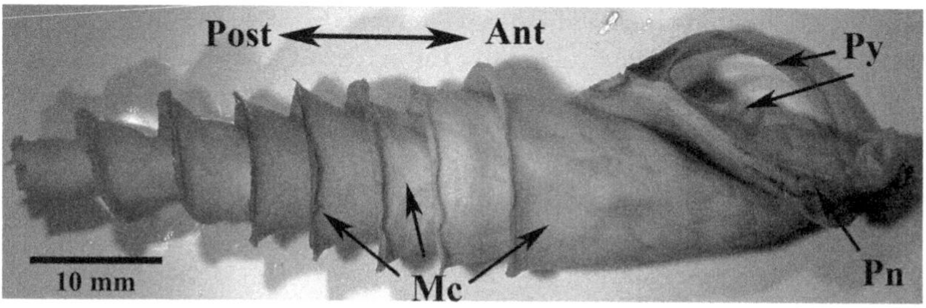

Fig. 3 Dorsal view of the spiral valve intestine of *N. forsteri* fixed in Bouins and external wall removed. Anterior to the right and left lateral side to up. There are nine full coils. The first spiral (back of this view) is about one-third of the total length of the spiral valve intestine. The direction of winding is clockwise from anterior to posterior. The spirals in the anterior region of the intestine overlap each other, making spiral cones, the apexes of which are in the anterior direction. Py: pyloric fold, Pn: pancreas, Mc: mucosal tissue, Ant: anterior direction, Post: posterior direction.

Beyond the pyloric fold and region containing rugae (see below), all three genera of lungfish have a similar arrangement of the spiral valve. The midgut, or post-pyloric intestine can be discussed in two sections, the medial axis and the mucosal layer. The medial axis in the midgut region is the free margin of the spiral valve, where the mucosal layer winds around it and segregates the posterior spleen from the intestinal lumen. As the midgut spiral valve merges with the hindgut, the spleen, which has been reduced in diameter, completely disappears. It is replaced by a double layer of mucosal epithelium separated by a thin layer of connective tissue in the penultimate coil of the spiral valve. Finally in the last coil, the medial axis and its attached mucosa vanish into the external wall leading into the non-spiral rectum (Rafn and Wingstrand 1981, personal observations).

When the spiral valve of *Neoceratodus* is unwound and laid flat, it has a hemi-annular structure (Figure 4), in which the short side is that of the medial axis and encases the posterior spleen and the long side is the attachment to the external wall. The mucosal layer is a long sheet of thin and delicate tissue, wrinkled between the long and short sides of the hemi-annular spiral valve. It is a double epithelial layer so that both sides of the mucosa are in direct contact with the nutritional contents of the lumen. As the contents of the lumen progress through the coils of the spiral valve, they pass both sides of the double mucosal layer. This mucosal epithelium is thicker towards the outer margins and the central axis of each coil.

PYLORIC FOLD

The distinct separation observed in most vertebrates between the pyloric region of the stomach and the duodenum is not so clear in *Neoceratodus*. Rather the pylorus is made up of a double layer of mucosa, referred to as the pyloric fold (Figure 5), which acts as an incomplete gate between the pre-pyloric stomach and the rest of the intestine, midgut. However, as there are no distinct *musculus sphincter pylori* in the pyloric fold (Rafn and Wingstrand 1981), this structure does not deserve

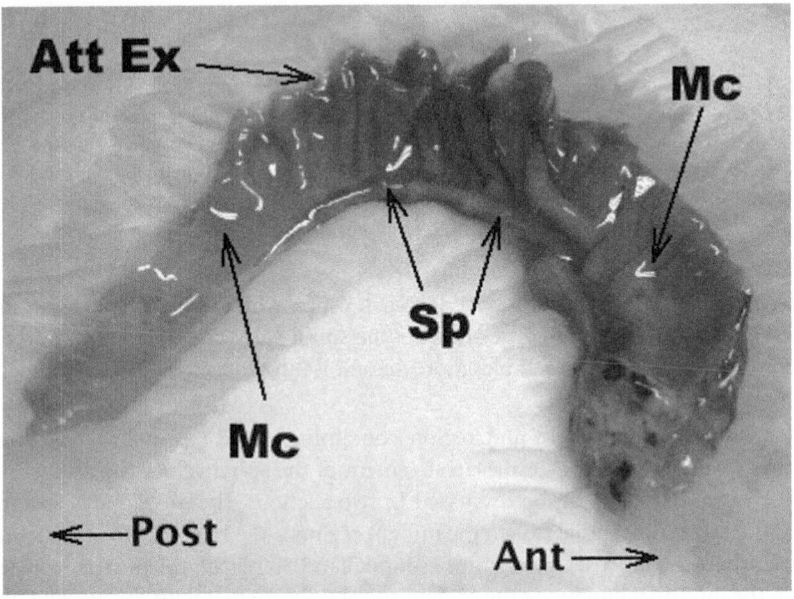

Fig. 4 Hemi-annular structure of the spiral valve of *N. forsteri* when unwound. The short side is the medial axis and encases the posterior spleen and the long side is the attachment to the external wall. The mucosal layer is wrinkled between the long and short sides of this hemi-annular structure. Att Ex: attachment of the mucosal tissue to the external wall, Mc: mucosal tissue, Sp: spleen, Ant: anterior direction, Post: posterior direction.

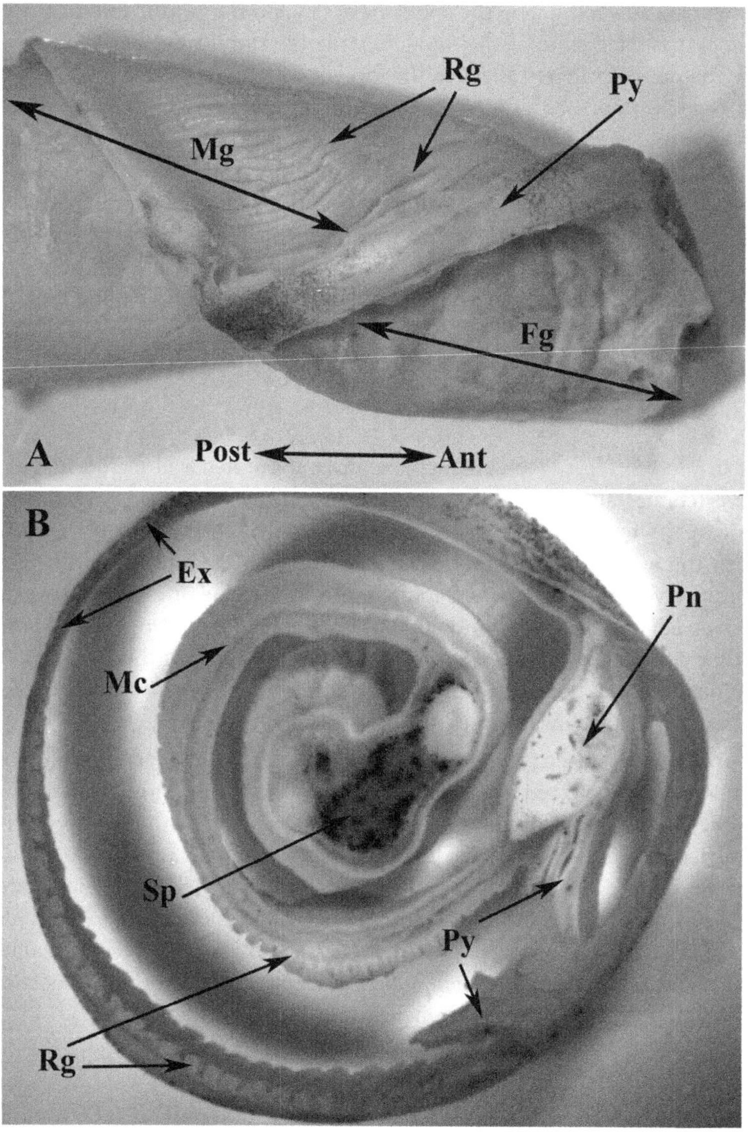

Fig. 5 A: The anterior part of the spiral intestine of *N. forsteri* fixed in Bouins, external wall removed. Fg: foregut, Mg: midgut, Rg: rugae, Py: pyloric fold, Ant: anterior direction, Post: posterior direction. B: Transverse section of the spiral valve in foregut region. Rugae are completely post-pyloric and are limited to the first coil of the midgut, between the pyloric fold and about halfway through the end of the first coil of the spiral valve (B). Rugae are parallel to the pyloric fold (A) and are dense in the pyloric fold side but gradually decrease in size and number until they completely disappear from about halfway through the end of the first coil (B). Rugae are present on the inner side of the external wall (B). Ex: external wall, Mc: mucosal tissue, Pn: pancreas, SP: spleen.

to be referred as a sphincter. The spiral valve beyond the pyloric fold or the post-pyloric part is referred to as the midgut, while the spiral valve before the pyloric fold or the pre-pyloric part is called the foregut. Due to its double layer structure, which based on our observations, extends in an anterior to posterior direction, the pyloric fold might have an important role in preventing the regurgitation of the digesta from midgut to foregut. The coiling process in *N. forsteri* begins in the pre-pyloric intestine (Figure 3). *Protopterus* and *Lepidosiren* also have a pyloric fold separating the gastric stomach from the intestine but unlike *Neoceratodus*, the spiralling does not begin until the region beyond the pyloric fold and the pyloric fold itself is a simple ring-shape at the hind end of the stomach (Jollie 1968; Rafn and Wingstrand 1981).

RUGAE AND STOMACH

In *Neoceratodus*, the rugae which characterize the inner wall of the stomach in most vertebrates and which allow considerable distension of this organ, are not present in the foregut ("stomach"). They are limited to the wall of the midgut between the pyloric fold and about halfway through the end of the first coil of the spiral valve. Rugae are parallel to the more anterior pyloric fold and are dense in the side adjacent to the pyloric fold but gradually decrease in size and number until they completely disappear from about halfway through the end of the first coil. Where present, the rugae can ramify into multiple branches as they progress from the medial axis towards the external wall (Figure 5).

The presence of rugae just beyond the pyloric fold implies that this region of the intestine may be subjected to some distension from contents, rather than the pre-pyloric region more usually associated with distension and storage in vertebrates. The entries of the common bile duct and pancreatic duct into this region, just beyond the pyloric fold (Rafn and Wingstrand 1981, personal observations), define this region with rugae as the duodenum (a distensible duodenum!). Although the presence of gastrin and cholecystokinin (CCK) have been identified in the anterior part of the intestine (Van Noorden and Joss, unpublished), the exact location of the stomach in *N. forsteri* remains unclear (Hassanpour and Joss 2009).

HISTOLOGY

The histology of the gastrointestinal tract of lungfish is that of a digestive/absorptive organ as it is in most vertebrates. The oesophagus is luminally composed of a thin epithelial layer, the lamina propria, and a very thick submucosa. The epithelial layer of the oesophagus, like the rest of the intestinal wall comprised pseudostratified columnar epithelium. Compared to the epithelium of the spiral valve, the goblet cells, which secrete mucus, are not abundant in the oesophageal epithelium. According to Rafn and Wingstrand (1981), unlike the epithelium of the midgut, the epithelial layer of the oesophagus is ciliated. The lamina propria is poorly developed in the oesophagus but the submucosa is very thick and mainly composed of adipose tissue (Rafn and Wingstrand 1981).

The mucosa of the post-pyloric intestine includes all of the spiral valve beyond the medial axis and external wall. However the epithelium of the medial axis and also the epithelium of the inner side of the external wall have histological similarities with the epithelium of the mucosal tissue. The mucosal layer comprised three distinct parts – the epithelium, lamina propria, and submucosa. The epithelium of the mid- and hind-gut is similar throughout, appearing as pseudostratified columnar epithelium, where it is assumed most of the absorption of nutrients is taking place. Three cell populations are present in this layer, epithelial cells, goblet cells, and intraepithelial lymphocytes (IEL). Based on our study in *N. forsteri*, a brush border and basement membrane are present in the epithelium of the intestine. The lamina propria is the layer where, in some regions, the lymphoid tissue is located. The submucosa is a relatively thick layer of connective tissue, which is shared between two epithelial layers of the mucosal tissue (Figure 6).

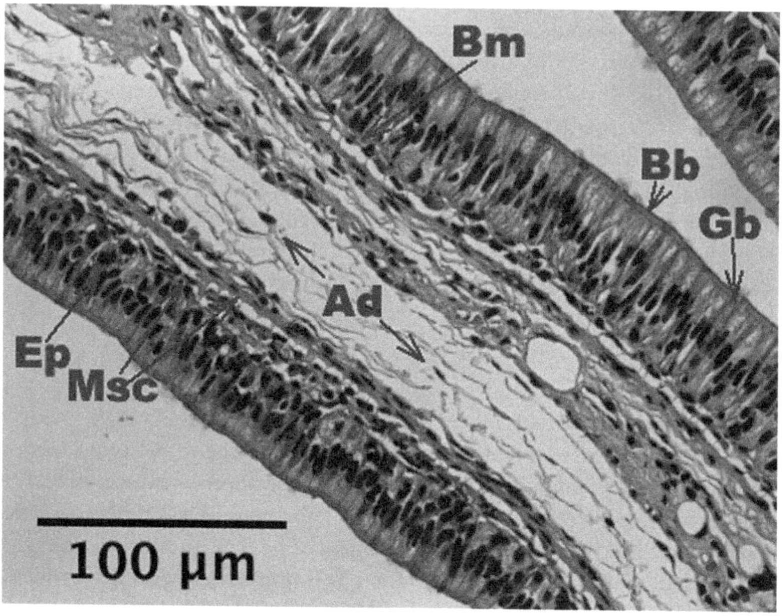

Fig. 6 Transverse section of the mucosal tissue of midgut in *N. forsteri*. The mucosal layer comprised three parts – the epithelium, lamina propria, and submucosa. Brush border and basement membrane are present. The submucosa, which is mainly made of adipose tissue, is shared between two epithelial layers of the mucosal tissue. Ad: adipose tissue, Bb: brush border, Bm: basement membrane, Ep: epithelium, Gb: goblet cell, Msc: smooth muscle (longitudinal and circular).

The ultrastructure of the cells lining the spiral valve of *Protopterus aethiopicus* has been described by Purkerson *et al.* (1975). These workers found ciliated cells in both the proximal and distal regions of the spiral valve. They argued that these ciliated epithelial cells, along with the intraepithelial capillaries, might indicate a respiratory function of the intestine in African lungfish. However, they did not rule out the role of cilia in the movement of the intestinal contents. According to Rafn and Wingstrand (1981) and based on our observations, cilia are present in the epithelium of foregut and pyloric fold in *N. forsteri*. Ciliated cells lining the intestine are a primitive feature, being found also in amphioxus, lampreys, several selachians, sturgeons, gars (*Lepisosteus*) and some anurans (Barrington 1942; Coujard and Coujard-Champy 1947).

SUBSTANCE P

The one unpublished study that localized immunocytochemically tachykinins present in mucosal cells is described in the chapter on the Endocrine System (Chapter 15). It is worth noting here that one of these tachykinins, Substance P (SP), has been sequenced from *Neoceratodus* and its effect on intestinal motility examined (Liu *et al.* 2002). The primary structure of this peptide is Lys-Pro-Arg-Pro-Asp-Glu-Phe-Tyr-Gly-Leu-Met.NH_2, which shows 64% identity with mammalian SP. In isolated preparations of lungfish foregut circular muscle (which almost certainly would have been the post-pyloric region of the gut containing rugae), lungfish SP produced slow, long-lasting tonic contractions. Midgut preparations responded to SP more rapidly and more complexly, demonstrating that lungfish SP may be a physiologically important regulator of spiral valve intestine motility (Liu *et al.* 2002; Jensen and Holmgren 1991).

LIVER

The liver comprises the usual three lobes, that on the animal's right being the largest. The middle lobe contains a large gallbladder. The cystic duct, from the gallbladder, and the hepatic duct join to form a common bile duct, which runs parallel to but completely separate from the pancreatic duct. Both these ducts (pancreatic and bile) empty into the intestine just beyond the pyloric fold (Rafn and Wingstrand 1981). This region can then be homologized with the duodenum of other vertebrates.

Bile, produced in the liver and stored in the gallbladder, has been used to determine the phylogenetic relationships between the three genera of lungfish, the coelacanth, amphibians and teleosts (Amos *et al.* 1977). The bile salts studied were those presumed to be used in the emulsification of lipids in the digesta. Essentially, the detailed analysis of the bile salts and alcohols found in the gallbladders of these species confirmed the most recent generally accepted relationships between them (Brinkman *et al.* 2003). Of further interest, however, were the similarities found between the bile salts of lungfish, especially *Neoceratodus*, and to some extent *Protopterus*, and the major bile salt of lampreys (5α-petromyzonol sulphate), perhaps confirming the basal place of lungfish in the evolution of gnathostomes (Haslewood and Tokes 1969).

PANCREAS

The arrangement of the pancreas and spleen is unusual. Without opening the intestine, they can be completely missed by the casual observer. In all three genera of lungfish, the pancreas is buried between two epithelial layers of the mucosa in the foregut. The pancreas is intermixed with the anterior spleen on the left side, which acts as the axis of coiling in the first spiral in *Neoceratodus* (Figure 3). In this genus, the pancreas is entirely pre-pyloric, but in *Protopterus* and *Lepidosiren* the pancreas extends posteriorly into the post-pyloric, spiral valve region.

The pancreas of *Neoceratodus* is saucer-shaped, being thickest at the base of the saucer. It is whitish or light grey in contrast to the spleen, which is black with some spots of white tissue distributed throughout. The pancreas of *Protopterus* and *Lepidosiren* is also hidden in the first coil of the spiral valve but, in these two genera, the spiral valve does not begin until beyond the pyloric fold. In the lepidosirenid lungfishes, the pancreas appears much darker than it is in *Neoceratodus* (Epple and Brinn 1975; Rafn and Wingstrand 1981) and is not so easily distinguished from the anterior spleen at the gross anatomical level. Histologically, the pancreas primarily contains the extensive sinuses and zymogen cells of the exocrine pancreas connecting to the pancreatic ducts. In *Neoceratodus*, the endocrine islets are dispersed amongst the exocrine cells (Figure 7), as they are in most other vertebrates, but in the lepidosirenid lungfishes, they are grouped together similarly to the "Brockman Bodies" of actinopterygian fish (Epple and Brinn 1987; chapter on Endocrine System – this volume).

HAEMOPOIESIS AND THE SPLEEN

In all three genera of living lungfishes, the spleen consists of two apparently identical parts, the anterior spleen, associated with the pancreas and the posterior spleen associated with the intestine beyond the gastric portion. Both the anterior and posterior spleen are long bars of splenic tissue located in the free margin of the spiral valve. In *Neoceratodus*, the posterior spleen is the longer and thicker of the two and progresses caudally beyond the region where the anterior and posterior spleens overlap. At this point, the posterior spleen is ventral to the anterior spleen. Depending on the condition of the lungfish there can be a layer of adipose tissue surrounding the posterior spleen to further separate it from the medial muscle layer separating it from the lumen of the spiral valve intestine. Based on our observation and those of Rafn and Wingstrand (1981), the posterior spleen in *N. forsteri* does not have its own capsule (Figure 8).

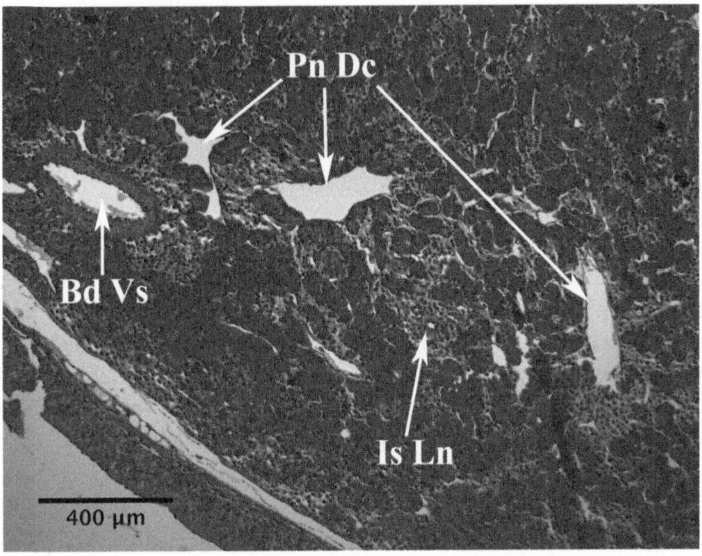

Fig. 7 A cross section of pancreas in *N. forsteri.* The dark-stained regions are associated with exocrine and the light-stained areas show endocrine related tissue. Is Ln: islet of Langerhans, Pn Dc: pancreatic ducts (lumen), Bd Vs: blood vessel.

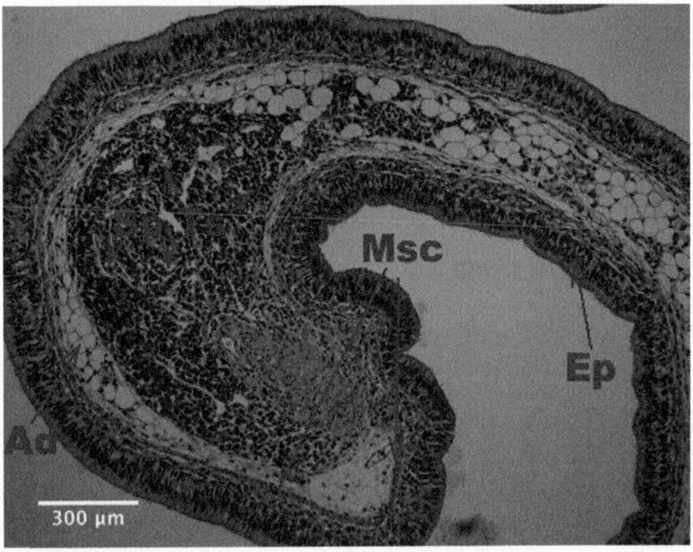

Fig. 8 Transverse section of the posterior spleen in *N. forsteri.* Observe that splenic tissue is a continuation of the mucosal tissue and is not encased in any capsule. Ad: adipose tissue, Ep: epithelium, Msc: smooth muscle (longitudinal and circular), Sp Ts: splenic tissue.

Histologically, there appears to be no difference between the two spleens. Rafn and Wingstrand (1981) distinguished three main types of tissue in the spleens, two of these lymphoid in nature and the third being a plexus of blood sinuses separated by lymphoid tissue. In adult lungfishes of all three genera, splenic tissue is characterized by its abundant pigment, black in *Neoceratodus* and *Lepidosiren* and brown in *Protopterus*. Although the accumulation of dark pigment in the spleen may sound unusual, it has also been shown to be characteristic of splenic tissue in the anuran amphibian, *Rana esculenta* (Scalia *et al.* 2004). Rafn and Wingstrand (1981) report that the pigment cells in all species are relatively enormous epithelial cells, which lie in connected strings, surrounded by smaller non-pigmented cells. Jordan and Speidel (1931) and Dustin (1934) both describe the pigment cells as eosinophil granulocytes that load with pigment as they degenerate. Rafn and Wingstrand (1981) showed histochemically that this pigment contains considerable quantities of iron, which suggests that a function of the spleen of lungfish is elimination of erythrocytes, as it is in most other vertebrates. Jordan and Speidel (1931) and Dustin (1934) have found that the spleen also functions in the formation of lymphocytes which were observed in large numbers in the blood sinuses of the spleen. Bielek and Strauss (1993) have described granulocytes (three types) in the spleen and kidney of *Lepidosiren*. Thus the spleen of lungfishes appears to function in production/storage and elimination of blood cells, erythrocytes and leucocytes as it does in other vertebrates. It is just its location, hidden as it is within the walls of the foregut and midgut, which is unusual. There is some evidence that very young stages of developing lungfish possess a small nodule of lymphoid tissue between the pancreas and the dorsal wall of the intestine which could be the beginnings of a spleen, but unfortunately there have been no new studies of the ontogeny of these organs in lungfish that might allow discussion of the relatedness of the lungfish spleen to the spleens of other vertebrates.

LYMPHOID TISSUE

In addition to the lymphoid tissue in the lungfish spleen, there are also discrete regions of the gut wall where nodes of lymphoid tissue are found (Figure 9A). It is too early to say whether any or all of these are the equivalent of the Peyer's patches found associated with the intestine in amniote vertebrates, described by Owen (1977). Peyer's patches are believed to survey the intestinal luminal contents for harmful microorganisms, generating immune responses within the mucosa. The description of actual Peyer's patches in fish was first made on paddlefish intestine (Weisel 1973; Petrie-Hanson and Peterman 2005; Peterman and Petrie-Hanson 2006). Similar aggregations of lymphoid tissue appear to be more extensive in the lepidosirenid lungfishes than they are in *Neoceratodus* (Rafn and Wingstrand 1981). Nevertheless, such gut-associated lymphoid tissue (GALT) can be recognized in *Neoceratodus*, first in the mucosal layer of the pyloric

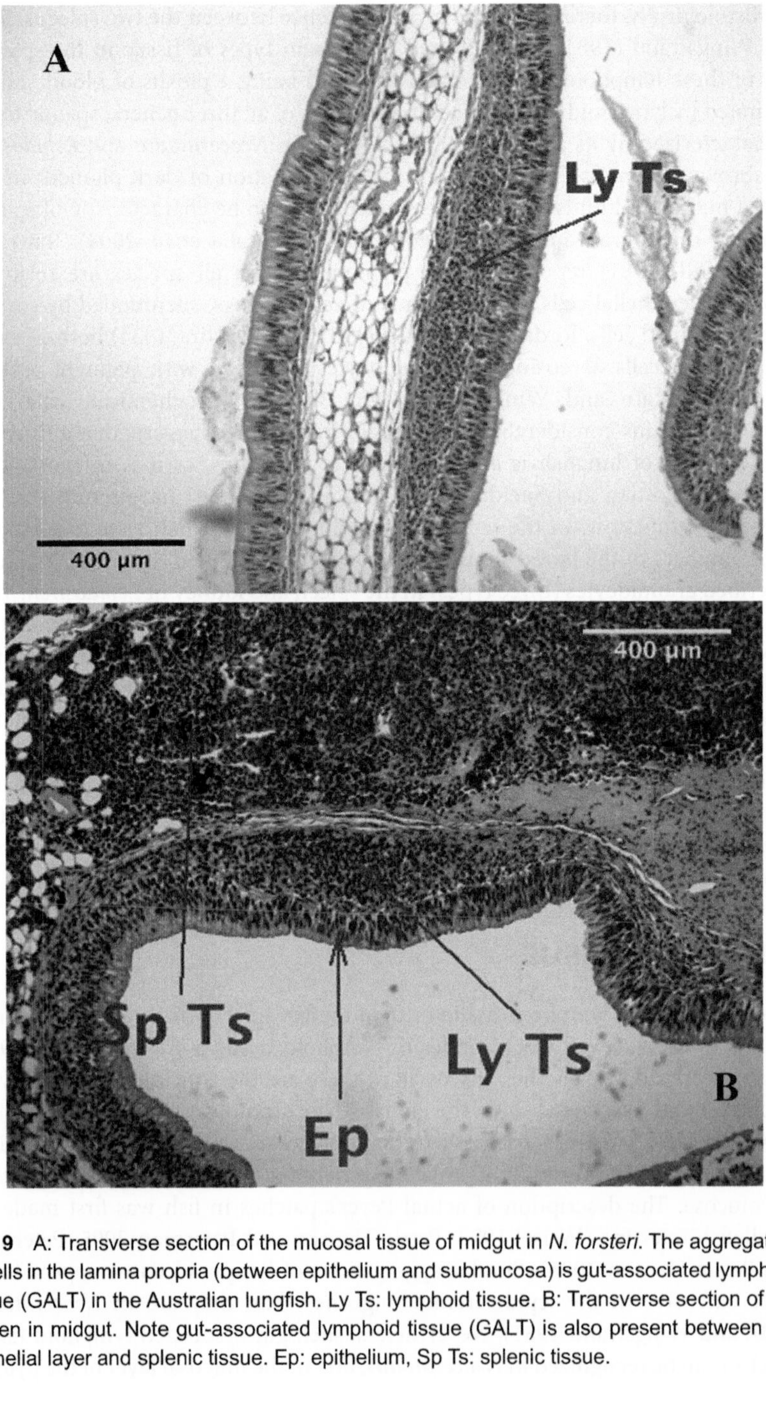

Fig. 9 A: Transverse section of the mucosal tissue of midgut in *N. forsteri*. The aggregation of cells in the lamina propria (between epithelium and submucosa) is gut-associated lymphoid tissue (GALT) in the Australian lungfish. Ly Ts: lymphoid tissue. B: Transverse section of the spleen in midgut. Note gut-associated lymphoid tissue (GALT) is also present between the epithelial layer and splenic tissue. Ep: epithelium, Sp Ts: splenic tissue.

fold (personal observations). There does not appear to be any lymphoid tissue found amongst the rugae of the foregut and early midgut, but beyond that, lying between the epithelial layer and the spleen's capsule, there appears to be some GALT (Figure 9B) in *Neoceratodus*. Therefore it is reasonable to regard the pylorus as the first 'battlefield' between lymphocytes residing there and microorganisms entering the fish via the gastrointestinal tract.

GUT PARASITES

The external wall of the spiral valve intestine of *Neoceratodus* commonly houses parasitic worms, most likely nematodes (Rafn and Wingstrand 1981; personal observations) (Figure 1, 2B/C). However, the only parasite recorded for *Neoceratodus* is a monogenean of the family Polystomatidae (*Concinnocotyla australiensis*), which lives exclusively in the oral cavity and gills of *Neoceratodus forsteri* and lays its eggs on the crushing surface and sides of the tooth plates. Larva or adult form of this parasite might transfer from one lungfish to another (Pichelin *et al.* 1991; Whittington and Pichelin 1991). We have observed parasitic infection in the spleen of *N. forsteri*, which we have not yet studied sufficiently to report in detail (Figure 10).

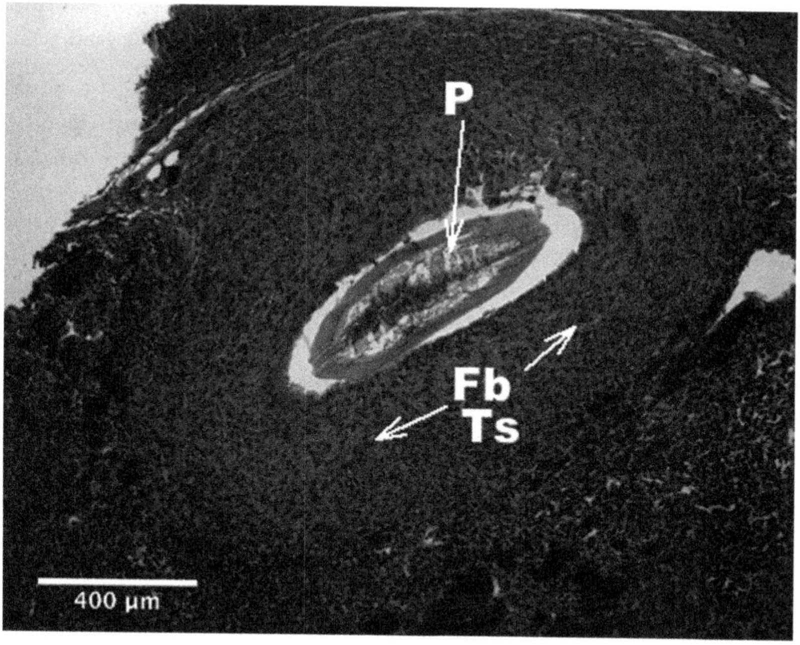

Fig. 10 Cross section of a parasite in the spleen of *N. forsteri*. P: parasite. Observe that this parasite has been surrounded by a thick layer of fibrous tissue (Fb Ts), which indicates an immunological response to this parasitic infection.

CONCLUSION

The spiral nature of the digestive system of lungfish is a primitive feature of many other living vertebrates such as elasmobranchs, non-teleost ray-finned fish and coelacanth. In *Neoceratodus*, the spiralling commences immediately after the oesophagus and is the only vertebrate described for which this occurs. It is thus a feature confirming the basal position of lungfish, in particular *Neoceratodus*, among the living osteichthyes.

References

Amos, B., Anderson, I.G., Haslewood, G.A.D. and Tokes, L. (1977). Bile salts of the lungfishes *Lepidosiren, Neoceratodus* and *Protopterus* and those of the coelacanth *Latimeria chalumnae* Smith. Biochemical Journal 161: 201-202.

Barrington, E.J.W. (1942). Gastric digestion in the lower vertebrates. Biological Reviews 17: 1-27.

Bemis, W.E. (1986). Feeding systems of living Dipnoi: anatomy and function. Journal of Morphology Supplement 1: 249-275.

Bielek, E. and Strauss, B. (1993). Ultrastructure of the granulocytes of the South American lungfish, *Lepidosiren paradoxa* – morphogenesis and comparison to other leukocytes. Journal of Morphology 218: 29-41.

Brinkman, H., Venkatesh, B. and Meyer, A. (2004). Nuclear protein-coding genes support lungfish and not coelacanth as the closest living relatives of land vertebrates. Proceedings of the National Academy of Sciences USA 101: 4900-4905.

Coujard, R. and Coujard-Champy, C. (1947). Recherches sur l'epithelium intestinal du Protoptere et sur evolution des enterocytes chez les vertebres. Archives d'Anatomie, d'Histologie et d'Embryologie 30: 69-97.

Dustin, P. 1934. Recherches sur les organes hématopoietiques du *Protopterus dolloi*. Archives of Biologie Paris 45: 1-26.

Epple, A. and Brinn, J.E. (1987). The Comparative Physiology of Pancreatic Islets, In: D.S. Farner (ed). Zoophysiology Series. Volume 21, Springer-Verlag, New York, USA.

Haslewood, G.A.D. and Tokes, L. (1969). Comparative studies of bile salts. Bile salts of the lamprey *Petromyzon marinus* L. Biochemical Journal 114: 179-204.

Hassanpour, M. and Joss, J. (2009). Anatomy and Histology of the Spiral Valve Intestine in Juvenile Australian Lungfish, *Neoceratodus forsteri*. The Open Zoology Journal 2: 62-85.

Jensen, J. and Holmgren, S. (1991). Tachykinins and intestinal motility in different fish groups. General and Comparative Endocrinology 83: 388-396.

Jollie, M. (1968). Chordate Morphology. Reinhold Book Co., N.Y., Amsterdam and London.

Jordan, H.E. and Speidel, C.C. (1931). Blood formation in the African lungfish, under normal conditions and under conditions of prolonged estivation and recovery. Journal of Morphology 51: 319-371.

Liu, L., Conlon, J.M., Joss, J.M.P. and Burcher, E. (2002). Purification, characterisation, and biological activity of a substance P-related peptide from the gut of the Australian lungfish, *Neoceratodus forsteri*. General and Comparative Endocrinology 125: 104-112.

McAllister, J.A. (1985). Reevaluation of the formation of spiral coprolites. The University of Kansas Paleontological Contributions. Paper 114.

Owen, R. L. (1977). Sequential uptake of horseradish peroxidase by lymphoid follicle epithelium of Peyer's patches in the normal unobstructed mouse intestine: an ultrastructural study. Gastroenterology 72(3): 440-51.

Parker, T.J. (1885). On the intestinal spiral valve in the genus *Raja*. Transactions of the Zoological Society of London 11: 49-61 (cited by McAllister 1985).

Peterman, A. F. and Petrie-Hanson, L. (2006). Ontogeny of American paddlefish lymphoid tissues. Journal of Fish Biology 69 (Supplement A): 72-88.

Petrie-Hanson, L. and Peterman, A.E. (2005). American paddlefish leukocytes demonstrate mammalian-like cytochemical staining characteristics in lymphoid tissues. Journal of Fish Biology 66: 1101-1115.

Pichelin, S., Whittington, I. and Pearson, J. (1991). *Concinnocotyla australiensis* (Monogenea: Polystomatidae) a new genus for the polystome from the Australian lungfish, *Neoceratodus forsteri*. Systematic Parasitology 18: 81-93.

Purkerson, M.L., Jarvis, J.U.M., Luse, S.A. and Dempsey, E.W. (1975). Electron microscopy of the intestine of the African lungfish, *Protopterus aethiopicus*. Anatomical Record 182: 71-90.

Rafn, S. and Wingstrand, K.G. (1981). Structure of intestine, pancreas, and spleen of the Australian lungfish, *Neoceratodus forsteri*. Zoologica Scripta 10: 223-239.

Scalia, M., Di Pierto, C., Poma, M., Ragusa, M., Sichel, G. and Corsaro, C. (2004). The Spleen Pigment Cells in Some Amphibia. Pigment Cell Research 17: 119-127.

Watt, M., Evans, C.S. and Joss, J.M.P. (1999). Use of electroreception during foraging by the Australian lungfish. Animal Behaviour 58: 1039-1045.

Weisel, G. F. (1973). Anatomy and Histology of the Digestive System of the Paddlefish, *Polyodon spathula*. Journal of Morphology 140: 243-256.

Whittington, I. and Pichelin, S. (1991). Attachment of eggs by *Concinnocotyla australiensis* (Monogenea: Polystomatidae) to the tooth plates of the Australian lungfish, *Neoceratodus forsteri*. International Journal for Parasitology 21: 341-346.

CHAPTER 14

The Lungfish Urogenital System

Jean Joss[*]
Department of Biological Sciences, Macquarie University,
Sydney, NSW 2109, Australia

ABSTRACT

This chapter briefly summarises the chapter of the same title by Wake in 'The Biology and Evolution of Lungfishes' (Bemis *et al.*.. 1987). New, largely unpublished, data is included on the development of the gonads in *Neoceratodus*, the location of the adrenal medulla equivalent and seasonal changes in mature ovaries and testes also of *Neoceratodus*.
Keywords: kidney, testes, ovaries, adrenal medulla, seasonal reproduction

INTRODUCTION

Marvalee Wake has provided excellent descriptions of the urogenital systems of all three genera of extant lungfish, including coverage of all descriptions previous to her own, in 'The Biology and Evolution of Lungfishes' (Bemis *et al.* 1987). In this chapter I will briefly summarise Wake's descriptions, adding observations of my own on *Neoceratodus* and any new information published since the mid 1980s that is not already included in the chapter on the endocrine system in this volume. The reader is directed to the appropriate section of the chapter on "Endocrine Systems", which contains the most recent functional studies on lungfish gonads, kidneys and adrenals.

KIDNEY

Lungfish kidneys are paired, lobed structures lying dorsally in the body cavity. In *Lepidosiren* they are nearly as long as the testes to which the kidneys are closely bound. In *Protopterus aethiopicus*, the kidneys are shorter and stouter and lie in

*E-mail: jjoss@rna.bio.mq.edu.au

the posterior half of the body cavity. The kidneys of *Neoceratodus* are even shorter and stouter and more posteriorly located than those of *Protopterus* and, unlike the lepidosirenid lungfishes, cannot be segregated into distinct anterior and posterior regions. The close association with the testes, confined to the posterior half of the lepidosirenid lungfish kidneys, occurs along the whole length of the *Neoceratodus* kidney. During the breeding season (from late winter to end of spring-early summer) nephric tubules along the entire length of the *Neoceratodus* kidney contain mature sperm (Jesperson 1969, Wake 1987, Joss *et al.* 1996). The cytology of the nephrons of all three genera closely resemble those of other "typical" freshwater fishes (Hickman and Trump 1969, Hentschel and Elger 1987)), in that they have well-developed encapsulated glomeruli and tubules that histologically appear unmodified for sperm transport. There is no evidence to suggest that the very dilute urine of lungfishes has any effect on sperm viability but this has not been directly investigated in any lungfish.

During development, the pronephric kidney comprises two functional nephrostomal units in larval lungfish, located on either side of the heart. Wake (1987) discusses the phylogenetic significance of this. The relationship to the adrenal medulla is discussed under the section, Adrenals, in this chapter.

Recently, Ojeda *et al.* (2006) described the microanatomy and ultrastructure of the kidney of *Protopterus dolloi*, as the first such description of any dipnoan kidney. They conducted this research with the view to understanding how a kidney adapted to coping with the removal of large amounts of water, can also function during long periods of estivation when water must be conserved. Their current report (2006) is of the freshwater phase of the kidney, which confirms and extends the older descriptions referred to above. Ojeda *et al.* (2006) found the arrangement of the nephrons within the kidney separate the proximal from the distal tubules in different zones with the position of renal corpuscles lying between these two zones. The proximal and distal tubules are in turn separated by an intermediate segment (IS) lying in the proximal tubule zone. An IS has also been described in *P. annectens* (Sawyer *et al.* 1982) and *Lepidosiren* (Guyton 1935) but not in *Neoceratodus* (Jespersen 1969). It is interesting that the nephrons of the lepidosirenid lungfish do have an IS, as do the nephrons of several other primitive freshwater fish, amphibians and reptiles. However, there are no functional studies to suggest that the IS may be used in a countercurrent fashion to concentrate the filtrate. It is quite possible that it is the lack of an IS that has made it possible for *Neoceratodus* to utilise all kidney nephrons in the transport of sperm but impossible for them to estivate. More of these detailed microanatomical descriptions and the accompanying functional studies of living lungfish kidneys are required to address these fascinating questions.

Apart from the functional studies by Sawyer *et al.* (1976, 1982) mentioned in the "Endocrine System" chapter there are no others published for lungfish kidneys, but the method of converting waste nitrogen generated from breakdown of amino acids has been studied. Two distinct pathways for urea synthesis leading to the liberation of urea in different cellular compartments can be distinguished in vertebrates. All fish, including coelacanths but excluding lungfish, synthesize urea using glutamine-dependent carbamoyl phosphate synthetase (CPS) and mitochondrial arginase. In contrast mammals, amphibians and lungfish have ammonia-dependent CPS and cytosolic arginase (Janssens and Cohen 1966, 1968a, b; Mommsen and Walsh 1989). This supports the hypothesis that the ornithine-urea cycle, a monophyletic trait in vertebrates, underwent two key changes in the fish immediately involved in the transition to tetrapods – a switch from glutamine- to ammonia-dependent CPS and a replacement of mitochondrial with cytosolic arginase.

GONADS

Ovary: Lungfish ovaries are large, elongate structures extending the length of the body cavity on either side of the intestine and strongly attached to the dorsal body wall. They are also tightly bound to the oviducts (Wake 1987). Wake makes no mention of the fact that the oocytes are contained in 'leaves' like a book with the spine of the book being the region that is tightly bound to the body wall. This may only be a feature of the ovary of *Neoceratodus*. The lines drawn on the ovary of *Neoceratodus* and not the on those of the other two genera in Figure 1 of Wake's (1987) chapter suggest that she may have recognized this unusual structure of the mature ovary of *Neoceratodus*. Each 'leaf' contains small, non-vitellogenic oocytes around the periphery. Growing oocytes that will later become vitellogenic are more centrally located (Fig. 1A). Wake's specimens of *Neoceratodus* were 520-560 mm in length. In my experience, this size is very rarely mature which would explain why the section of *Neoceratodus* ovary figured (5c) in her chapter contained no degenerating follicles from the previous season. In larger, mature females degenerating, unovulated, vitellogenic follicles are present throughout the year in large (>800 mm), mature female *Neoceratodus*. The more seasons a lungfish has had the more of these remains will be apparent alongside new degenerating vitellogenic follicles in summer, new season's follicles growing in autumn and winter and mature vitellogenic follicles in late winter/spring (Fig. 1B-D).

Testis: Lungfish testes are elongate structures bound to the kidney and dorsal body wall. In the lepidosirenid lungfishes the posterior part of the testis has been described as 'vesicular' and it is this region that invades the posterior portion of the kidneys (Kerr 1901; Wake 1987). During the breeding season the anterior testes contain tubules packed with mature sperm that are conveyed to the posterior nephrons of the kidney via the vesicular posterior portion of the testes, arranged as 'vasa efferentia' for this purpose. In *Neoceratodus*, there is no 'vesicular' posterior region to the testes. The entire testes which can be very large (~10% of body weight during spawning season, personal observation) are packed with mature sperm during the breeding season (Joss *et al.* 1996, Fig. 2A-C). The

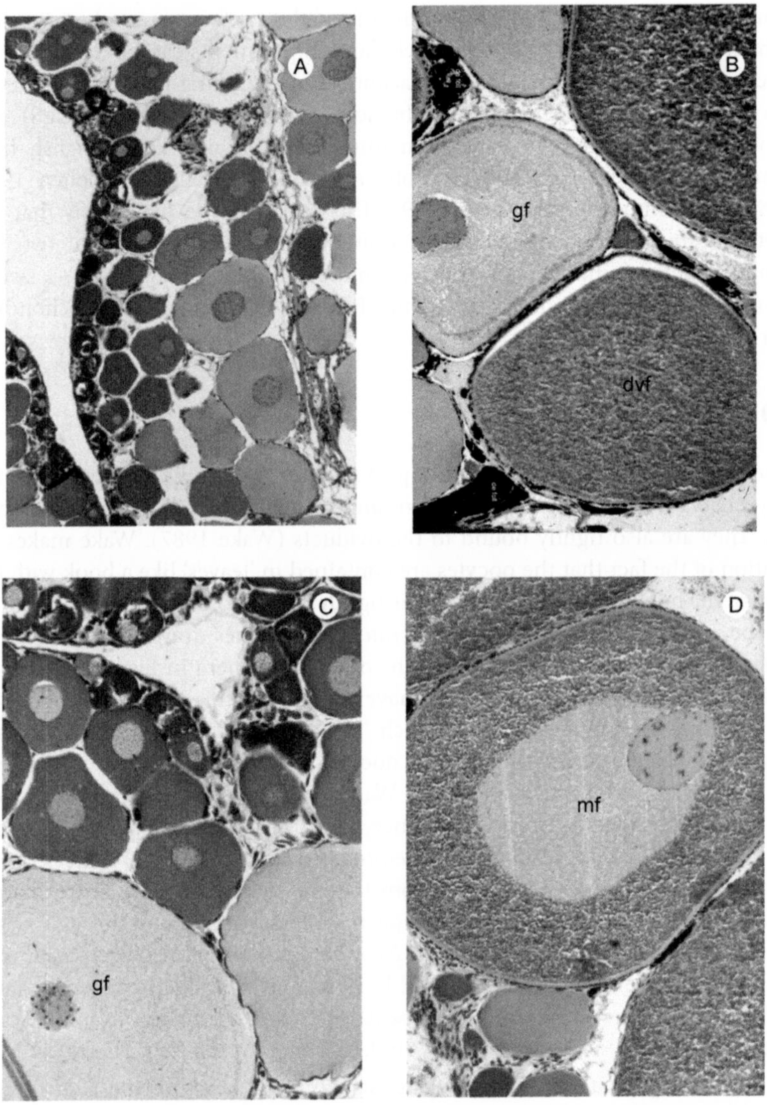

Fig. 1 Histological sections of the ovary of *Neoceratodus*, stained with haematoxylin and eosin, bar = 100 µm. **A.** Small section showing the intersection of two "leaves" with small oocytes on the periphery and larger growing oocytes more centrally located. **B.** Section of ovary in summer, showing degenerating vitellogenic follicles from the previous season and a degenerating vitellogenic follicle (dvf) and a growing follicle (gf). **C.** Section of ovary in autumn, showing growing follicles (gf). **D.** Section of ovary in spring, showing a mature vitellogenic follicle (mf).

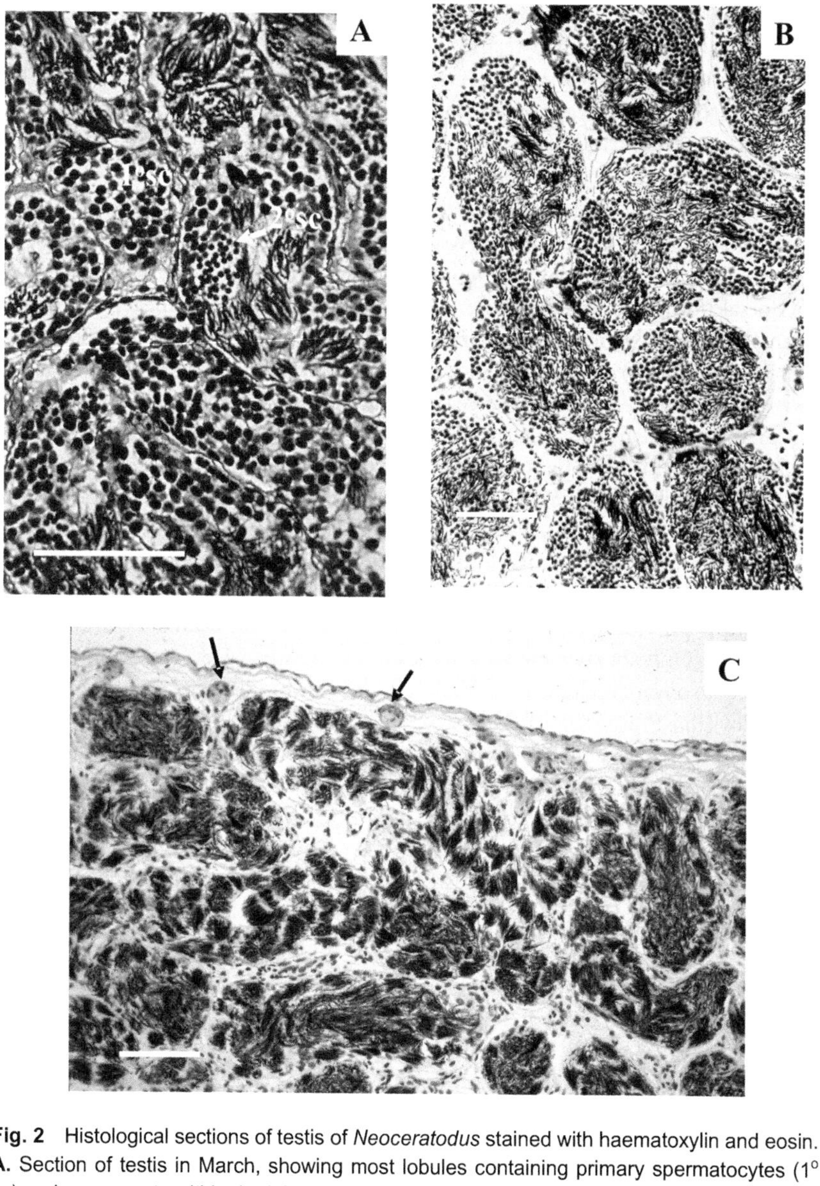

Fig. 2 Histological sections of testis of *Neoceratodus* stained with haematoxylin and eosin. **A.** Section of testis in March, showing most lobules containing primary spermatocytes (1° sc) and some nests within the lobules of secondary spermatocytes (2° sc and arrow). Bar = 20 μm. **B.** Section of testis in late April, showing secondary spermatocytes, spermatids and some tailed sperm. Bar = 100 μm. **C.** Section at the periphery of testis in October, showing all lobules containing mature sperm. Blood vessels can be seen at the periphery (arrows) and Leydig cells between the lobules containing sperm. Bar = 100 μm.

Color image of this figure appears in the color plate section at the end of the book.

structure of the mature testis in *Neoceratodus* is lobular according to Grier's (1992) re-classification of chordate testis types. As mentioned previously, sperm are transferred to the nephrons of the entire kidney for transport to the exterior for external fertilization of eggs oviposited by a female lungfish as part of an elaborate courtship ritual described by Grigg (1965) and confirmed by my own unpublished observations.

Gonad development: In *Neoceratodus*, gonadal development begins soon after hatching at stages 47-48 (stage when all yolk has been consumed and feeding has begun). The paired gonadal rudiments appear as gonadal ridges, evaginations of the dorsal body wall, one on either side of the medial, dorsal mesentery (route taken by migrating primordial germ cells to reach the gonads) in association with the archinephric (medullary) ducts and extending the entire length of the body cavity (Fig. 3A). By about a year-old developing gonads in all *Neoceratodus* appear to have begun ovarian differentiation. The cortical region has begun to thicken as a result of both somatic and germ cell proliferation and the medulla becomes vacuolated (Fig. 3B-C). This is characteristic of gonadal development for the next 15-20 months, after which testis differentiation appears to have begun in some individuals. Testicular differentiation is characterized by seminiferous tubules developing in the medulla and containing the primordial germ cells, while the cortex is providing only a very narrow (one-cell thick) boundary to the testis (Fig. 4A-C). Only 24 specimens were examined between 1-3 years of age and of these only two were recognized as males, the remaining 22 were all differentiating as ovaries (Raj-Prasad 2000). This strongly suggests that gonadal differentiation in *Neoceratodus* is either strongly skewed in favour of females or, more likely that all gonads begin differentiation first as ovaries and later, in male lungfish, change from ovarian to testicular differentiation. Unfortunately, we have not yet examined sufficient specimens to be sure which of these alternatives is correct. We also have not yet examined the ovaries or testes of immature/sub-adult *Neoceratodus* so there is a leap from the first 3 years to the early mature adults described by Wake (1987). However, we do have unpublished data on wild caught adult lungfish that suggests that male and female lungfish are represented approximately equally in wild populations.

ADRENALS

The lungfish equivalent of adrenal cortical tissue is located within the kidney, associated with the postcardinal veins. Wake (1987) and authors previous to 1987 were unable to locate a medulla equivalent. However Chopin and Bennett (1995) have described chromaffin positive cells within the atrial tissue of the heart of *Neoceratodus*, and given the close proximity of the pronephros to the heart during development, this tissue may be considered the lungfish equivalent of the adrenal medulla. In adult African lungfish, Abrahamsson *et al.* (1979a) also identified

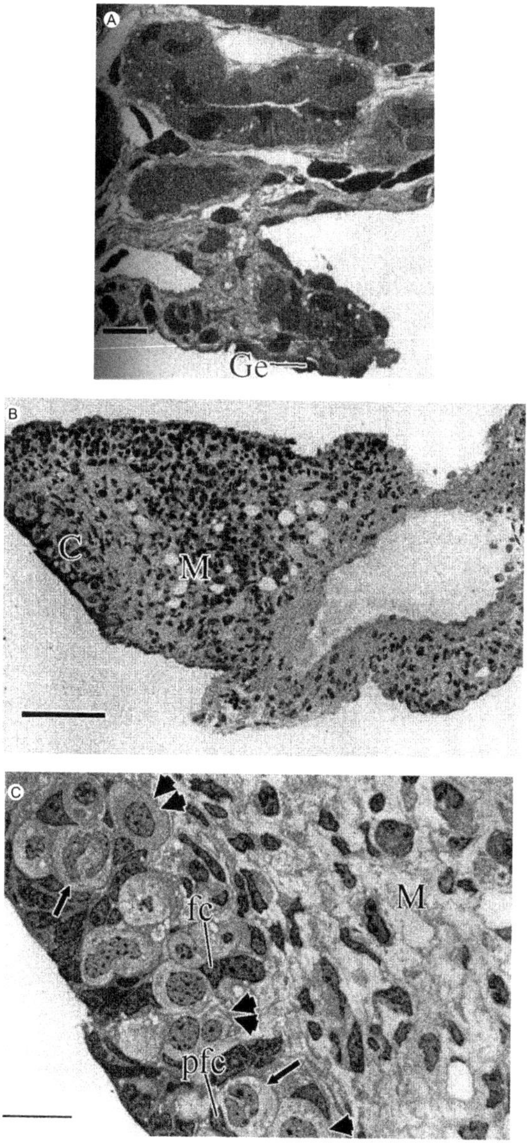

Fig. 3 Semithin araldite sections of developing gonads of *Neoceratodus* stained with methylene blue. **A.** Section of the dorsal body wall in region of the gonadal ridge of a 5-month-old juvenile. Note the germinal epithelium (Ge) surrounding the gonad anlagen. Bar = 30 µm. **B.** Section through developing gonad of a 30-month-old juvenile, differentiating into an ovary with a well-developed cortex (C) and degenerating medulla (M). Bar = 100 µm. **C.** Higher magnification of 3B in region of cortex, showing primary germ cells (arrows), primary oogonia (double arrowheads) and prefollicular (pfc) and follicular (fc) cells. The medulla (M) is to the right. Bar = 20 µm.

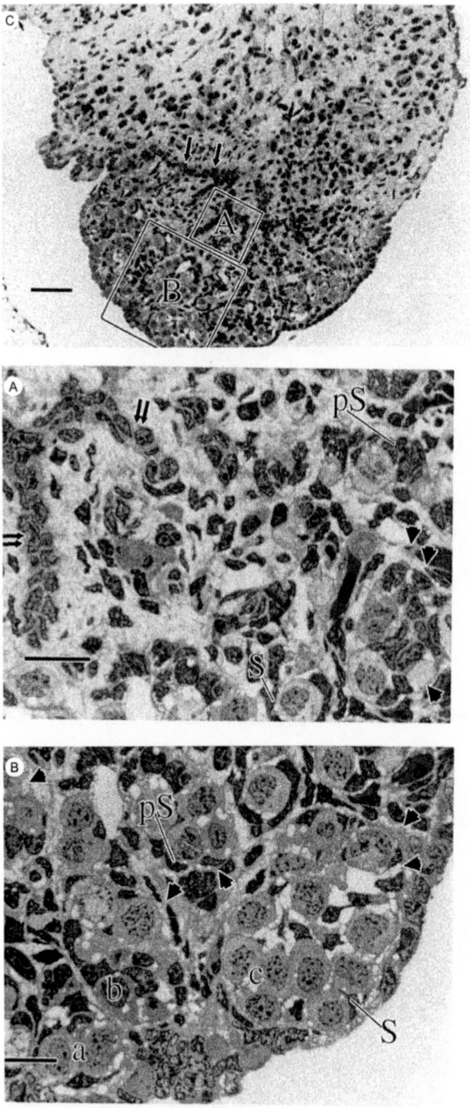

Fig. 4 Semithin araldite sections of developing gonads of *Neoceratodus* stained with methylene blue. **A.** Higher magnification of area marked 'A' in Figure 4C. Germ cells (double arrows) are in contact with pre-Sertoli (pS) and Sertoli (S) cells, and seminiferous tubules (double arrowheads) surrounding them. Bar = 20 µm. **B.** Higher magnification of area marked 'B' in Figure 4C. Seminiferous tubules (arrowheads) are beginning to form. In areas marked 'a' and 'b' they are at an earlier stage than in area marked 'c'. Bar = 20 µm. **C.** Section through developing gonad of a 30-month-old juvenile, differentiating into a testis. The cortex is only one-cell thick and the medulla contains seminiferous tubules (lobules). Bar = 100 µm.

chromaffin tissue in the walls of the atrium, the most anterior part of the left cardinal vein and the intercostal arteries.

SUMMARY

There has been very little recent interest in the urogenital system of lungfish. With the exception of Ojeda *et al's* (2006) description of the microanatomy of the *Protopterus* kidney, Wake's (1987) descriptions remain the most recent anatomical/histological information. This chapter adds some more recent (mostly unpublished) information for *Neoceratodus* only. This new data confirms the status of *Neoceratodus* as the most primitive of the three genera of living lungfish, most particularly because of the close association between the kidney and testis in males. Associations of this type are well known in vertebrates but *Neoceratodus*, as far as I am aware, is the only male vertebrate to utilize all of its kidney tubules for both urine and sperm transfer.

References

Abrahamsson, T., Holmgren, S., Nilsson, S. and Pettersson, K. (1979a). On the chromaffin system of the African lungfish, *Protopterus aethiopicus*. Acta Physiologica Scandanavia 107: 135-139.

Bemis, W.E., Burggren, W.W. and Kemp, N.E. (eds). (1987). The Biology and Evolution of Lungfishes. Alan R. Liss Inc. New York, USA.

Chopin, L.K. and Bennett, M.B. (1995). Cellular ultrastructure and catecholamine histofluorescence of the heart of the Australian lungfish, *Neoceratodus forsteri*. Journal of Morphology 223: 191-201.

Grier, H.J. (1992). Chordate testis: the extracellular matrix hypothesis. Journal of Experimental Zoology 261: 151-160.

Grigg, G.C. (1965). Spawning behaviour in the Queensland lungfish, *Neoceratodus forsteri*. Natural History 15: 75.

Guyton, J.S. (1935). The structure of the nephron in the South American Lungfish, *Lepidosiren paradox*. Anatomical Records 63: 213-229.

Hentschel, H. and Elger, M. (1987). The distal nephron in the kidney of the fishes. Advances in Embryology and Cell Biology 108: 1-151.

Hickman, C.P. Jr. and Trump, B.F. (1969). The Kidney. In: W.S. Hoar and D.J. Randall (eds). Fish Physiology, Volume. 1. Academic Press, New York, USA. pp. 91-239.

Janssens, P.A. and Cohen, P.P. (1966). Ornithine-urea cycle enzymes in the African lungfish, *Protopterus aethiopicus*. Science 152: 358-359.

Janssens, P.A. and Cohen, P.P. (1968a). Nitrogen metabolism in the African lungfish. Comparative Biochemistry and Physiology 24: 879-886.

Janssens, P.A. and Cohen, P.P. (1968b). Biosynthesis of urea in the aestivating African lungfish and in *Xenopus laevis* under conditions of water shortage. Comparative Biochemistry and Physiology 24: 887-898.

Jesperson, A. (1969). On the male urogenital organs of *Neoceratodus forsteri*. Biologiska Skrifter Danske Videnskabernes Selskab 16: 1-11.

Joss, J.M.P., Edwards, A. and Kime, D.E. (1996). *In vitro* biosynthesis of androgens in the Australian lungfish, *Neoceratodus forsteri*. General and Comparative Endocrinology 101: 256-263.

Kerr, J.G. (1901). On the male genito-urinary organs of the *Lepidosiren* and *Protopterus*. Proceedings of the Zoological Society of London 1901: 484-498.

Mommsen, T.P. and Walsh, P.J. (1989). Evolution of urea synthesis in the vertebrates: the piscine connection. Science 243: 72-75.

Ojeda, J.L., Icardo, J.M., Wong, W.P. and Ip, Y.K. (2006). Microanatomy and ultrastructure of the kidney of the African lungfish *Protopterus dolloi*. Anatomical Record 288A: 609-625.

Raj-Prasad, R.A. (2000). Germ cell migration, gonadal differentiation, and pituitary development in the Australian lungfish, *Neoceratodus forsteri*. PhD thesis, Macquarie University, Sydney, Australia.

Sawyer, W.H., Blair-West, J.R., Simpson, P.A. and Sawyer, M.K. (1976). Renal response of Australian lungfish to vasotocin, angiotensin II, and MaCl infusion. American Journal of Physiology 231: 593-602.

Sawyer, W.H., Uchiyama, M. and Pang, P.K.T. (1982). Control of renal functions in lungfishes. Federation Proceedings 41: 2361-2364.

Wake, M.H. (1987). Urinogenital morphology of dipnoans, with comparisons to other fishes and amphibians. In: W.E. Bemis, W.W. Burggren and N.E. Kemp (eds). The Biology and Evolution of Lungfishes. Alan R. Liss Inc., New York, USA. pp. 199-216.

CHAPTER 15

The Lungfish Endocrine System

Jean Joss[*]
Department of Biological Sciences, Macquarie University, Sydney, NSW 2109, Australia

ABSTRACT

This chapter deals with the endocrine system of lungfish under the headings of pituitary, urophysis, hypothalamus and brain peptides, neuropeptides, thyroid, parathyroid, ultimobranchial gland + PTHrp, gonads and reproduction, ovaries, testes, kidneys and adrenal glands, the renin-angiotensin system, gastrointestinal tract and pancreas. Each of the features of the vertebrate endocrine system have been dealt with in turn for lungfish. Since only living lungfish are considered, the three genera – *Neoceratodus, Protopterus* and *Lepidosiren* are indicated for each new piece of information given. The resurgence in interest in lungfish endocrine glands and their hormones over the last 15-20 years has confirmed their current phylogenetic position as sarcopterygian fish, more closely related to tetrapods than they are to other osteichthyans, particularly teleost fish. Interestingly, *Neoceratodus*, considered basal to the other two genera of lungfish, shows some similarities (e.g. morphology of pituitary, sequence of angiotensin II) with the basal actinopterygian fish such as sturgeon and paddlefish, while other hormone homology data, particularly for much larger peptide hormones, frequently finds closer similarity between lungfish and mammalian hormones than between lungfish and other tetrapods. However, the similarity between lungfish large peptide hormones and all tetrapod forms is always greater than between lungfish and teleost forms.

Keywords: hypothalamus, pituitary, thyroid, adrenals, gonads

*E-mail: jjoss@rna.bio.mq.edu.au

INTRODUCTION

At the International Symposium for Comparative Endocrinologists held in Hong Kong in 1981, James Atz urged us to direct our attention to species of particular evolutionary significance that have generally been neglected. Top of his list was the lungfish, *Neoceratodus forsteri*. I took up the challenge and at the following Symposium in 1985, I presented the first account of steroid hormones during the reproductive cycle of *Neoceratodus forsteri* and thus had begun my 'love affair' with lungfish which has involved collaborations with numerous other comparative endocrinologists over the intervening twenty years or so and will be the subject of this chapter.

Neoceratodus has attracted morphological and anatomical study since its discovery in 1870 (Krefft). Most of the endocrine studies to date have also been descriptions of the glands and more recently of the hormones, providing sequence data for many of them. Apart from the pioneering studies on osmoregulation by Pang *et al.* (1971) and Sawyer *et al.* (1976), there has been almost no experimental endocrinology on lungfish. This is not due to lack of interest but rather to their relative inaccessibility and to the difficulties of keeping such large fish in captivity, while avoiding the confounding affects of stress. As my own interest in these amazing fish grew, I took on the challenge of setting up spawning ponds at Macquarie University in Sydney. The success of this venture has made some functional studies more possible.

I will now discuss each of the major endocrine glands and summarise what is known to date about each in turn.

PITUITARY

The pituitary of vertebrates can be subdivided into anterior (adenohypophysis) and posterior (neurohypophysis) lobes. Lying between the two and generally in more intimate association with one or the other is an intermediate lobe. Developmentally, the anterior and intermediate lobes form as an evagination from the dorsal oral epithelium which grows up to meet with the posterior lobe which is forming as an evagination of the infundibular region of the hypothalamus. Within this general pattern there are very clear differences between the anatomy of fish pituitaries and those of tetrapods. Lungfish pituitaries however, more closely resemble those of tetrapods, particularly amphibians, than they do other fish pituitaries (Holmes and Ball 1974). This is more true for the lepidosirenid lungfishes (*Protopterus* and *Lepidosiren*) than it is for the ceratodontid and more ancient lungfish (*Neoceratodus*) the latter sharing more similarities with the pituitaries of some primitive actinopterygian fish, e.g. sturgeon (Joss *et al.* 1990). The regionalisation of cell types in the pars distalis of *Neoceratodus* is more fish-like and less like the tetrapod non-regionalised cells in lepidosirenid lungfishes.

Figure 1 gives an overview of pituitary anatomy in the two families of lungfish, a sturgeon, a teleost, a chondrichthyan, *Latimeria* and an amphibian (adapted from Gorbman *et al.* 1983).

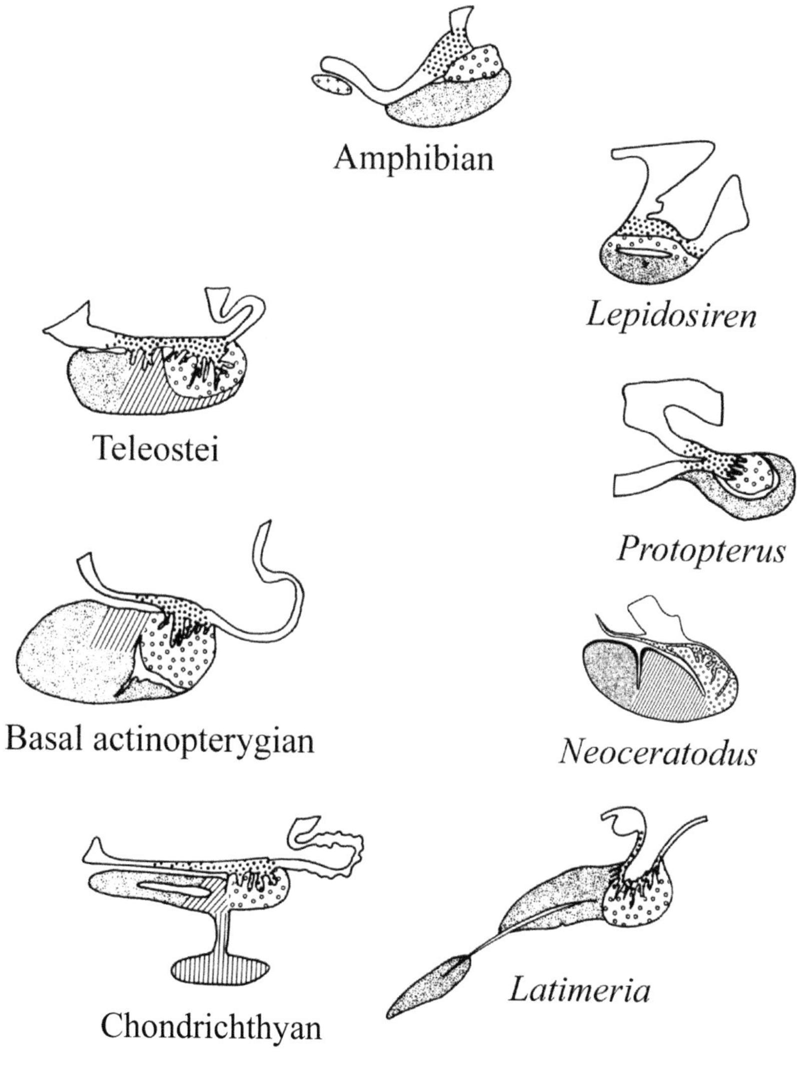

Fig. 1 Diagrammatic representation of vertebrate pituitary glands, showing generalised pituitaries from chondrichthyans, basal actinopterygians, teleosts, coelacanths, and the dipnoans, *Neoceratodus, Polypterus, Lepidosiren* and amphibians. Modified from Gorbman *et al.* (1983) and Joss *et al.* (1990). Fine stippling = rostral pars distalis; cross-hatching = proximal pars distalis; open circles = pars intermedia; coarse dots = pars nervosa.

Several studies have identified the cells of the pars distalis and pars intermedia of both families of lungfish by their cross reactivity with antibodies raised against pituitary hormones found in other vertebrates (Hansen *et al.* 1980; Joss *et al.* 1990; Hansen and Hansen 1994, 1998). These have shown clearly that *Neoceratodus* possesses a neurointermediate lobe, whereas the neuro- and intermediate regions of the pituitary in the lepidosirenid lungfishes are discretely separate. All of the usual pars distalis cell types are present, namely, prolactin cells, somatotropes, corticotropes, gonadotropes and thyrotropes. Melanotropes can be identified in the pars distalis as well as the neurointermediate lobe of *Neoceratodus* (Joss *et al.* 1990) whereas they are only present in the pars intermedia of the lepidosirenid lungfish (Hansen *et al.* 1980).

Lungfish do not appear to have a pars tuberalis but they do have a median eminence which most likely passes hypothalamic controlling hormones to the anterior pituitary. The pars nervosa occupies the central region of the neurointermediate lobe in *Neoceratodus* and in all three is made up of the nerve endings of magnocellular neurons from the hypothalamus (Schreibman 1986; Zambrano and Iturriza 1973).

There has been very little description of the development of lungfish pituitaries. Griffiths (1938) described the morphological changes during development of the *Neoceratodus* pituitary. My group has extended this description of *Neoceratodus* to include the ontogeny of the cell types in the anterior pituitary, identifying these by immunocytochemical and immunogold techniques (Joss *et al.* 1997). At hatching the anterior pituitary has separated from the roof of the mouth and has come to lie beneath the infundibular region of the hypothalamus. It comprises a hollow ball of cells, unidentifiable by antibody techniques. At approximately four weeks post-hatching (stage 48), four cell types can be distinguished by the size and arrangement of their secretory granules. Two of these can be identified by their affinity for antibodies to corticotropin and melanotropin. By stage 52 (~20 weeks), three more cell types can be distinguished, corresponding to prolactin cells, somatotropes and thyrotropes. This sequence of differentiation of anterior pituitary cell types, corticotropes first, followed by prolactin cells and thyrotropes with gonadotropes appearing much later is similar to that described for amphibians (Oota and Saga 1991; Ogawa *et al.* 1995; Miranda *et al.* 1996).

Two octapeptide hormones have been sequenced from the pars nervosa, arginine vasotocin (AVT) and an unusual form of mesotocin (MT), phenytocin in *Neoceratodus* (Hyodo *et al.* 1997). All of the anterior pituitary hormones have also been sequenced for at least one of the species of lungfish; the alpha subunit of the glycoprotein hormones in *Neoceratodus* (Arai *et al.* 1998) and the beta subunits of two gonadotropins (GtHs) and thyrotropin (TSH) in *Neoceratodus* (Querat *et al.* 2004); prolactin (PRL) in *Protopterus* (Noso *et al.* 1993), somatotropin (GH)in *Protopterus* (May *et al.* 1999), the relatively newly discovered somatolactin (SL) in *Protopterus* (May *et al.* 1999); and lastly adrenocorticotropic hormone (ACTH), which belongs to the proopiomelanocortin family (POMC) of hormones, in *Neoceratodus* (Dores *et al.* 1990), and *Protopterus* (Lee *et al.* 1999). POMC is also found with related proenkephalins (McDonald *et al.* 1991; Sollars *et al.* 2000) and prodynorphins (Dores *et al.* 2004) in the hypothalamus. The multiple forms of melanophore stimulating hormones (MSH) of the neurointermediate lobe, together with endorphin have also been characterised by Dores and his group (1988a, b) for *Neoceratodus*. The sequencing of all of these pituitary hormones shows a very clear alignment with tetrapod sequences rather than with other fish sequences with the possible exception of SL, which has not so far been found in any tetrapod (May *et al.* 1999).

Very few functional studies for the pituitary hormones have been conducted. Most of these have concentrated on the neurohypopophysial octapeptides and their role in osmoregulation and homeostasis. In 1982, Sawyer, Uchiyama and Pang reported that AVT is a vasopressor agent in lungfishes (*Protopterus* and *Neoceratodus*) that consistently increases the glomerular filtration rate and urine flow with increased blood pressure. This diuresis may be accompanied by natriuresis also, suggesting that all the effects of AVT on renal tubules of lungfish can be explained by its effect on blood pressure, not requiring any direct effect on the tubules themselves. These observations were later confirmed by Babiker and Rankin (1979) for *Protopterus* and further refined by Pang *et al.* (1983) working with *Lepidosiren*. In this latter publication, they tackled the seeming conundrum of AVT being diuretic in fish and antidiuretic in tetrapods, concluding that either 1) mesotocin, while having almost no effect on renal tubules in lungfishes, in high doses has a marked antidiuretetic effect on kidney tubules in amphibians, appearing to antagonize diuretic effects of vasotocin or 2) AVT receptors on the renal tubules of lungfish and amphibians may have evolved to ensure diuresis in a freshwater environment where it is so important and antidiuresis on land where desiccation is the major problem.

UROPHYSIS

Bern and colleagues (1973) surveyed the presence of a urophysis (caudal neurosecretory organ) in all major fish groups from agnathans through selachians, holocephalons, primitive actinopterygians, teleosts and lungfish and concluded from their study of *Lepidosiren* that there was so little bladder-contracting activity (urotensin II, UII) in all regions of the spinal cord as to make the existence of a urophysis dubious at best. Mathieu and colleagues (1999), using immunocytochemical localising techniques, confirmed the lack of UII in the pituitary and CNS of *Protopterus*. However, they did locate UI, colocalised with α-MSH in the melanotropes of the intermediate lobe of the pituitary. They went on to further characterise the lungfish UI with HPLC and cross-reactivity with mammalian CRH, teleost (white sucker) UI, amphibian sauvagine, and α-MSH, none of which cross-reacted with the lungfish UI, suggesting that it may act as a classic pituitary hormone or it may be involved in the control of melanotrope secretion.

HYPOTHALAMUS AND BRAIN PEPTIDES

In general, hypothalamic regulation of pituitary function in lungfishes appears to be neurovascular as it is in primitive ray-finned fishes, elasmobranchs and tetrapods, rather than neuroglandular as it is in teleost fishes, with the possible exception of *Lepidosiren* (Zambrano and Iturizza 1973). There is a distinct median eminence connected by a well-developed hypothalamo-pituitary portal system. The pars tuberalis, a tetrapod feature, is missing in all living lungfishes (Holmes and Ball 1974). Beyond this, there have been few other detailed anatomical descriptions of the endocrine hypothalamus in lungfish. Moreover, the only hormones that have been studied are proenkephalins and prodynorphins mentioned in the pituitary section, gonadotropin-releasing hormone (GnRH), thyrotropin-releasing hormone (TRH) and the neurohypophysial hormones, vasotocin and mesotocin (phenytocin in *Neoceratodus*) also mentioned above.

Two forms of GnRH have been identified by isolation and recognition using specific antibodies (Joss *et al.* 1994b; King *et al.* 1995). These are the chicken II and mammalian forms. While we have gone on to sequence the lungfish cGnRH II gene in *Neoceratodus*, the mGnRH gene has so far eluded us (Joss *et al.*, unpublished). Vallarino and colleagues (1998a; King *et al.* 1995) have reported on the distribution of cells and nerve fibres immunoreactive to these two forms of GnRH in *Protopterus*. cGnRH II and mGnRH cell bodies were located in quite separate locations within the preoptic nucleus and the ganglion of the terminal nerve. mGnRH fibres terminated in proximity to the median eminence. They were less abundant than the cGnRH II fibres which innervated the neural lobe of the pituitary, the telencephalon and the mesencephalon, suggesting that the chicken II form, as in other vertebrates, is most likely acting as a neuromodulator/transmitter, while the mammalian form is the releasing hormone stimulating the gonadotropes in the anterior pituitary. Further confirming this function for mGnRH is the finding that mGnRH administered to mature lungfish (*Neoceratodus*) via cardiac puncture in the low dose of 1.5 μg/kg, resulted in elevation of the circulating sex steroid hormones, testosterone in males and estradiol in females within 2 min (Figure 2). These steroid hormones returned to normal levels over the next hour, which suggests a very rapid response to mGnRH from pituitary gonadotropes, and rapid clearance (or degradation) of mGnRH , followed by a rapid response of the gonadal steroidogenic cells to the surge of gonadotropin (Joss *et al.* 1994b).

In 1973, Gorbman and Hyder reported the failure of TRH to stimulate thyroid function in the lungfish. Since then it has been clearly shown that many non-mammalian vertebrates use another releasing factor to stimulate their thyrotropes

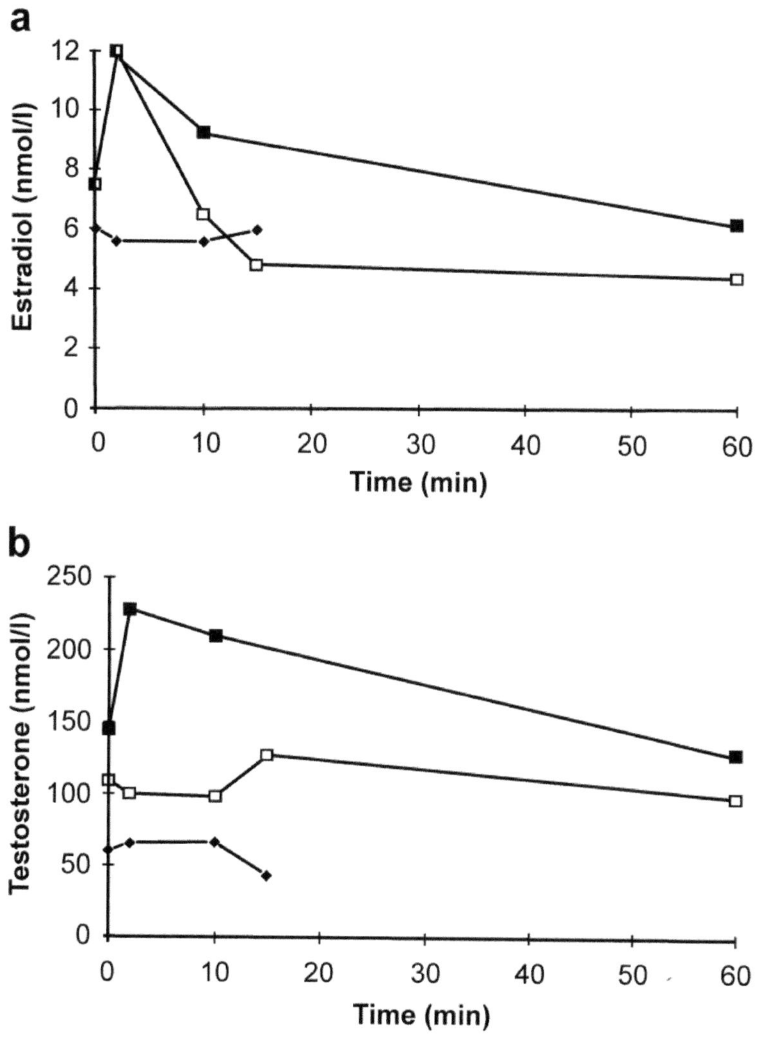

Fig. 2(a). Changes in serum estradiol-17β in female lungfish in response to a single cardiac injection of 1.5 µl/kg mGnRH administered in early spring (■), late spring (□) and early autumn (♦). Values represent means from two lungfish from each of the sampling times. **2(b).** Changes in serum testosterone in male lungfish in response to a single cardiac injection of 1.5 µl/kg mGnRH administered in early spring (3 fish) (■), late spring (4 fish) (□), and early autumn (3 fish) (♦). Adapted from Joss *et al.* (1994).

to release TSH, e.g. amphibians use corticotropin releasing factor/hormone (CRF/H) (Denver 1997, 2004). Although it has not been examined, it is very likely that CRH is also the releasing hormone for TSH in lungfish.

NEUROPEPTIDES

Dores and his co-workers have been exploring the evolutionary implications of the proopiomelanocortin/enkephalin/dynorphin family of peptides in the pituitary and brain of lungfish. The POMC family has been discussed above with the pituitary, the remaining two will be discussed here. The mammalian opioids, Met-enkephalin and Leu-enkephalin are derived from a common precursor. Wherever these two opioids have been immunolocated in non-mammalian brains, it has been assumed that they are encoded for in the same manner as in mammals. However, Sollars *et al.* (2000) showed quite emphatically that Met-enkephalin precursor in the brain of *Neoceratodus* only produces Met-enkephalins, five copies and two extra C-terminally extended forms of met-enkephalin. Distinct Met- and Leu-enkephalin-positive neurons had been detected in the CNS of *Protopterus* (Vallarino *et al.* 1998b). Sollars *et al.* (2000) went on to show that a full length prodynorphin cDNA could be cloned from the CNS of *Neoceratodus*. In addition to coding for α-neoendorphin, dynorphin A and dynorphin B, it also included coding for two Leu-enkephalins. A similar study with *Protopterus* found prodynorphin cDNA encoding for a Leu-enkephalin and a novel YGGFF sequence (Dores *et al.* 2004). These results clearly accounted for the separate location of Met- and Leu-enkephalin-positive neurons in *Protopterus* CNS (Vallarino *et al.* 1998b).

Vallarino and his co-workers have examined the brain of *Protopterus* for immunolocation of several other neuropeptides beyond the GnRHs and enkephalins mentioned above (Vallarino *et al.* 1998a):

Neuropeptide tyrosine (NPY) belongs to a family of peptides structurally related to and also including polypeptide YY and pancreatic polypeptide. NPY has been shown to have several functions in mammalian brains: stimulation of food intake, inhibition of sexual behaviour, control of cardiovascular function and regulation of neuroendocrine secretion (Vallarino *et al.* 1998a). Lungfish NPY was found to closely resemble frog NPY both in form and location sites in the brain. Like GnRH, it was found in the preoptic nucleus of the hypothalamus and also in various nuclei in the telencephalon along the ventral wall of the infundibulum, and the medial caudal tegmentum. Cell bodies of the terminal nerve were also strongly immunostained. NPY nerve fibres were most densely located in the medial and lateral subpallium, the hypothalamus and the tegmentum. Positive fibres were also found in the median eminence and the neural lobe of the pituitary suggesting that NPY in lungfish may be involved in neuroendocrine secretion as it is in mammals (Vallarino *et al.* 1998a).

α-MSH is formed as a cleavage product from the much larger precursor molecule, POMC. The POMC gene is expressed in pituitary corticotropes and melanotropes as well as in discrete populations of cells in the brain. In *Protopterus* brain, α-MSH immunoreactivity was located in the preoptic nucleus and caudal hypothalamus, as it is in amphibians (Vallarino *et al.* 1992).

The last neuropeptide in the lungfish brain to be mentioned is atrial natriuretic peptide (ANP) which is a member of the natriuretic peptides found in the heart and brain of mammals and other vertebrates known for their natriuretic, diuretic and hypotensive actions. The immunolocalisation of ANP in the lungfish brain closely resembles that found in amphibian brains and suggests that *Protopterus* ANP-related peptide may be involved in regulation of osmotic and behavioural adaptation, particularly during estivation, processing of olfactory information, regulation of pituitary peptide secretion, and modulation of melanotropin secretion, as previously demonstrated in amphibians (Vallarino *et al.* 1996). ANP has also been investigated in the heart of *Protopterus* (Larsen *et al.* 1994) who found it immunolocated to all areas of the heart – both atrial and ventricular myocardium. The intensity of secretory granules was greatest in the atria, suggesting an endocrine function for the myocardium in *Protopterus*.

The close similarities between *Protopterus* and amphibian neuropeptides of themselves are not strong evidence of a lungfish/amphibian clade to the exclusion of all other bony fish. Similar studies need to be performed on more basal actinopterygians to either confirm or rule out the possibility of the shared neuropeptide patterns between lungfish and amphibians being an ancestral character state.

THYROID

The thyroid gland of *Neoceratodus* is an unpaired gland, situated medially, just behind the tongue in the lower jaw and encapsulated by a thin layer of connective tissue, similar to chondrichthyan thyroid glands (Chavin 1976). The thyroid develops in the lungfish embryo as a thickening of the midventral pharyngeal endoderm. At hatching, it is barely apparent and several weeks later (stage 49) comprises 2-3 follicles of 9-12 cells each (Joss *et al.* 1997). Numbers of follicles increase slowly and quite variably between individuals (Figure 3), until the adult structure is achieved 15-20 years later.

Function of the thyroid axis has been shown to resemble that of larval amphibians. Circulating levels of T4 (thyroxin) and T3 (tri-iodothyronine) are very low to undetectable by most thyroid hormone assays (< 10 pg/ml, Joss, unpublished data). Deiodinating enzymes in the peripheral tissues are of the D2 and D3 type only, the latter being more active than the former throughout life. The D1 type, which converts T4 to T3 in the liver of mammals and most other vertebrates for release into the general circulation, has not been found by either molecular or physiological methods in *Neoceratodus* (Sutija and Joss 2005). Interestingly, this is also true for the few amphibians that have been studied (Becker *et al.* 1997). Feedback between the pituitary (and/or hypothalamus) and circulating levels of thyroid hormones could not be demonstrated (Joss *et al.* 1997). Taken together, these results from studying the function of the thyroid axis in *Neoceratodus* were

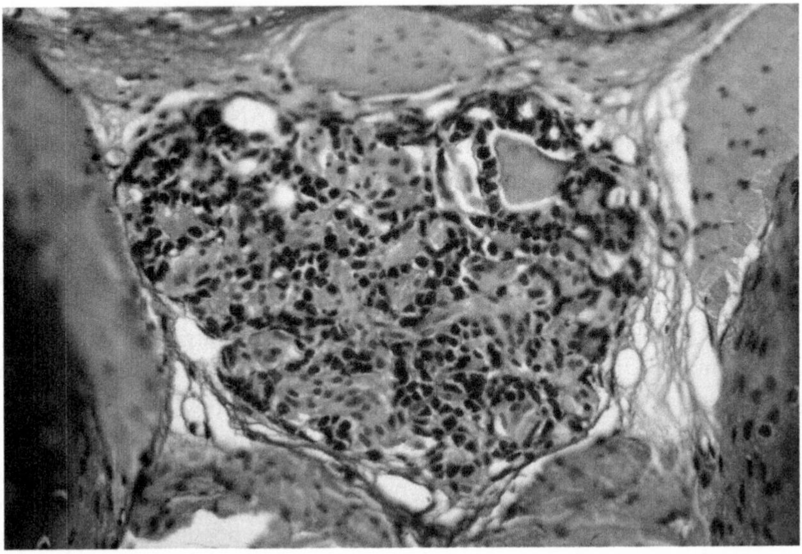

Fig. 3 Transverse section through the lower jaw of a juvenile lungfish showing follicles of the thyroid gland. The histological appearance is one of an inactive thyroid gland, with cuboidal follicle cells and colloid in the lumina of the follicles. Staining haematoxylin and eosin. Magnification x 20.

taken as a clear indication that it is a larval axis, even in reproductively mature lungfish. A strong case for non-inducible obligate neoteny in *Neoceratodus* and more than likely for all three living genera of lungfish has been made (Joss 2006).

A role for thyroid hormones in reawakening following aestivation in *Protopterus* spp. has been suggested (Fishman *et al.* 1987) but requires confirmation. Also for *Protopterus*, thyroxine (0.5 ml/fish, im. injection) has been shown to increase oxygen consumption for approximately one week (Mohsen and Godet 1960).

PARATHYROID, ULTIMOBRANCHIAL GLAND + PTHrp

A discrete parathyroid gland is not present in lungfish, or in any other fish. However, parathyroid hormone (PTH) has recently been found in several species of fish, including lungfish (Okabe and Graham 2004). Interestingly, its source was located in the gills, prompting the suggestion that the tetrapod parathyroid gland may have had its evolutionary forerunner in the gills of fish as might have been expected since both are products of the developing pharynx. The related peptide, PTH-rp has been described for most fish groups, including lungfish (Trivett *et al.* 1999) and is generally believed to be the more ancient of the two peptides. The distribution of PTH-rp is quite wide in the vertebrates studied, whereas PTH is much more discretely confined to its own gland, often associated with the thyroid in amniote vertebrates.

The ultimobranchial gland in all three genera of lungfish is located in the septum separating the peritoneal cavity from the pericardium. It does not appear to have aroused any recent interest. Pang and colleagues (1971), Pang and Sawyer (1975) demonstrated hypocalcaemic activity in the ultimobranchial gland of *Lepidosiren* but found no calcaemic activity in response to exogenously delivered PTH or calcitonin in the same species. They did find that PTH produced a distinct antidiuresis (Pang and Sawyer 1975). Calcium regulation in lungfish is an area in real need of further study, limiting as calcium must be in such obligate freshwater species as lungfish are today.

GONADS AND REPRODUCTION

The gonads of all vertebrates serve the dual functions of housing the gametes and producing several hormones of reproduction. The gonads of sarcopterygian fishes are no exception. However, while there are excellent descriptions of gonadal morphology and histology for all three genera of lungfish, there is much less known about their endocrinology. Wake's (1987) description of the gonads and associated kidneys and adrenals of lungfish provides the best overall description to date, even though it is now in excess of twenty years old. I refer readers to Wake's (1987) paper for lungfish gonadal morphology and will concentrate in this chapter on what is now known about gonadal endocrinology.

Lungfish are late maturers. *Neoceratodus* reaches adulthood at ~20 years of age, the males somewhat sooner than the females as judged by the relative size of *Neoceratodus* in their first year of spawning: males 3.5-6.0 kg; females: 5.5-6.0 kg (Joss, unpublished data).

OVARIES

In *Neoceratodus,* oocyte recrudescence begins in late summer and vitellogenesis progresses over winter and into spring. The exact timing of these events appears to be subject to age (size) of the fish, the larger the fish, the earlier, seasonal vitellogenesis begins. Lungfish ovarian follicles, wherein the oocytes grow, are surrounded by the usual inner granulosa layer, and outer thecal layers. Histochemical demonstration of the activity of the enzyme, 3βHSD, as an indication of steroidogenesis was negative for the granulosa layer and strongly positive for the inner thecal layer (Joss, unpublished data), suggesting that most steroidogenesis takes place in the thecal cells and the granulosa is primarily where androgens passed from the thecal cells to the granulosa are aromatized to estrogens. Circulating levels of estrogens (estradiol-17β) and progestogens (progesterone) are quite high and vary throughout the year; estrogen being highest (50 nmol/l) in late autumn

(May in Australia) when vitellogenesis is getting underway, lower (12 nmol/l) in spring when oviposition/spawning is occurring and lowest (5 nmol/l) in summer. Progesterone is highest (50 nmol/l) in female fish containing ripe eggs in their ovaries in spring (Joss, unpublished data). These data suggest that lungfish, like most other non-mammalian vertebrates, employ estrogen to stimulate production of yolk proteins by the liver and their transport in the blood to the ovary where they are sequestered by the growing oocytes during the process of vitellogenesis. Moreover, the high circulating level of progesterone found in spring may be promoting ovulation as it does in some amphibians (Follett and Redshaw 1974).

TESTES

In the testes of *Neoceratodus*, spermatogenesis is active during autumn months and mature sperm are stored in the tubules of the testes over winter and released into the kidney tubules in spring ready for spermiation and fertilization of the ova in late winter/early spring. The testes produce and release large quantities of the androgen, testosterone seasonally: fluctuations are highest (150 nmol/l) in autumn and spring and lowest (10 nmol/l) in summer (Joss *et al.* 1996). We went on to suggest that solubilised forms of androgens (androstenedione and testosterone glucuronide) produced by the testes in spring and not at other times of the year, could be released from mature male lungfish presumably to act as pheromones influencing the behaviour of both sexes prior to and during spawning.

The above studies on lungfish gonadal hormones and pheromones is still in its infancy and urgently requires more attention as *Neoceratodus* is becoming increasingly threatened by man-made activities in its very restricted natural habitat.

KIDNEYS AND ADRENAL GLANDS

The intimate relationship developed in many male anamniote vertebrates between the testes and the kidneys is most complete in male *Neoceratodus*. During the spawning season every tubule along the entire length of the kidney contains mature sperm (Wake 1987; Joss *et al.* 1996) in the process of being conveyed to the outside to fertilise ova released from their accompanying female. *Neoceratodus* is the only vertebrate of which I am aware that does not retain at least a portion of the kidney for conveying urine only. In *Lepidosiren* and *Protopterus*, only the posterior portion of the kidney is utilised for sperm transport from the testis to the archinephric duct of the kidney (Wake 1987), leaving the rest of the kidney tubules for transport of urine alone.

The two components of the mammalian adrenal gland, the cortex and the medulla, are not present as a compound gland in lungfish. The medulla, which in mammals differentiates in the embryo from neural crest cells, contains extensive autonomic innervation and secretes catecholamines. The equivalent of the medulla in lungfish (also known as chromaffin tissue because of its chromium salt staining reaction to catecholamines) is associated with the embryonic pronephros (head kidney) on either side of the heart which comes to lie in the walls of the atrium in adult lungfish. In adult African lungfish, Abrahamsson *et al.* (1979a) identified chromaffin tissue in the walls of the atrium, the most anterior part of the left cardinal vein and the intercostal arteries. In this distribution, lungfish more closely resemble the chromaffin tissue location of embryonic amniote vertebrates than they do of adults in which the chromaffin cells have migrated from the heart to be co-located with the adrenal cortex (Scheuermann 1993).

There have been several studies looking at the role of adrenaline (epinephrine) in carbohydrate metabolism in lungfish. Adrenaline is the dominant catecholamine secreted by the adrenal medulla in mammals and presumably all other vertebrates but studies of the catecholamine hormones of the mammalian adrenal medullary homologues in other vertebrates are rather sparse. Early studies using specific staining reactions identified dopamine (Scheuermann *et al.* 1981) but in 1993, Scheuermann had revised this to these cells having the ability to synthesise all three common catecholamines, dopamine, noradrenaline and adrenaline. Hanke and Janssens (1983) found that adrenaline stimulated glucose release from lungfish liver *in vitro*. They found that this was entirely due to glycogenolysis as adrenaline had no effect on gluconeogenesis. In a later study they confirmed that this action of adrenaline was mediated through β-adrenergic receptors, most likely β_2 type, on the hepatocytes. They suggest that the similarity between this and several amphibians (Janssens and Grigg 1988) compared to the α-receptor mediated response to adrenaline by mammalian hepatocytes, implies that α-adrenergic receptors may have evolved later as has the response of gluconoegenesis production of glucose.

The homologue of the mammalian adrenal cortex has attracted more attention than the medullary homologue, particularly for its role in osmoregulation. In *Neoceratodus,* the adrenocortical tissue is found within the kidneys (Figure 4) in close association with the post-cardinal veins as they traverse the length of each kidney in the outer cortical region which also contains the glomerulae (Call and Janssens 1975). In the lepidosirenid lungfish the adrenocortical cells are found as small cords located between renal and perirenal tissue, also adjacent to the post-cardinal veins (Janssens *et al.* 1965). This anatomical position more closely resembles that of urodele amphibians such as *Pleurodeles waltlii* (Certain 1961) than it does other fish groups.

The hormones secreted by adrenal cortical homologues are steroid hormones: corticosteroids and mineralocorticoids. Aldosterone, cortisol, corticosterone (and/or 11-deoxycorticosterone in *Neoceratodus*) circulate in the blood of lungfishes (Blair-West *et al.* 1977; Joss *et al.* 1994a; Norris 1996). Corticosterone levels are higher than cortisol levels, which confirms their phylogenetic position as aquatic

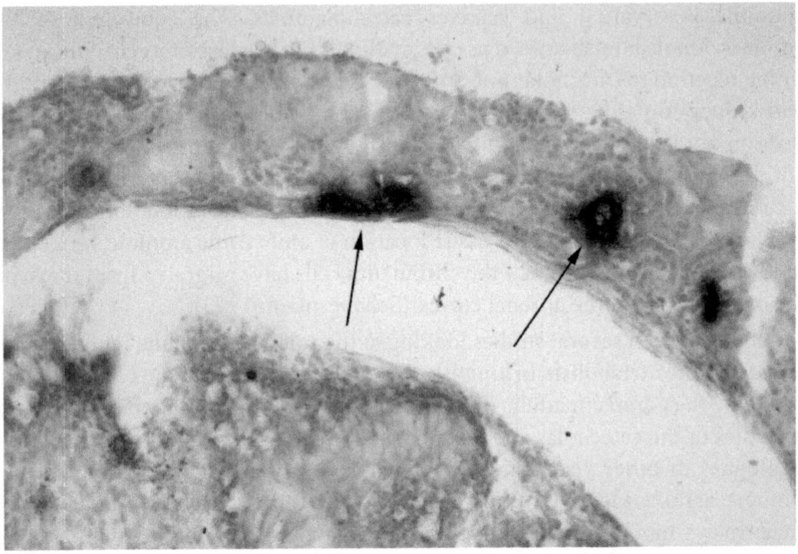

Fig. 4 Histochemical detection of the steroidogenic enzyme, 3β-steroid dehydrogenase, indicating adrenal cortical tissue (arrows) in the outer capsule of the kidney in an adult *Neoceratodus* (Joss, unpublished). Magnification × 20.

intermediates between the ray-finned fishes which secrete predominantly cortisol and the tetrapods which secrete corticosterone. The secretion of aldosterone is interesting, particularly as it appears to be stimulated by the renin-angiotensin system in *Neoceratodus* (Joss *et al.* 1994a) in a similar manner to that found in tetrapods (more in next section).

THE RENIN-ANGIOTENSIN SYSTEM

An intact renin-angiotensin system (RAS) has been demonstrated for all classes of gnathostome vertebrates. Within this system, angiotensin I is cleaved from angiotensinogen in the blood, by renin, an enzyme released from the kidneys in response to lowered blood pressure or sodium depletion. Not surprisingly, juxtaglomerular cells containing renin occur in lungfishes as they do in teleosts and tetrapods.

Angiotensin I produced by *Neoceratodus* has been shown to be the same sequence for the first 8 amino acid residues (angiotensin II sequence) as is found in teleosts and not one of the typical tetrapod or mammalian octapeptides (Joss *et al.* 1999). However, lungfish/teleost angiotensin II (octapeptide) behaves similarly to tetrapod angiotensins in that it stimulates aldosterone release from the adrenocortical cells in the kidneys of *Neoceratodus in vitro* while also increasing arterial blood pressure in a dose-dependent manner *in vivo* (Joss *et al.* 1999). The release of aldosterone in response to angiotensin II, to the exclusion of cortisol and/or corticosterone, is a typical tetrapod response, even though the form of angiotensin II is that typical of teleosts. Interestingly, the receptors in the adrenal cortex for angiotensin II in lungfish appear to be highly specific, as a release of aldosterone in response to tetrapod angiotensin II in *Neoceratodus* was not observed by Blair-West and colleagues (1977). Joss and colleagues (1999) found similar specificity in the blood vascular system of *Neoceratodus*, whereby only the lungfish/teleost form could elicit a strong vasopressor effect.

Masini and colleagues (1995) reported on a local RAS system in the heart of *Protopterus*.

GASTROINTESTINAL TRACT

The gastrointestinal tract of lungfishes is a wide straight organ occupying a large part of the abdominal cavity in immature lungfish. It contains a well-developed spiral valve. In *Neoceratodus* the spiralling begins far anteriorly as a deep groove on the right side of the foregut, just behind the glottis and may be clearly divided into pre-pyloric and post-pyloric regions. The prepyloric spiral valve is more poorly developed in the other two genera of lungfish which more closely resemble the spiral valves of other primitive fishes that all begin post-pylorically, i.e. following a distinct gastric or stomach region (Rafn and Wingstrand 1981).

There have been relatively few studies concerning endocrine regulation of gastrointestinal physiology in non-mammalian vertebrates in general and almost none in lungfish. An as yet unpublished (Van Noorden and Joss) immunocytochemical survey of several bioactive peptides/neuropeptides in the gut of *Neoceratodus* and *Protopterus* has found the two genera were similar in their immunoreactivities and distribution. Peptide immunoreactive endocrine cells and nerves were present in the stomach (pre-pyloris) and intestine. These cells were small and widely dispersed. They were mostly of the 'open' type extending from the basement membrane to the lumen with the immunoreactive material concentrated around the nucleus. Some bombesin-immunoreactive cells in the stomach were of the 'closed' type, having no extension to reach the lumen. In the stomach, endocrine cells were infrequent and positive for gastrin/cholecystokinin (CCK), bombesin and somatostatin 14. In the bursa entiana, endocrine cells were present along the sides of the folds but were concentrated at the tips in closer contact with the lumen. The predominant peptide immunoreactivity was to neuropeptide Y (NPY) and peptide YY (PYY), co-localised in the same cells. There were also less frequent cells immunoreactive for gastrin/CCK, bombesin, substance P (SP) and somatostatin. The spiral valve resembled the bursa in type of immunoreactive cells but frequency was greater. In the pyloris, gastrin/CCK, bombesin and somatostatin cells were present in greater numbers than in the pre-pyloric stomach. The bombesin cells were of the closed type, while very fine gastrin/CCK nerve fibres ran close to the base of the mucosal epithelium. Galanin-immunoreactive nerves are also present along the base of the mucosa and penetrating it, appearing to end on intra-epithelial capillaries. Vasoactive intestinal peptide (VIP) – immunoreactive nerves occur in the deeper muscle of the gut wall. The general conclusion from this study was that the gut endocrine system of lungfishes is less active than in other vertebrate species studied. The endocrine cells are both small and sparse and represent less types (though this could have been due to the antisera used).

For just one of the above peptides, SP, there is some data on function (Liu *et al.* 2002). SP has been isolated from the gut of *Neoceratodus* and sequenced. The primary structure of this peptide is Lys-Pro-Arg-Pro-Asp-Glu-Phe-Tyr-Gly-Leu-Met.NH_2, which shows 64% identity with mammalian SP. In isolated preparations of lungfish foregut circular muscle, lungfish SP produced a slow, long-lasting tonic contraction. Midgut preparations responded to SP more rapidly and more complexly, demonstrating that lungfish SP may be a physiologically important regulator of gut motility.

PANCREAS

The pancreas of lungfishes is hidden within the coils of their spiral valve intestine. In *Neoceratodus*, the pancreas is exclusively contained within the pre-pyloric spiral valve in contact with the gastric epithelium (Rafn and Wingstrand 1981 and see also Chapter on Digestive System in this volume), whereas in *Lepidosiren* and *Protopterus*, the bulk of the pancreas is more posteriorly located in the anterior region of the postpyloric spiral valve (Brinn 1973). The endocrine islet tissues of *Neoceratodus* look very much like those of the majority of tetrapods, being numerous and scattered throughout the exocrine tissue but having maximum frequency anteriorly. The islets of *Lepidosiren* and *Protopterus* are very large, surrounded by a thick capsule of connective tissue and concentrated into a small area of the posterior pancreas near the large blood vessels and excretory ducts. In this they more closely resemble the "Brockmann bodies" of actinopterygian fish (Epple and Brinn 1987).

Insulin-, glucagon- and somatostatin-like hormone-secreting cells have been identified immunocytochemically in the pancreas of *Neoceratodus* (Hansen *et al.* 1987). The insulin (B) cells are the most abundant and are all located within the islets of the pancreas. The glucagon (A) cells are far fewer, small groups on the periphery of the islets or distributed without the islets amongst the exocrine zymogen cells. The somatostatin (D) cells are even less than the glucagon cells, also found in the periphery of the islets. The pancreatic 'islets' of *Protopterus* have also been shown histologically to contain the three cell types found in *Neoceratodus* pancreatic islets (Gabe 1969; Brinn 1973). The fourth hormone (pancreatic polypeptide) has also been found in *Protopterus* (Scheuermann *et al.* 1991) by immunocytochemical techniques, which showed it to be co-localised with glucagon. Pancreatic polypeptide was not found by Van Noorden and Joss (unpublished), in their study of *Neoceratodus*.

There has been no functional study of the endocrine pancreas in any lungfish but the main islet hormone, insulin, has been sequenced for both *Protopterus* (Conlon *et al.* 1997) and *Neoceratodus* (Conlon *et al.* 1999). In both these genera, insulin is composed of two chains as it is in all other vertebrates sequenced. However, the sequence homology of *Neoceratodus* insulin more closely resembles human insulin and the insulins from phylogenetically older actinopterygians such as paddlefish and bichir than it does insulins of amphibians. Insulin from *Protopterus* differs from that of *Neoceratodus* in 13 positions, while both species possess extensions to the C-terminus of the A-chain and the N-terminus of the B-chain not found in tetrapod insulins, confirming that they form a monophyletic group and suggesting that the differences between the two genera have most likely occurred in the *Protopterus* lineage following an ancient divergence (Conlon *et al.* 1999).

SUMMARY

Study of the endocrine system in lungfish began with the location of each discrete gland and a description of its anatomy and histology for comparison with the same glands in other fish and tetrapods. There followed a brief and highly selective phase of investigating the function of the gland and its hormone(s). As the phylogenetic position of lungfish began to be more and more accepted as the closest living fish to the ancestor of the tetrapods and the molecular techniques for sequencing the hormones produced by these glands became available, so it moved into the new phase of lungfish hormone homologies, the details of which have not been recorded herein but rather the references to these details have been provided for the interested reader. What next? Is it time to return to functional studies? I fear that most comparative physiologists/endocrinologists have moved on to using their molecular techniques to look more closely at the biochemistry of hormone/transporter and hormone/receptor interactions and these using the usual model species. This is a great shame as our evolutionary colleagues are turning more and more away from model species to tackle the often more difficult phylogenetically key species, such as lungfish. The lungfish colony at Macquarie University has shown that it is possible to raise sufficient lungfish to support rewarding physiological studies, which must shed more light on the fish-tetrapod transition.

Acknowledgements

I would like to thank all my colleagues and my students, too numerous to name individually, who have contributed to the many endocrine studies referred to in this chapter. I would also like to thank Li Kershaw for the wonderful lungfish bibliography she has compiled over the last ten years, my husband, Greg, and my postdoctoral fellow, Rolf Ericsson, for assistance with the final production. Much of the work attributed to *Neoceratodus* has been funded by research grants from Macquarie University and the Australian Research Council.

References

Abrahamsson, T., Holmgren, S., Nilsson, S. and Pettersson K. (1979a). On the chromaffin system of the African lungfish, *Protopterus aethiopicus*. Acta Physiologica Scandinavica 107: 135-139.

Abrahamsson, T., Holmgren, S., Nilsson, S. and Pettersson, K. (1979b). Adrenergic and cholinergic effects on the heart, the lung and the spleen of the African lungfish. Acta Physiologica Scandinavica 107: 141-147.

Arai, Y., Kubokawa, K., Ishii, S. and Joss, J.M.P. (1998). Cloning of cDNA encoding the common alpha subunit precursor molecule of pituitary glycoprotein hormones in the Australian lungfish, *Neoceratodus forsteri*. General and Comparative Endocrinology 110: 109-117.

Babiker, M.M. and Rankin, J.C. (1979). Renal and vascular effects of neurohypophysial hormones in the African lungfish *Protopterus annectens* (Owen). General and Comparative Endocrinology 37: 26-34.

Becker, K.B., Stephens, K.C., Davey, J.C., Schneider, M.J. and Galton, V.A. (1997). The type 2 and type 3 iodothyronine deiodinases play important roles in coordinating development in *Rana catesbiana* tadpoles. Endocrinology 138: 2989-2997.

Bern, H.A., Gunther, R., Johnson, D.W. and Nishioka, R.S. (1973). Occurrence of urotensin II (bladder-contracting activity) in the caudal spinal cord of anamniote vertebrates. Acta Zoologica 54: 15-19.

Blair-West, J.R., Coghlan, J.P., Denton, D.A., Gibson, A.P., Oddie, C.J., Sawyer, W.H. and Scoggins, B.A. (1977). Plasma renin activity and blood corticosteroids in the Australian lungfish *Neoceratodus forsteri*. J. Endocrinology 74: 137-142.

Brinn, J. (1973). The pancreatic islets of bony fishes. American Zoologist 13: 653-665.

Call, R.N. and Janssens, P.A. (1975). Histochemical studies of the adrenocortical homologue in the kidney of the Australian lungfish, *Neoceratodus forsteri*. Cell and Tissue Research 156: 533-538.

Certain, P. (1961). Organogenèse des formations iinterrénales chez la Batracian urodèle *Pleurodeles waltlii* Michah. Bulletin biologique de la France et de la Belgique 95: 134-148.

Chavin, W. (1976). The thyroid of the sarcopterygian fishes (Dipnoi and Crossopterygii) and the origin of the tetrapod thyroid. General and Comparative Endocrinology 30: 142-155.

Conlon, J.M., Platz, J.E., Nielsen, P.F., Vaudry, H. and Vallarino, M. (1997). Primary structure of insulin from the African lungfish, *Protopterus annectens*. General and Comparative Endocrinology 107: 421-427.

Conlon, J.M., Basir, Y. and Joss, J.M.P. (1999). Purification and characterisation of insulin from the Australian lungfish, *Neoceratodus forsteri* (Dipnoi). General and Comparative Endocrinology 116: 1-9.

Denver, R.J. (1997). Environmental stress as a developmental cue: corticotrophin-releasing hormone is a proximate mediator of adaptive phenotypic plasticity in amphibian metamorphosis. Hormones and Behavior 31: 169-179.

Denver, R.J. (2004). Corticotropin-releasing hormone. In: L. Martini (ed). Encyclopaedia of Endocrinology and Endocrine Diseases, Academic Press, San Diego, California, USA. pp. 580-583.

Dores, R.M. and Joss, J.M.P. (1988a). Immunological evidence for multiple forms of a-melanotropin (a-MSH) in the pars intermedia of the Australian lungfish, *Neoceratodus forsteri*. General and Comparative Endocrinology 71: 468-474.

Dores, R.M., Stevenson, T.C. and Joss, J.M.P. (1988b). The isolation of multiple forms of b-endorphin from the intermediate pituitary of the Australian lungfish, *Neoceratodus forsteri*. Peptides 9: 801-808.

Dores, R.M., Adamczyk, D.L. and Joss, J.M.P. (1990). Analysis of ACTH-related and CLIP-related peptides partially purified from the pituitary of the Australian lungfish, *Neoceratodus forsteri*. General and Comparative Endocrinology 79: 64-73.

Dores, R.M., Sollars, C., Lecaude, S., Lee, J., Danielson, P., Alrubaian, J., Lihrman, I., Joss, J.M.P. and Vaudry, H. (2004). Cloning of prodynorphin cDNAs from the brain of Australian lungfish: implications for the evolution of the prodynorphin gene. Neuroendocrinology 79: 185-196.

Epple, A. and Brinn, J.E. (1987). The Comparative Physiology of Pancreatic Islets, In: D.S. Farner (ed). Zoophysiology Series. Volume 21, Springer-Verlag, New York, USA.

Fishman, A.P., Pack, A.I., Delaney, R. and Galante, R.J. (1987). Estivation in *Protopterus*. In: W.E. Bemis, W.W. Burggren and N.E. Kemp (eds). The Biology and Evolution of Lungfishes. Alan R. Liss Inc., New York, USA. pp. 237-248.

Follett, B.K. and Redshaw, M.R. (1974). The physiology of vitellogenesis. In: B. Lofts (ed). The Physiology of the Amphibia, Volume II. Academic Press, New York and London. pp. 219-308.

Gabe, M. (1969). Données Histologiques sur le pancreas endocrine de *Protopterus annectens* Owen. Archives d'anatomie microscopique et de morphologie expérimentale 58: 21-40.

Gorbman, A. and Hyder, M. (1973). Failure of mammalian TRH to stimulate thyroid function in the lungfish. General and Comparative Endocrinology 20: 588-589.

Gorbman, A., Dickhoff, W.W., Vigna, S.R., Clark, N.B. and Ralph, C.L. (1983). Comparative Endocrinology. Wiley, New York, USA.

Griffiths, M. (1938). Observations on the pituitary in Dipnoi and speculations concerning the evolution of the pituitary. Proceedings of the Linnean Society of New South Wales. 63: 89-94.

Hanke, W. and Janssens, P.A. (1983). The role of hormones in regulation of carbohydrate metabolism in the Australian lungfish, *Neoceratodus forsteri*. General and Comparative Endocrinology 51: 364-369.

Hansen, G.N. and Hansen, B.L. (1994). Immunocytochemical localization and characterization of mammalian prolactin- and somatotropin-like material in the pituitary of the Australian lungfish, *Neoceratodus forsteri*. Cell and Tissue Research 276: 117-121.

Hansen, G.N. and Hansen, B.L. (1998). Immunocytochemical localization and characterization of mammalian thyrotropin-like material in the pituitary of the Australian lungfish, *Neoceratodus forsteri*. Cell and Tissue Research 294: 515-523.

Hansen, G.N., Hansen, B.L. and Hummer, L. (1980). The cell types in the adenohypophysis in the South American lungfish, *Lepidosiren paradoxa*, with special reference to immunocytochemical identification of corticotrophin-containing cells. Cell and Tissue Research 209: 147-160.

Hansen, G.N., Hansen, B.L. and Jorgensen, P.N. (1987). Insulin-, glucagons- and somatostatin-like immunoreactivity in the endocrine pancreas of the lungfish, *Neoceratodus forsteri*. Cell and Tissue Research 248: 181-185.

Holmes, R.L. and Ball, J.N. (1974). The Pituitary Gland: A Comparative Account. Cambridge University Press, London and New York.

Hyodo, S., Ishii, S. and Joss, J.M.P. (1997). Australian lungfish neurohypophysial hormone genes encode vasotocin and [Phe2] mesotocin precursors homologous to tetrapod-type precursors. Proceedings of the National Academy of Sciences of the United States of America 94: 13339-13344.

Janssens, P.A. and Grigg, J.A. (1988). Binding of adrenergic ligands to liver plasma membrane preparations from the axalotl, *Ambystoma mexicanum*; the toad, *Xenopus laevis*; and the Australian lungfish, *Neoceratodus forsteri*. General and Comparative Endocrinology 71: 524-530.

Janssens, P.A., Vinson, G.P., Chester-Jones, I. and Mosley, W. (1965). Amphibian characteristics of the adrenal cortex of the African lungfish (*Protopterus* sp). Journal of Endocrinology 32: 373-382.

Joss, J.M.P. (2006). Lungfish evolution and development. General and Comparative Endocrinology 148: 285-289.

Joss, J.M.P., Beshaw, M., Williamson, S., Trimble, J. and Dores, R.M. (1990). The adenohypophysis of the Australian lungfish, *Neoceratodus forsteri* – an immunocytological study. General and Comparative Endocrinology 80: 274-287.

Joss, J.M.P., Arnold-Reed, D.E. and Balment, R.J. (1994a). The steroidogenic response to angiotensin II in the Australian lungfish, *Neoceratodus forsteri*. Journal of Comparative Physiology. B 164: 378-382.

Joss, J.M.P., King, J.A. and Millar, R.P. (1994b). Identification of molecular forms of a steroid hormone response to gonadotropin-releasing hormone in the Australian lungfish, *Neoceratodus forsteri*. General and Comparative Endocrinology 96: 392-4000.

Joss, J.M.P., Edwards, A. and Kime, D.E. (1996). *In vitro* biosynthesis of androgens in the Australian lungfish, *Neoceratodus forsteri*. General and Comparative Endocrinology 101: 256-263.

Joss, J.M.P., Rajasekar, P.S., Raj-Prasad, A. and Ruitenberg, K. (1997). Developmental endocrinology of the dipnoan, *Neoceratodus forsteri*. American Zoologist 37: 461-469.

Joss, J.M.P., Itahara, Y., Watanabe, T.X., Nakajima, K. and Takei, Y. (1999). Teleost-type angiotensin is present in Australian lungfish, *Neoceratodus forsteri*. General and Comparative Endocrinology 114: 206-212.

King, J.A., Millar, R.P., Vallarino, M. and Pierantoni, R. (1995). Loacalisation and characterisation of gonadotropin-releasing hormones in the brain, gonads and plasma of a Dipnoi (lungfish, *Protopterus annectans*). Regulatory Peptides 57: 163-174.

Larsen, T.H., Helle, K.B. and Satersdal, T. (1994). Immunoreactive atrial natriuretic peptide and dopamine beta-hydroxylase in myocytes and chromaffin cells of the heart of the African lungfish, *Protopterus aethiopicus*. General and Comparative Endocrinology 95: 1-12.

Lee, J., Lecaude, S., Danielson, P., Sollars, C., Alrubaian, J., Propper, C., Lihrman, I., Vaudry, H. and Dores, R.M. (1999). Cloning of proopiomelanocortin from the brain of the African lungfish, *Protopterus annectens*, and the brain of the western spadefoot toad, *Spea multiplicatus*. Neuroendocrinology 70: 43-64.

Liu, L., Conlon, J.M., Joss, J.M.P. and Burcher, E. (2002). Purification, characterisation, and biological activity of a substance P-related peptide from the gut of the Australian lungfish, *Neoceratodus forsteri*. General and Comparative Endocrinology 125: 104-112.

Masini, M.A., Devecchi, M., Napoli, L. and Uva, B.M. (1995). A local renin-angiotensin-like system in the heart of an African lungfish (*Protopterus annectens*). Comparative Biochemistry and Physiology 110A: 229-233.

Mathieu, M., Vallarino, M., Trabucchi, M., Chartrel, N., Vaudrey, H. and Conlon, J.M. (1999). Identification of an urotensin I-like peptide in the pituitary of the lungfish *Protopterus annectens*: immunocytochemical localization and biochemical charaterisation. Peptides 20: 1303-1310.

May, D., Alrubaian, J., Patel, S. and Dores, R.M. (1999). Studies on the GH/SL gene family: cloning of African lungfish (*Protopterus annectens*) growth hormone and somatolactin and toad (*Bufo marinus*) growth hormone. General and Comparative Endocrinology 113: 121-135.

McDonald, L.K., Joss, J.M.P. and Dores, R.M. (1991). The phylogeny of Met-enkephalin and Leu-enkephalin: studies on the holostean fish *Lepisosteus platyrhincus* and the Australian lungfish, *Neoceratodus forsteri*. General and Comparative Endocrinology 84: 228-236.

Miranda, L.A., Paz, D.A., Dezi, R. and Pisano, A. (1996). Immunocytochemical and morphometric study on the changes of TSH, PRL, GH and ACTH cells during the development of *Bufo arenarum*. Cell and Tissue Research 283: 125-132.

Mohsen, T. and Godet, R. (1960). Action of thyroxine on the rate of oxygen consumption of the lungfish (*Protopterus*). Nature 185: 108.

Norris, D.O. (1997). Vertebrate Endocrinology. 3rd Edition. Academic Press, San Diego, USA.

Noso, T., Nicoll, C.S. and Kawauchi, H. (1993). Lungfish prolactin exhibits close tetrapod relationships. Biochimica et Biophysica Acta 1164: 159-165.

Ogawa, K., Suzuki, E. and Taniguchi, K. (1995). Immunohistochemical studies on the development of the hypothalamo-hypophyseal system in *Xenopus laevis*. The Anatomical Record 241: 244-254.

Okabe, M. and Graham, A. (2004). The origin of the parathyroid; persistence of gills in tetrapods. Proceedings of the National Academy of Sciences of the United States of America 101: 17716-17719.

Oota, Y. and Saga, T. (1991). Chronological appearance of immunoreactivity for the different adenohypophyseal hormones in the pituitary of salamander larvae (*Hynobius nebulosis*). Zoological Science 8: 613-616.

Pang, P.K.T. and Sawyer, W.H. (1975). Parathyroid hormone preparations, salmon calcitonin, and urine flow in the South American lungfish, *Lepidosiren paradoxa*. Journal of Experimental Zoology 193: 407-412.

Pang, P.K.T., Clark, N.B. and Thomson, K.S. (1971). Hypocalcemic activities in the ultimobranchial bodies of lungfishes, *Neoceratodus forsteri* and *Lepidosiren paradoxa* and teleosts, *Fundulus heteroclitus* and *Gadus morhua*. General and Comparative Endocrinology 17: 582-585.

Pang, P.K.T., Furspan, P.B. and Sawyer, W.H. (1983). Evolution of neurohypophyseal hormone actions in vertebrates. American Zoologist 23: 655-662.

Querat, B., Arai, Y., Henry, A., Akama, Y., Longhurst, T.J. and Joss, J.M.P. (2004). Pituitary glycoprotein hormone b subunits in the Australian lungfish and estimation of the relative evolution rate of these subunits within vertebrates. Biology of Reproduction 70: 356-363.

Rafn, S. and Wingstrand, K.G. (1981). Structure of intestine, pancreas, and spleen of the Australian lungfish, *Neoceratodus forsteri* (Krefft). Zoologica Scripta 10: 223-240.

Sawyer, W.H. and Pang, P.K.T. (1975). Endocrine adaptation to osmotic requirements of the environment: endocrine factors in osmoregulation by lungfishes and amphibians. General and Comparative Endocrinology 25: 224-229.

Sawyer, W.B., Blair-West, J.R., Simpson, P.A. and Sawyer, M.K. (1976). Renal responses of Australian lungfish to vasotocin, angiotensin II, and NaCl infusion. American Journal of Physiology 231: 593-602.

Sawyer, W.H., Uchiyama, M. and Pang, P.K.T. (1982). Control of renal functions in lungfishes. Federation Proceedings 41: 2361-2364.

Scheuermann, D.W. (1993). Comparative morphology, cytochemistry and innervation of chromaffin tissue in vertebrates. Journal of Anatomy 183: 327-342.

Scheuermann, D.W., Stilman, C., Reinhold, C. and De Groodt-Lasseel, M.H.A. (1981). Microspectrofluorometric study of monoamines in the auricle of the heart of *Protopterus aethiopicus*. Cell and Tissue Research 217: 443-449.

Scheuermann, D.W., Adriaensen, D., Timmermans, J-P. and De Groodt-Lasseel, M.H.A. (1991). Immunohistochemical localization of polypeptide hormones in pancreatic endocrine cells of a dipnoan fish, *Protopterus aethiopicus*. Acta Histochemica 91: 185-192.

Schreibman, M.P. (1986). Pituitary gland. In: P.K.T. Pang and M.P. Schreibman (eds). Vertebrate Endocrinology: Fundamentals and Biomedical Implications.: Academic Press Inc., Orlando, USA. pp. 11-55.

Sollars, C., Danielson, P., Joss, J.M.P. and Dores, R.M. (2000). Deciphering the origin of Met-enkephalin and Leu-enkephalin in lobe-finned fish: cloning of Australian lungfish proenkephalin. Brain Research 874: 131-136.

Sutija, M. and Joss, J.M.P. (2005). Thyroid hormone deiodinases revisited: insights from lungfish: a review. Journal of Comparative Physiology. B 176: 87-92.

Trivett, M.K., Officer, R.A., Clement, J.G., Walker, T.I., Joss, J.M., Ingleton, P.M., Martin, T.J. and Danks, J.A. (1999). Parathyroid hormone-related protein (PTHrp) in cartilaginous and bony fish tissues. Journal of Experimental Zoology 284: 541-548.

Vallarino, M., Tranchand, Bunel, D. and Vaudry, H. (1992) Alpha-melanocyte-stimulating hormone (α-MSH) in the brain of the African lungfish, *Protopterus annectens*: immunohistochemical localization and biochemical characterization. The Journal of Comparative Neurology 322: 266-274.

Vallarino, M., Goula, D., Trabucchi, M., Masini, M.A., Chartrel, N. and Vaudry, H. (1996). Immunocytochemical localization of atrial natriuretic factor and autoradiographic distribution of atrial natriuretic binding sites in the brain of the African lungfish, *Protopterus annectens*. The Journal of Comparative Neurology 375: 345-362.

Vallarino, M., Trabucchi, M. and Vaudry, H. (1998a). Neuropeptides in the lungfish brain: phylogenetic implication. Annals of the New York Academy of Sciences 839: 53-59.

Vallarino, M., Thoumas, J-L., Masini, M.A., Trabucchi, M., Chartrel, N. and Vaudry, H. (1998b). Immunocytochemical localization of enkephalins in the brain of the African lungfish, *Protopterus annectens*, provides evidence for differential distribution of Met-enkephalin and Leu-enkephalin. Journal of Comparative Neurology 396: 275-287.

Wake, M.H. (1987). Urinogenital morphology of dipnoans, with comparisons to other fishes and amphibians. In: W.E. Bemis, W.W. Burggren and N.E. Kemp (eds). The Biology and Evolution of Lungfishes. Alan R. Liss Inc., New York, USA. pp. 199-216.

Zambrano, D. and Iturriza, F.C. (1973). Hypothalamo-hypophysial relationships in the South American lungfish, *Lepidosiren paradoxa*. General and Comparative Endocrinology 20: 256-273.

CHAPTER 16

The Central Nervous System of Lungfishes

R. Glenn Northcutt*

Laboratory of Comparative Neurobiology, Scripps Institution of Oceanography and Department of Neurosciences, School of Medicine, University of California, San Diego, La Jolla, CA 92093-0201, USA

ABSTRACT

The central nervous system of lungfishes consists of a spinal cord and brain. The spinal cord is cylindrical and is composed of a central canal, surrounded by a core of gray matter, surrounded, in turn, by white matter. The gray matter is divided into dorsal, lateral, medial, and ventrolateral cellular columns. The white matter consists of dorsal, lateral, and ventral funiculi, which comprise both ascending and descending axonal tracts. The South American lungfish, *Lepidosiren paradoxa* (Fitzinger 1837) and African lungfishes, such as the Spotted African lungfish, *Protopterus dolloi* (Boulenger 1900) all have slender brains, with both the hindbrain and the telencephalic hemispheres being wider than the diencephalon and midbrain. The Australian lungfish, *Neoceratodus forsteri* (Krefft 1870) also has a slender brain, but the cerebellum and optic tectum are relatively larger than those structures in other lungfishes. Although there is an extensive descriptive literature on the brain in each genus of lungfishes, there are few immunohistochemical or hodological studies. The literature is reviewed, and the anatomy of the brain and spinal cord of the Spotted African lungfish is described in detail, beginning with the spinal cord and continuing rostrally through the telencephalon and olfactory bulbs. There is general agreement on the basic anatomy of the hindbrain and midbrain in all three lungfish genera, but the organization of the forebrain is poorly understood and the subject of disagreement. Most of the controversy on its organization centers on differing interpretations of the pallial-subpallial border, the extent of the

*E-mail: rgnorthcutt@ucsd.edu

striatopallidal systems, and the organization of the amygdala. Each of these areas of disagreement is reviewed, and a new model of telencephalic organization in lungfishes is proposed. In this model, the telencephalon of lungfishes is far more similar to that of amphibians than has previously been suspected. The chapter ends with a series of proposals concerning future directions for neurobiological research on these fishes.
Keywords: diencephalon, cerebellum, hindbrain, midbrain, prosencephalon, spinal cord

INTRODUCTION

One or more parts of the central nervous system (CNS) have been described in each genus of lungfishes by several researchers: 1) *Lepidosiren* (Elliot Smith 1908; Nieuwenhuys and Meek 1990; Zambrano and Iturriza 1972, 1973; Thors and Nieuwenhuys 1979; Northcutt 1986a; Nieuwenhuys 1998); 2) *Neoceratodus* (Beauregard 1881; Sanders 1889; Wilder 1887; Bing and Burckhardt 1904, 1905; Holmgren and van der Horst 1925; Keenan 1928; Jeener 1930; Nieuwenhuys and Hickey 1965; Nieuwenhuys 1967a; von Bartheld *et al.* 1990; Nieuwenhuys and Meek 1990; Whitt 1969; Northcutt 1986a); and 3) *Protopterus* (Burckhardt 1892; Kölliker 1896; Jeener 1930; Gerlach 1933; Holmgren 1922, 1959; Rudebeck 1945, Dorn 1957; Schnitzlein 1966; Schnitzlein and Crosby 1967, 1968; Nieuwenhuys 1967a, b, 1969; Capanna and Clairambault 1973; Clairambault and Capanna 1973; Clairambault *et al.* 1974a, b, c; von Bartheld and Meyer 1988; von Bartheld 1992; von Bartheld *et al.* 1990; Northcutt 1986a, 2008). The descriptive neuroanatomy of the two genera of lepidosirenid lungfishes (*Lepidosiren* and *Protopterus*) is far more complete than that of *Neoceratodus*. The most recent neuroanatomical description of the entire CNS of a lungfish is that for *Lepidosiren* (Nieuwenhuys 1998), and there is no recent description of the entire CNS of *Neoceratodus* or *Protopterus*. In order to partially rectify this problem, the CNS of the Spotted African lungfish, *Protopterus dolloi* Boulenger 1900, will be emphasized in this chapter.

Although there is a fairly extensive descriptive literature on the CNS of lungfishes, far less is known about the development or immunohistochemistry of the CNS in these fishes. Bing and Burckhardt (1905) briefly described the development of the brain of *Neoceratodus* and Whiting *et al.* (1992) described the development of the brain and cranial nerves. Rudebeck (1945) provided the most detailed description of the development of the telencephalon of *Protopterus*, whereas Bergquist (1932) undertook a similar description of the development of the diencephalon of *Neoceratodus*. Thus, there are no recent descriptions of the development of the entire brain or the expression of any of the genes involved in the development of the CNS of lungfishes.

There are also few immunohistochemical studies of the CNS in lungfishes. Reiner and Northcutt (1987) described the distribution of a number of neuroactive substances (substance P, leucine-enkephalin, avian pancreatic polypetid and

LANT6) in the forebrain of *Protopterus annectens*. The telencephalic distribution of acetylcholinesterase (AChE) in *Neoceratodus* and *Protopterus* was described by von Bartheld *et al.* (1990). Vallarino *et al.* (1992, 1995) described the distribution of α-melanocyte-stimulating hormone and neuropeptid Y, respectively in *P. annectens*. The distribution of the peptide FMRFamide in the terminal nerve fibers of *Neoceratodus* has been described by Fiorentino *et al.* (2002). Finally, messenger RNAs encoding for two gutamic acid decarboxylases (GAD65 and GAD67) have been characterized in *P. annectens* and their distribution in the brain described by Trabucchi *et al.* (2008).

Further, there are only a handful of experimental studies on the CNS in lungfishes. The primary olfactory and terminal nerve projections in *P. dolloi* were described by von Bartheld and colleagues (von Bartheld and Meyer 1988; von Bartheld *et al.* 1988). The projections of the terminal nerve have also been described in *Neoceratodus* (Schober *et al.* 1994), as have the projections of the paraventricular organ in *P. dolloi* (von Bartheld and Meyer 1990). The retinal projections to the diencephalon and midbrain have been described in *Lepidosiren* (Northcutt 1977), in *Neoceratodus* (Northcutt 1980), and in *Protopterus* (Clairambault and Flood 1975; Northcutt 1977). The oculomotor and mesencephalic trigeminal neurons have been localized in *Neoceratodus* and *Protopterus* (von Bartheld, 1992). Finally, the central projections of the lateral line nerves in lepidosirenid lungfishes (Northcutt 1983a) and the descending projections to the spinal cord in *Protopterus* (Ronan and Northcutt 1985) have been established.

The gross brain structure of lungfishes (Fig. 1) is reviewed and compared to the brains of other lobe-finned fishes (Fig. 2). The major divisions of the brain (Figs. 3-10) and what is known about their connections are then described, beginning with the spinal cord and continuing rostrally through the telencephalon and olfactory bulbs. Finally, the chapter ends with a short section on future directions for neurobiological research on lungfishes.

GROSS BRAIN STRUCTURE

The brains of *Lepidosiren* and *Protopterus* are very similar. In both taxa, the brains are fairly slender, with both the hindbrain and the telencephalic hemispheres being wider than the diencephalon and midbrain (Fig. 1A). In both taxa, the cerebellum and the optic tectum are poorly developed and are similar in relative size to those structures in salamanders, such as *Ambystoma* (Fig. 2B). Although the brain of *Neoceratodus* (Fig. 1B) is also elongated, with the hindbrain and telencephalic hemispheres being wider than the diencephalon and midbrain, both the cerebellum and the optic tectum of *Neoceratodus* are relatively larger than these structures in lepidosirenid lungfishes. In many ways, the cerebellum and optic tectum of *Neoceratodus* are similar in appearance to these structures in the coelacanth, *Latimeria* (Fig. 2A).

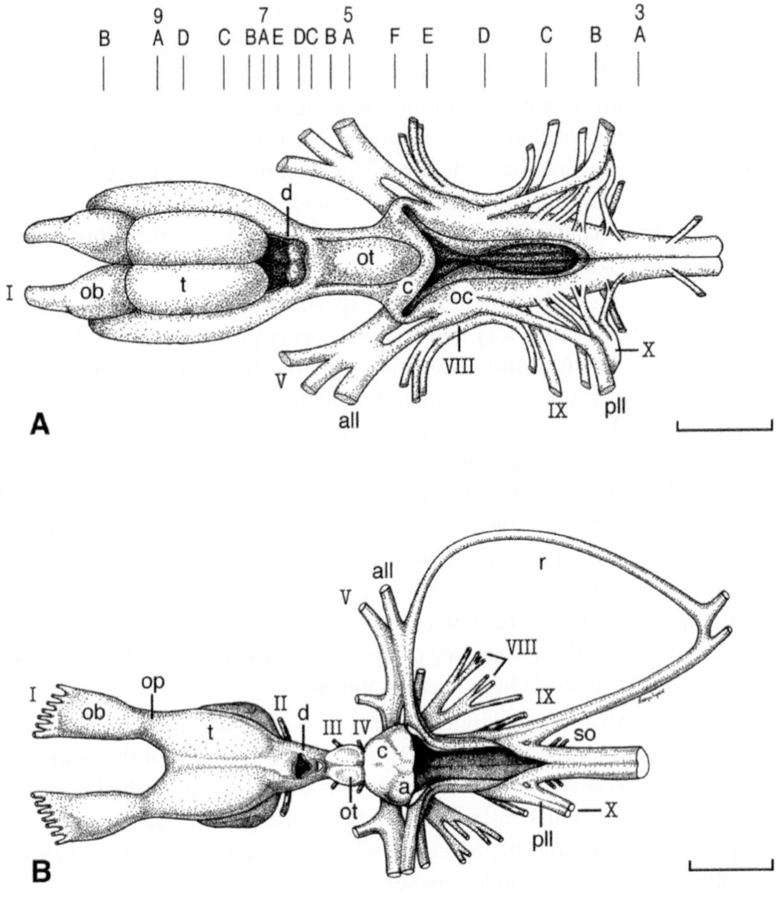

Fig. 1 Dorsal views of the brain of *Protopterus* (A) and *Neoceratodus* (B). The relative positions of the transverse sections from *Protopterus* in Figures 3, 5, 7, and 9 are indicated by labeled lines (A-F, A-E, A-D, and A-B) at the top of the figure. Bar scales equal 5 mm (A) and 1 cm (B).

The olfactory bulbs are large in all three genera of lungfishes, but they are sessile and arise dorsally from the rostral poles of the telencephalic hemispheres in lepidosirenid lungfishes (Fig. 1A), whereas in *Neoceratodus* they are connected directly to the telencephalic hemispheres via short hollow peduncles composed of the secondary olfactory tracts (Fig. 1B). Although the attachment of the olfactory bulbs to the telencephalic hemispheres in lungfishes differs, the telencephalic hemispheres in all three taxa are characterized by an expanded floor, or subpallium, which greatly increases both the height and width of the telencephalic hemispheres (Fig. 1).

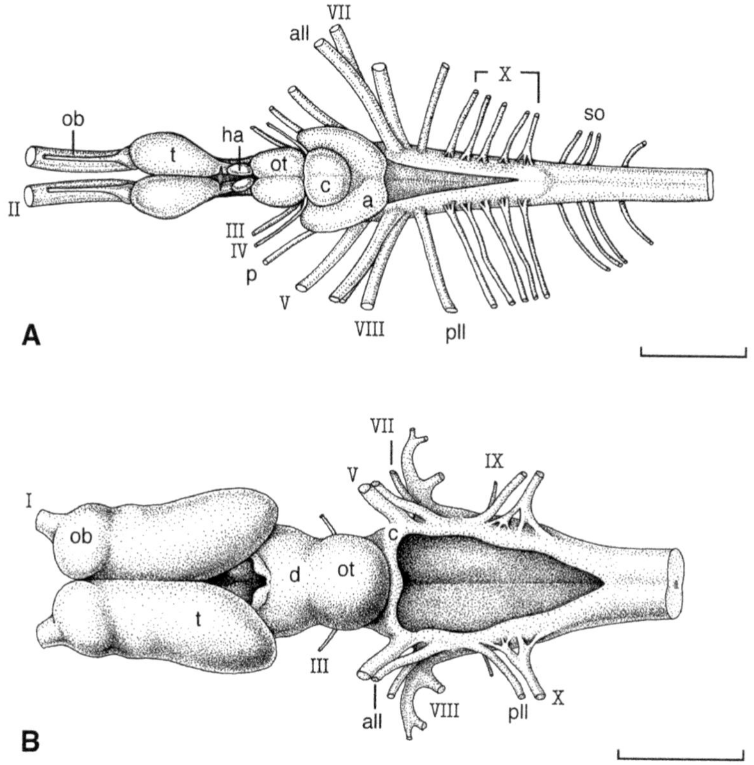

Fig. 2 Dorsal view of the brain of *Latimeria* (A) and *Ambystoma* (B). Bar scales equal 1 cm (A) and 3 mm (B).

SPINAL CORD

The neuroanatomy of the spinal cord has been described in all three genera of lungfishes: 1) *Lepidosiren* (Nieuwenhuys, 1998); 2) *Neoceratodus* (Keenan 1928); and 3) *Protopterus* (Burckhardt 1892; Kölliker 1896). In each of the genera, the central canal is surrounded by a distinct ependymal layer (Figs. 3A, 4A) which, in turn, is surrounded by gray matter, then white matter. The gray matter is divided into dorsal, lateral, medial and ventrolateral columns. The dorsal column consists of small darkly staining, densely packed neurons (Figs. 3A, 4A), among which single, larger multipolar neurons are occasionally seen. Burckhardt (1892) thought that these larger cells within the dorsal cell column might be homologous to the dorsal cells of lampreys and the Rohon-Beard cells of larval amphibians, both of which are primary sensory neurons. This is unlikely, however, as centrally located primary sensory neurons are spherical and usually unipolar in appearance rather than being multipolar.

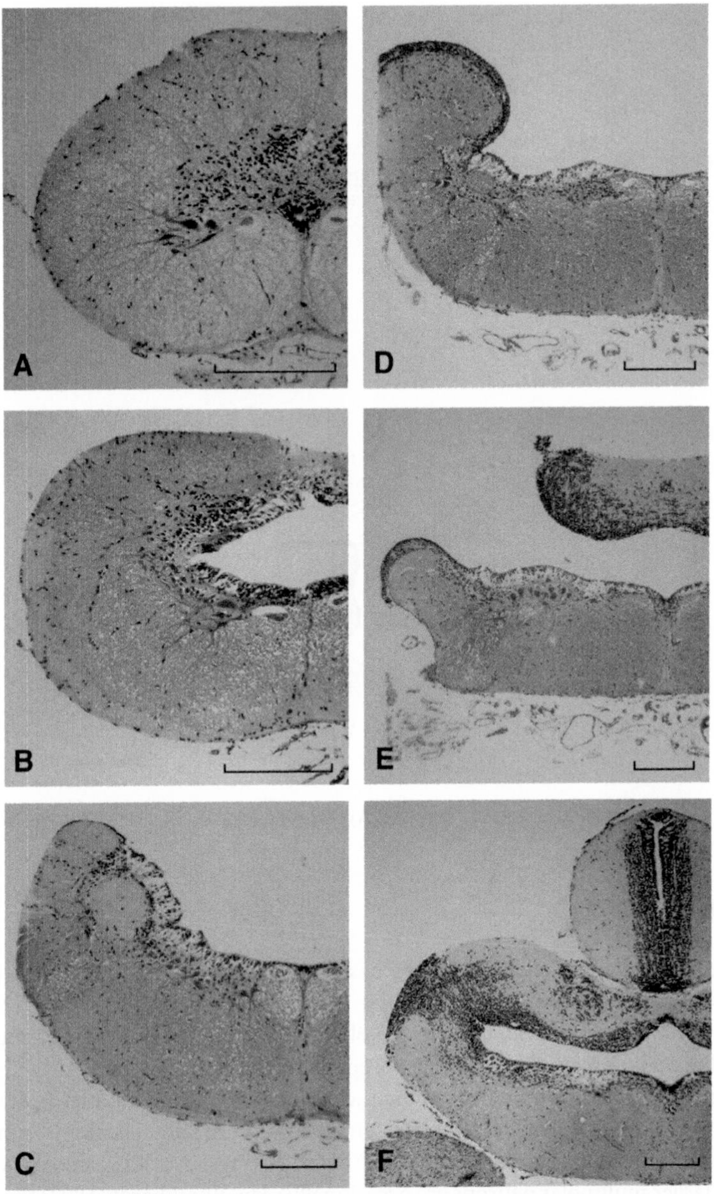

Fig. 3 Photomicrographs of cresyl violet stained transverse sections through one side, revealing cell bodies in the spinal cord (A), medulla oblongata (B-E) and cerebellum (F) of the Spotted African lungfish, *Protopterus dolloi*. Boundaries of cell groups and tracts are indicated by solid and broken lines, respectively. Larger neurons are indicated by solid ovals. Bar scales equal 500 μm.

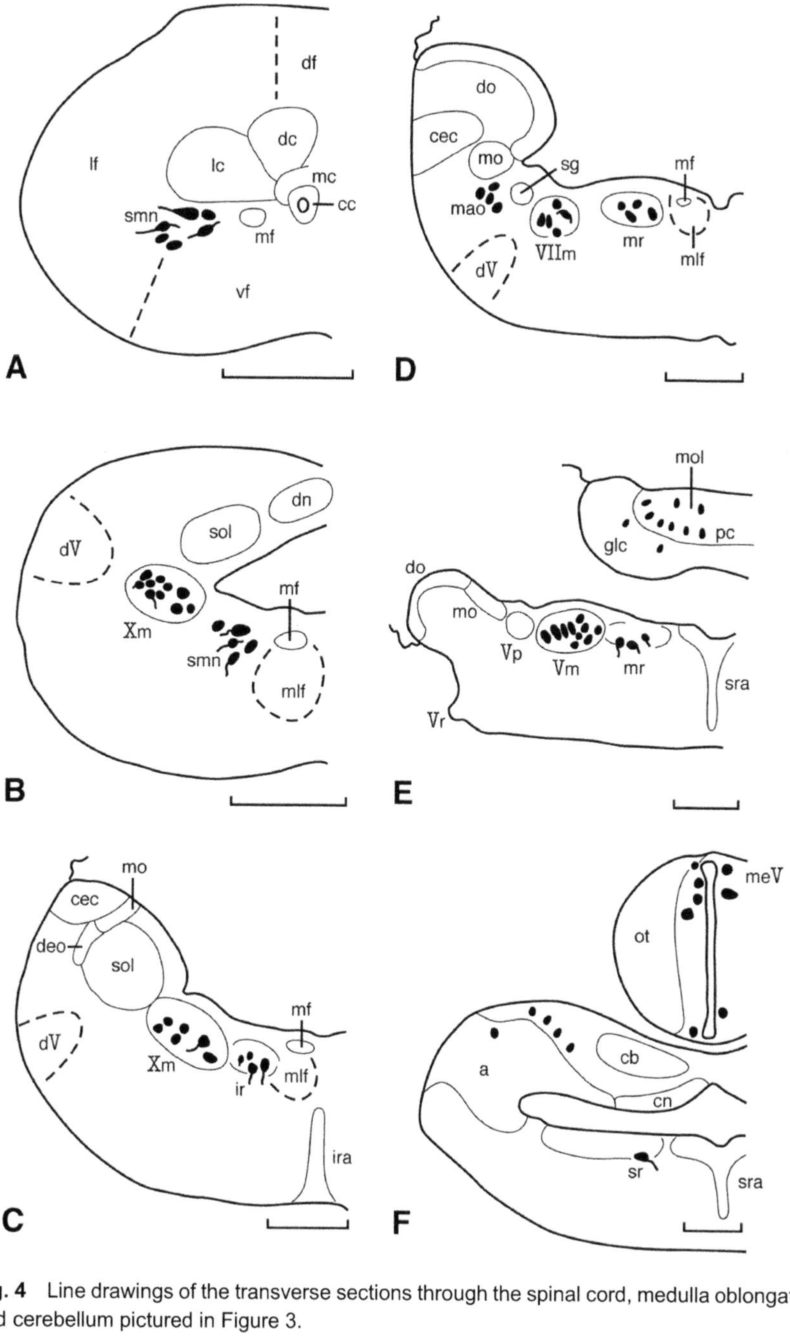

Fig. 4 Line drawings of the transverse sections through the spinal cord, medulla oblongata, and cerebellum pictured in Figure 3.

The neurons of the lateral cell column are larger than those of the medial cell column, and they stain more lightly and are also more scattered (Figs. 3A, 4A). In many cases, their cell bodies are sufficiently large to allow tracing of their dendrites into the lateral white funiculus; these cells, as well as the other cell classes in the spinal cord, must form most of their synapses in the peripheral white funiculi. The cells of the medial cell column consist of one or two rows of small multipolar neurons, located adjacent to the ependymal layer, and the more ventrally located Mauthner axon (Figs. 3A, 4A). Nothing else is known about the cells of the medial column. In contrast, the cells of the ventrolateral cell column comprise the largest neurons in the spinal cord and constitute the primary somatic motor neurons, whose axons exit the spinal cord via the ventral roots of the spinal nerves to innervate the striated muscles of the body wall and fins. Many of the somatic motor neurons are bipolar, with a lateral dendrite that extends into the lateral funiculus and a medial dendrite that extends into the ventral funiculus.

As in other vertebrates, the peripheral white matter of the spinal cord consists of axonal tracts passing up and down the spinal cord, as well as the dendrites of the more centrally located neurons of the cellular columns. The peripheral matter is divided into dorsal, lateral and ventral funiculi, whose exact borders are only relative. Generally, the diameter of the axons in each funiculus increases from the dorsal funiculus, through the lateral funiculus, to the ventral funiculus. In this regard, the largest axons in the spinal cord are the Mauthner axons, which run in the ventral funiculus (Figs. 3A, 4A) and arise from a pair of giant neurons located in the medulla oblongata at the level of entry of the eighth cranial nerve. The axons of the Mauthner cells decussate almost immediately after issuing from the Mauthner cell bodies and become invested in a substantial myelin sheath in most anamniotic vertebrates. Lungfishes, however, appear unique in that accessory axons, apparently the axons of cells within the adjacent reticular nuclei of the medulla oblongata, occur within the myelin sheath of the Mauthner axons and form an axonal complex (Kölliker 1896; Holmgren and van der Horst 1925; Nieuwenhuys 1998).

Unfortunately, there are few experimental data on the location of most of the cells whose axons form the funiculi. Similarly, almost nothing is known about the nature of the information carried by these axons. At present, there is a single study on the origins of the descending spinal projections in lepidosirenid lungfishes (Ronan and Northcutt 1985). When gelfoam pledgets, saturated with horseradish peroxidase, were inserted unilaterally in the rostral spinal cord, large numbers of retrogradely labeled neuronal cell bodies were seen in the medulla oblongata, midbrain, and diencephalon (Fig. 11). More cell bodies were labeled ipsilaterally than contralaterally, except for the red nucleus, which was only labeled contralaterally. This study demonstrated that the neurons of the reticular nuclei provide the largest source of descending projections to the spinal cord, followed by the magnocellular octaval nucleus and the nucleus of the medial longitudinal fasciculus. Although it was impossible to determine the funicular topography within the spinal cord, most of the descending fibers were probably restricted to the lateral and ventral funiculi, as this is the characteristic route for octaval and reticular centers in other vertebrates.

HINDBRAIN

Traditionally, the nuclei and tracts of the brainstem (Thors and Nieuwenhuys 1979; Nieuwenhuys 1998), which is herein defined as the hindbrain and the floor of the midbrain in lungfishes, as in most other vertebrates (Herrick 1948; Nieuwenhuys *et al.* 1998), have been grouped into four longitudinal zones, separated primarily by a series of longitudinal sulci. There is a great deal of variation in the sulcal pattern of the brainstem among vertebrates, and distinct functional groups may or may not be grouped together in a single longitudinal zone. For this reason, a functional grouping, based on somatic sensory, octavolateralis, visceral sensory, visceral motor, and somatic motor modalities will be adopted in this chapter.

Somatic sensory nuclei and tracts. Although no physiological studies have been conducted on the centers and pathways carrying pain, temperature and touch information from the skin and body wall of lungfishes, information from other vertebrates suggest there should be at least three pathways carrying this information: 1) a dorsal column system, 2) an ascending spinoreticular system, and 3) the trigeminal system. In most vertebrates, touch and proprioceptive information regarding the state of muscle contraction is carried by dorsal funicular fibers (Figs. 3A, 4A), which terminate in the dorsal funicular (dorsal column) nucleus. In lungfishes, a putative homologue of this nucleus can be identified at the level of the obex (Figs. 3B, 4B). If indeed, this cell group is a homologue of the dorsal column nucleus, the axons of its cells should decussate as internal arcuate fibers and then ascend as the medial lemniscus and terminate in the reticular formation, cerebellum, and midbrain. Similarly, pain and temperature information should be carried by the entering sensory fibers of the other dorsal root ganglionic neurons of the spinal nerves, fibers which terminate in the cells of the dorsal horn (dorsal and lateral cell columns) and, after decussating, ascend as spinoreticular, spinocerebellar, and spinotectal pathways. Although experimental studies on ascending spinal projections in lungfishes have not been undertaken, ascending spinal projections in teleost fishes have been demonstrated to these centers (Hayle 1973a, b; Murakami and Ito 1985; Oka *et al.* 1986).

Pain, temperature, and touch information from the skin and deeper tissues of the head is carried by sensory fibers of the trigeminal nerve and possibly fibers of the facial, glossopharyngeal, and vagal nerves. The sensory fibers of the trigeminal nerve, which arise from its sensory ganglion, enter the rostral lateral wall of the hindbrain (Fig. 1) and either terminate in the ipsilateral principal sensory nucleus (Figs. 3E, 4E) or descend along the lateral wall of the hindbrain as the descending trigeminal tract (Figs. 3B-D, 4B-D) to terminate on neurons that are scattered within the tract and constitute a "nucleus" of the descending trigeminal tract. In most vertebrates, fibers carrying touch information terminate in the principal sensory nucleus, whereas fibers carrying pain and temperature information terminate in the nucleus of the descending trigeminal tract. There are no experimental data on the primary trigeminal projections in lungfishes or on their higher order connections.

There is one additional class of trigeminal sensory information in lungfishes and other vertebrates, proprioceptive information, which is conveyed by the mesencephalic trigeminal neurons. These extremely large primary sensory neurons carry information from one or more of the extrinsic eye muscles, as well as the jaw muscles. They are not located within the sensory ganglion of the trigeminal nerve but, rather, constitute approximately 500 neurons scattered throughout the optic tectum (Figs. 3F, 4F, 5A, 6A) in *Protopterus* and *Neoceratodus* (von Bartheld 1992).

Octavolateral sensory system. In lungfishes and other anamniotic vertebrates, the octavolateralis system is composed of the inner ear and lateral line organs, which are innervated by the eighth, or octaval, nerve and the lateral line nerves, respectively. The inner ear has been described in detail in all three genera of lungfishes (Platt *et al.* 2004; Jørgensen *et al.*, this volume) and consists of semicircular canals, which detect angular acceleration, a dorsally situated utricule, and a saccule and a lagena. In lungfishes, unlike in most vertebrates, the saccule and lagena are housed in a single chamber ventral to the utricule. Although it is generally assumed that the utricule detects linear acceleration and that the saccule and lagena detect sound, there is increasing evidence that all three otolith organs have multiple functions (Popper and Platt 1993; Popper and Fay 1999). Unfortunately, there are no physiological or behavioral studies on either acoustic or vestibular functions in lungfishes. Similarly, there are no experimental anatomical studies on the central projections of the end organs of the inner ear in lungfishes.

In other fishes, the octaval nerve projects to a series of octaval nuclei that form the ventral tier of the octavolateral column, which forms the dorsolateral wall of the hindbrain (Fig. 1). In a rostral to caudal sequence, the octaval nuclei in these fishes consist of an anterior nucleus, a magnocellular nucleus, a descending nucleus, and a posterior nucleus (McCormick 1999). Comparable octaval nuclei almost certainly exist in lungfishes, as a magnocellular nucleus (Figs. 3D, 4D) is easily recognized in *Protopterus,* as is a more caudal descending nucleus (Figs. 3C, 4C). Anterior and posterior octaval nuclei also probably exist in *Protopterus,* but the cellular differentiation within the octaval column is too poor to allow these nuclei to be distinguished from more dorsally located nuclei, and experimental studies are needed.

In other fishes, all of the end organs of the inner ear project to all four octaval nuclei. Thus, many of the cells in each nucleus must give rise to vestibular circuits (McCormick 1999). Some neurons within the anterior and descending nuclei do

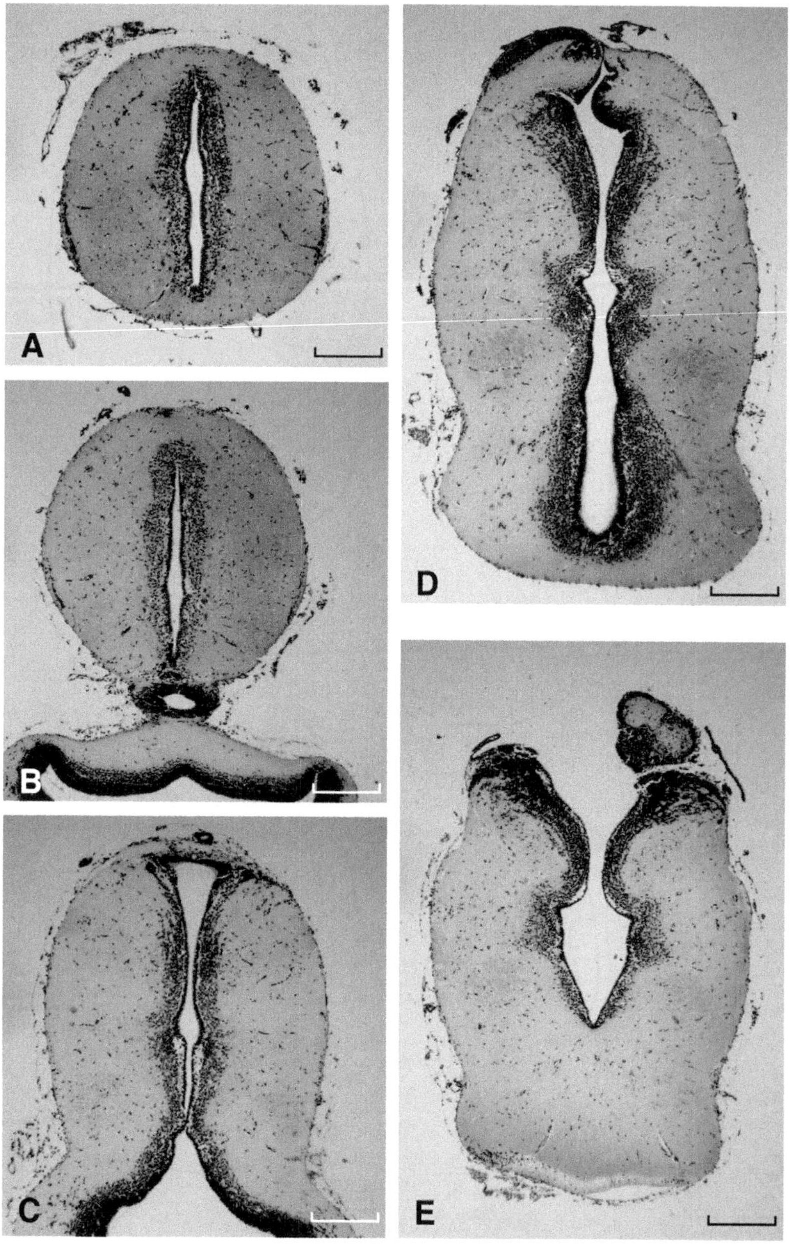

Fig. 5 Photomicrographs of transverse sections through the midbrain (A), pretectum (B, C) and diencephalon (D, E) of the Spotted African lungfish. Bar scales equal 500 µm.

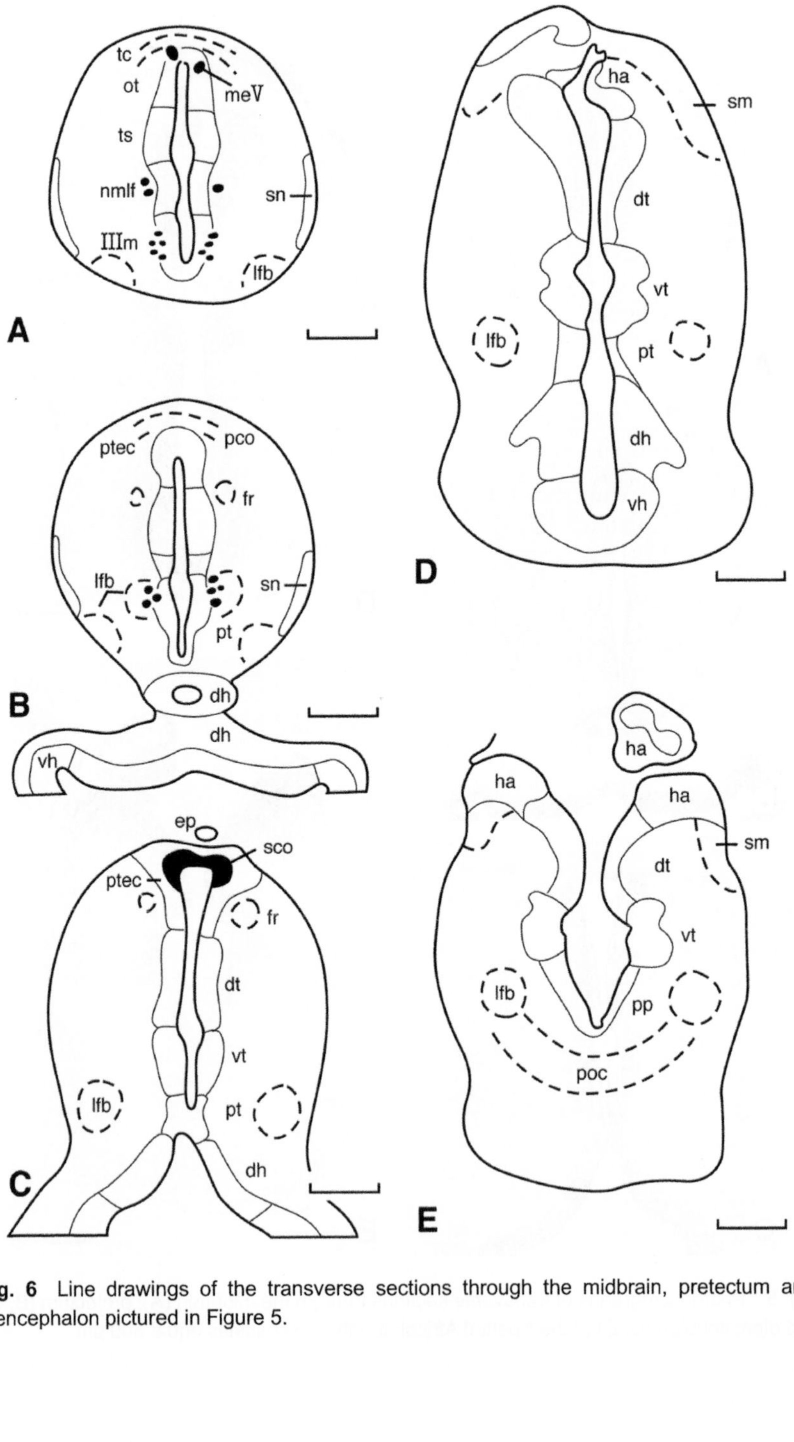

Fig. 6 Line drawings of the transverse sections through the midbrain, pretectum and diencephalon pictured in Figure 5.

project to the torus semicircularis, however, and are believed to carry auditory information. In teleost fishes where our understanding of ascending auditory projects is most complete, there is now clear evidence that auditory information reaches telencephalic levels after being relayed in the diencephalon (McCormick 1999; Northcutt 2006).

In lungfishes, the peripheral lateral line system consists of mechanoreceptive neuromasts and electroreceptive ampullary organs. The former are housed in grooves or appear as canals distributed on the head and trunk; the latter are also distributed on the head and trunk (Northcutt 1986b). Both neuromasts and ampullary organs are innervated by up to six pairs of lateral line nerves that are not part of the twelve traditionally recognized cranial nerves and are therefore not numbered but are named after the embryonic placodes that give rise to the lateral line nerves and the receptors they innervate (Northcutt 1989, 1992; Northcutt and Bemis 1993; Northcutt and Brändle 1995; O'Neill *et al.* 2007). In most fishes there are three pairs of preotic and three pairs of postotic lateral line nerves that innervate the neuromasts of the head and trunk. In lungfishes, however, only the development of the lateral lines (Kemp 1999), not the development of the lateral line nerves, has been investigated. For this reason, all of the preotic and postotic lateral line nerves will be collectively referred to as anterior and posterior lateral line nerves, respectively (Fig. 1). In this nomenclature, the neuromasts and electroreceptors on the head are innervated by the anterior lateral line nerve, whereas the neuromasts on the trunk are innervated by the posterior lateral line nerve. In lungfishes, where electroreceptive ampulllary organs also occur on the trunk, the electroreceptors are innervated by a recurrent ramus (Fig. 1B) of the anterior lateral line nerve, rather than by the posterior lateral line nerve (Northcutt 1986b).

The octavolateral column in lungfishes retains a primitive pattern of organization, which also characterizes the octavolateral column in lampreys, cartilaginous fishes, and chondrostean bony fishes. In all of these fishes, the octavolateral column is divided into a dorsal octavolateral nucleus (Figs. 3D,E, 4D,E), a medial or intermediate octavolateral nucleus (Figs. 3C-E, 4C-E), and a ventral octaval series of four nuclei that have already been described. The anterior lateral line nerve, which conveys mechanoreceptive information from neuromasts located on the head and electroreceptive information from the entire body, divides into dorsal and ventral roots as it enters the rostral medulla (Fig. 1). The dorsal root of the anterior lateral line nerve, which consists of all fibers carrying electroreceptive information, terminates in the ipsilateral dorsal octavolateral nucleus, whereas the ventral root of the anterior lateral line nerve, which consists of all fibers carrying mechanoreceptive information from the neuromasts on the head, terminates in the ipsilateral medial octavolateral nucleus (Northcutt 1983a). Electroreceptive information within the dorsal octavolateral nucleus is represented somatotopically with information from head and trunk electroreceptors mapped rostromedially and caudolaterally, respectively.

The posterior lateral line nerve carries only mechanoreceptive information from the trunk neuromasts and enters the medulla more caudally, closely associated with the vagal nerve (Fig. 1). At this point, fibers of the posterior lateral line nerve divide into ascending and descending branches, which terminate within the medial octavolateral nucleus. Some portion of the ascending mechanoreceptive fibers of the anterior and posterior lateral line nerves continue rostrally and terminate in the ipsilateral auricle and ipsilateral body of the cerebellum (Northcutt 1983a). The course of the secondary lateral line projections in lungfishes is not known, but in other fishes that retain the primitive pattern of organization for the octavolateral column the secondary lateral line projections ascend bilaterally in a lateral lemniscus and terminate in the torus semicircularis and optic tectum. From these centers, higher order projections are relayed through the thalamus to several telencephalic areas.

Visceral sensory system. This sensory system consists of free nerve endings in the endothelium of the oropharynx that convey pain and temperature information from the oropharynx and chemosensory information from taste buds that are scattered in the endothelium of the oropharynx. Lungfishes, at least lepidosirenid lungfishes, are among the few fishes that also have external taste buds scattered in the epidermis over the head (Fahrenholz 1929; Cordier 1936; Pfeiffer 1968). All of these visceral sensory receptors in lungfishes are believed to be innervated by fibers of the facial, glossopharyngeal and vagal nerves, as they are in other vertebrates. The visceral sensory fibers of the facial, glossopharyngeal and vagal nerves are distinguished from other fibers of these nerves, and from fibers of the lateral line nerves, by their small size and light myelinization. In each case, these visceral fibers terminate primarily, if not solely, within the core of the solitary nucleus of the hindbrain (Figs. 3B,C, 4B,C). Some of the visceral fibers that enter the solitary nucleus appear to descend within the nucleus and decussate through the commissural infirma located at the level of the obex. These decussating fibers presumably terminate in the contralateral nucleus solitarius. Although no experimental studies have reported on the distribution of visceral sensory fibers terminating within the solitary nucleus of lungfishes, it is probable that fibers carrying information from taste buds terminate in the more rostral part of the nucleus, whereas general visceral sensory fibers terminate in the more caudal part of the nucleus, as this pattern of termination is fairly common in other vertebrates.

An ascending secondary visceral tract can be traced rostrally from each solitary nucleus (Figs. 3D, 4D) and may terminate in a secondary gustatory nucleus just rostral to the cerebellar nucleus identified in *Lepidosiren* (Nieuwenhuys 1998) and *Neoceratodus* (Holmgren and van der Horst 1925). In teleosts, where there is an extensive experimental literature on the ascending projections of the solitary nucleus (reviewed in Wullimann 1998), bilateral ascending projections have been traced to a secondary gustatory nucleus in the isthmal region, then seen to relay through a number of diencephalic centers to the telencephalon. Similar ascending projections may exist in lungfishes, but these await experimental confirmation.

Visceral motor system. The visceral motor system in the hindbrain of vertebrates has traditionally been divided into a group of general visceral motor nuclei, which are the sites of first order visceral motor neurons of the parasympathetic nervous system, and a group of special visceral motor nuclei, which are the sites of the motor neurons that innervate the jaw muscles and muscles of the branchial arches. The motor centers that innervate jaw and branchial arch muscles, which are striated, were traditionally considered visceral motor centers, because these muscles were believed to develop from lateral plate mesoderm, which forms the smooth muscle of the gut in the trunk. In 1983, Noden transplanted segments of paraxial head mesoderm between quail and chick embryos and demonstrated that jaw and branchial arch muscles, also termed branchiomeric muscles, arise from paraxial mesoderm, not lateral plate mesoderm. The hindbrain motor centers that innervate branchiomeric muscles should therefore be considered special somatic motor centers, not special visceral motor centers, and these special somatic motor centers will be described in the next section on the medulla oblongata.

Generally, so-called general visceral motor centers consist of fairly small neurons that are closely associated with the much larger neurons of the somatic motor centers of the oculomotor, facial, glossopharyngeal and vagal nerves, but it is almost impossible to identify these neurons unless the rami of the cranial nerves that contain visceral motor fibers are experimentally labeled. Because such experiments have not been conducted in any of the three genera of lungfishes, it is presently impossible to locate these visceral motor centers in lungfishes.

Somatic motor system. This group of motor centers is divided into the branchiomeric motor nuclei that innervate the jaw muscles and branchial muscles of the gill arches, and a traditionally recognized series of somatic motor nuclei that innervate the extrinsic eye muscles and the hypobranchial muscles. The neurons of the branchiomeric motor nuclei are located in the lateral half of the basal plate of the brainstem and constitute some of the largest motor neurons in the lungfishes and other vertebrates. The trigeminal motor nucleus (Figs. 3E, 4E, 11), which innervates the jaw muscles derived from the first branchiomere, is the most rostral of the branchiomeric motor centers and forms a fan-shaped nucleus of large multipolar neurons in the hindbrain at the level of the entry and exit of the sensory and motor roots of the trigeminal nerve. The facial motor nucleus (Figs. 3D, 4D), which innervates the muscles of the second (hyoid) branchiomere, occurs more caudally within the hindbrain, and its cells are smaller and more scattered. The glossopharyngeal and vagal motor nuclei, which innervate the branchial muscles of the functional gill arches, form a single column of cells (Figs. 3B,C, 4B,C, 11) located in the caudal half of the hindbrain. The motor neurons of this column are comparable in size to those of the facial motor nucleus and are also more scattered than those of the trigeminal motor nucleus.

The extrinsic eye muscles arise from the rostral paraxial mesoderm of the head. In those fishes that have distinct somite-like structures termed head cavities, the superior rectus, internal rectus, inferior rectus, and inferior oblique eye muscles arise from the first head cavity and are all innervated by neurons of the oculomotor nucleus. The second head cavity gives rise to the superior oblique eye muscle which is innervated by neurons of the trochlear nucleus. The second and third head cavities both contribute to the formation of the external rectus eye muscle, which is innervated by the abducent nucleus. In most vertebrates, it is relatively easy to identify the oculomotor, trochlear, and abducent motor nuclei within the brainstem due to the relative large size of their neurons. Although this appears to be true for *Neoceratodus* (Holmgren and van der Horst 1925), which has relatively well developed eyes, it is not true for the lepidosirenid lungfishes with their reduced eyes. There is general agreement regarding the location of the oculomotor nucleus in *Lepidosiren* (Thors and Nieuwenhuys 1979; Nieuwenhuys 1998) and *Protopterus* (Gerlach 1933), but Clairambault *et al.* (1974c) misidentified the trigeminal motor nucleus as the oculomotor nucleus. In *Protopterus*, the oculomotor nucleus occupies the rostral midbrain tegmentum (Figs. 5A, 6A) and consists of medium sized neurons that are scattered among the smaller neurons of the walls and floor of the periventricular gray. Approximately 100 motor neurons have been identified in *P. dolloi* by labeling the extrinsic eye muscles with the fluorescent tracer true blue (von Bartheld 1992). In the Spotted African lungfish, the oculomotor nucleus can be divided into dorsolateral (approximately 70 neurons), ventromedial (approximately 10 neurons) and ventrolateral (approximately 10 neurons) subdivisions. The cells of the dorsolateral subdivision only innervate ipsilaterally located extrinsic eye muscles, whereas the ventrolateral subdivision only innervates contralateral extrinsic eye muscles. On the other hand, the ventromedial subdivision innervates extrinsic eye muscles bilaterally.

Nieuwenhuys (1998) was able to identify a small trochlear nerve in *Lepidosiren*, but failed to locate the motor nucleus. Burckhardt claimed that the trochlear motor nucleus is located in the cerebellum in *Protopterus* but von Bartheld (1992) has experimentally shown that the trochlear nucleus consists of approximately only 20 motor neurons, which occupy the ventrolateral periventricular wall and floor of the midbrain tegmentum approximately 100 μm caudal to the caudal pole of the oculomotor nucleus. Given its close association with the oculomotor nucleus, neurons of the trochlear motor nucleus almost certainly have been considered part of the oculomotor nucleus in most studies. In *Lepidosiren*, neither an abducent motor nucleus nor an abducent nerve has been identified (Nieuwenhuys 1998), but in *P. dolloi* an abducent motor nucleus has been experimentally identified as a collection of some 30 motor neurons located slightly caudal to the Mauthner cells at mid-medullar levels.

The hypobranchial muscles arise from the postotic myotomes of the head, and the differentiating muscle cells migrate laterally and ventrally to form the hypobranchial muscles, which function in fishes to open the lower jaw, elevate the floor of the mouth, and extend the gill pouches. In lungfishes these muscles are innervated by the somatic motor neurons of the rostral spinal cord (Figs. 3A, 4A). The axons of these motor neurons issue the spinal cord as two pairs of spino-occipital nerves (Fig. 1). In amniotes, the collection of nerves that innervate these muscles is called the hypoglossal nerve, and that portion of the somatic motor column that innervates the hypobranchial muscles is called the hypoglossal motor nucleus. The descriptor "spino-occipital" nerves is normally used in fishes, because of the variable number of nerves that innervate the muscles, whether or not the nerves issue from the skull, and the fact that they frequently possess dorsal root ganglia and have sensory functions which are lacking in other somatic motor nerves, with the exception of those nerves that contain the peripheral processes of mesencephalic trigeminal neurons.

CEREBELLUM

In all three genera of lungfishes, the cerebellum is divided into a plate-like corpus or body, paired auricles, and a single pair of cerebellar nuclei (Figs. 1, 3E,F, 4E,F). Each auricle consists of an upper leaf, which is continuous with the body of the cerebellum, and a lower leaf which is continuous with the more caudal octavolateral columns (Figs. 1, 3F, 4F). In lepidosirenid lungfishes much of the body of the cerebellum is covered by the optic tectum (Figs. 1A, 3F, 4F), but in *Neoceratodus*, the optic tectum does not overlap any part of the cerebellum, even though the optic tectum is relatively better developed than that in lepidosirenid lungfishes.

Histologically, the body of the cerebellum in all lungfishes consists of an outer molecular layer, a middle layer of Purkinje cells, and an inner layer of granular cells (Figs. 3E, 4E). Unlike the cerebellum in other vertebrates, however, the Purkinje cells do not form a distinct lamina but are more scattered, and a small number of these cells occur in both the molecular and granular layers (Figs. 3E, 4F). The auricles of the cerebellum consist almost entirely of granular cells, but a small number of Purkinje cells are also scattered among the granular cells of the upper leaf of the auricle in *Protopterus* (Figs. 3F, 4F).

Near its rostral border, the body of the cerebellum is reduced to a small island of granular cells with an occasional Purkinje cell (Figs. 3F, 4F), and the periventricular layer, which more caudally forms the granular cell layer, transforms into a cerebellar nucleus that protrudes into the fourth ventricle (Figs. 3F, 4F).

Almost nothing is known about cerebellar connections in lungfishes. Holmgren and van der Horst (1925) tried to describe these connections in non-experimental material and suggested that lateral line fibers reach the auricle and the corpus. This has been partly substantiated in *Protopterus*, where the primary projections of the lateral line nerves have been traced by labeling them with horseradish peroxidase (Northcutt 1983a). Rami of the lateral line nerves that carry information from mechanoreceptive neuromasts do reach the ipsilateral auricle (eminentia granularis) and corpus of the cerebellum, but the lateral line rami that carry information from electroreceptive ampullary organs only terminate in the ipsilateral dorsal octavolateral nucleus. Holmgren and van der Horst (1925) predicted that spinal and trigeminal information should reach the corpus. Although these putative connections have not been examined with experimental methods, they probably exist. Tectocerebellar connections have also been postulated, but if lungfishes are like other anamniotic taxa, the pretectum, not the optic tectum, will be found to project to the cerebellum. A presumed efferent bundle, the superior cerebellar tract, or brachium conjunctivum, has been claimed to exit the cerebellar nucleus in *Neoceratodus* (Holmgren and van der Horst 1925) and *Protopterus* (Niewenhuys 1998), and is believed to terminate in the midbrain tegmentum, but this tract also awaits confirmation by experimental methods.

Immediately rostral to the cerebellum, the width of the brain narrows dramatically in all three genera of lungfishes, forming a transitional region between the medulla oblongata and the midbrain (Fig. 1), termed the isthmus. The isthmus is perhaps best known as a brain region responsible for organizing the cerebellum and midbrain developmentally, and its walls form a putative homologue of the secondary gustatory nucleus (described in the visceral sensory section of the medulla) and a more dorsally located superficial isthmal nucleus (Figs. 5A,B, 6A,B). The superficial isthmal nucleus consists of small, darkly staining granular cells that occupy a submeningeal position throughout the length of the optic tectum and are considered to be isthmal in origin due to their close association with the most dorsomedial and rostral granular cells of the body of the cerebellum (Figs. 3F, 4F). The only known input to the superficial isthmal nucleus is a direct retinal input to its most rostral portion, which has been experimentally established in all three genera of lungfishes (Clairambault and Flood 1975; Northcutt 1977, 1980).

The isthmal floor consists of a median interpeduncular nucleus flanked laterally by a periventricular central rhombencephalic gray formation. The interpeduncular nucleus in lepidosirenid lungfishes is almost impossible to identify in cresyl violet stained material, as the nucleus contains very few cells. In Bodian silver material, however, it is very distinct as a midline neuropil located immediately rostral to the cerebellar auricles. The bulk of the interpeduncular neuropil appears to consist of the fibers of the fasciculus retroflexus (Figs. 5C, 6C) which is the main descending pathway of the habenular nuclei. The periventricular central rhombencephalic gray forms a transitional zone between the floor of the rostral medulla and the midbrain tegmentum. At present, nothing is known about its connections or functions.

MIDBRAIN

The midbrain of lepidosirenid lungfishes is oval in transverse sections (Figs. 5A, 6A), and its lateral walls are flanked by a narrow vertically oriented ventricle. With the exception of the superficial isthmal nucleus, almost all of the midbrain's neurons occupy a periventricular position, and individual cell groups are best distinguised by differences in the thickness of the periventricular cellular layer and the adjacent ventricular sulci. The dorsolateral walls form the optic tectum and the torus semicircularis, and the ventrolateral walls form the midbrain tegmentum. In general, the dorsolateral periventricular layer is thicker than the ventrolaterally situated tegmentum. Caudally, the caudal pole of the midbrain extends over the corpus of the cerebellum (Figs. 3F, 4F), and its rostral border is marked dorsally by the posterior commissure of the pretectum (Figs. 5B, 6B) and ventrally by the larger cells of the oculomotor nucleus (Figs. 5A, 6A). In contrast to the midbrain of lepidosirenid lungfishes, the midbrain of *Neoceratodus* is more heart-shaped when viewed in transverse sections, the roof or optic tectum is bilobed, and the optic tectum is distinctly laminated. Beyond these differences, however, the same cell groups and boundaries can be recognized in all three genera of lungfishes.

In lepidosirenid lungfishes, the border between the optic tectum and torus semicircularis is marked by a slight constriction of the periventricular cell plate (Figs. 5A, 6A), which coincides with the ventral extent of the retinal terminal field in the outer third of the tectal wall (Northcutt 1977). The tecto-toral border in *Neoceratodus* is even easier to detect due to the lamination of the tectum (Northcutt 1980).

The piriform-shaped neurons of the tectal periventricular layer In lepidosirenid lungfishes send single apical dendrites into the more superficial tectal neuropil, where these dendrites branch two or three times in the outer half of the tectal neuropil (Clairambault and Flood 1975). No other experimental information exists on afferent projections to the optic tectum, but given the experimental information from other groups of fishes (Northcutt 1982, 1983b), the inner half of the tectal neuropil is a likely site for the termination of ascending lateral line and spinal pathways.

Descriptions of lungfish brains based on non-experimental Bodian preparations (Holmgren and van der Horst 1925; Nieuwenhuys 1998), and the experimental data available on other groups of fishes (Northcutt 1983b), suggest that the axons of most tectal cells collect in the tectal neuropil and form both ascending and descending tectofugal pathways. The ascending tectofugal pathway probably issues ipsilaterally and passes ventral to the entering fibers of the optic tract to terminate in the ipsilateral pretectum, dorsal thalamus, and ventral thalamus. A small number of these ascending tectal efferents probably decussate in the postoptic comissure (Figs. 5E, 6E) and turn caudally to terminate in the contralateral diencephalon. The descending tectofugal fibers almost certainly exit in the deeper layers of the tectal neuropil and course ventrally into the midbrain tegmentum, where many of them decussate to the contralateral tegmentum. Thus both crossed and uncrossed tectobulbar and tectospinal pathways must exist, as reported by Ronan and Northcutt (1985), who labeled neurons in the optic tectum of both *Lepidosiren* and *Protopterus* after injections of horseradish peroxidase ipsilaterallly into the rostral spinal cord.

Again, there is no experimental information on the connections of the torus semicircularis of lungfishes. Based on information from other groups of fishes (Wullimann 1998), the torus semicircularis of lungfishes almost certainly receives both auditory and mechanoreceptive lateral line inputs and projects ipsilaterally to the caudal dorsal thalamus. It may or may not possess descending torofugal projections to one or more of the primary octaval nuclei, but there is no evidence that any projections from the torus reach spinal levels (Ronan and Northcutt 1985).

DIENCEPHALON

In lungfishes, the diencephalon, like the midbrain, primarily consists of an extensive periventricular cellular plate bordering a slender, vertically oriented ventricle (Figs. 5B-E, 6B-E). Unlike the midbrain, however, the diencephalic periventricular cellular plate is better differentiated, so that differences in the thickness of the cellular plate and the packing density of its cells offer valuable clues for the recognition of the different centers. Developmental studies also provide an additional set of criteria for distinguishing different diencephalic cell groups and their topography (Bergquist and Kallén 1954; Vaage 1969; Puelles and Rubenstein 1993). Today there is a general consensus that the diencephalon in all vertebrates consists of three genetically determined regions of cell proliferation, whose boundaries restrict cell lineage and facilitate axonal outgroups termed neuromeres. The diencephalon can be divided into a caudal neuromere, termed the synencephalon, and a rostral neuromere, termed the parencephalon, which is further divided into anterior and posterior subdivisions. The synencephalon is divided into a dorsal pretectum and a ventral nucleus of the medial longitudinal fasciculus (Figs. 5A, 6A). The posterior parencephalon forms a dorsal epithalamus, an intermediate dorsal thalamus, and a ventral posterior tubercle (Figs. 5C-E, 6C-E). The anterior parencephalon consists of the ventral thalamus, which more accurately should be termed the prethalamus (Figs. 5B-E, 6B-E). Although the hypothalamus has traditionally been considered a diencephalic region, neuromeric models indicate that the preoptic area, hypothalamus, and telencephalon form a developmental unit termed the secondary prosencephalon. For this reason, the preoptic area and hypothalamus will be described in the section on the secondary prosencephalon.

Synencephalon. Caudally, the pretectum of lungfishes consists of a periventricular cellular plate, which is capped dorsally by the posterior commissure (Figs. 5B, 6B). More rostrally, the periventricular cellular plate is displaced dorsally by the subcommissural organ (Figs. 5C, 6C). At this level, a small number of migrated neurons are scattered in the neuropil. Throughout its length, the pretectum is bordered ventrally (which is actually rostrally, developmentally) by the dorsal thalamus, from which it can be distinguished by the broader periventricular cellular plate of the dorsal thalamus (Figs. 5C, 6C). The primary retinal projections that terminate in the outer half of the neuropil in lepidosirenid lungfishes (Clairambault and Flood 1975; Northcutt 1977), and terminate more deeply in *Neoceratodus* (Northcutt 1980), are the only connections confirmed experimentally. In other fishes, however, the pretectum forms extensive connections with the hypothalamus and cerebellum (Wullimann 1998).

The nucleus of the medial longitudinal fasciculus forms the ventrolateral walls of the synencephlon and consists of large neurons that begin rostral to the oculomotor nucleus but continue caudally to overlap the rostral one-third of this nucleus. The nucleus of the medial longitudinal fasciculus is the only group of cells in the midbrain whose projections have been experimentally determined, at least in part. The axons of these cells form one of the major components of the medial longitudinal fasciculus, which can be traced through the medulla (Figs. 3B-D, 4B-D) to the spinal cord, where they form many of the axons in the ventral funiculus (Figs. 3A, 4A). Injections of horseradish peroxidase into one side of the rostral spinal cord of lepidosirenid lungfishes indicate that the cells of the medial longitudinal fasciculus project both ipsilaterally and contralaterally. It is probable that many of these cells terminate on the somatic motor cells that innervate the extrinsic eye muscles and on the cells of the reticular formation.

Posterior parencephalon. The epithalamus is the most dorsally situated segment of the posterior parencephalon. The epithalamus is composed of the habenular nuclei and their associated pathways – the stria medullaris and the habenular commissure (Figs. 5D,C, 6D,E) – and an evaginated epiphysis (Figs. 5C, 6C). The habenular nuclei comprise a shell of densely packed small neurons surrounding a centrally located neuropil composed of the fibers of the stria medullaris as well as the dendrites and axons of the more peripheral cells of the habenular nuclei (Figs. 5D, 6D). In both *Lepidosiren* and *Protopterus*, the right habenular nucleus is larger than the left nucleus, but this does not appear to be true for *Neoceratodus* (Holmgren and van der Horst 1925). Both habenular nuclei are bridged by the fibers of the habenular commissure, which is formed in part by at least some of the fibers of the stria medullaris. Throughout their rostrocaudal length, the habenular nuclei are bordered ventrally by the dorsal thalamus, but there is no break between the periventricular cellular plates of the nuclei. Rather, the cells of the habenular nuclei gradually grade over into those of the dorsal thalamus. Generally, however, the dorsal thalamus as a whole can be distinguished from the habenular nuclei by the increased thickness of its periventricular cellular plate, particularly in its dorsal half. There are no experimental data on the connections of the habenular nuclei in lungfishes, and the data reported for other vertebrates are highly variable. It is possible, however, that the habenular nuclei receive secondary olfactory projections that arise in the olfactory bulbs, as well as afferents from the lateral pallium, septal nuclei, dorsal pallidum, preoptic area, and hypothalamus. The main efferent pathway that arises in the habenular nuclei is the fasciculus retroflexus (Figs. 5C, 6C), which terminates in the interpeduncular nucleus in the rostral floor of the midbrain, as described in the section on the midbrain.

The epiphysis or pineal gland of lungfishes consists of a stalk and a terminal vesicle, the latter lined by sensory cells similar to those reported in amphibians (Holmgren 1959). The stalk of the epiphysis consists of fibers which almost certainly are both afferent and efferent in nature. Although there are no experimental studies of this structure in lungfishes, the epiphysis probably has efferents to the suprachiasmatic nucleus and hypothalamus, it may secrete melatonin, and it is probably involved in circadian rhythms.

The largest part of the posterior parencephalon is formed by the dorsal thalamus (Figs. 5B-E, 6B-E). The dorsal thalamus is bordered dorsally by the habenular nuclei and the pretectum. Ventrally, the dorsal thalamus is bordered by the ventral thalamus, whose periventricular cellular plate is wider than that of the dorsal thalamus. Although Schnitzlein and Crosby (1968) recognized numerous subdivisions, including a number of migrated cell groups, other workers (Clairambault *et al.* 1974c; Northcutt 1977; Nieuwenhuys 1998) have not supported these claims and believe that the dorsal thalamus exhibits no migrated cell groups. Generally, the dorsal thalamus can be divided into rostral and caudal subdivisions based on differences in the organization of its periventricular cellular plate. Rostrally, the periventricular plate protrudes into the ventricle and is subdivided, albeit poorly, into three or four cellular laminae (Figs. 5D,E, 6D,E). More caudally, the periventricular cellular plate of the dorsal thalamus protrudes only slightly into the ventricle, and the cellular plate is far more uniform (Figs. 5C, 6C).

The primary retinal projections are the only afferents to the dorsal thalamus that have been determined experimentally (Clairambault and Flood 1975; Northcutt 1977, 1980). Although retinal projections to the dorsal thalamus of both lepidosirenid lungfishes are reported to be solely contralateral (Clairambault and Flood 1975; Northcutt 1977), recent unpublished work in my laboratory, with the use of more sensitive tracers, indicates that the retinal projections in both lepidosirenid lungfishes are bilateral, as has been reported for *Neoceratodus* (Northcutt 1980). In all three genera of lungfishes, the retinal projections to the dorsal thalamus are heavier contralaterally than ipsilaterally. Furthermore, the retinal terminals do not form a continuous sheet, but rather comprise a series of punctuate terminal fields that are mainly confined to the rostral dorsal thalamus. In *Protopterus,* where the terminal fields were plotted in detail, only one small terminal field was seen to occur in the caudal subdivision of the dorsal thalamus (Northcutt 1977). Much needed experimental studies will almost certainly reveal that the dorsal thalamus of lungfishes receives tectal and toral inputs, and if the thalamic efferents in lungfishes are similar to those of ray-finned fishes (Northcutt 2006, 2008), most thalamic efferents will project primarily to the ipsilateral septum and striatum via the lateral forebrain bundle (Figs. 5C-E, 6C-E).

The posterior tubercle (Figs. 5B-D, 6B-D) forms the most ventral cell group in the posterior parencephalon. Rostrally the nucleus is recognized in *Protopterus* and the other lungfishes as a narrow periventricular cellular plate interposed between the more dorsally located ventral thalamus and the more ventrally located dorsal hypothalamus. The posterior tubercle is easily distinguished from both these cell groups by its more narrow cellular plate (Figs. 5D, 6D). The posterior tubercle increases in height as it is traced caudally, and comes to be characterized by large neurons (Figs. 5B, 6B) that are positive for tyrosine hydroxylase and are probably dopaminergic (Reiner and Northcutt 1987). Although the connections of the posterior tubercle in lungfishes have not been experimentally determined, the tubercle receives a substance P positive input from the striatum (Reiner and Northcutt 1987), and the posterior tubercle probably projects back upon the striatum and to the spinal cord as in other fishes (Wullimann and Vernier 2007).

Anterior parencephalon. The ventral thalamus (Figs. 5B-E, 6B-E) is the sole cell group in this subdivision of the diencephalon, and it is characterized, particularly rostrally, by the Figure 8-like configuration of its periventricular cellular plate (Figs. 5D,E, 6D,E). More caudally, it gradually grades into the posterior tubercle ventrally (Figs. 5C, 6C). Retinal input is the only connection that has been experimentally determined for the ventral thalamus. In lepidosirenid lungfishes, the retinal projection terminates throughout most of the peripheral neuropil of the ventral thalamus (Clairambault and Flood 1975; Northcutt 1977); whereas in *Neoceratodus,* it terminates among the cell bodies of the dorsal segment of the periventricular cellular plate of the ventral thalamus, as well as in its more superficial neuropil (Northcutt 1980). The ventral thalamus of lungfishes probably also receives a bilateral tectal input and probably projects back upon the optic tectum, as is the case in other fishes (Northcutt 1982, 1983b).

SECONDARY PROSENCEPHALON

The preoptic area, the hypothalamus, and the telencephalon, with its olfactory bulbs, constitute the secondary prosencephalon in lungfishes and other vertebrates (Puelles and Rubenstein, 1993).

Preoptic area and hypothalamus. The postoptic commissure decussates through the suprachiasmatic ridge (Figs. 5E, 6E), which divides the floor of the diencephalic ventricle into a rostral preoptic area and a more caudal hypothalamus (Figs. 5B-E, 6B-E, 7A, 8A). The preoptic area forms a broad periventricular cellular plate (Figs. 7A, 8A); rostrally, it begins immediately caudal to the lamina terminalis, which consists of the bed nucleus of the stria terminalis and the anterior commissure (Figs. 7B, 8B); caudally, it continues to form the rostral wall and dorsal surface of the suprachiasmatic ridge (Figs. 5E, 6E). Throughout most of its rostrocaudal extent, the preoptic area consists of small, tightly packed neurons, whose apical dendrites extend into a lateral neuropil (Figs. 7A, 8A), but just caudal to the level illustrated in Figure 7A, a group of much larger cells, the magnocellular preoptic nucleus, begins to replace the small cells of the preoptic cellular plate dorsally. More caudally, the large cells of the magnocellular preoptic nucleus are replaced, in turn, by a second cellular plate of small neurons (Figs. 5E, 6E). Thus, the large cells of the magnocellular preoptic nucleus essentially divide the periventricular cellular plate of the preoptic area into anterior (Figs. 7A, 8A) and posterior (Figs. 5E, 6E) periventricular preoptic nuclei. Once again, the retinal efferents are the only known projections to the preoptic area (Northcutt 1980). In *Neoceratodus*, substantial retinal projections occur bilaterally in the anterior periventricular preoptic nucleus, as well as more caudally in a lateral segment of the posterior periventricular preoptic nucleus that represents a putative homologue of the suprachiasmatic nucleus in other vertebrates. In lepidosirenid lungfishes a retinal projection to the preoptic area was not reported by Northcutt (1977), and Clairambault and Flood (1975) were uncertain regarding such a projection in *Protopterus*. In any case, the preoptic area almost certainly has reciprocal connections with a number of telencephalic and hypothalamic areas in lungfishes, as it does in other fishes (Wullimann 1998; Northcutt 2006).

The hypothalamus of lungfishes begins rostrally along the caudal surface of the suprachiasmatic ridge (Figs. 5D, 6D) and continues as a flattened hollow lobe beneath the midbrain floor (Figs. 5B, 6B) to end caudally in contact with the pituitary. Throughout most of its rostrocaudal extent, the hypothalamus is divided into dorsal and ventral nuclei (Figs. 5B-D, 6B-D), but different authors (Holmgren and van der Horst 1925; Northcutt 1977; Nieuwenhuys 1998) disagree as to the exact boundaries of these two nuclei. In *Protopterus*, the dorsal hypothalamic nucleus (Figs. 5D, 6D) is a wedge-shaped cell group, whose width is greater than that of the ventral hypothalamic nucleus, whose cells are darker and more densely packed. More caudally, the cells of the dorsal hypothalamic nucleus are still more scattered, but the width of the cellular plate decreases (Figs. 5C, 6C). Finally, the dorsal hypothalamic nucleus ends dorsally as a caudal recess that separates from the more ventral part of the dorsal hypothalamic and ventral hypothalamic nuclei (Figs. 5B, 6B). No hypothalamic connections have been experimentally demonstrated in lungfishes, with the exception of the efferents of the cells of the paraventricular organ in *P. dolloi* (von Bartheld and Meyer 1990). The paraventricular organ in *Protopterus* consists of approximately 300 cerebrospinal fluid-contacting neurons located in both caudal divisions of the dorsal hypothalamic nucleus. These neurons were inadvertently labeled with cobaltous lysine during an attempt to trace retinal efferents. Label that leaked out of the injected eye appears to have entered the blood stream and then pinocytosed from the ventricular cerebrospinal fluid by the cells of the paraventricular organ. The cells of this organ reportedly terminate in the medial wall of the telencephalon, the preoptic area, the habenular nuclei, the anterior dorsal thalamus, the optic tectum, the superficial isthmal nucleus, the middle reticular formation, and even the spinal cord. Von Bartheld and Meyer suggested that the cells of the paraventricular organ are involved in neuroendocrine control and, perhaps, estivation. It is not known whether or not these cells occur in *Neoceratodus*. The hypothalamo-hypophyseal connections are not covered in this review, but Nieuwenhuys (1998) provides an excellent summary.

Telencephalon and olfactory bulbs. The telencephalon of lungfishes consists of a caudal telencephalon impar (Figs. 7A,B, 8A,B), which is that part of the telencephalon not involved in the embryonic evagination that produces the telencephalic hemispheres (Figs. 7C,D, 8C,D, 9A, 10A), and a pair of olfactory bulbs (Figs. 9B, 10B), which form as secondary evaginations from the rostrodorsal poles of the telencephalic hemispheres.

Unfortunately, there is no consensus regarding division of the cell groups of the telencephalon impar or the hemispheres in lungfishes, nor is there any agreement regarding their possible homologues in other vertebrates. Most of the disagreements (Figs. 12-15) stem from differing interpretations of three features: 1) the border between the medial pallium and the septal nuclei; 2) the extent of the striatopallidal systems; and 3) the organization of the amygdala.

Medial pallial-septal border. The traditional view regarding this border is that the entire medial telencephalic wall in lungfishes is homologous to the septal nuclei (Elliot Smith 1908; Holmgren and van der Horst 1925; Kuhlenbeck 1929; Gerlach 1933; Rudebeck 1945; Whitt 1969; von Bartheld *et al.* 1990; Nieuwenhuys 1969, 1998). Proponents of this hypothesis are said to believe the restricted pallial hypothesis (Fig. 12B), i.e., that the pallium of lungfishes is restricted to the dorsal and dorsolateral roof of the telencephalic hemisphere. Proponents of an alternative view, the extended pallial hypothesis (Fig. 12C), believe that the medial wall of the telencephalic hemispheres in lungfishes consists of a dorsal medial pallium and ventral septal nuclei, as it does in other vertebrates (Schnitzlein and Crosby 1967; Clairambault and Capanna 1973; Northcutt 1986a, 2008; Reiner and Northcutt 1987).

In this chapter, the extended pallial hypothesis is accepted, and the medial pallium is interpreted as a pallial division that is characterized by extensive cell migrations. It is believed to occupy the ventromedial rostral pole of the telencephalic hemisphere (Figs. 9B, 10B). More caudally, it expands to occupy most of the medial hemispheric wall (Figs. 9A, 10A); dorsally, it borders the dorsal pallium; and ventrally, it borders the lateral septal nucleus. Both the dorsal pallium and the lateral septal nucleus are formed by densely packed periventricular cellular plates. At mid-hemispheric levels (Figs. 7D, 8D), the lateral septum has expanded, and the medial pallium is restricted to the dorsal half of the medial hemispheric wall. Near the caudal pole of the telencephalic hemisphere, the medial pallium continues to occupy the dorsal half of the medial wall, and both lateral and medial septal nuclei can be recognized (Figs. 7C, 8C). At this level, the medial ependymal thickening marks the position of the medial pallial–subpallial border. The interventricular foramen forms immediately caudal to this area, interconnecting

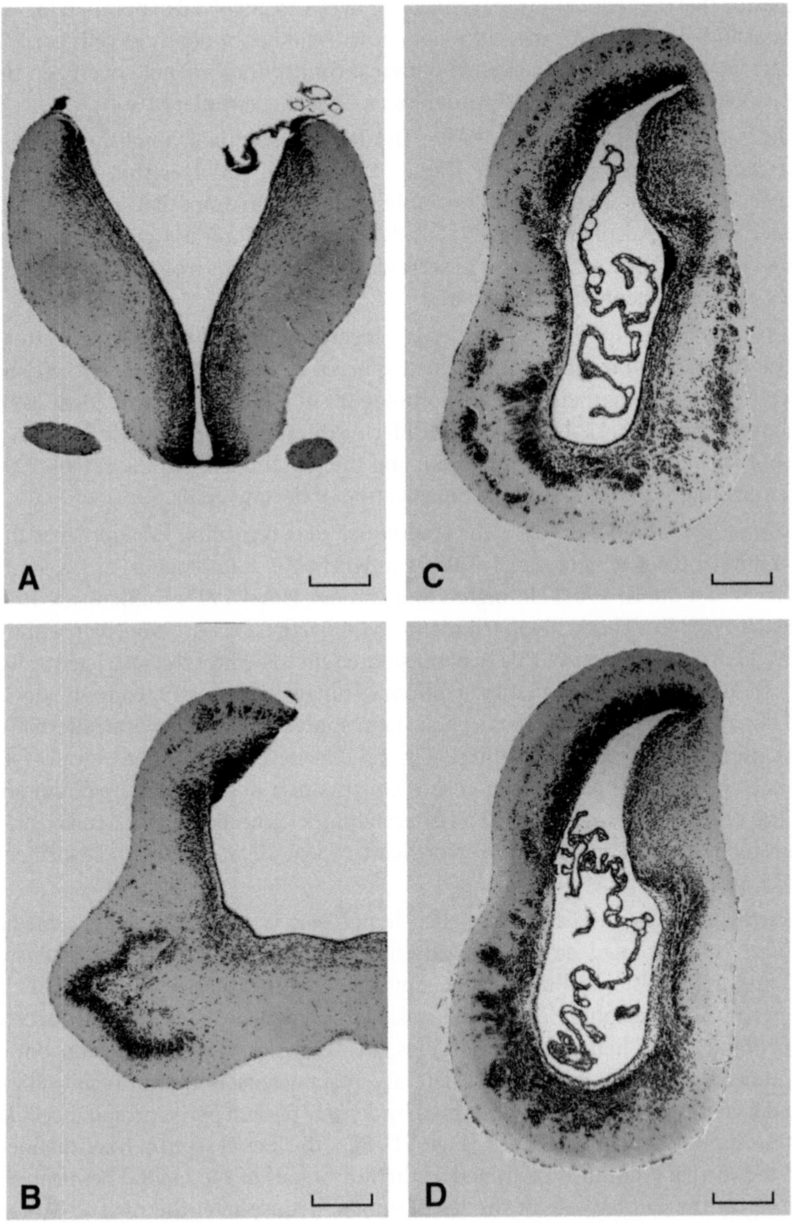

Fig. 7 Photomicrographs of transverse sections through the telencephalon impar (A, B) and caudal telencephalic hemisphere (C, D) of the Spotted African lungfish. Bar scales equal 500 μm.

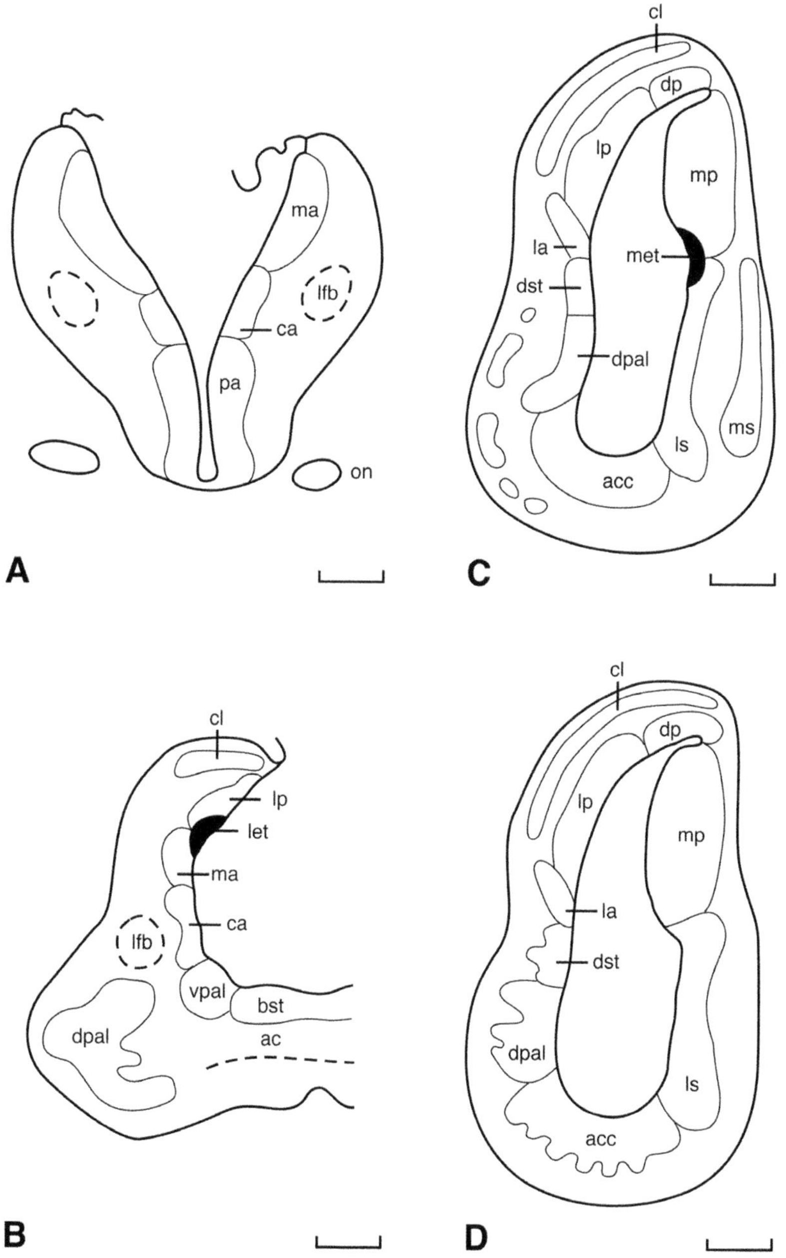

Fig. 8 Line drawings of the transverse sections through the telencephalon impar and caudal telencephalic hemisphere pictured in Figure 7.

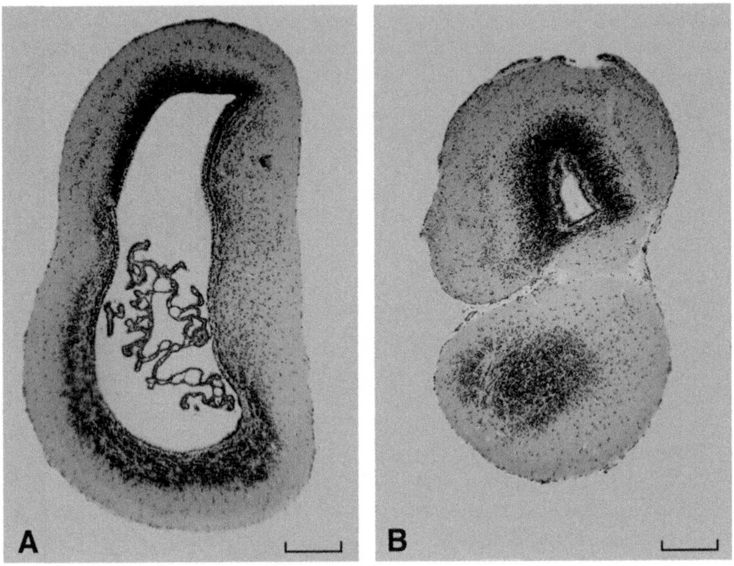

Fig. 9 Photomicrographs of transverse sections through the rostral telencephalic hemisphere (A) and olfactory bulb (B) of the Spotted African lungfish. Bar scales equal 500 μm.

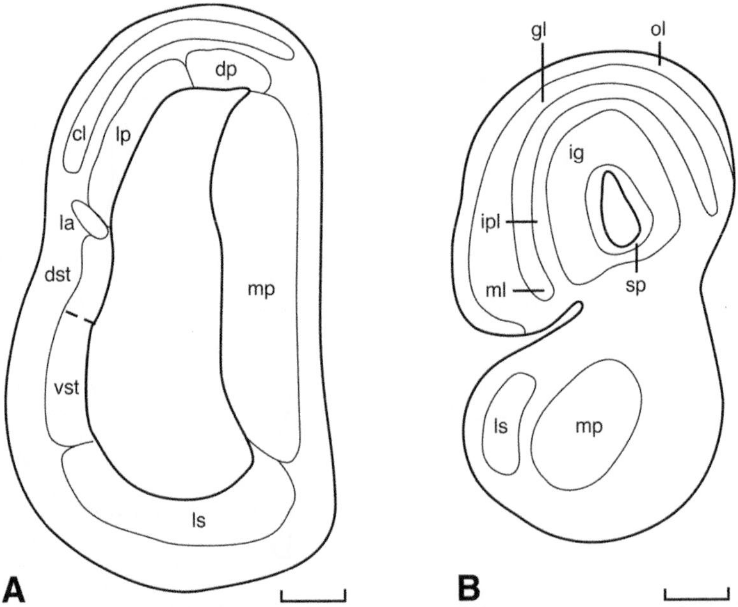

Fig. 10 Line drawings of the transverse sections through the rostral telencephalon and olfactory bulb pictured in Figure 9.

the lateral ventricles of the telencephalic hemispheres with the unpaired ventricle of the telencephalon impar (Figs. 7A,B, 8A,B, 14A). The caudal end of the lateral septal nucleus forms the roof of the interventricular foramen (Fig. 14A,C), and it and the caudal medial pallium are immediately replaced by the tela choroidea (Figs. 7B, 8B).

The extended pallial hypothesis was initially accepted (Northcutt 1986a) for two reasons: the marked similarity in the positions (1) and the relative sizes (2) of the cell groups in the medial hemispheric wall in lepidosirenid lungfishes and amphibians. Subsequently, the distribution of substance P in the medial hemispheric wall of *Protopterus* (Fig. 14B) also supported the extended pallial hypothesis (Reiner and Northcutt 1987). The distribution of substance P is one of the most robust indicators of the medial pallial-subpallial border in a large number of vertebrate groups (Marín *et al.* 1998a). Although there are no published data on the connections of the medial pallium or the septal nuclei in lungfishes, there is an extensive literature on the differences in the connections of these two cell groups in amphibians, which are the logical outgroup for comparison. These differences have been reviewed in Northcutt (2008), and experimental studies on lungfishes, currently in progress in my laboratory, indicate that the cell groups identified as the medial pallium and septal nuclei in the extended pallial hypothesis are homologous to the same named cell groups in amphibians.

The major challenge to the extended pallial hypothesis is incorporated into seven points raised by von Bartheld *et al.* (1990): 1) The position of the interventricular foramen in lungfishes is inconsistent with a large medial pallium; 2) Comparison of the telencephalic ependymal thickenings between lungfishes and amphibians is inconsistent with a large medial pallium; 3) In amphibians, the septal nuclei receive hypothalamic input, but the medial pallium does not; 4) In amphibians, the medial pallium receives only a small secondary olfactory input; 5) In amphibians, the terminal nerve projects to the septal nuclei but not to the medial pallium; 6) In amphibians, acetylcholinesterase (AChE) is high in the septal nuclei but low in the medial pallium; and 7) The chemical markers described by Reiner and Northcutt (1987) in *Protopterus* indicate that the medial pallial-subpallial border occurs at the sulcus hippocampi, which lies at the dorsalmost point of the medial hemispheric wall (Fig. 14B). Each of these points can be refuted as follows:

(1) Von Bartheld *et al.* (1990) interpreted the interventricular foramen in *Protopterus* as occurring at the dorsomedial angle of the medial hemispheric wall (their Figs. 13H and 15), in a position where the pallium thins caudally into a tela choroidea (Fig. 14C), but the interventricular foramen occurs more ventrally (Fig. 14A,C), as was indicated by Gerlach (1933) and Nieuwenhuys (1998). If the interventricular foramen occurred dorsally, as suggested by von Bartheld and colleagues, all of the cell group identified as the medial pallium in the extended pallial hypothesis would

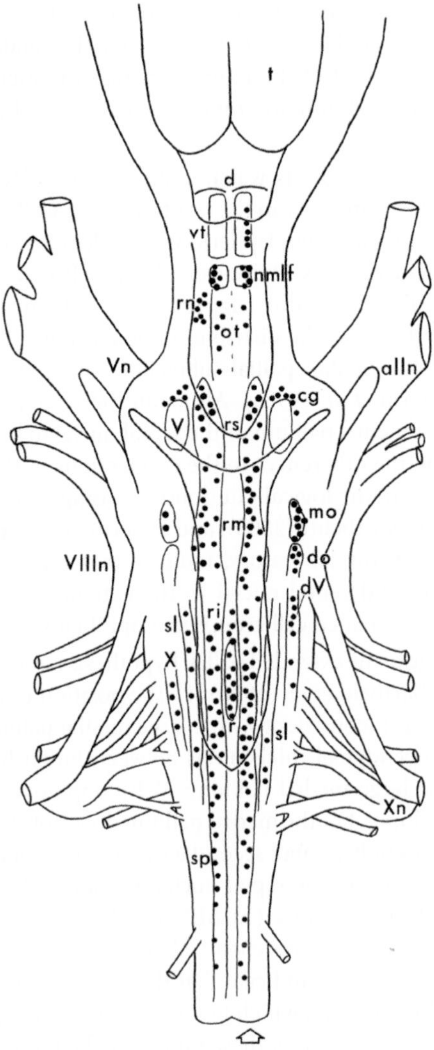

Fig. 11 A line drawing of the brain of an African lungfish, in dorsal view, showing the descending spinal projections in lepidosirenid lungfishes. Neurons retrogradely labeled by introduction of HRP into the spinal cord are indicated by smaller black dots, larger neuronal somata by larger black dots. Open arrow at the bottom indicates the side of the spinal cord that was injected with HRP. (From Ronan and Northcutt, 1985). Abbreviations: alln, anterior lateral line nerve; cg, granule cells of the cerebellar corpus; d, diencephalon; do, descending octaval nucleus; dV, descending trigeminal tract; mo, magnocellular octaval nucleus; nmlf, nucleus of the medial longitudinal fasciculus; ot, optic tectum; r, raphe; ri, inferior reticular nucleus; rm, middle reticular nucleus; rn, red nucleus; rs, superior reticular nucleus; sl, solitary nucleus and tract; sp, spinal gray matter; t, telencephalon; V, trigeminal motor nucleus; Vn, trigeminal nerve; vt, ventral thalamus; X, vagal motor nucleus; Xn, vagal nerve.

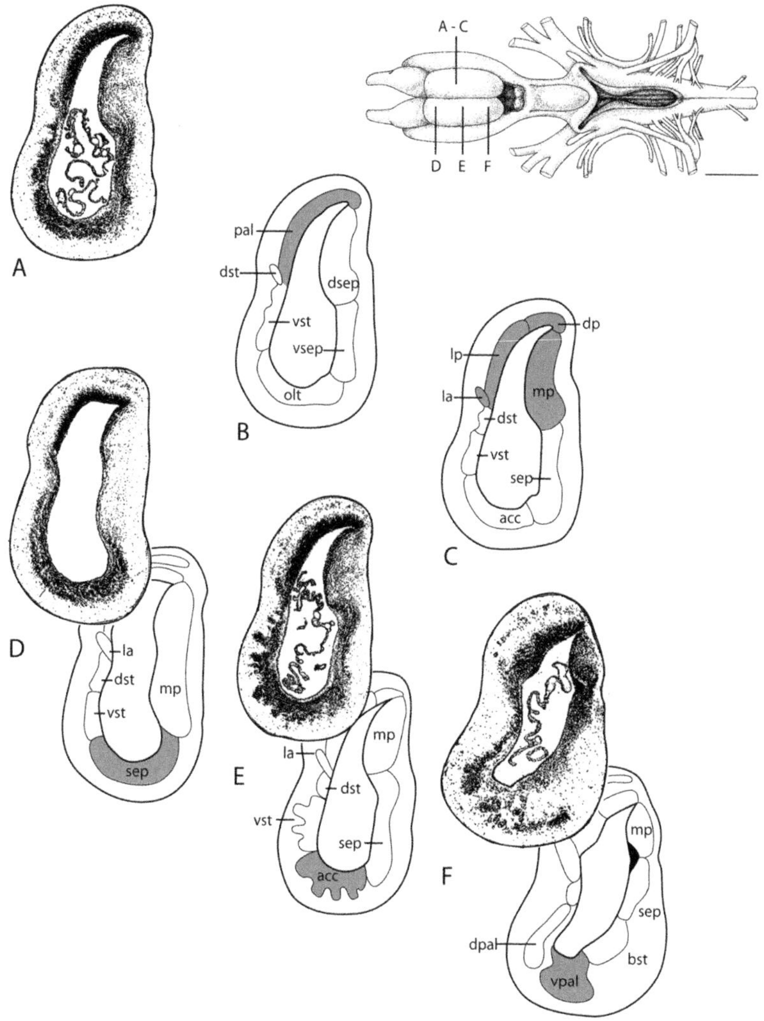

Fig. 12 Hypotheses regarding a restricted pallium vs. an extended pallium in the African lungfish (A-C), and the two hypotheses regarding the ventral striatopallidal system (D-F). The hypothesis of a restricted pallium, illustrated by gray shading in B, holds that the pallium is restricted to the dorsolateral roof of the hemisphere, while the entire medial wall is occupied by the septum; the hypothesis of an extended pallium, illustrated by gray shading in C, holds that there are four divisions of the pallium, and, together, they occupy the dorsomedial hemispheric wall, as well as the dorsolateral roof. One hypothesis regarding the ventral striatopallidal system holds that the olfactory tubercle (indicated by gray shading in D-F) extends the entire length of the cerebral hemisphere, and claims that there is no ventral pallidum. The second hypothesis holds that the floor of the hemisphere comprises a rostral septum (D), a more ventral striatum (nucleus accumbens), and a ventral pallidum. (From Northcutt 2008).

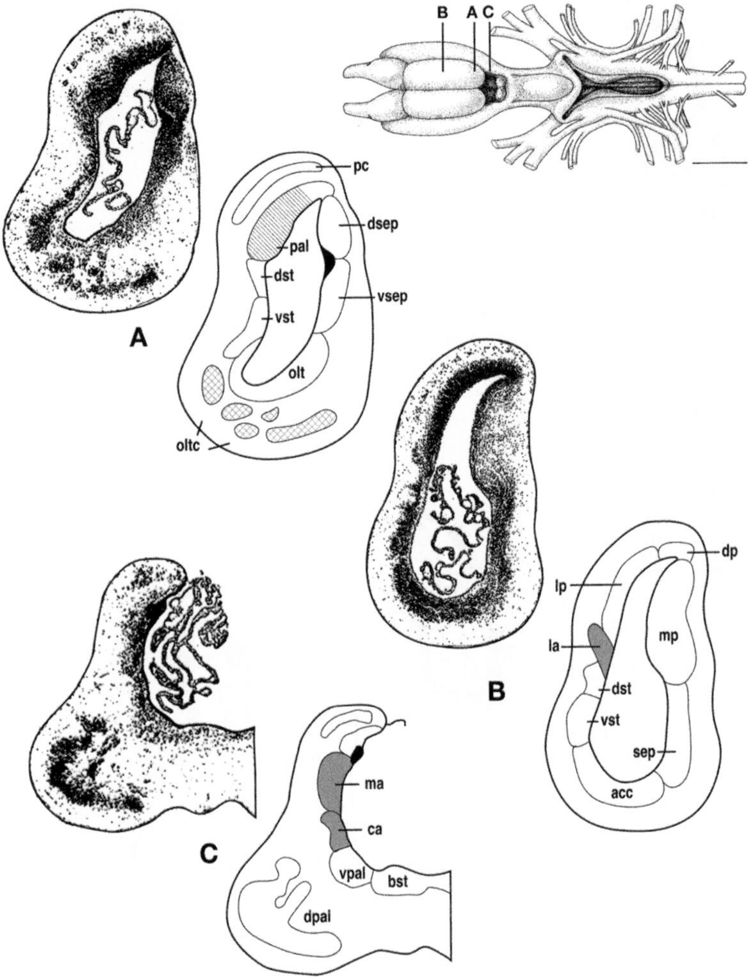

Fig. 13 Interpretations of amygdalar organization in the African lungfish. The cell groups in the lateral wall and floor of the hemisphere have been subject to differing interpretations by different researchers: The caudal dorsolateral pallium (oblique lines in A) was termed amygdalar nuclei by Schnitzlein and Crosby (1967); Cell clusters in the caudal floor (cross-hatching in A) were interpreted as the amygdala by Clairambault and Capanna (1973); Cell groups forming the dorsal half of the telencephalon impar (gray areas in C) were said to be the amygdala by Rudebeck (1945). A new hypothesis, proposed herein, holds that the amygdala has three divisions: a lateral nucleus (gray area in B), a central nucleus, and a medial nucleus (gray areas in C). (From Northcutt 2008).

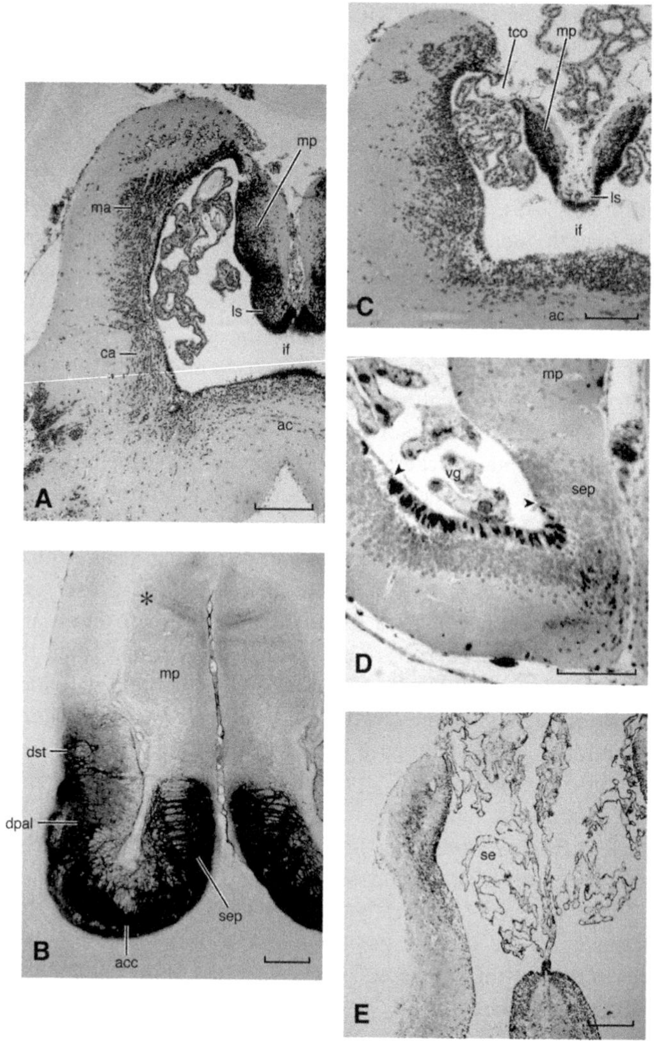

Fig. 14 Photomicrographs of transverse sections through the telencephalic hemisphere of the Spotted African lungfish (A-C), an axolotl, *Ambystoma mexicanum* (D), and an Australian lungfish (E). A) A caudal section through the interventricular foramen, whose roof is formed by the lateral septal nucleus; B) A section through a mid-telencephalic level, showing robust distrubtion of substance P in the subpallium and a terminal field (asterix) in the medial pallium; C) A section through the interventricular foramen, below the lateral septum, with incursion of the tela choroidea into the pallium; D) The rostral telencephalon of an axolotl that received an injection of BrdU, which labels the proliferating cells of the ventral proliferation zone; E) The position of the septum ependymale, which may or may not separate the pallium from the subpallium. Bar scales equal 250 µm (C, D) and 500 µm (A, B, E).

lie ventral to the interventricular foramen, as is the case for most of the septal nuclei in amphibians (Fig. 15). This is clearly not the case, however.

(2) In amphibians, dorsal and ventral ependymal thickenings characterize much of the rostrocaudal length of the hemispheres, with the dorsal ependymal thickening lying adjacent to much of the pallium, and the ventral ependymal thickening lying adjacent to the striatum and septum (Fig. 15). Von Bartheld and colleagues noted that what appear to be homologous thickenings in lungfishes occur medially and laterally (Figs. 7C, 8C, 15C), as though they had been rotated in lungfishes. If the medial ependymal thickening of lungfishes is homologous to the ventral ependymal thickening in amphibians, the cell groups bordered by the medial ependymal thickening in *Protopterus* should be on the border of the striatum and the septum, not the border of the septum and the medial pallium (Fig. 15C). A reasonable first question to ask, however, is: what are the ependymal thickenings? In juvenile lungfishes, the thickenings mark positions in the ependyma where the cells lining the ventricle are not cuboidal, as they are throughout most the ependyma, but elongated and oval-shaped. They grade into several layers of cells, which, in turn, grade into what appear to be neuronal cell bodies. Closer examination of the elongated oval-shaped ependymal cells reveals that many of them exhibit mitotic figures. Thus they appear to be germinal zones that are still actively producing new cells, probably including both glial and neuronal cells. Bromodeoxyuridine (BrdU) is a synthetic nucleotide and analogue of thymidine, commonly used to detect proliferating cells. Injections of BrdU into larval axolotls, *Ambystoma mexicanum* Shaw and Nodder 1798, indicate that the dorsal and ventral ependymal thickenings are, in fact, germinal zones (Fig. 14D). If sections through the telencephalic hemispheres of axolotls and lepidosirenid lungfishes are compared at the larval stage (Fig. 15), rather than a later juvenile stage, there are no differences in the relative positions of the dorsal and ventral germinal zones in the rostral telencephalon (Fig. 15A,B). More caudally, the ventral germinal zone in lungfishes does shift more dorsally, but this appears to be due to the more caudal and dorsal shift in the position of the septum, and to the relative position of interventricular foramen formation in the two groups of animals (Fig. 15C,D). Furthermore, the caudalmost segment of the so-called lateral germinal zone in lungfishes has no direct homologue in amphibians. Much of the amygdala in lungfishes borders the "lateral" germinal zone, which occurs within the ependyma of the telencephalon impar, as opposed to the caudal part of the ventral germinal zone, which lies adjacent to the amygdala, much of which is included in the evaginating hemispheres in amphibians (Fig. 15). Thus, at least rostrally, the germinal zones occupy comparable positions, and "rotation" is most likely related to differences in the extent of hemispheric evagination.

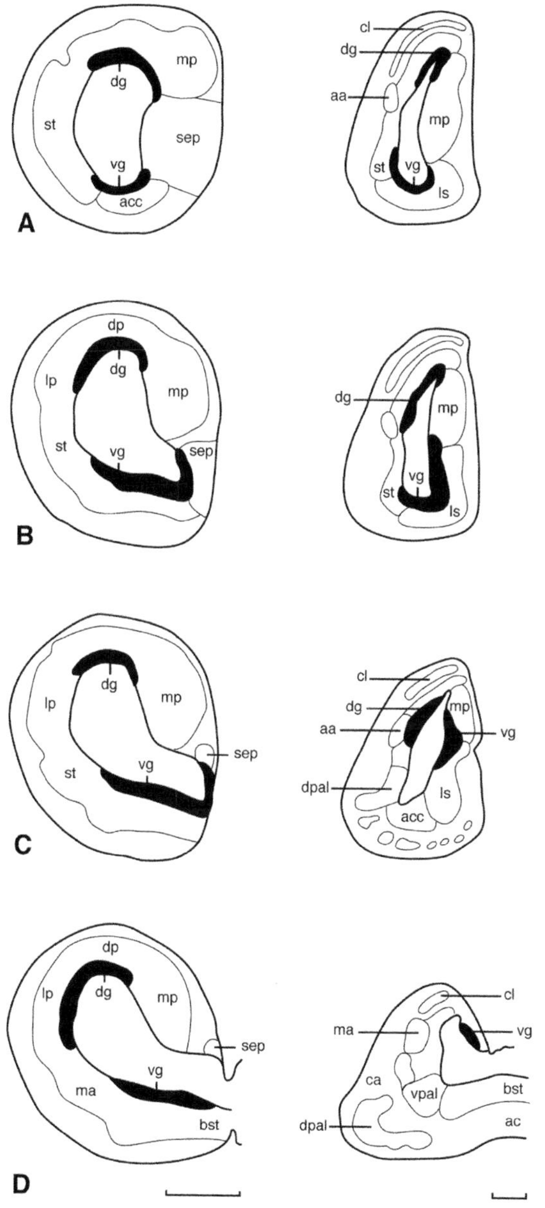

Fig. 15 Camera lucida drawings from rostral (A) to caudal (D) through one telencephalic hemisphere of a larval axolotl (left) and a larval South American lungfish (right). The portions of the ependyma shown in black indicate the dorsal and ventral germinal zones in each animal. Bar scales equal 500 μm.

(3) Von Bartheld and colleagues noted that the cells of the hypothalamic paraventricular organ in lungfishes project to the medial hemispheric wall, which must be homologous to the septum of amphibians, since the hypothalamus does not project to the medial pallium in amphibians. Although hypothalamic projections to the medial pallium of amphibians had not been reported in 1990, when von Bartheld *et al.* framed their objections to the extended pallial hypothesis, such projections have since been documented in the fire-bellied toad (Roth *et al.* 2004).

(4) Von Bartheld and colleagues noted that the medial pallium of amphibians receives only a sparse projection from the olfactory bulbs. They quoted Reiner and Northcutt (1987) as stating that the olfactory bulbs in *Protopterus* project to the dorsal and lateral pallium, but they noted that Reiner and Northcutt did not mention the medial pallium, and they thus apparently assumed that Reiner and Northcutt were denying projections to that area. Therefore, the possible absence of a small olfactory projection to the medial hemispheric wall suggested to von Bartheld *et al.* that the entire medial hemispheric wall must be septal. The olfactory bulbs *do* project to the septum and the medial pallium in lepidosirenid lungfishes (unpublished observations), as they do in many vertebrates (Northcutt 2006).

(5) Von Bartheld and colleagues claimed that the terminal nerve in amphibians projects to the septal nuclei but not to the medial pallium. In lungfishes, the terminal nerve projects heavily to the medial hemispheric wall, an area that von Bartheld *et al.* concluded is consistent with the septal nuclei but not the medial pallium. In fact, the terminal nerve projects heavily to both the septal nuclei and the medial pallium in amphibians (Muske and Moore 1994; Northcutt and Muske 1994) and in elasmobranch fishes (Lovejoy *et al.* 1992).

(6) Von Bartheld and colleagues noted that in amphibians, acetylcholinesterase (AChE) is high in the septal nuclei but low in the medial pallium, and they claimed that a band of AChE positive fibers and terminals in the neuropil of the dorsomedial hemispheric wall in *Protopterus* indicates that this region is part of the septum. Their own study, however, indicates that the concentration of AChE is much higher in the ventral half of the medial wall, as it is in amphibians, where there is no question that it occurs within the septal nuclei. It is true, however, that AChE is very low to nonexistent in the medial pallium of amphibians. As noted by Reiner and Northcutt (1987), a band positive for substance P also occurs in the same region of the dorsomedial wall in *Protopterus,* and it is very probable that the AChE and substance P activity is related to a tract of axons whose cells of origin are located in the ventromedial hemispheric wall. If this is the case, this tract may be evidence of a septo-hippocampal pathway, which

occurs in all vertebrates where the connections of the septal nuclei have been determined experimentally.

(7) Von Bartheld and colleagues claimed that the distribution of the chemical markers examined by Reiner and Northcutt (1987) indicate that the medial pallial-septal border occurs dorsally rather than half way down the medial hemispheric wall in *Protopterus*. If this were the case, then the entire medial hemispheric wall should be interpreted as the septal nuclei. None of the chemical markers examined by Reiner and Northcutt (1987) provided an absolute indication of pallial versus subpallial origins. The relative distribution of substance P within the telencephalon of most vertebrates, however, provides a robust indicator of the pallial or subpallial origin of a telencephalic cell group. While it is true that moderate amounts of substance P do occur within the dorsomedial hemispheric wall of *Protopterus*, there is no question that the major differences in substance P distribution occur half way down both the lateral and medial hemispheric walls (Fig. 14B), not far dorsally as claimed by von Bartheld *et al.* (1990). This distribution is consistent with the extended pallial hypothesis, not the restricted pallial hypothesis.

The telencephalic hemispheres of *Neoceratodus* appear very different from those of the lepidoserenid lungfishes. In *Neoceratodus*, most of the dorsomedial hemispheric walls consists of a highly folded ependymal tela, termed the septum ependymale (Fig. 14E). As indicated by the nomenclature, this tela is usually regarded as separating the dorsolateral pallium from the more ventromedial septum (Holmgren and van der Horst 1925; Nieuwenhuys and Hickey 1965; Whitt 1969; von Bartheld *et al.* 1990; Nieuwenhuys 1998). If this is true, the telencephalic anatomy would support the restricted pallial hypothesis. It is possible, however, that the septum ependymale is not septal but pallial in nature. If the tela seen in the caudal pallium of *Protopterus* (Fig. 14C) were to be expanded rostrally, the telencephalic hemispheres of lepidosirenid lungfishes and *Neoceratodus* would appear far more similar. In this context, Nieuwenhuys and Hickey (1965) and Whitt (1969) indicated at least two, if not three, distinct cytological fields in the ventromedial hemispheric wall of *Neoceratodus*, and it is possible that the most dorsal cellular field is actually part of the medial pallium. If this were the case, hemispheric organization in *Neoceratodus* would support the extended pallial hypothesis. Resolution of this issue awaits new immunohistochemical and hodological studies.

Striatopallidal systems. In all terrestrial vertebrates, these systems occupy the ventrolateral wall and floor of the telencephalon and form dorsal and ventral subdivisions (Heimer *et al.* 1995; Marín *et al.* 1998a,b). The dorsal striatopallidal system consists of a dorsal striatum (caudate and putamen in mammals) and a dorsal pallidum (globus pallidus in mammals). The ventral striatopallidal system consists of the ventral striatum (nucleus accumbens and the olfactory tubercle in mammals) and a ventral pallidum.

One or more cell groups of the ventrolateral wall in lungfishes have been considered homologous to the dorsal striatum, but different studies have recognized different cell groups (Rudebeck 1945; Schnitzlein and Crosby 1967; Whitt 1969; Clairambault and Capanna 1973; Reiner and Northcutt 1987; Nieuwenhuys 1998). Herein, the dorsal striatum is considered to occupy most of the rostral ventrolateral hemispheric wall in *Protopterus* and is divided into dorsal and ventral subdivisions (Figs. 9A, 10A). Both subdivisions consist of a periventricular cellular plate flanked by a lateral neuropil, but the cells of the dorsal subdivision are more scattered than those of the ventral subdivision. More caudally (Figs. 7D, 8D), the dorsal subdivision can still be recognized, but the ventral subdivision has been replaced by the rostral pole of a cell group that exhibits pronounced, finger-like cellular protrusions that extend into the lateral neuropil. This new cell group is interpreted as the rostral pole of the dorsal pallidum. Surprisingly, only Schnitzlein and Crosby (1967) have previously recognized a dorsal pallidum, and they named a more ventral and caudal cellular group that has been interpreted as the olfactory tubercle in most studies. Further caudally, the finger-like cellular protrusions are no longer connected to the periventricular cellular plate but, instead, form isolated cellular islands within the neuropil (Figs. 7C, 8C). At this same level, the ventralmost part of the lateral periventricular cellular plate extends laterally and forms a distinct cortical-like plate, which constitutes the caudal pole of the dorsal pallidum (Figs 7B, 8B).

Current concepts of a ventral striatopallidal system have been developed only in the last 10 years. Thus, when Elliot Smith (1908) noted similarities between the olfactory tubercle of mammals and the extensive, folded, cortical-like periventricular sheet in the hemispheric floor of lepidosirenid lungfishes, it was reasonable for him to interpret this cellular sheet as a homologue of the mammalian olfactory tubercle. This interpretation was adopted in most subsequent studies (Kuhlenbeck 1929; Rudebeck 1945; Nieuwenhuys and Hickey 1965; Nieuwenhuys 1969, 1998). Given our current understanding of the ventral striatopallidal system, however, we need to ask at least three questions: 1) What biological pressures could have caused lungfishes to evolve an olfactory tubercle that accounts for approximately half of the total hemispheric volume; 2) If the olfactory tubercle of lungfishes is so extensive, where is an equally extensive nucleus accumbens located in lungfishes; and 3) If the olfactory tubercle of lungfishes is so extensive, where is its major target, the ventral pallidum? No one has attempted to answer the first question, but Nieuwenhuys (1998) approached the second question by suggesting that the ventral subdivision of the dorsal striatum (as described herein) might be homologous to both the dorsal and ventral striatal "moieties" of mammals. Finally, as regards, the third question, no previous study has proposed a homologue of the ventral pallidum.

A closer examination of the so-called olfactory tubercle of lungfishes (Fig. 12D-F) suggests that it is not a single, uniform cellular field. Recently, it was suggested that the compact periventricular cellular plate in the rostral telencephalon of lepidosirenid lungfishes (Figs. 9A, 10A, 12D) is homologous to the lateral septum of terrestrial vertebrates (Northcutt, 2008). Further, the more caudal cell group, which consists of a periventricular cellular plate and migrated cellular patches (Figs. 7C, 8C, 12E), may be homologous to nucleus accumbens and the olfactory tubercle of terrestrial vertebrates, and that the most caudal segment of the floor (Figs. 7B, 8B, 12F) may be homologous to the ventral pallidum of terrestrial vertebrates. Certainly the distribution of LANT6-positive cells in this caudal segment of the floor (Reiner and Northcutt 1987) is consistent with this interpretation. At present, there are no experimental data on any of the lungfish cell groups proposed as putative homologues of striatopallidal nuclei in other vertebrates, but the connections of these cell groups should be comparable to those established in amphibians (Marín *et al.* 1997a,b; Endepols *et al.* 2004) and other terrestrial vertebrates (Marín *et al.* 1998a).

Amygdalar organization. As with many other telencephalic neural centers in lungfishes, there is no consensus regarding the organization of the amygdala. Several of the early studies (Elliot Smith 1908; Holmgren and van der Horst 1925; Gerlach 1933) failed to recognize any amygdalar nuclei. Rudebeck (1945) was the first to suggest that the dorsal half of the telencephalon impar (Figs. 7A, 8A, 13C), which he termed the superior preoptic nucleus, was homologous to the massa cellularis reuniens, which, at that time, was thought to be the amygdalar homologue in amphibians (Herrick 1927, 1933). On the other hand, Schnitzlein and Crosby (1967) recognized anterior, basolateral, and corticomedial amygdalar nuclei, which they believed formed the caudal dorsolateral pallium (Fig. 13A) in *Protopterus.* Whitt (1969) recognized the same amygdalar nuclei in a comparable position in *Neoceratodus.* In other studies, however, these cells have been identified as part of the general pallium. Finally, Clairambault and Capanna (1973) described as the amygdala in *Protopterus* a single cell group in the caudal telencephalon that has been identified in most other studies as an olfactory tubercular cortex (Fig. 13A).

In the last 10 years, remarkable progress has been made in resolving amygdalar organization in amphibians and other tetrapods (Bruce and Neary 1995; Marín *et al.* 1998b; Moreno *et al.* 2004; Moreno and González 2003, 2004, 2005, 2007a,b; Laberge *et al.* 2005). Today, at least three amygdalar cell groups are recognized in terrestrial vertebrates: central, lateral, and medial nuclei, as well as a closely associated bed nucleus of the stria terminalis. In tetrapods, the bed nucleus of the stria terminalis is located in the lamina terminalis, dorsal to the anterior commissure. A comparable nucleus can also be recognized in lungfishes (Figs. 7B, 8B, 13C). A central amygdalar nucleus composed of large cells that are positive for NADPH diaphorase replaces the dorsal subdivision of the dorsal striatum in amphibians caudally. A comparable group of large neurons, positive for NADPH diaphorase (Northcutt 2009) , also exists in a similar position in *Protopterus* (Figs. 7A,B, 8A,B, 13A).

The other amygdalar nuclei in terrestrial vertebrates are part of the dorsolateral telencephalic wall and occur ventral to the lateral pallium, which is the primary target of the secondary olfactory projections. Traditionally, the lateral pallium of amphibians was divided into dorsal and ventral divisions, but recent gene studies (Brox *et al.* 2003, 2004; Moreno *et al.* 2004) indicate that the ventral division is actually a separate major pallial division, which is now termed the ventral pallium. In amphibians, the ventral pallium forms the lateral amygdalar nucleus, whose lateral neuropil is rich in NADPH diaphorase-positive fibers and terminals (Marín *et al.* 1998b). A similarly located cell group can also be recognized in *Protopterus* (Figs. 7C,D, 8C,D, 9A, 10A, 13B), and it is also rich in NADPH diaphoarase (Northcutt 2009), as well as in substance P and tyrosine hydroxylase (Reiner and Northcutt 1987).

In lepidosirenid lungfishes, the lateral amygdalar nucleus can be recognized rostrally as an oval-shaped cluster of larger densely packed cells located adjacent to a deep bundle of secondary olfactory fibers (Figs. 9A, 10A). More caudally, the lateral amygdalar nucleus becomes elongated and forms cell clusters along its dorsolateral border (Figs. 7C,D, 8C,D). As the nucleus reaches the telencephalon impar, it is replaced (Figs. 7A,B, 8A,B) by the medial amygdalar nucleus, which occupies much of the telencephalon impar (Figs. 7A,B, 8A,B). In amphibians, a similar transformation occurs. The lateral amygdalar nucleus is replaced caudally by a reversed C-shaped nucleus, which receives a major input from the accessory olfactory bulb, and it is identified as a medial amygdalar nucleus. At present there are no experimental data on the amygdalar pathways in lungfishes.

The olfactory bulbs. The olfactory bulbs in all three genera of lungfishes are large (Fig. 1) and well differentiated histologically. Each bulb consists of a series of more or less concentric laminae: 1) an outer layer of primary olfactory fibers, which are the axons of the olfactory sensory receptors; 2) a glomerular layer, which consists of the terminal endings of the primary olfactory fibers and the apical dendritic branches of the more deeply lying mitral cells; 3) a layer of mitral cell bodies; 4) the internal plexiform layer, which is primarily the axons of the mitral cells that leave the olfactory bulb as the secondary olfactory tracts; 5) the internal granular layer which contains most of the neurons located within the olfactory bulb; and 6) a subependymal fiber plexus of uncertain origin. Rudebeck (1945) studied the cells in the olfactory bulb of *P. annectens* using the Golgi method, which allowed him to describe the cytology of the cell types in the different laminae. He noted that a sizable number of neurons occur scattered among the entering primary olfactory fibers, as well as between the individual glomeruli of the glomerular layer. Many of these cells were observed to have a dendrite that ramifies within the glomeruli, as well as axons forming an external olfactory tract that may be unique to lungfishes.

Rudebeck (1945) also noted that most granular cells within the inner granular layer have one or more apical dendrites that extend into the more superficial zones, including apical processes that end within the glomeruli. Surprisingly, many of the granular cells of the internal granular layer have unmyelinated axons that enter the internal plexiform layer and leave the olfactory bulb with the myelinated axons of the mitral cells.

Most studies indicate that the caudal end of the olfactory bulb is marked by a shallow transverse sulcus, termed the external limiting bulbar sulcus (Rudebeck 1945) or the circular sulcus (Schnitzlein and Crosby 1967). Clairambault and Capanna (1973) claimed that the olfactory bulb extends throughout the rostral one-third of the hemisphere, and they identified the rostral cortical layer (Figs. 9A, 10A) of most researchers as mitral cells. Their conclusion was rejected by Derivot (1984) and Northcutt (1986a). The nature of this area was finally resolved when von Bartheld *et al.* (1988) injected horseradish peroxidase into the olfactory organ of *P. dolloi* and demonstrated that the labeled primary olfactory fibers cannot be traced caudal to the external limiting bulbar sulcus.

Schnitzlein and Crosby (1967) described a rostromedial division of the olfactory organ in *Protopterus* as a vomeronasal organ and claimed that the most medial fascicle of the olfactory nerve represents a vomeronasal nerve. They believed that this fascicle could be traced to a specialized segment of the dorsomedial bulb, which they termed an accessory olfactory bulb. Neither Derivot (1984) nor Northcutt (1986a) were able to confirm these observations. Similarly, von Bartheld *et al.* (1988) did not recognize a vomeronasal organ or nerve in their experimental study of the projections of the olfactory organ and nerve in *P. dolloi*. At present, there is no evidence for a morphologically distinct vomeronasal organ or nerve or an accessory olfactory bulb in lungfishes. It is possible, however, that some parts of the olfactory organ and olfactory bulb do represent a cryptic vomeronasal system in these fishes, and the olfactory epithelium and bulb should be examined with specific vomeronasal histochemical markers to resolve this issue.

The most complete description of the secondary olfactory projections in lungfishes is that of Rudebeck (1945), who described six different olfactory tracts leaving the olfactory bulb to terminate in most pallial and subpallial areas in *P. annectens*. Today, these projections can best be viewed as constituting lateral and medial olfactory tracts. The lateral olfactory tract consists of a dorsal division, which terminates in the dorsal and lateral pallium, including the anterior and lateral amygdalar nuclei, and a ventral division, which terminates in the dorsal striatum and nucleus accumbens. On entering the pallium, the dorsal division of the lateral olfactory tract divides into external and internal subdivisions, which terminate external and internal to the cellular layer of the pallial divisions. The medial olfactory tract, however, does not divide. It exits the dorsomedial olfactory bulb and terminates primarily in the dorsal third of the medial pallium. These connections have been experimentally confirmed in *P. dolloi* in my laboratory (unpublished observations). In addition to these projections, some fibers of the lateral olfactory tract decussate in the anterior and habenular commissures and project contralaterally to the dorsal and lateral pallial divisions, as well as bilaterally to the preoptic area and dorsal hypothalamus.

Lungfishes, like many vertebrates, exhibit a terminal nerve that enters the forebrain in close association with the olfactory nerve. Unlike that of other fishes, however, the terminal nerve of lungfishes possesses anterior and posterior roots (Pinkus 1894; Sewertzoff 1902; Holmgren and van der Horst 1925; Rudebeck 1945; von Bartheld 2004). The anterior root enters the forebrain in association with the olfactory nerve, whereas the posterior root, which has also been termed the nervus praeopticus, enters the telencephalon in the lamina terminalis.

The projections of the anterior root of the terminal nerve were traced in *P. dolloi* after the olfactory epithelium was labeled with horseradish peroxidase (von Bartheld *et al.*, 1988). The labeled fibers enter the telencephalic hemispheres after coursing in the dorsomedial quadrant of the overlying olfactory bulb, and they terminate along the dorsoventral extent of the medial wall of each hemisphere. Some fibers decussate in the anterior commissure and appear to terminate bilaterally in the overlying bed nucleus of the stria medullaris, whereas other labeled fibers were traced into the diencephalon, where they appear to terminate in the preoptic area. Fibers also appear to decussate in the postoptic commissure and to terminate bilaterally in the dorsal hypothalamic nucleus and the posterior tubercle. The anterior root of the terminal nerve in *Neoceratodus* appears to have similar projections, based on a visualization of the fibers of this root with NADPH diaphorase (Schober *et al.* 1994).

Although the projections of the posterior root of the terminal nerve have not been experimentally traced in either of the lepidosirenid lungfishes, Schober *et al.* (1994) visualized these fibers in both *Protopterus* and *Neoceratodus* with NADPH diaphorase. In both genera, fibers of the posterior root enter the lamina terminalis and appear to terminate in the preoptic area. Few fibers appear to decussate in the anterior commissure; rather, most pass dorsally, adjacent to the periventricular gray of the ventricle, and join fibers of the anterior root. At present, it is assumed that the more caudally directed fibers of the posterior root terminate in many, if not all, of the same centers reached by fibers of the anterior root. It is known, however, that other chemical markers have revealed both similarities and differences in the fibers of the terminal nerve. In *Neoceratodus*, fibers in both roots are positive for FMRFamide (Fiorentino *et al.* 2002), and in *Protopterus* the fibers of the posterior root are positive for AChE and GnRH, but the fibers of the anterior root are not (von Bartheld *et al.* 1988, 1990).

FUTURE DIRECTIONS

Essentially, every aspect of the central nervous system of lungfishes needs continued study. Although there are a small number of older descriptive studies of the general development of the brain and cranial nerves, a wealth of new information could be revealed on the pallial-subpallial border and other aspects of forebrain organization by examining the expression patterns of numerous genes involved in patterning the development of the brain, such as *Dlx2, Islet-1, lhx, NKx2.1*, and *Tbr2* (Mueller *et al.* 2008; Moreno and González 2007a). Due to current breeding programs, the Australian lungfish, *Neoceratodus*, may offer an excellent opportunity for carrying out such studies, and it would be of particular interest, given the unique anatomy of its telencephalic hemispheres. Certainly any resolution of the current hypotheses regarding pallial organization in lungfishes must include experimental confirmation of the pallial-subpallial border in *Neoceratodus*.

New histochemical studies that include *Neoceratodus* are also needed. Studies using NADPH diaphorase and other indicators of nitric oxide activity, as well as γ-amino butyric acid (GABA) have been particularly efficacious in resolving amygdalar and striatopallidal organization in amphibians (Márin *et al.* 1998b; Mühlenbrock-Lenter *et al.* 2005; Moreno and González 2007a,b) and would greatly increase our understanding of these systems in lungfishes. A new generation of histochemical studies should also focus on other parts of the brain and spinal cord, which have been largely ignored in earlier studies.

Today, comparative neuroanatomists have an extensive armamentarium of experimental techniques, involving highly sensitive tracers that allow far greater resolution of neuronal connections than has been possible any time in the past. In spite of the availability of these powerful tools, fewer experimental studies are being done today than 20 years ago. It is time to implement a program that addresses, with modern experimental methods, how major neural centers, such as the spinal cord, cerebellum, optic tectum, thalamus, etc. are "wired" in lungfishes. Such information is necessary to provide the basis for us to begin to think about the evolution of these centers and to design neurophysiological and behavioral experiments to investigate their functions. It is truly amazing that there is almost no information on the connections and physiological properties of any part of the nervous system in lungfishes except for their Mauthner neurons. Implementation of a new experimental program at this time is particularly important, as the interpretation of their central nervous system proposed herein suggests that brain organization in lungfishes, particularly forebrain organization, is far more similar to that of amphibians than was previously suspected. Brain organization in the living lungfishes can, therefore, provide critical information on the earliest lobe-finned fishes.

What is equally remarkable, almost nothing is known about the behavior of lungfishes, 200 years after their discovery by the Western World. These unique and fascinating animals offer the best, perhaps only, opportunity for us to glimpse a largely extinct world when lobe-finned fishes gave rise to terrestrial vertebrates.

Acknowledgments

I thank the editors for inviting me to contribute to this volume on the biology of lungfishes, surely one of the most fascinating groups of living vertebrates. I am grateful to Jo Griffith for her expertise in illustrating, to Susan Commerford for literature retrieval and word processing, and to Mary Sue Northcutt for assistance with numerous aspects of the writing and illustration of this manuscript. I also thank Rudolf Nieuwenhuys for his helpful comments on this chapter. This project was supported, in part, by the National Science Foundation [1BN-0236018].

List of Abbreviations

a – cerebellar auricle
ac – anterior commissure
acc – nucleus accumbens (ventral striatum)
all – anterior lateral line nerve
bst – bed nucleus of the stria terminalis
c – cerebellum
ca – central amygdala
cb – cerebellar body
cc – central canal
cec – cerebellar crest
cl – cortical layer
cn – cerebellar nucleus
d – diencephalon
dc – dorsal cell column
deo – descending octaval nucleus
df – dorsal funiculus
dg – dorsal germinal zone
dh – dorsal hypothalamus
dn – dorsal column nucleus
do – dorsal octavolateral nucleus
dp – dorsal pallium
dpal – dorsal pallidum
dsep – dorsal septal nucleus
dst – dorsal subdivision of the striatum
dt – dorsal thalamus
dte – diencephalic tela

dV – descending trigeminal tract
ep – epiphysis
fr – fasciculus retroflexus
gl – glomerular layer of the olfactory bulb
glc – granular layer of the body of the cerebellum
ha – habenular nuclei
if – interventricular foramen
ig – internal granular layer of the olfactory bulb
ipl – internal plexiform layer of the olfactory bulb
ir – inferior reticular nucleus
ira – inferior raphe
la – lateral amygdala
lc – lateral cell column
let – lateral ependymal thickening
lf – lateral funiculus
lfb – lateral forebrain bundle
lp – lateral pallium
ls – lateral septal nucleus
ma – medial amygdala
mao – magnocellular octaval nucleus
mc – medial cell column
met – medial ependymal thickening
meV – mesencephalic trigeminal neuron
mf – Mauthner axon
ml – mitral layer of the olfactory bulb
mlf – medial longitudinal fasciculus
mo – medial octavolateral nucleus
mol – molecular layer of the cerebellum
mp – medial pallium
mr – middle reticular nucleus
ms – medial septal nucleus
nmlf – nucleus of the medial longitudinal fasciculus
ob – olfactory bulb
oc – octavolateral column
ol – olfactory nerve
olt – olfactory tubercle
oltc – olfactory tubercular cortex

on – optic nerve
op – olfactory peduncle
ot – optic tectum
otr – optic tract
p – profundal nerve
pa – anterior periventricular preoptic area
pal – pallium
pc – Purkinje cell
pco – posterior commissure
pll – posterior lateral line nerve
poc – postoptic commissure
pp – posterior periventricular preoptic area
pt – posterior tubercle
ptec – pretectum
r – recurrent ramus of anterior lateral line nerve
rb – rostral body
sco – subcommissural organ
se – septum ependymale
sep – septum
sg – secondary gustatory tract
sm – stria medullaris
smn – somatic motor nucleus
sn – superficial isthmal nucleus
so – spino-occipital nerves
sol – solitary nucleus
sp – subependymal fiber plexus
sr – superior reticular formation
sra – superior raphe
st – striatum
t – telencephalon
tc – tectal commissure
tco – tela choroidea
ts – torus semicircularis
vg – ventral germinal zone
vf – ventral funiculus
vh – ventral hypothalamus
vpal – ventral pallidum

vsep – ventral septal nucleus
vst – ventral subdivision of the striatum
vt – ventral thalamus
I – olfactory nerve
II – optic nerve
III – oculomotor nerve
IIIm – oculomotor nucleus
IV – trochlear nerve
V – trigeminal nerve
Vr – trigeminal motor root
Vp – principal trigeminal sensory nucleus
Vm – trigeminal motor nucleus
VI – abducent nerve
VII – facial nerve
VIIm – facial motor nucleus
VIII – octaval nerve
IX – glossopharyngeal nerve
X – vagal nerve
Xm – vagal motor nucleus

References

Beauregard, H. (1881). Encéphale et nerfs craniens du Ceratodus forsteri. J. de l'anat. et de la Physiol. Journal de l'anatomie et de la physiologie normales et pathologiques de l'homme et des animaux 17: 1-13.

Bergquist, H. (1932). Zur Morphologie des Zwischenhirns bei niederen Wirbeltieren. Acta Zoologica 13: 57-304.

Bergquist, H. and Källén, B. (1954). Notes on the early histogenesis and morphogenesis of the central nervous system in vertebrates. Journal of Comparative Neurology 100: 627-660.

Bing, R. and Burckhardt, R. (1904). Das Zentralnervensystem von *Ceratodus forsteri*. Anatomische Anzeiger 25: 588-599.

Bing, R. and Burckhardt, R. (1905). Das Centralnervensystem von *Ceratodus forsteri*. Jenaische Denkschriften 4: 513-584.

Brox, A., Puelles, L., Ferreiro, B. and Medina, L. (2003). Expression of the genes GAD67 and distal-less-4 in the forebrain of *Xenopus laevis* confirms a common pattern in tetrapods. Journal of Comparative Neurology 461: 370-393.

Brox, A., Puelles, L., Ferreiro, B. and Medina, L. (2004). Expression of the genes Emxl, Tbr1, and Eomes (Tbr2) in the telencephalon of *Xenopus laevis* confirms the existence of four pallial divisions in tetrapods. Journal of Comparative Neurology 474: 562-577.

Bruce, L. and Neary, T.J. (1995). The limbic system of tetrapods: a comparative analysis of cortical and amygdalar populations. Brain, Behavior and Evolution 46: 224-234.

Burckhardt, R. (1892). Das Centralnervensystem von Protopterus annectens. Friedlander. Berlin, Germany.

Capanna, E. and Clairambault, P. (1973). Some considerations on the forebrain of the bipulmonate Dipnoi. Rendiconti Academia Nazionale Lincei 55: 603-608.

Clairambault, P. and Capanna, E. (1973). Suggestion for a revision of the cytoarchitectonics of the telencephalon of *Protopterus*, *Protopterus annectens* (Owen). Bolletinio di Zoologia 40: 149-171.

Clairambault, P. and Flood, P. (1975). Les centres visuals primaries de *Protopterus dolloi* Boulenger. Journal fur Hirnforschung 16: 497-509.

Clairambault, P., Capanna, E., Chanconie, M. and Pinganaud. G. (1974a). Tipologia neurale del septum telencefalico di un Dipnoi lepidosireniforme (*Protopterus dolloi* Boulenger). Rendiconti Academia Nazionale Lincei 56: 423-431.

Clairambault, P., Capanna, E., Chanconie, M. and Pinganaud, G. (1974b). Typologie neuronique du complexe strioamygdaloide de *Protopterus dolloi* Boulenger. Rendiconti Academia Nazionale Lincei 56: 1017-1025.

Clairambault, P., Capanna, E., Chanconie, M. and Pinganaud, G. (1974c). Architectutal pattern of the diencephalon and mesencephalon of the African lungfish, *Protopterus dolloi*. Bolletinio di Zoologia 41: 107-122.

Cordier, R. (1936). Les organs sensoriels cutanés du Protoptère. Bulletin de la Classe des Sciences 22: 474-483.

Derivot, J.H. (1984). Functional anatomy of the peripheral olfactory system of the African lungfish, *Protopterus annectens* Owen: macroscopic, microscopic, and morphometric aspects. American Journal of Anatomy. 169: 177-192.

Dorn, E. (1957). Über das Zwischenhirn-Hypophysen-System von *Protopterus annectens*. Zeitschrift für Zellforschung und Mikroskopische Anatomie 46: 108-114.

Elliot Smith, G. (1908). The cerebral cortex in *Lepidosiren*, with comparative notes on the interpretation of certain features of the forebrain in other vertebrates. Anatomischer Anzeiger 33: 513-540.

Endepols, H., Roden, K., Luksch, H., Dicke, U. and Walkowiak, W. (2004). Dorsal striatopallidal system in anurans. Journal of Comparative Neurology 468: 299-310.

Fahrenholz, C. (1929). Über die "Drüsen" und die Sinnesorgane in der Haut der Lungenfische. Zeitschrift für Zellforschung und Mikroskopische Anatomie 16: 55-74.

Fiorentino, M., D'Aniello, B., Joss, J., Polese, G. and Rastogi, R.K. (2002). Ontogenetic organization of the FMRFamide immunoreactivity in the nervus terminalis of the lungfish, *Neoceratodus forsteri*. Journal of Comparative Neurology 450: 115-121.

Gerlach, J. (1933). Uber das Gehirn von *Protopterus annectens*. Ein Beitrag zur Morphologie des Dipnoerhirnes. Anatomischer Anzeiger 75: 305-448.

Hayle, T.H. (1973a). A comparative study of spinal projections to the brain (except cerebellum) in three classes of poikilothermic vertebrates. Journal of Comparative Neurology 149: 463-476.

Hayle, T.H. (1973b). A comparative study of spinocerebellar systems in three classes of poikilothermic vertebrates. Journal of Comparative Neurology 149: 477-496.

Heimer, L., Zahm, D.S. and Alheid, G.F. (1995). Basal ganglia. In: G. Paxinos (ed.). The Rat Nervous System. 2nd Edition. Academic Press, San Diego, USA. pp. 579-628.

Herrick, C.J. (1927). The amphibian forebrain. IV. The cerebral hemispheres of *Amblystoma*. Journal of Comparative Neurology 43: 231-325.

Herrick, C.J. (1933). The amphibian forebrain. VI. *Necturus*. Journal of Comparative Neurology 58: 1-288.

Herrick, C.J. (1948). The Brain of the Tiger Salamander. University of Chicago Press, Chicago, USA.

Holmgren, N. (1922). Points of view concerning forebrain morphology in lower vertebrates. Journal of Comparative Neurology 34: 391-440.

Holmgren, N. and van der Horst, C.J. (1925). Contribution to the morphology of the brain of *Ceratodus*. Acta Zoologica 6: 59-165.

Holmgren, U. (1959). On the pineal area and adjacent structures of the brain of the dipnoan fish, *Protopterus annectens* (Owen). Breviora Museum of Comparative Zoology 108: 1-7.

Jeener, R. (1930). Evolution des centres diencéphaliques périventriculaires des Téléostomes. Proceedings of the Koninklijke Nederlands Akademie Wetenschappen [Amsterdam] B 33: 1-16.

Keenan, E. (1928). The phylogenetic development of the substantia gelatinosa Rolandi, Part I. Fishes. Koninklijke Nederlands Akademie Wetenschappen [Amsterdam] 31: 837-854.

Kemp, A. (1999). Ontogeny of the skull of the Australian lungfish *Neoceratodus forsteri* (Osteichthyes: Dipnoi). Journal of Zoology London 248: 97-137.

Kölliker, A. (1896). Nervensystem des Menschen und der Thiere. In: Handbuch der Gewebelehre des Menschen, Volume 2. Engelmann, Leipzig, Germany. p. 334.

Kuhlenbeck, H. (1929). Über die Grundbestandteile des Endhirns im Lichte der Bauplanlehre. Anatomischer Anzeiger 67: 1-51.

Laberge, F., Mühlenbrock-Lenter, S., Grunwald, W. and Roth, G. (2005). Evolution of the amygdala: new insights from studies in amphibians. Brain, Behavior and Evolution 67: 177-187.

Lovejoy, D.A., Ashmead, B.J., Coe, I.R. and Sherwood, N.M. (1992). Presence of gonadotropin-releasing hormone immunoreactivity in dogfish and skate brains. Journal of Experimental Zoology 263: 272-283.

Marín, O., Smeets, W.J.A.J. and González, A. (1997a). Basal ganglia organization in amphibians: afferent connections to the striatum and the nucleus accumbens. Journal of Comparative Neurology 378: 16-49.

Marín, O., Smeets, W.J.A.J. and González, A. (1997b). Basal ganglia organization in amphibians: efferent connections of the striatum and the nucleus accumbens. Journal of Comparative Neurology 380: 23-50

Marín, O., Smeets, W.J.A.J. and González, A. (1998a). Evolution of the basal ganglia in tetrapods: a new perspective based on recent studies in amphibians. Trends in Neuroscience 21: 487-494.

Marín, O., Smeets, W.J.A.J. and González, A. (1998b). Basal ganglia organization in amphibians: chemoarchitecture. Journal of Comparative Neurology 392: 285-312.

McCormick, C.A. (1999). Anatomy of the central auditory pathways of fish and amphibians. In: R.R. Fay and A.N. Popper (eds). Comparative Hearing: Fish and Amphibians. Springer, New York, USA. pp. 155-217.

Moreno, N. and González, A. (2003). Hodological characterization of the medial amygdala in anuran amphibians. Journal of Comparative Neurology 466: 389-408.

Moreno, N. and González, A. (2004). Localization and connectivity of the lateral amygdala in anuran amphibians. Journal of Comparative Neurology 479: 130-148.

Moreno, N. and González, A. (2005). Central amygdala in anuran amphibians: neurochemical organization and connectivity. Journal of Comparative Neurology 489: 69-91.

Moreno, N. and González, A. (2007a). Regionalization of the telencephalon in urodele amphibians and its bearing on the identification of the amygdaloid complex. Frontiers in Neuroanatomy 1: 1-12.

Moreno, N. and González, A. (2007b). Development of the vomeronasal amygdala in anuran amphibians: hodological, neurochemical, and gene expression characterization. Journal of Comparative Neurology 503: 815-831.

Moreno, N., Bachy, L., Rétaux, S. and González, A. (2004). LIM-hemeodomain genes as developmental and adult genetic markers of *Xenopus* forebrain functional subdivisions. Journal of Comparative Neurology 472: 52-72.

Mueller, T., Wullimann, M.F. and Guo, S. (2008). Early teleostean basal ganglia development visualized by zebrafish Dlx2a, Lhx6, Lhx7, Tbr2 (eomesa), and GAD67 gene expression. Journal of Comparative Neurology 507: 1245-1257.

Mühlenbrock-Lenter, S., Endepols, H., Roth, G. and Walkowiak, W. (2005). Immunohistological characterization of striatal and amygdalar structures in the telencephalon of the fire-bellied toad *Bombina orientalis*. Neuroscience 134: 705-719.

Murakami, T. and Ito, H. (1985). Long ascending projections of the spinal dorsal horn in a teleost, *Sebastiscus marmoratus*. Brain Research 346: 168-170.

Muske, L. and Moore, F.L. (1994). Antibodies against different forms of GnRH distinguish different populations of cells and axonal pathways in a urodele amphibian, *Taricha granulosa*. Journal of Comparative Neurology 345: 139-147.

Nieuwenhuys, R. (1967a). Comparative anatomy of olfactory centres and tracts. Progress in Brain Research 23: 1-64.

Nieuwenhuys, R. (1967b). Comparative anatomy of the cerebellum. Progress in Brain Research 25: 1-93.

Nieuwenhuys, R. (1969). A survey of the structure of the forebrain in higher bony fishes. Annals of the New York Academy of Sciences 167: 31-64.

Nieuwenhuys, R. (1998). Lungfishes. In: R. Nieuwenhuys, H.J. ten Donkelaar and C. Nicholson (eds). The Central Nervous System of Vertebrates, Volume 2. Springer, Berlin, Germany. pp. 939-1006.

Nieuwenhuys, R. and Hickey, M. (1965). A survey of the forebrain of the Australian lungfish *Neoceratodus forsteri*. Journal fur Hirnforschung 7: 433-452.

Nieuwenhuys, R. and Meek, J. (1990). The telencephalon of sarcopterygian fishes. In: E.G. Jones and A. Peters (eds). Cerebral Cortex, Volume 8A: Comparative Structure and Evolution of Cerebral Cortex, Part I. Plenum, New York, USA. pp. 75-106.

Nieuwenhuys, R., ten Donkelaar, H.J. and Nicholson, C. (1998). The Central Nervous System of Vertebrates, 3 volumes. Springer, Berlin, Germany.

Noden, D.M. (1983). The embryonic origins of avian cephalic and cervical muscles and associated connective tissues. American Journal of Anatomy 168: 257-276.

Northcutt, R.G. (1977). Retinofugal projections in the lepidosirenid lungfishes. Journal of Comparative Neurology 174: 553-574.

Northcutt, R.G. (1980). Retinal projections in the Australian lungfish. Brain Research 185: 85-90.

Northcutt, R.G. (1982). Localization of neurons afferent to the optic tetum in longnose gars. Journal of Comparative Neurology 204: 325-335.

Northcutt, R.G. (1983a). The primary lateral line afferents in lepidosirenid lungfishes. Society for Neuroscience Annual Meeting Abstracts 9: 1167.

Northcutt, R.G. (1983b) Evolution of the optic tectum in ray-finned fishes. In: R.E. Davis and R.G. Northcutt (eds). Fish Neurobiology, Volume 2. University of Michigan Press, Ann Arbor, Michigan, USA. pp. 1-42.

Northcutt, R.G. (1986a). Lungfish neural characters and their bearing on sarcopterygian phylogeny. Journal of Morphology Supplement 1: 277-297.

Northcutt, R.G. (1986b). Electroreception in nonteleost bony fishes. In: T.H. Bullock and W.F. Heiligenberg (eds). Electroreception. Wiley, New York, USA. pp. 257-285.

Northcutt, R.G. (1989). The phylogenetic distribution and innervation of craniate mechanoreceptive lateral lines. In: S. Coombs, P. Görner and H. Münz (eds). Mechanosensory Lateral Line: Neurobiology and Evolution. Springer, New York, USA. pp. 17-78.

Northcutt, R.G. (1992). Distribution and innervation of lateral line organs in the axolotl. Journal of Comparative Neurology 325: 95-123.

Northcutt, R.G. (2006). Connections of the lateral and medial divisions of the goldfish telencephalic pallium. Journal of Comparative Neurology 494: 903-943.

Northcutt, R.G. (2008). Forebrain evolution in bony fishes. Brain Research Bulletin 75: 191-205.

Northcutt, R.G. (2009). Telencephalic organization in the Spotted African lungfish, *Protopterus dolloi*: A new cytological model. Brain, Behavior and Evolution 73: 59-80.

Northcutt, R.G. and Bemis, W.E. (1993) Cranial nerves of the coelacanth, *Latimeria chalumnae* [Osteichthyes: Sarcoptergii: Actinistia], and comparisons with other Craniata. Brain, Behavior and Evolution, 42 (Supplement 1): 1-76.

Northcutt, R.G. and Muske, L.E. (1994). Multiple embryonic origins of gonadotropin-releasing hormone (GnRH) immunoreactive neurons. Developmental Brain Research 78: 279-290.

Northcutt, R.G. and Brändle, K. (1995). Development of branchiomeric and lateral line nerves in the axolotl. Journal of Comparative Neurology 355: 427-454.

Oka, Y., Satou, M. and Ueda, K. (1986). Ascending pathways from the spinal cord in the himé salmon (landlocked red salmon, *Oncorhynchus nerka*). Journal of Comparative Neurology 254: 104-112.

O'Neill, P., McCole, R.B. and Baker, C.V.H. (2007). A molecular analysis of neurogenic placode and cranial sensory development in the shark *Scyliorhinus canicula*. Developmental Biology 304: 156-181.

Pfeiffer, W. (1968). Die Fahrenholzschen Organe der Dipinoi und Brachiopterygii. Zeitschrift für Zellforschung und Mikroskopische Anatomie 90: 127-147.

Pinkus, F. (1894). Über einen nocht nicht beschriebenen Hirnnerven des *Protopterus annectens*. Anatomische Anzeiger 9: 562-566.

Platt, C., Jørgensen, J.M. and Popper, A.N. (2004). The inner ear of the lungfish *Protopterus*. Journal of Comparative Neurology 471: 277-288.

Popper, A.N. and Platt, C. (1993). Inner ear and lateral line of bony fishes. In: D.H. Evans (ed). The Physiology of Fishes. CRC Press, Boca Raton, FL, USA. pp. 99-136.

Popper, A.N. and Fay, R.R. (1999). The auditory periphery in fishes. In: R.R. Fay and A.N. Popper (eds). Comparative Hearing: Fish and Amphibians. Springer, New York, USA. pp. 43-100.

Puelles, L. and Rubenstein, J.L.R. (1993). Expression patterns of homeobox and other putative regulatory genes in the embryonic mouse forebrain suggest a neuromeric organization. Trends in Neuroscience 16: 472-479.

Reiner, A. and Northcutt, R.G. (1987). An immunohistochemical study of the telencephalon of the African lungfish, *Protopterus annectens*. Journal of Comparative Neurology 256: 463-481.

Ronan, M.C. and Northcutt, R.G. (1985). The origins of descending spinal projections in lepidosirenid lungfishes. Journal of Comparative Neurology 241: 435-444.

Roth, G., Mühlenbrock-Lenter, S., Grunwald, W. and Laberge, F. (2004). Morphology and axonal projection pattern of neurons in the telencephalon of the fire-bellied toad *Bombina orientalis*: an anterograde, retrograde, and intracellular biocytin labeling study. Journal of Comparative Neurology 478: 35-61.

Rudebeck, B. (1945). Contributions to forebrain morphology in dipnoi. Acta Zoologica 26: 10-157.

Sanders, A. (1889). Contributions to the anatomy of the central nervous system in *Ceratodus forsteri*. Annals of the Magazine of Natural History 3: 157-188.

Schnitzlein, H.N. (1966). The olfactory tubercle of the African lungfish, *Protopterus*. Alabama Journal of Medical Sciences 3: 39-45.

Schnitzlein, H.N. and Crosby, E.C. (1967). The telencephalon of the lungfish, *Protopterus*. Journal fur Hirnforschung 9: 105-149.

Schnitzlein, H.N. and Crosby, E.C. (1968). The epithalamus and thalamus of the lungfish, *Protopterus*. Journal fur Hirnforschung 10: 351-371.

Schober, A., Meyer, D.L. and von Bartheld, C.S. (1994). Central projections of the nervus terminalis and the nervus praeopticus in the lungfish brain revealed by nitric oxide synthase. Journal of Comparative Neurology 349: 1-19.

Sewertzoff, A.N. (1902). Zur Entwicklungsgeschichte des *Ceratodus forsteri*. Anatomische Anzeiger 21: 593-608.

Thors, F. and Nieuwenhuys, R. (1979). Topological analysis of the brain stem of the lungfish *Lepidosiren paradoxa*. Journal of Comparative Neurology 187: 589-612.

Trabucchi, M., Trudeau, V.L., Drouin, G., Tostivint, H., Ihrmann, I., Vallarino, M. and Vaudry, H. (2008). Molecular characterization and comparative localization of the mRNAs encoding two glutamic acid decarboxylases (GAD65 and GAD 67) in the brain of the African lungfish, *Protopterus annectens*. Journal of Comparative Neurology 506: 979-988.

Vaage, S. (1969). The segmentation of the primitive neural tube in chick embryos. Advances in Anatomical Embryology and Cell Biology 41: 7-87.

Vallarino, M., Bunel, D.T. and Vaudry, H. (1992). Alpha-melanocyte-stimulating hormone (α-MSH) in the brain of the African lungfish, *Protopterus annectens*: immunohistochemical localization and biochemical characterization. Journal of Comparative Neurology 322: 266-274.

Vallarino, M., Tranchand-Bunel, D., Thoumas, J-L., Masini, M.A., Conlon, J.M., Fournier, A., Pelletier, G. and Vaudry, H. (1995). Neuropeptide tyrosine in the brain of the African lungfish, *Protopterus annectens*: immunohistochemical localization and biochemical characterization. Journal of Comparative Neurology 356: 537-551.

von Bartheld, C.S. (1992). Oculomotor and sensory mesencephalic trigeminal neurons in lungfishes: phylogenetic implications. Brain, Behavior and Evolution 39: 247-263.

von Bartheld, C.S. (2004). The terminal nerve and its relation with extrabulbar "olfactory" projections: lessons from lampreys and lungfishes. Microscopy Research and Technique 65: 13-24.

von Bartheld, C.S. and Meyer, D.L. (1988). Central projections of the nervus terminalis in lampreys, lungfishes, and bichirs. Brain, Behavior and Evolution 32: 151-159.

von Bartheld, C.S. and Meyer, D.L. (1990). Paraventricular organ of the lungfish *Protopterus dolloi*: morphology and projections of the CSF-contacting neurons. Journal of Comparative Neurology 297: 410-434.

von Bartheld, C.S., Claas, B., Münz, H. and Meyer, D.L. (1988). Primary olfactory projections and the nervus terminalis in the African lungfish. Implications for the phylogeny of cranial nerves. American Journal of Anatomy 182: 325-334.

von Bartheld, C.S., Collin, S.P. and Meyer, D.L. (1990). Dorsomedial telencephalon of lungfishes: a pallial or subpallial structure? Criteria based on histology, connectivity, and histochemistry. Journal of Comparative Neurology 294: 14-29.

Whiting, H.P., Bannister, L.H., Barwick, R.E. and Bone, Q. (1992). Early locomotor behaviour and the structure of the nervous system in embryos and larva of the Australian lungfish, *Neoceratodus forsteri*. Journal of Zoology [London] 226: 175-198.

Whitt, G.G. (1969). The telencephalon of the Australian lungfish, *Epiceratodus*. Alabama Journal of Medical Sciences 6: 165-191.

Wilder, B.G. (1887). The dipnoan brain. American Naturalist 21: 544-548.

Wullimann, M.F. (1998). The central nervous system. In: D.H. Evans (ed). The Physiology of Fishes, 2nd edition. CRC Press, Boca Raton, FL, USA. pp. 245-282.

Wullimann, M.F. and Vernier, P. (2007). Evolution of the nervous system in fishes. In: J. Kaas and T.H. Bullock (eds). Evolution of Nervous Systems, Non-mammalian Vertebrates. Volume 2, Academic Press, San Diego, USA. pp. 39-60.

Zambrano, D. and Iturriza, F.C. (1972). Histology and ultrastructure of the neurohypophysis of the South American lungfish, *Lepidosiren paradoxa*. Zeitschrift fur Zellforschung 131: 47-62.

Zambrano, D. and Iturriza, F.C. (1973). Hypothalamic-hypophysial relationships in the South American lungfish *Lepidosiren paradoxa*. General and Comparative Endocrinology 20: 256-273.

CHAPTER 17

Vision in Lungfish

Justin Marshall[1,*], Shaun Collin[2], Nathan Hart[2] and Helena Bailes[1,3]

[1]Sensory Neurobiology Group, School of Biomedical Sciences and Queensland Brain Institute, The University of Queensland, Brisbane, QLD 4072, Australia

[2]The University of Western Australia, School of Animal Biology & The Oceans Institute, Crawley, WA 6009, Australia

[3]Faculty of Life Sciences, Michael Smith Building, Oxford Road, Manchester, M13 9PT, England

ABSTRACT

All three extant genera of lungfish, Australian, African and South American, appear to possess unremarkable, even 'degenerate eyes' when viewed externally. The eyes of the Australian lungfish, *Neoceratodus forsteri*, are slightly larger than those of the other species (seven African species in the genus *Protopterus* and the single South American species *Lepidosiren paradoxa*). *N. forsteri* seems to be the most visually-oriented of the extant lungfishes. All three genera of lungfish, however, possess remarkable and beautiful retinal adaptations, including, coloured oil droplets, multiple cone spectral sensitivities and large photoreceptor inner segments, making them more closely aligned in design to modern amphibians and other terrestrial animals, than to teleosts. The tetrapod-like retinal features of *N. forsteri* provide the capability for tetrachromatic colour vision and add to the debate on the phylogenetic origin(s) of lungfish. They also suggest that the complex colour vision system of vertebrates on land, exemplified by birds, may have first evolved in the aquatic environment or at least close to the time when aquatic life emerged onto land. Other ocular adaptations in dipnoans include a non-spherical lens, the anatomical mechanism for accommodation, a mobile pupil and giant retinal cells. This eye design suggests a need to increase light flux, rather than for a reliance on high spatial acuity, a conclusion supported by the relatively low ganglion cell densities. Future work should certainly aim at a better understanding of the visual biology, behaviour and ecology of all lungfish, especially in light of their disappearing habitat worldwide. Both African and South American species also need a full description of their visual system

*Corresponding author: E-mail: justin.marshall@uq.edu.au

before they are properly consigned to being 'less well developed', than *N. forsteri.*

Keywords: Vision, colour, visual ecology, colour signals, neurobiology

INTRODUCTION

An ability to breathe air, the possession of lobe-fins, and an evolutionary position possibly closer to amphibians than to fish, places the lungfishes (Dipnoi) in an iconic position within the Osteichthyes. Several aspects of eye design in lungfishes suggest that they possess a visual system much more like that of a land dwelling vertebrate such as a bird, reptile or amphibian. The potential for complex tetrachromatic colour vision, spectral coverage from the UV to deep in the red region of the spectrum and coloured filters to tune the light incident on the retina has recently been realised (Robinson 1994; Bailes *et al.* 2006a). Although Bailes and colleagues have provided a recent and thorough examination of the visual system of the Australian lungfish, *Neoceratodus forsteri*, components of this surprisingly un-fish-like visual system have been known for more than a century (Krefft 1870; Gunther 1871; Schiefferdecker 1886; Kerr 1902; Grynfeltt 1911; Walls 1942). Robinson (1994) revived interest in lungfish vision by a more complete description of the most surprising aspect of their vision, the coloured oil droplets and ellipsoids housed within the photoreceptors, and by suggesting the potential for tetrachromatic colour vision in *N. forsteri.* These sorts of coloured filters, while present in the eyes of many land vertebrates, are rare among fish (Walls 1942; Douglas and Marshall 1999), although the coelacanth *Latimeria chalumnae*, a close relative of the lungfish, is one example where a large oil droplet is associated with one of its three large single cones (Locket 1973).

A gaping hole in our understanding, that will unfortunately permeate this chapter, is the lack of work on the visual systems of the African and South American species of lungfishes. In general, due to their small eye size (Fig. 1) and comparatively reduced complexity in a number of ocular features, these species are thought to be less 'visual' than the Australian lungfish. It is, however, possible that this is an expression of our ignorance rather than a reflection of the real situation. Wherever possible in this chapter we refer to known work on *Lepidosiren paradoxa* and *Protopterus* species, but the majority of what is summarised here concerns *N. forsteri.*

Retinal morphology and the optics of the eyes of *N. forsteri*, along with visual ecology and some scant observations on visual behaviour, indicate that this visual system is designed for increased sensitivity rather than for high acuity. This conclusion is supported by the findings of very large or "monstrous" photoreceptors described by Walls (1942), a general feature shared by all species of lungfishes. Conversely, the relatively small eyes, in particular those of the Lepidosirenidae (this includes both African and South American species) suggest that sensitivity per se is not of importance to the sensory biology of the lungfishes and that other

Fig. 1 The heads and eyes of lungfish. (a) Juvenile Australian lungfish, *Neoceratodus forsteri*, at around two years of age (25 cm total length). The body proportions are more similar to the adult at this stage. Note relatively large eye and lateral line pores. (Photograph Justin Marshall and Helena Bales). (b) South American lungfish, *Lepidosiren paradoxa*. Note relatively small eye and pigmented lateral line canals on head. (Photographic source – www.opencage.info/ Image 800_8532). Size unknown. (c) African lungfish *Protopterus annectens*. Note relatively small eye in this species also. (Photographic source – www.Photovault.com). Size unknown. (d) Same fish as (a) demonstrating eye/body proportions.

Color image of this figure appears in the color plate section at the end of the book.

factors such as the presence of dense coloured oil droplets in their inner segments may have driven the apparent need for sensitive photoreceptors (Partridge 1989).

Several aspects of lungfish vision add to the debate on the relationship of lungfish to teleost fish and urodele amphibians (Robinson 1994; Carroll 1988, 1997; Joss 1998; Pough *et al.* 1999; Takezaki *et al.* 2004; Bailes *et al.* 2006a, 2007a,b). While this is the subject of another chapter (Chapter 20), it is worth noting in passing here the striking resemblance between what we think of as an advanced terrestrial vertebrate colour vision system and that of the lungfish visual system (Bailes 2006; Bailes *et al.* 2006a; Hart *et al.* 2008). Molecular analysis of the visual pigments or opsin genes also places lungfishes closer to the amphibians than teleost fishes (Bailes *et al.* 2007b).

The complexity of the lungfish colour vision system contributes to the discussion on the evolution of colour vision in vertebrates (Tresize and Collin 2005; Lamb *et al.* 2007), suggesting that the capacity for colour discrimination first evolved in water and was retained after emerging onto land, although specific classes of photoreceptors have subsequently been lost. The suggestion of neoteny in lungfish (Joss 1998) could also point towards the retention of visual characteristics evolved on land in a now largely aquatic animal.

Prior to the recent work of Bailes and colleagues, fewer than ten publications existed on lungfish eyes. Gunther (1871), Schiefferdecker (1886), Kerr (1902) and Grynfeltt (1911) provided early anatomical descriptions. Walls (1942) reviewed this early work and went on to provide what is still considered the best source of comparative morphological study on the eyes of the extant species. More recently, Munk (1969), Locket (1970), Ali and Anctil (1973, 1976), Pow (1994) and Robinson (1994) have contributed morphological studies, largely based on individual species. A common theme of these investigations is that the retina and choroid in all species described are considered thin, while the cells of all retinal layers are very large. Walls (1942) implies that the eyes of the *L. paradoxa* and *Protopterus* sp. are degenerate, at least when compared to *N. forsteri*, while Ali and Anctil (1976) rank them according to complexity in the ascending order South American, African and Australian species. Walls (1942) states:

"The dearth of knowledge about *Lepidosiren* is of no great importance, since this form is in the same family as *Protopterus*. But *Neoceratodus* deserves a thorough investigation, for this large fish has none of the appearances of degeneracy characteristic of the Lepidosirenidae. Its relatively large eye may have, in particular, a mechanism of accommodation: and its cone oil droplets may be coloured in life. But the animal is reputedly nocturnal (in captivity, at least), and may not have retained such things even though some diurnal ancestor may have had them."

GROSS MORPHOLOGY OF THE EYE

In all species, the eyes are laterally-placed in the head and possess circular-shaped pupils (Fig. 1; Walls 1942; Bailes *et al.* 2007a). The eyes are not remarkably coloured or protruding and the iris is generally pigmented to match the pale brown colour of the body. Eyes in members of the Lepidosirenidae are small, i.e. 1.5 mm in a 50 cm fish in *L. paradoxa* (Ali and Anctil 1973) and 2 mm in a 30 cm fish in *Protopterus dolloi* (Pfeiffer 1968). The eyes of *N. forsteri* are proportionally larger i.e. 14 mm in a 127 cm fish, and the axial length of the eye increases with total body length according to a linear relationship (Bailes *et al.* 2007a). Compared to an "average teleost fish", the axial length of the eye in *N. forsteri* is, however, relatively small (Howland *et al.* 2004; Bailes *et al.* 2007a).

CORNEA, LENS AND PUPIL

In *N. forsteri*, the non-refractile cornea consists of two layers; inner definitive (scleral) cornea and an outer secondary spectacle (Walls 1942), the outer layer being evenly pigmented yellow, which is especially dense in the adults (Fig. 2). As the eye grows, the the lens and globe become increasingly elliptical in shape (Fig. 2) but retains an even magnification from centre to periphery. Functionally, the result is a decrease in focal ratio and an increase in illumination on the retina, suggesting increased sensitivity is needed (Bailes *et al.* 2007a). Aspherical lenses are known in deep-sea teleosts, most elasmobranchs and lampreys with a variety of explanations given (Munk 1984; Sivak 1990; Collin *et al.* 1999; Douglas *et al.* 2002). The lens of the adult lungfish is also yellow, with an increasing concentration of pigment

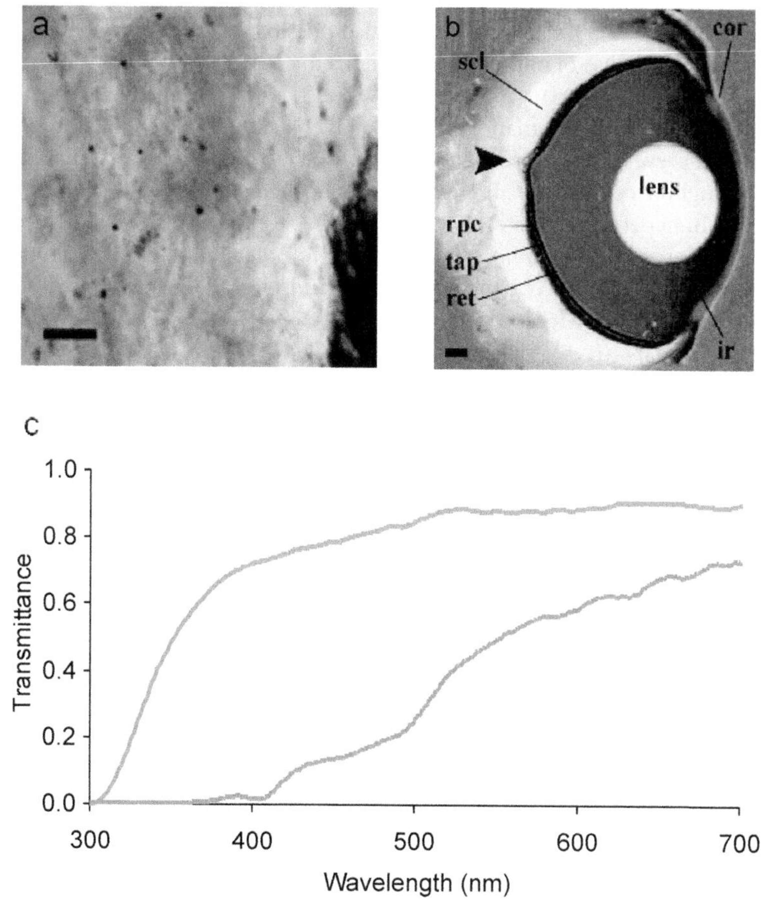

Fig. 2 Some optical features of the eye of *Neoceratodus forsteri*. (a) Freshly dissected cornea of adult lungfish with transmitted light to show intense yellow pigment. Scale 250 µm. (b) Frozen hemisected left eye of adult (127 cm total length), close to the geometrical centre. Note slight flattening of the lens. scl – sclera, ir – iris, cor – cornea, tap – tapetum, rpe – retinal pigment epithelium, ret – retina. Retinal indentation (arrow) close to optic nerve head. Scale - 1 mm.

Color image of this figure appears in the color plate section at the end of the book.

located towards the lens centre (Bailes 2006). Possible functions of a yellow lens and cornea are discussed in sections on the retina and colour vision.

A muscular attachment has been found to secure the lens of *N. forsteri* anteriorly, suggesting that the lens could accommodate and focus on near or far objects in the environment (Bailes 2006; Bailes *et al.*, 2007a). However, this has yet to be shown directly in *Protopterus annectens* and other species (Walls 1942; Munk 1964).

The optical performance of the lungfish lens and eye appears rather poor with the likelihood of both spherical aberration (more marked in the periphery of the lens) and hyperopia reducing the ability to discriminate fine detail (Bailes *et al.* 2006b; Bailes *et al.* 2007a). However it is also possible that accommodatory lens movement and the use of multiple focal lengths to correct for spherical aberration (Malkki and Kröger 2005) are present in *N. forsteri*, both mechanisms providing improved spatial (and chromatic) vision (Bailes *et al.* 2007a). The spatial resolving power of the eye is discussed in more detail below along with the changes in retinal ganglion cell density. It is worth noting here, however, that if the eye does remain hyperopic at rest, this will be an over-estimate due to the defocused image. The larger eyes of *N. forsteri* compared to other lungfish species, may in fact be optimized both for enhanced sensitivity and/or spatial resolution within its environment (Land 1981; Howland *et al.* 2004).

Most of the teleosts described with pupillary responses are benthically-oriented (Walls 1942; Rubin and Nolte 1982; Douglas *et al.* 1998; Douglas *et al.* 2002). The finding of pupillary movement in *N. forsteri* reveals that this ability is a primitive characteristic for vertebrates, a suggestion confirmed by the finding of a mobile pupil in the southern hemisphere lamprey, *Mordacia mordax* (Collin *et al.* 2004). In lungfish, pupillary contraction is noted in *Protopterus annectens* (Steinach 1890, Walls 1942), where on release from aestivation, the eye has a slit-shaped pupil that rounds up after several hours. Whether this slit is only present when the fish lies within the protection of its mud cocoon is not known, but apparently no intraocular eye muscles are involved in the slow papillary response (Walls 1942). Other animals with slit pupils narrow the slit with increasing light intensity (Walls 1942). The pupillary response in *N. forsteri* is also slow, taking at least 30 minutes, but being mostly complete after 10 minutes (Bailes *et al.* 2007a). During constriction of the adult pupil, the retinal illumination changes by a factor of two and the F-number of the eye changes from 0.47 to 0.64 for the dilated and constricted states, respectively (Bailes 2006). Contraction is described as similar to that observed in amphibians, almost all of which are known to show a slow pupillary response (Walls 1942; Cornell and Hailman 1984), one factor that initially suggested the possibility of a close relationship between lungfishes and amphibians (Walls 1942).

GENERAL DESCRIPTIONS OF RETINAL LAYERS AND SURROUNDING TISSUES

General descriptions exist for the retinae of all three lungfish genera, based mainly on light and some electron microscopical studies. Walls (1942) describes a *Protopterus* species, probably mostly *P. aethiopicus*. Walls describes 'huge' retinal pigment epithelial (RPE) cells making the pigment epithelium as thick as the sclera and its rudimentary (thin) choroid. "All retinal elements are monstrous," he proports, where the outer nuclear layer possesses two rows of nuclei of rod and cone soma. The inner nuclear layer contains four rows of nuclei, the outer plexiform layer is very thin and the inner plexiform layer is thick. Horizontal cells are "slenderly fibrous" or possibly not present at all. A single row of ganglion cells exists.

Ali and Anctil (1973, 1976) describe the retina of *Lepidosiren* as essentially similar to that of *Protopterus*. The retinal cells are both large and sparse and the photoreceptor, bipolar and ganglion cell populations are among the largest in any fish. No retinal mosaic was observed, as often seen in teleost fishes, and indeed is not considered present in the other two species although no nearest neighbour analysis was undertaken. After light and dark adaptation, no retinomotor movements (the exchange in retinal position of rods and cones on a diurnal cycle) were noted, in agreement with similar observations for *Protopterus* (Pfeiffer 1968). Walls (1942), however, makes a reference to the possibility of retinomotor movements in *Lepidosiren*, in connection with light flux regulation, but without any reference. In the outer plexiform and inner nuclear layers of the *Lepidosiren* retina, apparently horizontal and amacrine cells could not be distinguished from bipolar cells, but their presence was not eliminated. Retinal cell distribution was largely uniform in the four eye quadrants (dorsal, ventral, nasal, temporal) with the number of rods about the same as the number of cones over the retina .

Locket (1970), Pow (1994), Robinson (1994) and Bailes *et al.* (2006a) provide a more detailed description of the retinal morphology in *Neoceratodus forsteri*. Again, all cell types are large although the entire retina is thin (Walls 1942) where the nuclear layers are placed between two thin plexiform layers. Bailes *et al.* (2006a) note up to four sub-layers in both outer and inner nuclear layers. As in the other species, there is no retinal vasculature and no retinomotor movement of photoreceptors. The choroid has large blood vessels and a rete mirabile within a meshwork of cells containing reflective material that comprise a tapetum lucidum, as identified with transmission electron microscopy (Bailes *et al.* 2006a). The tapetum elicits a reddish-pink reflex from the dorsal two thirds of the eye, although the red colour probably is the result of the large red cone oil droplets housed within the inner segments of a sub-population of photoreceptors. Below the tapetum lucidum lies a darkly-pigmented, non-reflective strip of tissue. Vitread to the choroid, large RPE cells extend processes between all photoreceptor types to the external limiting membrane. Using immunohistochemical methods, Pow (1994) identified two types of horizontal cells, bipolar cells (displaced to the outer nuclear layer and lying within the inner nuclear layer), seven different types of amacrine cells within the inner nuclear layer (four were noted by Bailes *et al.* 2006a), two types of glial cells and both normal and displaced types of ganglion cells. Bailes *et al.* (2006b) identified displaced amacrine cells in the ganglion cell layer by retrograde labeling and four types of ganglion cells. Pow (1994) also demonstrated the retinal distribution of retinal neurotransmitters, such as glutamate and GABA, along with other amino acids.

Previous work, summarised by Walls (1942), Locket (1970) and Pow (1994) describe bipolar cell processes called Landolt's clubs in *Protopterus* and *Neoceratodus*. These structures contain two centrioles and a cilium and are also observed in urodele amphibians and elasmobranchs (summarised in Locket 1970). Their function, however, is speculative.

PHOTORECEPTOR MORPHOLOGY AND DISTRIBUTION

The presence of oil droplets, including the rods of the lepidosireniform species, and the large size of all retinal cells are the two key observations arising from the early body of work on lungfish (Fig. 3; Gunther 1871; Schiefferdecker 1886; Kerr 1902; Grynfeltt 1911; Walls 1942; Munk 1969; Ali and Anctil 1973). Oil droplets, while present in amphibians, birds reptiles and non-placental mammals, are rare in fish (Walls 1942; Douglas and Marshall 1999). Notable exceptions are the relative of lungfishes, the coelacanth, *Latimeria chalumnae* and some species of early ray-finned fishes such as chondrosteans (sturgeon and paddle-fish; Walls 1942; Locket 1973). The lamprey, *Geotria australis* also contains coloured inclusions within three of its five photoreceptor sub-types. These receptors contain yellow inclusions localized within the endoplasmic reticulum of the myoid region (Collin *et al.* 2003). Oil droplet-like inclusions are also part of the retina of the downstream and upstream migrants of *Mordacia mordax* (Collin *et al.* 2004) and *G. australis* (Collin and Trezise 2004), respectively but these do not contain any coloured filters and are ellipsosomes formed from mitochondria. Double cones are present in *Protopterus* sp. (Fig. 3) but, despite previous suggestions (Munk 1964, Pow 1994), they probably do not appear in *N. forsteri* (Bailes *et al.* 2006a) or in *L. paradoxa* (Walls 1942; Ali and Anctil 1973, 1976). The unequal double cones of *Protopterus* sp. possess an oil droplet in the principle (larger) member of the double cone (Fig. 3). Unfortunately, the colours of oil droplets of the lepidosireniforms are not well known as these are largely described from sectioned material only, where the oil droplets may have faded. In the older literature,they are usually described as colourless (summarised in Walls 1942), but this may not be accurate as there are accounts of red oil droplets in both *Protopterus* (Locket 1999) and *Lepidosiren* (Nicol *et al.* 1972). It would be particularly interesting to learn the colour of the rod oil droplet (Fig. 3) as this and other photoreceptor features have led to the suggestion that these rods are in fact modified cones (Walls 1942).

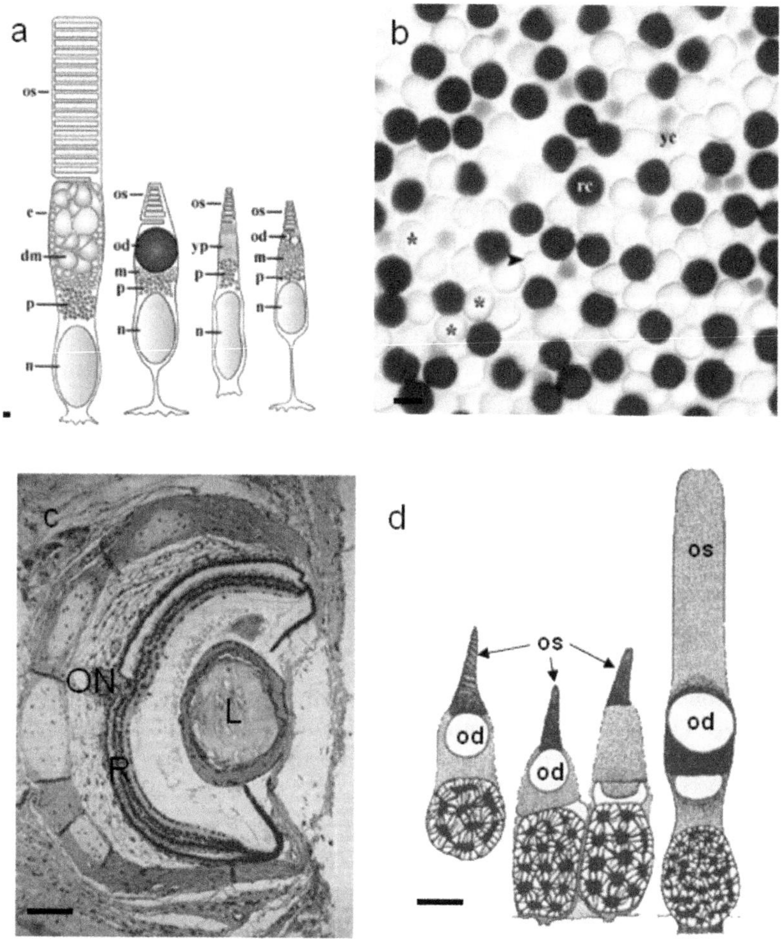

Fig. 3 Photoreceptors and retina of lungfish. (a) Schematic diagram of four morphologically distinct photoreceptors in *Neoceratodus forsteri* from left to right, rod, red cone, yellow cone, clear cone (after Bailes 2006). dm - distended mitochondria, e – ellipsosome, os – outer segment, p – parabaloid, n – nucleus, od – oil droplet, m – myoid, n – nucleus (for explanation of all terms see Walls 1942; Bailes 2006). Scale 10 μm. (b) Retinal whole-mount of fresh *Neoceratodus forsteri* retina showing all four morphological photoreceptor types. Large clear photoreceptors are rods (asterisks), red and yellow photoreceptor inclusions clearly distinguish these cone types. Arrow marks one clear cone type, being notably smaller than the rods. Scale 10 μm. (c) Section through the eye of *Lepidosiren paradoxa*. (From Ali and Anctil 1973, with kind permission). Note thin retinal layers. R – retina, L – lens, ON – optic nerve. scale 250 μm. (d) Schematic diagram of the photoreceptors of *Protopterus aethiopicus*. Note large oil droplet in the rod (far left). Apparent double cone (centre) and single cone (left) (after Walls 1942). Abbreviations as (a). Scale 10 μm.

Color image of this figure appears in the color plate section at the end of the book.

Through a combination of techniques including whole-mount and electron microscopy, five morphological types of photoreceptor, four cone types and a rod, have been identified and extensively characterised in *N. forsteri* (Fig. 3; Robinson 1994; Pow 1994; Bailes *et al.* 2006a). The large rods (around 18 × 40 μm outer segments) make up around 53% of the photoreceptor population. The most abundant cone type (34% of the total population) contains a red oil droplet in the inner segment, with an outer segment close to 15 μm long with basal diameter of 6 μm. The next most frequently found cone (10% of the photoreceptors) possesses a granular, yellow elipsoid pigment and outer segments around 4 × 10 μm in dimension. The third, and least abundant cone type (3% of the photoreceptors), has clusters of clear oil droplets in the ellipsoid and the outer segments are around 4 × 3 μm in size. Both molecular biology and microspectrophotometry (MSP) identify a fifth type of photoreceptor in juvenile lungfish with a spectral sensitivity peaking in the ultraviolet (UV). Morphologically, these cells appear much like the clear oil droplet category just described, although a little smaller (Table 1).

The large rods of lungfish are more like those of amphibians than teleosts in size and 'doom' the animal 'forever to low visual acuity' (Walls 1942). Teleosts, on the other hand, tend to possess many tiny rods and larger cones (Ali and Anctil 1976). Spatial resolving power calculated based on cone spacing is equally poor at around 3.3 cycles per degree in adults (Bailes *et al.* 2006a). In a range of marine teleosts, cone-density based spatial resolving power varied from 4.2 to 14.3 cycles per degree (Tamura 1957). More accurate estimates of spatial resolving power come from ganglion cell mapping as detailed below (Collin and Pettigrew 1989; Bailes *et al.* 2006b).

The distribution of all *N. forsteri* photoreceptors has been mapped in some detail (Bailes 2006). For all cone types, there is an increase in density mostly in the dorso-temporal retina but also in the ventro-nasal retina (Fig. 4, Bailes *et al.* 2006a). The rods are concentrated within a weak horizontal streak of increased density, matching that of the ganglion cells. This distribution of photoreceptors has been taken to imply an interest in objects in front of the fish and possibly an interest in a local horizon such as the sand/water interface.

Large photoreceptors generally imply a need for higher sensitivity rather than high spatial resolution (Land 1981). *N. forsteri* is reported to be active nocturnally or during crepuscular periods (Grigg 1965a; Kemp 1986) and this may explain the size of both cones and rods in these fishes.

GANGLION CELLS AND VISUAL ACUITY

A single row of ganglion cells is described in both *Protopterus* and *Lepidosiren* (Walls 1942, Ali and Anctil 1976). *Lepidosiren* has one of the highest ratios of photoreceptors to retinal ganglion cells of any vertebrate and contains a total of only 1,500 ganglion cells in its retina (Ali and Anctil 1973, Northcutt 1977).

Table 1 Photoreceptor spectral properties of juvenile and adult *N. forsteri*

Juvenile	*Rod*	*Clear UVS*	*Clear SWS*	*Yellow MWS*	*Red LWS*
Visual pigment mean λ-max	540	374	481	558	624
Filter mean λ-cut		<330	<330	534	563
Spectral sensitivity peak	535	383	484	584	631
Opsin type	Rh1	SWS1	SWS2	Rh2	LWS
Adult	*Rod*		*Clear SWS*	*Yellow MWS*	*Red LWS*
Visual pigment mean λ-max	540		481	558	624
Filter mean λ-cut			<330	534	591
Spectral sensitivity peak	574		517	584	656
Opsin type	Rh1		SWS2	Rh2	LWS

Values are to the nearest nm wavelength. λ-cut wavelength of intercept at the maximum measured absorptance by a line tangent to the absorptance curve of the filter at half maximum measured absorptance (Hart and Vorobyev 2005; Bailes *et al.* 2006a). Visual pigment data for the UVS cone is from a juvenile fish only. This cone type probably does not exist in adults. Peak spectral sensitivities are calculated from the filtering effect of the oil droplets and ocular media on the visual pigment. Note effect of more yellow cornea and larger oil droplets in the adult is to shift spectral sensitivities to longer peak values. Opsin localisation in the retina is assumed from spectral sensitivity rather than *in-situ* labeling and therefore are likely rather than certain (Bailes *et al.* 2006a).

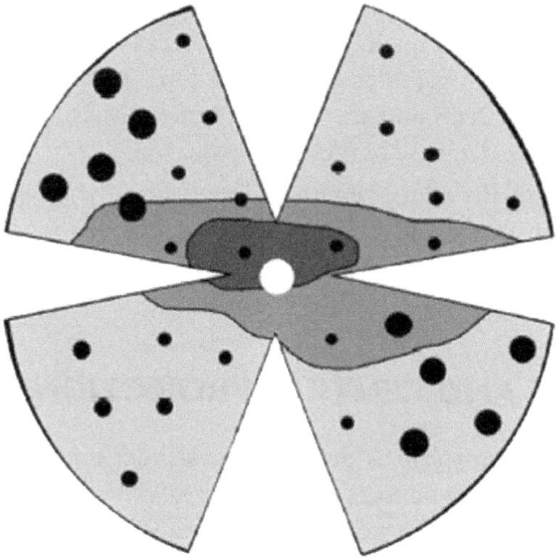

Fig. 4 Schematic representation of the topographic distribution of ganglion cells and cone photoreceptors derived from a retinal wholemount in *Neoceratodus forsteri*. The schematic retinal cup is flattened by making 4 cuts, dorsal (uppermost), ventral, nasal (right hand side) and temporal. The optic nerve head is represented by the clear area in the centre. Ganglion cell density is plotted as iso-density contours and shaded areas and shows a horizontal streak of increased density (darkest grey – 1.2 × 103 cells per mm, light grey – 0.048 × 103 cells per mm). Cone density, which is lowest towards the middle of the retina (1.85 × 103 cells per mm) and highest in dorso-temporal retina (3.49 × 103 cells per mm) is represented by black circles. Scale 1 mm.

This is thought to be an indication that the retina has stopped developing during growth and has remained larval-like or neotenic. In contrast, Bailes *et al.* (2006b) describe around 50,000 ganglion cells in *N. forsteri* divided into four categories, along with one type of amacrine cell that 'invades' the ganglion cell layer. They go on to map their distribution in detail (Fig. 4) demonstrating that a horizontal streak of higher ganglion cell density, matching that of the rods, lies across the meridian of the eye. Spatial resolving power based on ganglion cell spacing is low, averaging 1.7 cycles per degree. As the ganglion cells represent the final input to the brain, these estimates, rather than estimates based on rod or cone densities, are considered more accurate. Compared to highly visual teleosts, whose peak ganglion cell density can reach more than one hundred times that observed in *N. forsteri*, lungfishes clearly do not need to resolve fine detail (Collin and Pettigrew 1988). Ganglion cell density in *N. forsteri* is comparable to other riverine fish with benthic lifestyles, such as catfish (Dunn-Meynell and Sharma 1987). The horizontal streak may help stabilise the eye on the local horizon, the sand/water interface, and may even help in prey localisation as the fish sweeps the substrate for food (Bailes 2006). However, more in situ behavioural observations are needed to reliably begin to predict the importance of specific parts of the visual field to the survival of this species. Other aquatic vertebrates, such as teleosts, elasmobranchs, anurans and cetaceans also exhibit horizontal streaks and increased interest in the local horizon is often given as the driving force behind such retinal design (Hughes 1977; Collin, 2008 and see http://optometrists.asn.au/ceo/retinalsearch for a new comparative database of retinal maps).

Although, as with the other genera of lungfishes, the ganglion cell layer in *N. forsteri*, is relatively thin (7-10 μm), the presence of four different types of ganglion cells implies a degree of visual complexity not previously expected (Walls 1942; Bailes *et al.* 2006b).

OPTIC NERVE AND CENTRAL PROJECTIONS

The optic nerve in *Protopterus* is described as slender and simple, while in the other two genera, the optic nerve is divided into fascicles (Walls 1942). Northcutt (1977) notes that "the visual system of lepidosirenid lungfish is one of the most reduced visual systems in living vertebrates, as reflected by the small number of optic fibres." Bailes *et al.* (2006b) describe the optic nerve in *N. forsteri* as unpleated and divided into around 15 large fascicles. Each large fascicle is divided into smaller fascicles of axons surrounding a group of glial cells. In one adult examined (125 cm total length), there were 74,100 axons in the optic nerve of which around 26% were unmyelinated. In a comparison with the retinal ganglion cells counted in retinal wholemount, 62% of the axons in the optic nerve were derived from ganglion cells. The remaining axons may represent a retinopetal population of cells or be the result of axonal branching (Bailes *et al.* 2006b).

Retinofugal projections (projections to the visual centres of the brain) have been described for all genera of lungfishes by Northcutt (1977, 1980). In common with the retina and eye, *Protopterus* and *Lepidosiren* both possess an apparently simple pattern of projections with no ipsilateral retinal projections. However, *N. forsteri* exhibits both contralateral and ipsilateral projections, the latter often associated with stereopsis or at least binocular cross-talk (Pettigrew 1991; Northcutt 1980). The complexity of the larger eye of the Australian lungfish is also reflected in the neurons and brain areas concerned with vision, with a large differentiated dorsal thalamus and pre-tectum and a clearly laminated optic tectum (Northcutt 1980; Robinson 1997). These features imply a more complex visual repertoire. The lungfish brain is closer in structure to both the coelacanth (*Latimeria chalumnae*) brain and the amphibian brain than that of teleost fishes (Northcutt 1977, 1980, 1986).

VISUAL PIGMENTS, OIL DROPLETS AND SPECTRAL SENSITIVITIES

In almost every fish examined, morphologically-distinct photoreceptors contain different visual pigments and usually, therefore, possess different spectral sensitivities (Losey *et al.* 1999; Kusmic and Gualtieri 2000; Bowmaker and Loew 2008). Microspectrophotometry (MSP, Dartnall 1975) of the outer segments of the retina in adult *N. forsteri* (individuals 105 and 110 cm total length) has revealed a rod and three cone types with a maximal absorbance, or λ-max, of 534 nm (rod), 481 nm (short wavelength sensitive, clear oil droplet cone), 560 nm (medium wavelength sensitive, yellow ellipsoid cone) and 624 nm (long wavelength sensitive, red oil droplet cone) (Fig. 5). In juveniles (24, 28, 32 cm total length), a fourth type of cone was found with maximal absorbance in the ultra violet UV region of the spectrum at 375 nm (Fig. 5). This cone type shares the morphology of the short wavelength sensitive cone in the adult, including the presence of several clear oil droplets, although these UV sensitive photoreceptors are probably slightly smaller (Bailes 2006a; Hart *et al.* 2008). Such diversity and a large spectral range of photoreceptors argues against the idea that these animals are crepuscular/nocturnal, slow moving and not very reliant on vision. Rather, this sort of complexity invokes a diurnal existence with visual demands requiring a finely tuned colour vision system.

In common with several fish species, most notably the salmonids (Hawryshyn *et al.* 1989; Beaudet et al 1997; Bowmaker and Loew 2008), it seems most likely that the UV cone is lost in adulthood. MSP can be a haphazard technique in terms of photoreceptor sampling and can miss photoreceptor types all together, especially if they are located in discrete retinal regions and are particularly rare and/or small. It is therefore possible that UV cones persist in the adult, however the fact that UV wavelengths are also blocked from reaching the retina by an

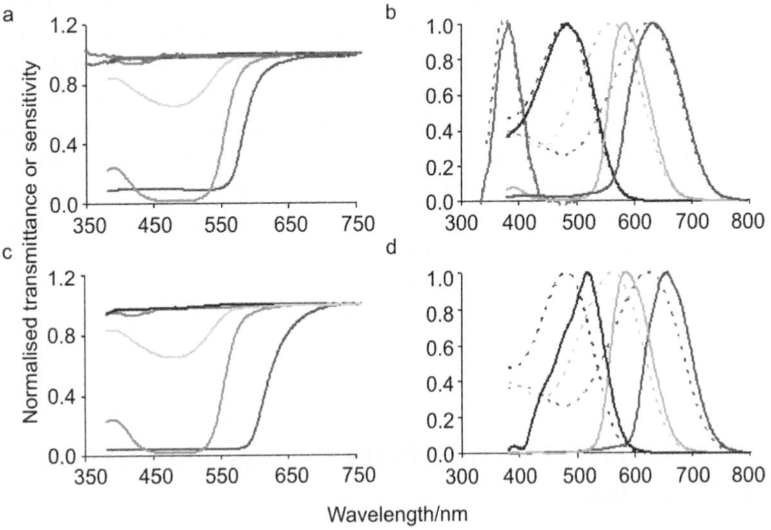

Fig. 5 Oil droplet and visual pigment spectral characteristics (a) Transmittance of oil droplets and yellow pigment in juvenile *Neoceratodus forsteri* photoreceptors relative to maximum transmission at 750 nm. The red oil droplet absorbs strongly below its λ-mid at 585 nm (red line), the yellow ellipsoid pigment λ-mid is at 556 nm (orange line) and the clear oil droplets of both UV (purple line) and short wavelength sensitive clear cones (green line) have negligible absorbance from 350-750 nm. (b) Visual pigment sensitivities of four cones (dotted curves) and the actual photoreceptor spectral sensitivities resulting from filtering by oil droplets and ocular media (Fig. 2) in juvenile *Neoceratodus forsteri*. Visual pigment curves are a nomogram based on peak absorbance (Table 1) and a vitamin A2 template, except for the UV sensitive photoreceptors which do not fit a template well and are just averaged direct measurements (Hart *et al.* 2008). Note loss of spectral sensitivity overlap and peak sensitivities pushed to longer wavelengths. (c) As (a) but for adult *Neoceratodus forsteri*. The red oil droplet is larger and absorbs more light, its λ-mid pushed longer at 623 nm. Yellow and blue pigments are the same as the juvenile. (d) Visual pigment sensitivities of three cones (dotted curves) and the actual photoreceptor spectral sensitivities resulting from filtering by oil droplets and ocular media (Fig. 2) in adult *Neoceratodus forsteri*. Visual pigment curves are a nomogram based on peak absorbance (Table 1) and a vitamin A2 template. Note loss of spectral sensitivity overlap and peak sensitivities pushed to longer wavelengths. Unlike the juvenile (b), the denser yellow adult ocular media (Fig. 2) absorbs strongly below 400 nm.

Color image of this figure appears in the color plate section at the end of the book.

increasingly yellowing cornea and lens in the adult, supports the notion that these photoreceptors are no longer useful to the mature fish (Figs. 2, 5, 7; Hart *et al.* 2008). The behavioural function of an apparent shift from potential tetrachromacy to potential trichromacy (Fig. 7) needs a clearer understanding of the environmental factors driving this change and of the behavioural ecology of these lungfishes. Possession of UV sensitivity may indicate a more surface biased existence, where

UV is still abundant even in green/brown water, or that a UV-specific task needs to be performed such as planktivory (Losey *et al.* 1999; Loew *et al.* 1993).

The yellow cornea and the yellow lens, particularly notable in adult *N. forsteri*, modifies the spectrum of light reaching the retina (Figs. 2, 5). The cornea and lens of juveniles filters out light below 330 nm, while in the adult, this moves up to around 400 nm. Variable UV filtering by ocular media is common in fishes (Kondrachev *et al.* 1986; Douglas and McGuigan 1989; Douglas and Thorpe 1992; Thorpe and Douglas 1993; Thorpe *et al.* 1993; Douglas and Marshall 1999; Siebeck and Marshall 2001; Siebeck *et al.* 2003) and is often thought to prevent the potentially damaging UV wavelengths from reaching the delicate retinal tissue. Increasing the density of yellow, short wavelength absorbing pigment within the ocular media during ontogeny is common in fishes (Thorpe and Douglas 1993) and indeed other animals including humans. Yellowing may be a result of an ageing eye as well as screening this damaging irradiation from reaching an increasingly large and therefore sensitive retina (Douglas and McGuigan 1989; Whitmore and Bowmaker 1989; Thorpe and Douglas 1993; Douglas and Marshall 1999).

All adult visual pigments contain a vitamin A2-based chromophore and this is typical of many freshwater fishes (Lythgoe 1979; Kusmic and Gualtieri 2000; Bowmaker and Loew 2008). The A2 containing visual pigments display longer wavelength sensitivities than their A1 equivalent and this appears to be of ecological benefit due to the largely longer wavelength dominated 'green' world of streams and lakes as opposed to the blues of oceanic waters (Lythgoe 1979). The UV sensitive cones of the juvenile lungfish, break this trend and indeed the 375 nm λ-max visual pigment in the juveniles was not a good fit to the A2 chromophore templates (template fitting is used to determine λ-max and chromophore type, Govardovski *et al.* 2000). A1 templates also did not fit the UV visual pigment absorbance curve well, implying secondary filtering or another factor as yet unknown may determine the absorbance in these photoreceptors (Fig. 5).

Each of the four morphological types of cone photoreceptor in *N. forsteri* contains a different oil droplet visual pigment combination (Table 1; Bailes *et al.* 2006a). Oil droplets, and other coloured inclusions in the eyes of many vertebrates, act as filters, tuning the spectral sensitivity provided by the visual pigment alone. The transmittance of oil droplets and the ocular media (cornea and lens) of *N. forsteri* have also been measured using MSP and spectrophotometry (Table 1; Fig. 5, 6; Hart *et al.* 2008). Both the clear and the intensely coloured red oil droplet and yellow ellipsoid of *N. forsteri* are similar to the coloured and clear oil droplets and ellipsoidal inclusions of birds and reptiles and the coloured retinal inclusions of the lamprey *Geotria australis* (Partridge 1989; Douglas and Marshall 1999; Collin *et al.* 2003; Bailes *et al.* 2006a; Hart *et al.* 2008). The oil droplets of amphibians, non-placental mammals and sturgeons are mostly colourless and this is often said to be the case for *Lepidosiren* and *Protopterus* (Walls 1942, see above). However, red oil droplets in both *Protopterus* (Locket 1999) and *Lepidosiren*

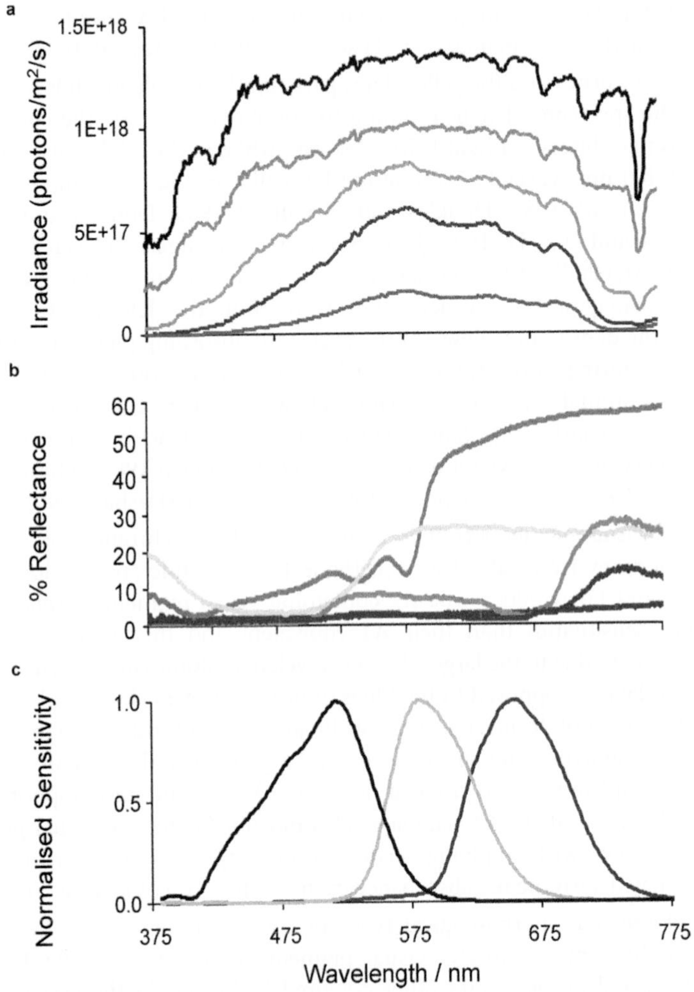

Fig. 6 Aspects of visual ecology in *Neoceratodus forsteri*. (a) Irradiance of light in the Mary River (Queensland, Australia), known habitat of *Neoceratodus forsteri* at: the surface (blue line), 0.05 m (bright green line), 0.5 m (mid-green line), 1 m dark green line, 1.25 m (kaki line). (b) Reflectance of objects in the environment of *Neoceratodus forsteri*: gravid female belly (orange line), non gravid female belly (yellow line), macrophytes from Mary River (green lines – see Hart *et al.* 2008 for details), submerged log from Mary River (brown line). These and other objects are plotted in the colour space of adult and juvenile in Fig. 7. (c) Adult spectral sensitivities (as Fig. 5b) included here for direct comparison with illumination in habitat and to complete complement of components needed for visual ecology characterisation; spectral sensitivities, illuminant on objects of interest and reflectance from objects of interest. Using these 3 factors is the first step towards the colour vision system models of Fig. 7 (Kelber *et al.* 2003).

Color image of this figure appears in the color plate section at the end of the book.

(Nicol *et al.* 1972) are now thought likely from tapetal examinations. Clearly, a careful examination of these retinae is needed and of particular interest will be the colour, if any, of the large oil droplets reported in the rods of these species (Fig. 3; Kerr 1902; Rochon-Duvigneaud 1941; Walls 1942; Ali and Anctil 1977).

Coloured photoreceptor inclusions generally act as long-wavelength-pass cut-off filters, absorbing wavelengths of light most strongly below their so called λ-cut (Table 1; Liebman and Granada 1975; Bowmaker and Knowles 1977; Partridge 1989; Hart and Vorobyev 2005). The functional result of this is examined in more detail in the visual ecology section below. Here, we just demonstrate the way in which the spectral sensitivities of the photoreceptors are sharpened and shifted to longer wavelengths (Fig. 5). In short, the spectral tuning by these filters improves the spectral discrimination ability of the owner's colour vision system (reviewed in Kelber *et al.* 2003). The composition of oil droplets and ellipsoids is comprehensively described elsewhere and is generally thought to be lipid and/or carotenoid based, as are the yellow pigments in the cornea and lens (Muntz 1972; Walls and Judd 1933; Wald and Zussman 1937; Meyer 1965; Johnston and Hudson 1976).

One result of filtering, by both ocular media and coloured photoreceptor inclusions, is a reduction in sensitivity of sometimes 90% or more (Douglas and Marshall 1999. The dimensions of the large and therefore most sensitive cone (and in the lepidosirenid lungfish, rod), may be a response to the quantal flux reduction imposed by the oil droplets and ellipsoids, along with a need to retain complex colour vision in the potentially light limited world of lungfish. If the currently observed activity patterns of *N. forsteri* are indeed crepuscular or nocturnal as suggested (Kemp 1986), any colour vision system that needed to retain colour discrimination in these challenging light conditions would certainly need to operate with 'monstrous' photoreceptors in order to catch enough photons (Land 1981; Warrant 1999; Warrant and Locket 2004). Night colour vision has recently been shown in a number of invertebrates (Kelber and Henique 1999; Kelber *et al.* 2003; Warrant 1999) and photoreceptor enormity is certainly in evidence there.

MOLECULAR BIOLOGY OF THE VISUAL PIGMENTS

Molecular characterisation the visual pigments of *N. forsteri* reveals the presence of all five known vertebrate opsin families: Rh1 (medium wavelength sensitive 1, found in rods), Rh2 (medium wavelength sensitive 2, found in cones), LWS (long wavelength sensitive, found in cones), SWS1 (UV/violet/short wavelength sensitive 1, found in cones) and SWS2 (blue/short wavelength sensitive 2, found in cones) (Yokoyama 2000a,b, 2002; Bowmaker and Hunt 2006; Bailes *et al.* 2007b). Table 1 shows the distribution of these within the 4 morphologically-distinct photoreceptors. The opsin is the protein part of the visual pigment molecule and its tertiary structure is one of the factors determining the spectral tuning of the visual pigment (Bowmaker and Hunt 1999; Bowmaker and Loew 2008).

This sort of opsin diversity is now being found in several fish lineages and interestingly, fish seem to be able to 'choose' when to express a particular opsin within a photoreceptor, both in an ontogenetic and a phylogenetic context (Bowmaker and Hunt 1999; Spady *et al.* 2006). It is, at least in part, the spectral environment and visual ecology of the species that determines the loss, duplication or retention of these genes during evolution (Bowmaker and Hunt 1999; Bowmaker and Loew 2008; Trezise and Collin 2005).

Interestingly, it was the use of molecular techniques that first suggested the presence of a SWS1 UV/violet sensitivity in lungfish and this was later confirmed with microspectrophotometry (MSP) (Bailes 2006; Bailes *et al.* 2007b; Hart *et al.* 2008). From both molecular and MSP evidence, it now seems likely that adults loose the fourth UV sensitive cone type and that this cone is retained in juveniles only (Bailes *et al.* 2007b; Hart *et al.* 2008). Similar loss of UV sensitivity during ontogeny is known in several fish species, notably the salmonids, and reflects changes in visual demands often associated with migrations and differing visual environments during different life stages (Hawryshyn *et al.* 1989; Archer and Hirano 1996; Beaudet *et al.* 1997; Bowmaker and Hunt 1999; Bowmaker and Loew 2008).

Phylogenetic analysis of the sequences of amino acids of the opsins from a range of vertebrates, places lungfish opsins closer to amphibian opsins than to those of teleost fish, again lending support to this evolutionary scenario (Bailes *et al.* 2007b). Partial sequences of Rh1 and LWS genes have been found in African lungfish, *Protopterus* sp. (Venkatesh ***et al.*** 2001; H.J. Bailes, W.L. Davies, A.E.O. Tresize, S.P. Collin – unpublished data) but more work is needed on this species and in *Lepidosiren* to determine the complement of opsin families present and their level of expression. As *N. forsteri* is generally regarded as plesiomorphic and the lepidosirenids derived and neotenic, determining the relationship of opsins within the three lungfish genera would add useful data to the current debate on their phylogenetic relationship and that of their nearest neighbours (Bailes *et al.* 2007b).

HABITAT, BEHAVIOUR AND VISUAL ECOLOGY

The visual behaviour of all lungfish species is poorly understood and it has even been stated that vision is of little importance to them (Dean 1906; Walls 1942; Kemp 1986; Simpson *et al.* 2002). Other sensory systems, including the lateral line and electroreception are well developed and some focus has been placed on these modalities (Watt *et al.* 1999; Jørgensen and Kemp, pers comm and see Chapters 18 and 20). *N. forsteri* is said to be mostly sedentary showing an increase in activity in the late afternoon and during the night, however the majority of these observations are from captive individuals (Grigg 1965a, Kemp 1986). While they can move fast if necessary (Kemp 1986), nothing is known about visual guidance of such behaviour. The diet of adult *N. forsteri* includes molluscs, fish, crustaceans and amphibians as well as some plant matter and this could suggest a visual component to prey capture (Kemp 1986). Olfaction and gustation are likely to play a significant role in feeding behaviour as the fish have been observed snuffling around in the weeds looking for food (Kemp 1986, Kind 2002). Live food is preferred by juveniles (Kemp 1986), although this observation is based largely on captive breeding programmes rather than field observations and, due to their poor spatial resolution, it is unlikely that they visually pursue prey (Bailes 2006; Bailes *et al.* 2006b). Courtship and spawning may include visual input since these behaviours are relatively elaborate and involve circling near the surface followed by spawning over apparently carefully selected macrophyte beds containing *Vallisneria gigantea, Hydrilla verticillata* and *Nitella* sp. (Grigg 1965a; Kemp 1984; Brooks and Kind 2001). Gravid females also become bright orange on the belly and this is likely to be a strong visual signal, especially in mid to long wavelength dominated fresh water systems (Lythgoe 1966, 1968a, b; Loew and McFarland 1990; Bowmaker and Loew 2008). A more detailed examination of how such colours may appear to *N. forsteri* is included below (Fig. 7).

An extensive survey of Australian lungfish movements and habits has all but failed to find any juveniles and, aside from a few sightings, this is a lasting mystery (Kind 2002). While it has been suggested that juveniles may spend much of their first few years buried in mud, Kemp (1986) suggests that the 'absence' of juveniles is due to inadequate capture mechanisms or a small fish's ability to avoid nets and escape. Kind's survey did determine that the average depth which adults inhabit was around 2 m but with considerable variation (Kind 2002). While a walk up the Mary or Fraser rivers might give the impression that *N. forsteri* lived in predominantly green, murky water, these fish are known from a wide variety of water types, use very shallow tributaries for spawning and are seen around surface waters (Grigg 1965a; Kemp 1986; Brookes and Kind 2001). That is, they clearly spend some time in broad spectrum, brightly-lit (possibly clear?) habitats in which a complex colour vision system, including UV wavelengths, may be useful.

The photic habitat of *N. forsteri* in the Mary River (Queensland) has been quantified, and in common with many algae/chlorophyll-laden freshwater systems, is largely green (Fig. 6). The spectral range of this species is clearly influenced by its light availability, as is the case for most fish (Fig. 6; Lythgoe 1966, 1968a,b, 1979, Bowmaker and Loew 2008). How UV may be of importance to juveniles, or even if it is truly lost in adults remains unresolved, although colours in the lungfish habitat, including body colours, contain a UV component (Fig. 6). This reflection in the UV is not remarkable as it can be found in other, apparently specific UV, signalling systems (Losey *et al.* 1999; Losey 2003; Hausmann *et al.* 2003; Marshall *et al.* 2003a, b; Siebeck 2004).

Protopterus is also found in a wide variety of natural habitats, both lentic and lotic, and is known to be an omnivorous carnivore (Greenwood 1986). Both

African and South American lungfishes are very poorly documented in terms of any behaviour that could be called visually driven. Unlike their Australian relative, they are capable of aestivation during drought and all lungfish species can flop about out of water, breathing air (Grigg 1965b; Burggren and Johansen 1986). *N. forsteri* is even said to 'walk' using its lobed fins but given the body weight of adults at several 10s of kilograms and the rudimentary musculature of these fins, this is not likely (Grigg 1965a, b; Kemp 1986). There are no reliable reports of any lungfish using their eyes for visually guided behaviour out of water, so the tenuous hope of lungfish complex colour vision being specifically for any out of water behaviour is questionable.

In short, we do not know much about where any species of lungfish spends its time and almost nothing about what it is doing while it is there. Given our new knowledge of the colour vision of Australian lungfish and the recent characterisation of the visual ecology of this animal (Bailes 2006; Hart *et al.* 2008), it is possible to predict detection and discrimination capabilities of potentially important colours to lungfish. As part of this, the function of the oil droplets and yellow ellipsoid pigments also becomes clear (Figs. 5, 7). These predictions using established visual models are now discussed in the context of oil droplet function and likely tri- or tetrachromacy in lungfish.

As previously demonstrated in birds (with visual systems also consisting of four cone types for colour vision, each with unique combinations of visual pigment and oil droplet), oil droplets function to expand the colour space of the animal (Fig. 7; Govardovski 1983; Vorobyev 2003 and see Kelber *et al.* 2003 for a good review of colour models and visual systems). In brief, by narrowing the spectral sensitivity of the naturally broad-band visual pigment contained in the cone to which they are associated, and by spacing the spectral sensitivities evenly in the available spectrum (Fig. 5), oil droplet filtering increases the distance of the loci spectra occupying each animal's colour space. As colour discrimination is, broadly speaking, proportional to distance between loci, this potentially allows finer spectral discrimination (Kelber *et al.* 2003 and Fig. 7) as well as improved colour constancy (Osorio *et al.* 1997; Neumeyer *et al.* 2002).

While we are not sure what colours may be of importance to lungfish, Fig. 7 plots the spectra of objects found in habitat of the Australian lungfish, i.e. plants, rocks, sand, logs and the lungfish itself, in the modelled colour space of both juvenile and adult *N. forsteri* (Bailes 2006; Hart *et al.* 2008). For the adult, this is done with oil droplets present and theoretically removed to show the contribution of these coloured retinal filters. Object spectra chosen include potential spawning-site plants and the colour of the female's ventral surface. The latter is known to change colour from a dull yellow to bright orange as the female becomes more gravid (Kemp 1984, 1986) and this may act as a colour signal to the males. While this is currently supposition, it is interesting that the position in *N. forsteri* colour space occupied by a ripening female's belly colours allows them to be particularly well discriminated (Fig. 7).

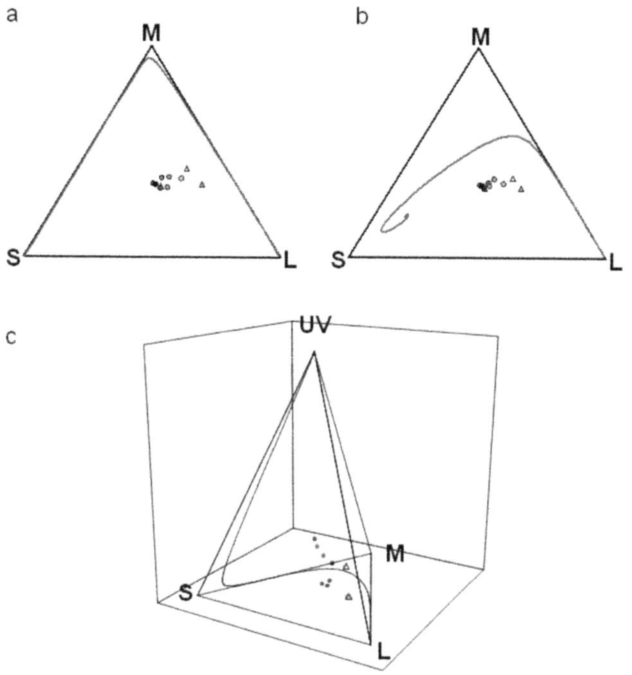

Fig. 7 Modelled colour vision performance in *Neoceratodus forsteri* examining coloured objects from Mary River environment (see Kelber *et al.* 2003 for review and references therein for methods). (a) Colour space based on Maxwell triangle for three cone sensitivities in adult *N. forsteri*. Although behavioural proof is required, this suggests trichromacy. The centre of the triangle is the achromatic point (occupied by flat spectra of e.g. white, grey and black) and areas towards the edges of the triangle are occupied by increasingly saturated spectral chromaticity loci up to the blue line – the monochromatic locus. The shape of the monochromatic locus is set by the spectral sensitivity characteristics and a comparison of (a) and (b) here demonstrates the way filtering by coloured ocular inclusions (oil droplets, ellipsoid pigment and ocular media) expands this space, thus improving the potential colour vision. Each symbol in the triangle (and tetrahedron in (c)) is the position occupied by one reflectance spectrum (Fig. 6b). The corners of the triangle; S, M and L plot the spectral loci of maximal stimulation of the short wavelength, medium wavelength and long wavelength photoreceptors respectively. Thus, for example, red spectra will plot near the L corner. Colour coding same as spectra in Fig. 6b. Female lungfish belly, gravid – orange triangle, female lungfish belly, non-gravid yellow triangle, macrophytes – green circles, log – brown circle, rocks and sand – grey circles (not plotted in Fig. 6b). (b) As (a) except for the hypothetical adult lungfish spectral sensitivities in the absence of filtering by coloured ocular inclusions (Govardovski 1983). Note the contraction of colour space delimited by the monochromatic locus (blue line) and the lesser separation of the loci of the coloured objects from lungfish habitat. (c) Three dimensional tetrahedral colour space of juvenile, potentially tetrachromatic, *N. forsteri*. Conventions and colours the same as triangles, UV, denotes the chromatic locus of the UV sensitive cone (Fig. 5b).

Color image of this figure appears in the color plate section at the end of the book.

Required input parameters to this model are environmental light in photons, reflectance measurements of colours and spectral sensitivity (calculated from visual pigment absorbance and the filtering effects of both oil droplets and cornea; Figs. 5, 6). These, along with the relative photoreceptor number, allow an accurate assessment of the performance of different colour vision systems (Vorobyev and Osorio 1989; Kelber *et al.* 2003). A comparison of the juvenile and adult (Fig. 7) does not shed light on any added advantage of potential tetrachromacy for the juveniles, however, given the paucity of knowledge regarding visual behaviour and habitat of either, this is no surprise.

It is clear that the general envelope and long wavelength bias of the cones in adult *N. forsteri* are well matched to the relatively long wavelength light prevalent in the freshwater habitat (Fig. 6) as is the case for many fish species, both freshwater and marine (Levine and MacNichol 1979; Lythgoe 1979; Loew and McFarland 1990).

Whether the adult and juvenile *N. forsteri* possessed truly trichromatic and tetrachromatic colour vision would require behavioural proof (Neumeyer 1992; Vorobyev *et al.* 1998; Kelber *et al.* 2003). Both forms of colour vision are known in freshwater fish (Neumeyer 1991, 1992, 1998), but they lack the added complexity of oil droplet filtering, a visual attribute of the terrestrial vertebrates. The dimensionality and details of the lepidosireniform lungfish colour vision (if any) are unknown, although the presence of (probably) coloured oil droplets certainly makes work on these species an exciting future project. Why Australian lungfish, with apparently poor spatial resolution and a desire to increase sensitivity with its large photoreceptors then 'invests' in the same sort of complexity for colour vision as a bird is another mystery. Colour vision often has rather poor spatial resolution (Vorobyev *et al.* 2001), but this does limit the sorts of task performed to those involving large colourful objects. Perhaps judging the quality of a big female with a bright orange belly and the correctly coloured patch of water-weed to lay eggs on are what has driven this colour vision system? On the other hand, it may be a neotenic survivor from a previously truly terrestrial amphibian. While this is unlikely to be resolved, visual systems are metabolically expensive and the fact that lungfish retain apparently complex colour vision suggests that they still use it. After the recent revelations regarding this remarkable fish's visual sense, it is clearly even more imperative that we both find out more and work to prevent the disappearance of lungfish through habitat destruction (Chapter 20).

Acknowledgements

The authors would like to thank Ann Trezise, Wayne Davies, Mike Bennett, Steve Robinson and Jack Pettigrew for much help, encouragement and collaboration during the recent studies on the Australian lungfish. The Australian Research Council funded many aspects of this work.

References

Ali, M.A. and Anctil, M. (1973). Retina of the South American lungfish *Lepidosiren paradoxa* Fitzinger. Canadian Journal of Zoology 51: 969-972.

Ali, M.A. and Anctil, M. (1976). Retinas of Fishes: An Atlas. Springer, Berlin, Heidelberg, New York.

Archer, S. and Hirano, J. (1996). Absorbance spectra and molecular structure of the blue-sensitive rod visual pigment in the conger eel (*Conger conger*). Proceedings of the Royal Society of London Series B 263: 761-767.

Bailes, H.J. (2006). The visual system of the Australian lungfish *Neoceratodus forsteri* (Kreft 1870). PhD Thesis, University of Queensland, Australia.

Bailes, H.J., Robinson, S.R., Trezise, A.E.O. and Collin, S.P. (2006a). Morphology, characterisation and distribution of retinal photoreceptors in the Australian lungfish *Neoceratodus forsteri* (Krefft, 1870). The Journal of Comparative Neurology 494: 381-397.

Bailes, H.J., Trezise, A.E.O. and Collin, S.P. (2006b). The number, morphology and distribution of retinal ganglion cells and optic axons in the Australian lungfish *Neoceratodus forsteri* (Krefft, 1870). Visual Neuroscience 23: 257-273.

Bailes, H.J., Trezise, A.E. and Collin, S.P. (2007a). The optics of the growing lungfish eye: lens shape, focal ratio and pupillary movements in *Neoceratodus forsteri* (Krefft, 1870). Visual Neuroscience 24(3): 377-87.

Bailes, H.J., Davies, W.L., Trezise, A.E. and Collin, S.P. (2007b). Visual pigments in a living fossil, the Australian lungfish *Neoceratodus forsteri*. Evolutionary Biology 7: 200.

Beaudet, L., Novales Flamarique, I. and Hawawryshyn, C. (1997). Cone photoreceptor topography in the retina of sexually mature Pacific salmonid fishes. Journal of Comparative Neurology 383: 49-59.

Bowmaker, J.K. and Knowles, A. (1977). The visual pigments and oil droplets of the chicken retina. Vision Research 17: 755-764.

Bowmaker, J.K. and Hunt, D.M. (1999). Molecular Biology of Photoreceptor Spectral Sensitivity. In: M.A. Archer, M.B.A. Djamgoz, E.R. Loew J.C. Partridge and S. Vallerga (eds). Adaptive Mechanisms in the Ecology of Vision. Kluwer Academic Publishers. Dordrecht, The *Netherlands*. pp. 439-462.

Bowmaker, J.K. and Loew, E.R. (2008). Vision in Fish. In: A.I. Basbaum, A. Kaneko, G.M. Shepherd and G. Westheimer (eds). The Senses: A Comprehensive Review. Volume 1. Academic Press, San Diego, USA. pp. 53-76.

Brooks, S. and Kind, P. (2001). Ecology and demographics of lungfish (*Neoceratodus forsteri*) and general fish communities in the Burnett River, Queensland, with reference to the impacts of Walla Weir and future infrastructure development. Department of Primary Industries Report: Queensland Government, Brisbane, Australia.

Burggren, W.W. and Johansen, K. (1986). Circulation and Respiration in Lungfishes (Dipnoi). Journal of Morphology pp. 217-236.

Carroll, R.L. (1988). Vertebrate Paleontology and Evolution. W.H. Freeman, New York, USA.

Carroll, R.L. (1997). Patterns and Processes of Vertebrate Evolution. Cambridge University Press, Cambridge, England.

Collin, S.P. (2008) A database of retinal topography maps. Clinical and Experimental Optometry 91: 85-95.

Collin, S.P. and Pettigrew, J.D. (1988a). Retinal topography in reef teleosts. I. Some species with well-developed areae but poorly-developed streaks. Brain, Behavior and Evolution 31: 269-282.

Collin, S.P. and Pettigrew, J.D. (1988b). Retinal topography in reef teleosts. II. Some species with prominent horizontal streaks and high-density areae. Brain, Behavior and Evolution 31: 283-295.

Collin, S.P. and Pettigrew, J.D. (1989). Quantitative comparison of the limits on visual spatial resolution set by the ganglion cell layer in twelve species of reef teleosts. Brain, Behavior and Evolution 34: 184-192.

Collin, S. P. and Trezise, A. E. O. (2004). The origin of colour vision in vertebrates. Clinical and Experimental Optometry 87: 217-223.

Collin, S.P., Potter, I.C. and Braekevelt, C.R. (1999). The ocular morphology of the southern hemisphere lamprey *Geotria australis* Gray, with special reference to optical specialisations and the characterisation and phylogeny of photoreceptor types. Brain, Behavior and Evolution 54: 96-118.

Collin, S.P., Hart, N.S., Shand, J. and Potter, I.C. (2003). Morphology and spectral absorption characterisitics of retinal photoreceptors in the southern hemisphere lamprey (*Geotria australis*). Visual Neuroscience 20: 119-130.

Collin, S.P., Hart, N.S., Wallace, K.M., Shand, J. and Potter, I.C. (2004). Vision in the southern hemisphere lamprey, *Mordacia mordax*: spatial distribution, spectral absorption characteristics and optical sensitivity of a single class of retinal photoreceptor. Vision Neuroscience 21: 765-773.

Conant, E.B. (1986). Bibliography of Lungfishes, 1811-1985. Journal of Morphology Supplement 1: 305-373.

Cornell, E.A. and Hailman, J.P. (1984). Pupillary responses of two *Rana pipiens*-complex anuran species. Herpetologica 40: 356-366.

Dartnall, H.J.A. (1975). Assessing the fitness of visual pigments for their photic environments. In: M.A. Ali (ed). Vision in Fishes. Plenum Press, New York, USA. pp. 543-563.

Dean, B. (1906). Notes on the living specimens of the Australian lungfish, *Ceratodus forsteri*, in the Zoological Society's collection. Proceedings of the Zoological Society of London 1906: 168-178.

Douglas, R.H. and McGuigan, C.M. (1989). The spectral transmission of freshwater teleost ocular media—an interspecific comparison and a guide to potential ultraviolet sensitivity. Vision Research 29: 871-897.

Douglas, R.H. and Thorpe, A. (1992). Short-wave absorbing pigments in the ocular lenses of deep-sea teleosts. Journal of the Marine Biological Association of the U.K. 72: 93-112.

Douglas, R.H. and Marshall, N.J. (1999). A review of vertebrate and invertebrate ocular filters. In: S.N. Archer, M.B.A. Djamgoz, E.R. Loew, J.C. Partridge and S. Vallerga (eds). Adaptive Mechanisms in the Ecology of Vision. Kluwer Academic Publishers, Dordrecht, The Netherlands. pp. 95-162.

Douglas, R.H., Harper, R.D. and Case, J.F. (1998). The pupil response of a teleost fish, *Porichthys notatus*: description and comparison to other species. Vision Research 38: 2697-2710.

Douglas, R.H., Collin, S.P. and Corrigan, J. (2002). The eyes of suckermouth armoured catfish (Loricariidae, subfamily Hypostpmus): pupil response, lenticular longitudinal spherical aberration and retinal topography. Journal of Experimental Biology 205: 3425-3433.

Dunn-Meynell, A. and Sharma, S.C. (1987). Visual system of the channel catfish (*Ictalurus punctatus*) II. The morphology associated with the optic papillae and retinal ganglion-cell distribution. Journal of Comparative Neurology 257: 166-175.

Govardovskii, V.I. (1983). On the role of oil drops in colour vision. Vision Research 23: 1739-1740.

Govardovskii, V.I. and Zueva, L.V. (1987). Photoreceptors and visual pigments in sturgeons. Journal of Evolutionary Biochemistry and Physiology 23: 685-686.

Govardovskii, V.I., Fyhrquist, N., Reuter, T., Kuzmin, D.G. and Donner, K. (2000). In search of the visual pigment template. Visual Neuroscience 17: 509-528.

Greenwood, P.H. (1986). The Natural History of African Lungfishes. Journal of Morphology Supplement 1: 163-179.

Grigg, G.C. (1965a). Spawning behaviour in the Queensland lungfish, *Neoceratodus forsteri*. Australian Natural History 15: 75.

Grigg, G.C. (1965b). Studies on the Queensland lungfish, *Neoceratodus forsteri* (Krefft). III. Aerial respiration in relation to habits. Australian Journal of Zoology 13: 413-421.

Grynfeltt, E. (1911). Études anatomiques et histologiques sur l'oeil du *Protopterus annectens*. Bulletin Mensuel de l'Académie des Sciences et des Lettres de Montpellier 1911: 210

Günther, A. (1871). Description of *Ceratodus*, a genus of ganoid fishes. Philosophical Transactions of the Royal Society of London B 161: 511.

Hart, N.S. and Vorobyev, M. (2005). Modeling oil droplet absorption spectra and spectral sensitivities of bird cone photoreceptors. Journal of Comparative Physiology A 191: 381-392.

Hart, N.S., Bailes, H.J., Vorobyev, M., Marshall, N.J. and Collin, S.P. (2008). Visual ecology of the Australian lungfish (*Neoceratodus forsteri*). BMC Ecology 8: 21.

Hausmann, K., Arnold, N. J., Marshall, N. J. and Owens, I. P. F. (2003). UV signals in birds are special. Proceedings of the Royal Society of London B 270: 61-67.

Hawryshyn, C.W., Arnold, M.G., Chaisson, D.J. and Martin, P.C. (1989). The ontogeny of ultraviolet photosensitivity in rainbow trout (*Salmo gairdneri*). Visual Neuroscience 2: 247-254.

Howland, H.C., Merola, S. and Basarab, J.R. (2004). The allometry and scaling of the size of vertebrate eyes. Vision Research 44: 2043-2065.

Hughes, A. (1977). The topography of vision in mammals of contrasting lifestyles: comparative optics and retinal organization. In: F. Crescitelli (ed). Handbook of Sensory Physiology: Volume.VIII/V. Springer-Verlag, Berlin, Germany. pp. 615-756.

Johnston, D. and Hudson, R.A. (1976). Isolation and composition of carotenoid-containing oil droplets from cone photoreceptors. Biochimica et Biophysica Acta 424: 235-245.

Joss, J.M.P. (1998). Are extant lungfish neotenic? Clinical Experimental Pharmacology and Physiology 25: 733-735.

Kelber, A. and Henique, U. (1999). Trichromatcic colour vision in the hummingbird hawkmoth, *Macroglossum stellatarum* L. Journal of Comparative Physiology A 184: 535-541.

Kelber, A., Vorobyev, M. and Osorio, D. (2003). Animal colour vision—behavioural tests and physiological concepts. Biological Reviews 78: 81-118.

Kemp, A. (1984). Spawning of the Australian lungfish, *Neoceratodus forsteri* (Krefft) in the Brisbane River and in Enoggera Reservoir, Queensland. Memoirs of the Queensland Museum 21: 391-399.

Kemp, A. (1986). The Biology of the Australian Lungfish, *Neoceratodus forsteri* (Krefft 1870). Journal of Morphology Supplement 1: 181-198.

Kerr, J.G. (1902). The development of *Lepidosiren paradoxa*. III. Development of the skin and its derivatives. Quarterly Journal of Microscopical Science 46: 417-406.

Kind, P.K. (2002). Movement patterns and habitat use in the Queensland lungfish *Neoceratodus forsteri* (Krefft 1870). PhD Thesis, The University of Queensland, Brisbane, Australia.

Kondrashev, S.L., Gamburtseva, A.G., Gnjubkina, V.P., Orlov, O.J. and My, P.T. (1986). Colouration of corneas in fish: a list of species. Vision Res. 26: 287-290.

Krefft, G. (1870). Description of a giant amphibian allied to the genus *Lepidosiren* from the Wide Bay district, Queensland. Proceedings of the Zoological Society of London 1870: 221-224.

Kusmic, C. and Gualtieri, P. (2000). Morphology and spectral sensitivities of retinal and extraretinal photoreceptors in freshwater teleosts. Micron 31: 183-200.

Land, M.F. (1981). Optics and Vision in Invertebrates. In: H. Autrum (ed). Handbook of Sensory Physiology. Volume VII/6, Springer-Verlag, Berlin, Germany. pp. 472-592.

Lamb, T.D., Collin, S.P. and Pugh, E.N. Jr. (2007). Evolution of the vertebrate eye: opsins, photoreceptors, retina and eye cup. Nature Reviews Neuroscience 8(12): 960-976.

Levine, J.S. and MacNichol, E.F. (1979). Visual pigments in teleost fishes: Effects of habitat, microhabitat, and behaviour on visual system evolution. Sensory Processes 3: 95-131.

Liebman, P.A. and Granda, A.M. (1975). Super dense carotenoid spectra resolved in single cone oil droplets. Nature 253: 370-372.

Locket, N.A. (1970). Landolt's club in the retina of the African lungfish, *Protopterus aethiopicus* Heckel. Vision Research 10: 299-306.

Locket, N.A. (1973). Retinal structure in *Latimeria chalumnae*. Philosophical Transactions of the Royal Society of London B 266: 493-521.

Locket, N.A. (1977). Adaptations to the deep-sea environment. In: F. Crescitelli (ed). Handbook of Sensory Physiology. Springer, Berlin, Germany. pp. 67-192.

Locket, N.A. (1999). Vertebrate photoreceptors. In: S.N. Archer, M.B.A. Djamgoz, E.R. Loew, J.C. Partridge and S.Vallerga (eds). Adaptive Mechanisms in the Ecology of Vision. Kluwer Academic Publishers, Dordrecht, The Netherlands. pp. 163-196.

Loew, E.R. and McFarland, W.N. (1990). The underwater visual environment. In: R.H. Douglas, and M.B.A. Djamgoz (eds). The Visual System of Fishes. Chapman and Hall, London, UK. pp. 1-43.

Loew, E.R., McFarland, W.N., Mills, E.L. and Hunter, D. (1993). A chromatic action spectrum for planktonic predation by juvenile yellow perch, *Perca flavescens*. Canadian Journal of Zoology 71(2): 384-387.

Losey, G.S.J. (2003). Crypsis and communication functions of UV-visible coloration in two coral reef damselfish, *Dascyllus aruanus* and *D. reticulatus*. Animal Behaviour 66: 299-307.

Losey, G.S., Cronin, T.W., Goldsmith, T.H., Hyde, D., Marshall, N.J. and McFarland, W.N. (1999). The UV visual world of fishes: a review. Journal of Fish Biology 54: 921-943.

Losey, G.S., McFarland, W.N., Loew, E.R., Zamzow, J.P., Nelson, P.A. and Marshall, N.J. (2003). Visual biology of Hawaiian coral reef fishes. I. Ocular transmission and visual pigments. Copeia 2003: 433-454.

Lythgoe, J.N. (1966). Visual Pigments and Underwater Vision. In: R. Bainbridge, C.C. Evans and O. Rackman (eds). Light as an Ecological Factor. Blackwell, Oxford, UK. pp. 375-392.

Lythgoe, J.N. (1968a). Red and yellow as conspicuous colors underwater. Underwater Association Report 1: 51-53.

Lythgoe J.N. (1968b). Visual pigments and visual range underwater. Vision Research 8: 997-1012.

Lythgoe, J.N. (1972). The adaptation of visual pigments to the photic environment. In: H.J.A. Dartnall (ed). Handbook of Sensory Physiology: Photochemistry of Vision. Volume VII/1, Springer-Verlag, Berlin, Germany. pp. 567-603.

Lythgoe, J.N. (1979). The Ecology of Vision. Clarendon, Oxford, UK.

Malkki, P.E. and Kröger, R.H.H. (2005). Visualization of chromatic correction of fish lenses by multiple focal lengths. Journal of Optics A 7: 691–700.

Marshall, N.J., Jennings, K., McFarland, W.N., Loew, E.R. and Losey, G.S. (2003a). Visual Biology of Hawaiian Coral reef Fishes. III. Environmental Light and an Integrated Approach to the Ecology of Reef fish Vision. Copeia 3: 467-480.

Marshall, N.J., Jennings, K., McFarland, W.N., Loew, E.R. and Losey, G.S. (2003b). Visual Biology of Hawaiian Coral Reef Fishes. II. Colors of Hawaiian Coral Reef Fish. Copeia 3: 455-466.

Meyer, D.B., Cooper, T.G. and Gernez, C. (1965). The structure of the eye: Retinal oil droplets. In: J.W. Rohen (ed). The Structure of the Eye. Schattauer, Stuttgart, Germany. pp. 521-533.

Munk, O. (1969). On the visual cells of some primitive fishes with particular regard to the classification of rods and cones. Videnskabelige Meddellelser Dansk Naturhistorisk Førening 132: 25-30.

Munk, O. (1984). Non-spherical lenses in the eyes of some deep-sea teleosts. Archiv Fische Wissenschaft 34: 145-153.

Muntz, W.R.A. (1972). Inert absorbing and reflecting pigments. In: H.J. Dartnall (eds). Handbook of Sensory Physiology: Photochemistry of Vision. Berlin-Heidelberg-New York, Springer. VII/I: 529-565.

Neumeyer, C. (1991). Evolution of colour vision. In: J.R. Cronly-Dillon and R.L. Gregory (eds). Vision and Visual Dysfunction: Evolution of the Eye and Visual System. Macmillan Press, London, UK. pp. 284-305.

Neumeyer, C. (1992). Tetrachromatic color vision in goldfish: evidence from color mixture experiments. Journal of Comparative Physiology A 171: 639-649.

Neumeyer, C. (1998). Color Vision in Lower Vertebrates. In: W.G.K. Backhaus, R. Kliegl and J.S. Werner (eds). Color Vision—Perspectives from Different Disciplines. Walter de Gruyter & Co., Berlin, Germany. pp. 149-162.

Neumeyer, C., Dorr, S., Fritsch, J. and Kardelky, C. (2002). Colour constancy in goldfish and man: influence of surround size and lightness. Perception 31: 171-187.

Nicol, J.A.C., Arnott, H.J. and Best, A.C.G. (1972). *Tapeta lucida* in bony fishes (Actinopterygii): a survey. Canadian Journal of Zoology 51: 69-81.

Northcutt, G.R. (1977). Retinofugal projections in the lepidosirenid lungfishes. Journal of Comparative Neurology 174: 55-574.

Northcutt, R.G. (1980). Retinal projections in the Australian lungfish. Brain Research 185: 85-90.

Northcutt, R.G. (1986). Lungfish Neural Characters and Their Bearing on Sarcopterygian Phylogeny. Journal of Morphology Supplement 1: 277-297.

Osorio, D., Marshall, N.J. and Cronin, T.W. (1997). Stomatopod photoreceptor spectral tuning as an adaptation for colour constancy in water. Vision Research 37(23): 3299-3309.

Partridge, J.C. (1989). The visual ecology of avian cone oil droplets. Journal of Comparative Physiology A 165: 415-426.

Pettigrew, J.D. (1991). Evolution of binocular vision. In: J.R. Cronly-Dillon and R.L. Gregory (eds). Evolution of the Eye and Visual System. CRC Press, Boca Raton pp. 271-283.

Pfeiffer, W. (1968). Retina und Retinomotorik der Dipnoi und Brachiopterygii. Zeitschrift für Zellforschung und Mikroskopische Anatomie 89: 62-72.

Pough, F.H., Janis, C.M. and Heiser, J.B. (1999). Vertebrate Life. Prentice-Hall Inc., Upper Saddle River, New Jersey, USA.

Pow, D.V. (1994). Taurine, amino-acid transmitters, and related molecules in the retina of the Australian lungfish *Neoceratodus forsteri*—a light-microscopic immunocytochemical and electron-microscopic study. Cell & Tissue Research 278: 311-326.

Ramsey, E.P. (1876). On the habits of *Ceratodus*. Proceedings of the Zoological Society of London 1876: 698-699.

Robinson, S.R. (1994). Early vertebrate color-vision. Nature 367: 121-121.

Robinson, S.R. (1997). Relationships between Müller cells and neurons in a primitive tetrapod, the Australian lungfish. Visual Neuroscience 14: 795-800.

Rochon-Duvigneaud, A. (1941). L'oeil de *Lepidosiren paradoxa*. Comptes Rendus Hebdomadaires des Séances de l'Académie des Sciences 212: 307-309.

Rubin, L. and Nolte, J. (1982). Autonomic innervation and photosensitivity of the sphincter pupillae muscle of two teleosts: *Lophius piscatorius* and *Opsanus tau*. Current Eye Research 1: 543-551.

Schiefferdecker, P. (1886). Studien zur vergleichenden Histologie der Retina. Archiv fur Mikroskopische Anatomie 28: 305-396.

Siebeck, U.E. (2004). Communication in coral reef fish: the role of ultraviolet colour patterns in damselfish territorial behaviour. Animal Behaviour 68: 273-282.

Siebeck, U.E. and Marshall, N.J. (2001). Ocular media of coral reef fish—can coral reef fish see ultraviolet light? Vision Research 41: 133-149.

Siebeck, U. E., Collin, S.P., Ghoddussi, M. and Marshall, N.J. (2003). Occlusable corneas in toadfishes: Light transmission, movement and ultrastructure of pigment during light and dark adaptation. Journal of Experimental Biology 206: 2177-2190.

Simpson, R., Kind, P. and Brooks, S. (2002). Trials of the Queensland Lungfish. Nature Australia 2002: 36-43.

Sivak, J.G. (1980). Accommodation in vertebrates: a contemporary survey. In: J.A. Zadunaisky and H. Davson (eds.). Current Topics in Eye Research. Academic Press, New York, USA. pp. 281-330.

Sivak, J.G. (1990). The optical variability of the fish lens. In: R.H. Douglas and M.B.A. Djamgoz (eds). The Visual System of Fish. Chapman & Hall, London, UK. pp. 63-80.

Somiya, H. (1987). Dynamic mechanism of visual accommodation in teleosts: structure of the lens muscle and its nerve control. Proceedings of the Royal Society of London, Series B 230: 77-91.

Spady, T.C., Parry, J.W., Robinson, P.R., Hunt, D.M., Bowmaker, J.K. and Carleton, K.L. (2006). Evolution of the cichlid visual palette through ontogenetic subfunctionalization of the opsin gene arrays. Molecular Biology and Evolution 23(8): 1538-1547.

Steinach, E. (1890). Untersuchungen zur vergleichenden physiologie der iris. Pflügers Archiv für die Gesamte Physiologie des Menschen und der Tiere 47: 289-340.

Takezaki, N., Figueroa, F., Zaleska-Rutczynska, Z. and Takahata, N. (2004). The phylogenetic relationship of tetrapod, coelacanth, and lungfish revealed by the sequences of 44 nuclear genes. Molecular Biology and Evolution 21: 1512-1524.

Tamura, T. (1957). On the relation between the intensity of illumination and the shifting of cones in the fish retina. Bulletin of the Japanese Society of Scientific Fisheries 22: 742-746.

Thorpe, A. and Douglas, R.H. (1993). Spectral transmission and short-wave absorbing pigments in the fish lens—II. Effects of age. Vision Research 33: 301-307.

Thorpe, A., Douglas, R.H. and Truscott, R.J.W. (1993). Spectral transmission and short-wave absorbing pigments in the fish lens—I. Phylogenetic distribution and identity. Vision Research 33: 289-300.

Trezise, A.E.O. and Collin, S.P. (2005). Opsins: Evolution in waiting. Current Biology 15: R794-R796.

Venkatesh, B., Mark, E.V. and Brenner, S. (2001). Molecular synapomorphies resolve evolutionary relationships of extant jawed vertebrates. Proceedings of the National Academy of Sciences of the USA 98: 11382-11387.

Vorobyev, M. (2003). Coloured oil droplets enhance colour discrimination. Proceedings of the Royal Society of London 270: 1255-1261.

Vorobyev, M. and Osorio, D. (1998). Receptor noise as a determinant of colour threshold. Proceedings of the Royal Society of London, series B 265: 351-358.

Vorobyev, M., Osorio, D., Bennett, A.T.D., Marshall, N.J. and Cuthill, I.C. (1998). Tetrachromacy, oil droplets and bird plumage colours. Journal of Comparative Physiology A 183: 621-633.

Vorobyev, M., Brandt, R., Peitsch., D., Laughlin, S.B. and Menzel, R. (2001). Colour thresholds and receptor noise: behaviour and physiology compared. Vision Research 41: 639-653.

Wald, G. and Zussman, H. (1937). Carotenoids of the chicken retina. Nature 140: 197.

Walls, G.L. and Judd, H.D. (1933). The intra-ocular colour filters of vertebrates. British Journal of Ophthalmology 17: 641-675; 705-725.

Walls, G. (1942). The Vertebrate Eye and its Adaptive Radiation. Hafner Publishing Company, New York, USA.

Warrant, E. J. (1999). Seeing better at night: lifestyle, eye design and the optimum strategy of spatial and temporal summation. Vision Research 39: 1611-1630.

Warrant, E. J. and Locket, N.A. (2004). Vision in the deep sea. Biological Reviews 79: 671-712.

Watt, M., Evans, C.S. and Joss, J.M.P. (1999). Use of electroreception during foraging by the Australian lungfish. Animal Behavior 58: 1039-1045.

Whitmore, A.V. and Bowmaker, J.K. (1989). Seasonal variation in cone sensitivity and shortwave absorbing visual pigments in the rudd *Scardinius erythrophthalmus*. Journal of Comparative Physiology A 166: 103-115.

Yokoyama, S. and Shi, Y.S. (2000a). Genetics and evolution of ultraviolet vision in vertebrates. FEBS Letters 486: 167-172.

Yokoyama, S. (2000b). Molecular evolution of vertebrate visual pigments. Progress in Retinal Eye Research 19: 385-419.

Yokoyama, S. (2002). Molecular evolution of color vision in vertebrates. Gene 300: 69-78.

CHAPTER 18

The Lateral Line System in Lungfishes: Mechanoreceptive Neuromasts and Electroreceptive Ampullary Organs

Jørgen Mørup Jørgensen*

Department of Zoophysiology, Buliding 1131, Biological Institute, University of Aarhus, DK- 8000 Århus C, Denmark

ABSTRACT

The lateral line system in lungfishes, or dipnoans, consists of two different sensory modalities: mechanoreception and electroreception. Neuromasts, which are responsible for mechanoreception, and electrosensory ampullae, which are responsible for electroreception, are both formed ontogenetically from placodal material.

The neuromasts of the mechanosensory system are found in several canals, which vary among the three genera, and in lines on the head and trunk. Neuromasts also occur as superficial organs in the epidermis, where they are mostly organized in so-called pit lines. Neuromasts consist of sensory hair cells and supporting cells, which together are surrounded by mantle cells. Each hair cell has an apical bundle composed of a single kinocilium and its associated stereovilli, and each has both afferent and efferent contacts.

The electrosensory organs occur as ampullae spread over the head and trunk, with a concentration on the snout. They are similar to mechanoreceptive neuromasts but have a more narrow apical part, with a single cilium and no stereovilli.

There is considerable variation in the three genera of lungfishes and their lateral line systems. A special feature for the lepidosirenid lungfishes (*Protopterus* and *Lepidosiren*) is their ability to aestivate in dry periods.

Keywords: hair cells, mechanoreceptors, neuromasts, ampullary organs, electroreceptors

*E-mail: joergen.moerup.joergensen@biology.au.dk

INTRODUCTION

There are three living genera of lungfishes, or dipnoans: *Protopterus* (the African lungfish), *Lepidosiren* (the South American lungfish), and *Neoceratodus* (the Australian lungfish). In these fishes, the lateral line system comprises two different sensory modalities: mechanoreception and electroreception. The organs responsible for mechanoreception are called neuromasts, and those responsible for electroreception are termed electrosensory ampullary organs, or ampullae. Both types of receptors are formed ontogenetically from placodal material, as demonstrated experimentally in the axolotl by Northcutt *et al.* (1994, 1995).

The mechanosensory neuromasts of the lateral line system consist of sensory hair cells, supporting cells, and mantle cells, and they occur within canals as well as superficially in lines on the head and trunk. The electrosensory organs of the lateral line system occur as ampullae, which are concentrated on the snout but also spread over the head and trunk. The sensory epithelium of these ampullae is similar to that of the mechanoreceptive neuromasts, but the electrosensory ampullae have a more narrow apical part, with a single cilium and no stereovilli.

The three genera of lungfishes vary in several respects. One notable difference is that the lepidosirenid lungfishes (*Protopterus* and *Lepidosiren*) are able to aestivate during dry periods. When their habitat literally dries up annually, adult individuals burrow into the mud and wait for the rainy reason to recreate their floodplains. This is known as their aestivation phase, as opposed to the preceding active or aquatic phase. In the West African lungfish, *Protopterus annectens*, the neuromasts remain intact, but their apical parts, with the sensory hair cell bundles, become enclosed in small cavities. Also during aestivation, the electroreceptive ampullae in *Protopterus annectens* become overgrown with epidermal cells, leaving a small ampullary lumen with the apical cilia of the sensory cells.

The mechanoreceptive neuromasts and the electroreceptive ampullary organs of the lateral line system in lungfishes will be described in greater detail below in sections on their morphology, their distribution in the three genera, and their development. Unless noted otherwise, all descriptions are based on adult, i.e. sexually mature, lungfishes.

MECHANORECEPTIVE NEUROMASTS – MORPHOLOGY

The individual mechanosensory organs, or neuromasts, are similar in morphology, whether they occur superficially on the skin or in canals. Neuromasts are composed of sensory hair cells, which are mixed with supporting cells in the middle of an onion-shaped structure enclosed by several layers of mantle cells (Fig. 5). Neuromasts in lungfishes are similar to those described in numerous other fishes and in amphibians (Flock 1965; Jørgensen and Flock 1973; Flock and Jørgensen 1974; Hetherington and Wake 1979).

Each neuromast contains 12-20 hair cells. These are pear-shaped and 40-60 μm long. Their diameter at the apical end is 2.5-3.3 μm, and at the level of the nucleus it is 7-10 μm. Transmission electron microscopy shows that each hair cell in a lungfish neuromast consists of a single eccentric kinocilium and a bundle of stereovilli. The kinocilium has a normal 9+2 microtubular arrangement and arises from the basal body of the cell, below which a centriole is found. The kinocilium is taller than the stereovilli, whose height decreases according to their distance from the kinocilium. The stereovilli have darkly staining tips. They contain numerous microfilaments that form a rod-like structure in the narrow transition from the base of the neuromast to the hair cell body. This rod-like structure spreads and forms rootlets, which extend into the cuticular plate that occupies the apical end of the cell, except for an area around the basal body of the kinocilium. The cuticular plate has a uniform appearance, dominated by fine granulated material and fine microfilaments. Mitochondria are widespread in all parts of the hair cells and are especially numerous above the nucleus of the cell.

Basally, hair cells form contacts with both afferent and efferent nerve endings (Fig. 6). A dark sphere surrounded by synaptic vesicles in the hair cell characterizes the afferent endings, and opposite to this is a rounded nerve ending with mitochondria. Efferent nerve endings are filled with vesicles and are only seen infrequently in histology preparations of the neuromasts.

Supporting cells reach from the basal lamina to the apical surface of the epithelium and separate the hair cells. Most of the widened basal lamina contains a rough endoplasmic epithelium and a few Golgi complexes. Apically, numerous vesicles are found in the supporting cells.

Mantle cells are arranged like the outer leaves of an onion, enclosing the hair cells and the supporting cells. Most of the mantle cells reach from the basal lamina to the apical surface of the organ. They contain a well-developed rough endoplasmic reticulum, mitochondria and numerous microfilaments. Apically they have short microvilli.

MECHANORECEPTIVE NEUROMASTS – DISTRIBUTION

As noted by Kingsbury (1895), lungfishes vary greatly in the complexity of their lateral line systems, and, indeed, the three genera differ greatly in the extent to which their mechanosensory organs are enclosed in canals, occur in open furrows or grooves, or appear superficially in the skin.

In adult *Neoceratodus*, the Australian genus, many of the mechanoreceptive neuromasts of the lateral line system are enclosed in canals (Fig. 1), although some also occur in lines or as scattered superficial neuromasts. There are well organized supraorbital and infraorbital canals, an oral canal, and a preopercular canal, as well as an anterior commissure and a supratemporal commissure. A trunk canal stretches along the side for most of the fish's length.

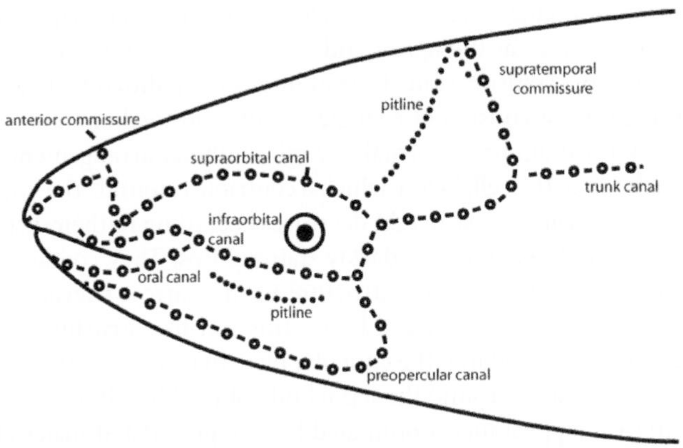

Fig. 1 Drawing of the head of an adult Australian lungfish (*Neoceratodus forsteri*, 105 cm TL), showing the canals of the lateral line system.

In the African genus, *Protopterus*, some mechanoreceptive neuromasts are enclosed in canals, while others appear in superficial grooves. As seen in Fig. 2, the supraorbital and infraorbital neuromasts form both a canal and a line. All other lines of neuromasts occur in shallow grooves, including the three trunk lines.

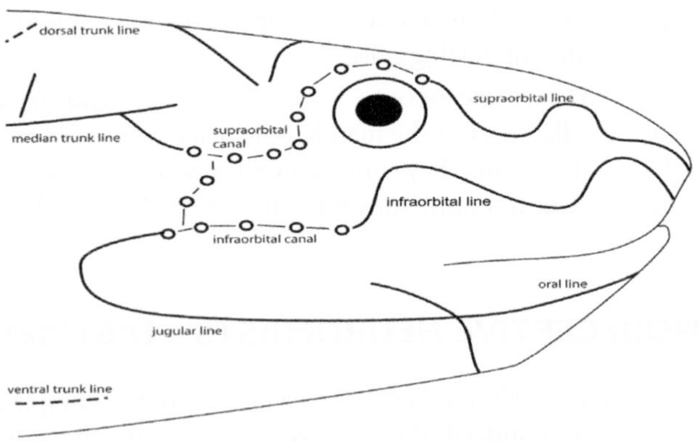

Fig. 2 Drawing of the head of an African lungfish (*Protopterus annectens*, 30 cm TL), showing the canals of the lateral line system.

In the South American lungfish, *Lepidosiren*, all of the mechanoreceptors are superficial, as none of the mechanoreceptive neuromasts are enclosed in canals. There are lines of shallow furrows on the head, with the neuromasts regularly distributed in a row, as seen in Fig. 3. On the body, a main trunk line extends

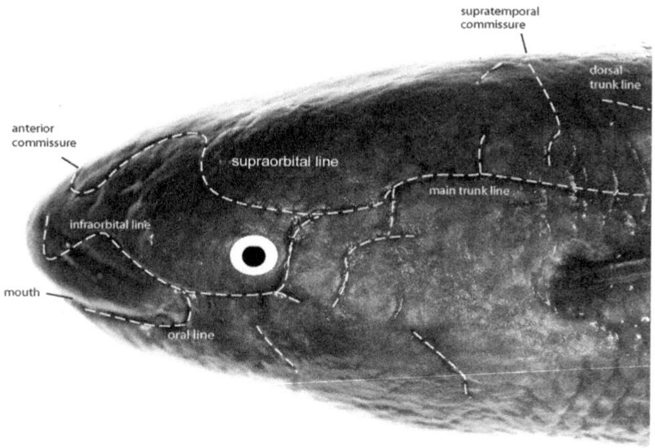

Fig. 3 Photograph of the head of a South American lungfish (*Lepidosiren paradoxa*, 40 cm TL), with the position of the sensory lines of the lateral line system indicated by dashed lines.

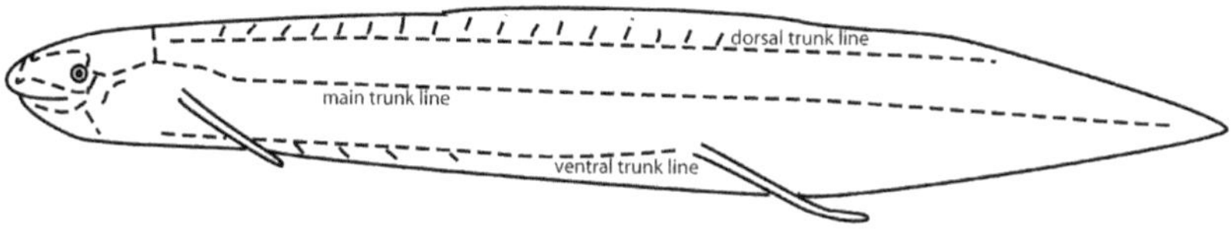

Fig. 4 Drawing a South American lungfish (*Lepidosiren paradoxa*, 40 cm TL) in lateral view, showing the trunk lines of the lateral line system.

midlaterally for the entire length of the fish and is flanked by shorter dorsal and ventral trunk lines, as seen in Fig. 4.

MECHANORECEPTIVE NEUROMASTS — DEVELOPMENT

Not a great deal is known about development in any of the three genera of lungfishes, and most of what we do know about the ontogeny of the lateral line system dates to 1949 and Pehrson's study of the development of this system in the head of lungfishes, with additional insights from Kemp's (1999) study of the ontogeny of the skull of the Australian lungfish.

Neoceratodus. Pehrson's descriptions of the ontogeny of the lateral line system in this genus were based on embryos of 8.6 mm and larger, whereas Kemp's study used material from pre-hatching through adults. Both types of sensory receptors of the lateral line system derive from placodes in *Neoceratodus*, as they do in all vertebrates examined. The first indication of the canals of a lateral line system

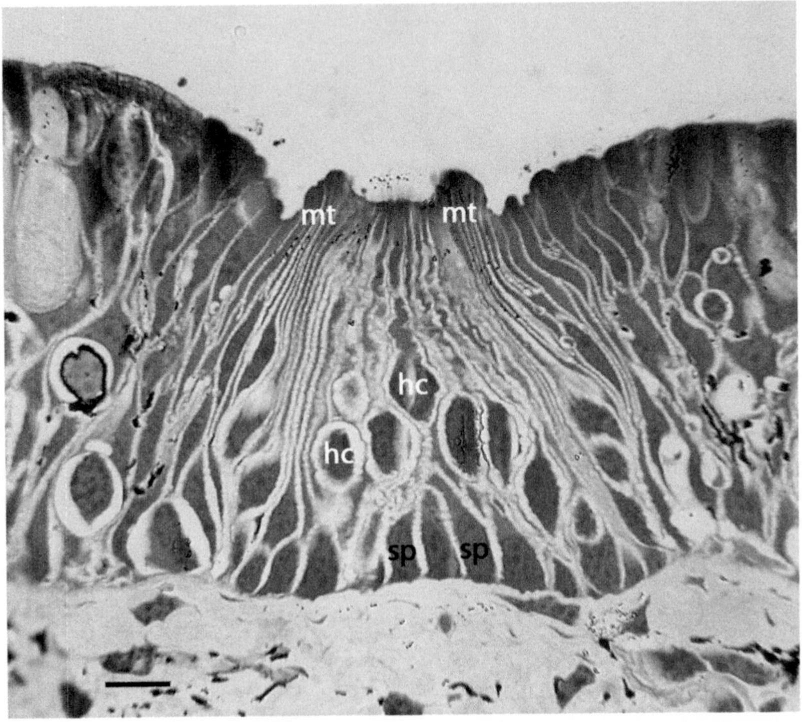

Fig. 5 Photomicrograph of a transverse section of a neuromast in the lateral line system in a South American lungfish (*Lepidosiren paradoxa*, 28 cm TL), the central hair cells (hc) and supporting cells (sp), both of which are enclosed by mantle cells (mt). Scale bar = 20 µm.

is a series of grooves in stage 50 larvae (Fig. 7). Later these grooves deepen and close, so that only the pores connecting the canals with the surrounding water are finally visible from the outside. In the bottom of the groove, neuromasts are seen very early. Rows of neuromasts in pit lines appear, lying superficially on the skin, and numerous ampullary organs are found close to the canals, as reported by Northcutt (1986).

Protopterus. The distribution of the main lines of the lateral line system of *Protopterus* were shown in the active or aquatic phase as early as 1895 by Kingsbury.

Inside the canals there are several neuromasts, each between two pores, as in some chondrichthyans (Jørgensen and Pickles 2002), instead of the single neuromast found in most teleosts, as previously noted by Webb and Northcutt (1997). The ontogenetic development of the lateral line system in *Protopterus annectens* was further described by Pehrson (1949), based on embryos from 10.1 to 55 mm in length. In the aestivation of *Protopterus annectens*, the canal and

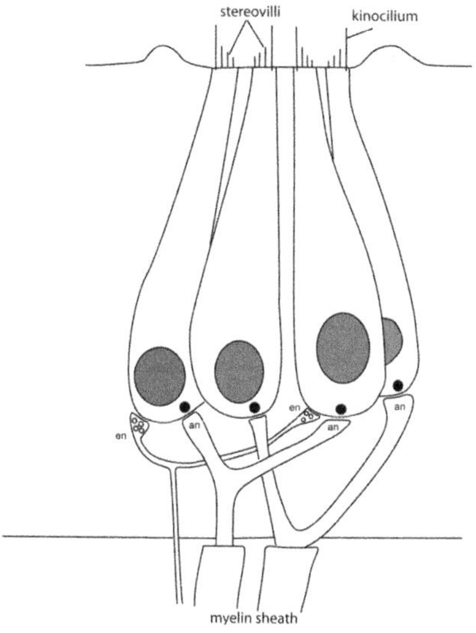

Fig. 6 Schematic drawing of a neuromast, showing the hair cells and afferent (an) and efferent (en) nerve fibres. The drawing is not based on a precise reconstruction from serial sections, but on the author's impression from many different light and transmission electron microscopic sections.

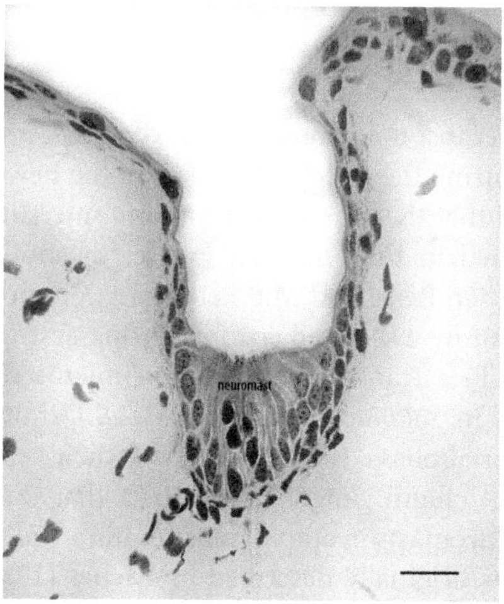

Fig. 7 Photomicrograph of a transverse section through a canal in the developing lateral line system of a larval Australian lungfish (*Neoceratodus forsteri*, stage 50), showing a neuromast in the furrow. Scale bar = 50 μm.

superficial neuromasts are covered by a layer of epidermal cells, leaving a small cavity at the apical part of the sense organ (Figs. 8 and 9). Transmission electron microscopy shows that the sensory hair cells and other parts of the neuromast remain intact (Fig. 10). Even the darkly staining synaptic body opposite the afferent nerve endings is still apparent.

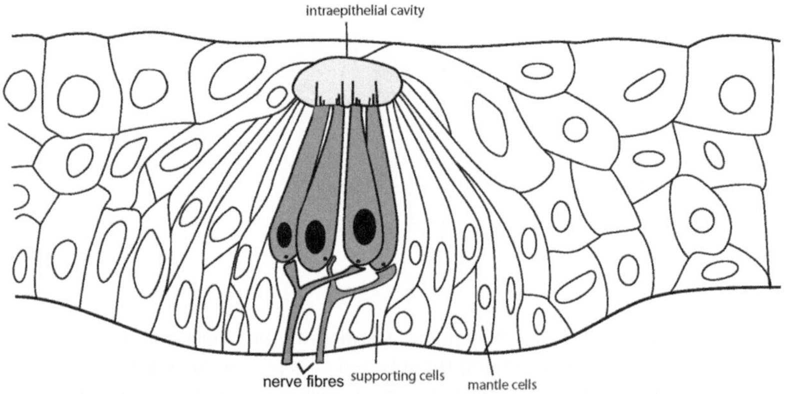

Fig. 8 Schematic drawing of a neuromast in the skin of an aestivating African lungfish (*Protopterus annectens*). The top of the hair cells is intact, though hidden in an intraepithelial cavity.

This is interesting, because many amphibians experience profound disintegration of their neuromasts during the terrestrial phase. In the newt *Triturus viridescens*, Dawson (1936) found that the mantle, sensory and interstitial (supporting) cells become undifferentiated in the terrestrial phase. In another light microscopy study of *Siren intermedia*, Reno and Middleton (1973) reported that most of the neuromasts are covered by a secreted cocoon during aestivation, and only a few of them disintegrate. In a transmission electron microscopical examination of the lateral line organs in *Salamandrina terdigiatata*, Delfino *et al.* (1984) found that during the terrestrial phase, the hair cells lose their sensory hairs, but retain the basal body of the cilium. Interesting also is that during the ontogenetic development of the neuromasts in amphibians, an intraepidermal cavity is formed above the hair cells, as originally described by Escher (1925). In this cavity the sensory hairs develop, undisturbed by external factors. So it is that during lungfish aestivation their neuromasts are again in a situation reminiscent of that in which they originally developed.

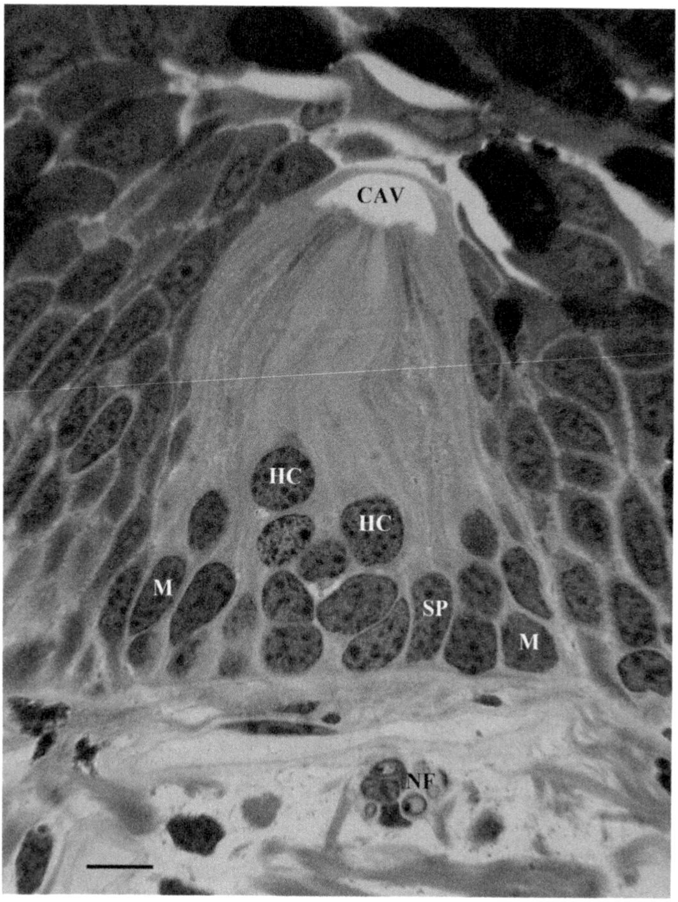

Fig. 9 Photomicrograph of a neuromast in the epidermis of an aestivating African lungfish (*Protopterus annectens*, 28 cm TL). The apical parts of the hair cells (HC) and supporting cells (SP) are seen with the sensory hairs in the intraepidermal cavity (CAV). Mantle cells (M) lie as the outer layers of an onion. Below the organ nerve fibers (NF) can be seen below the neuromast. Bar scale = 10 μm.

ELECTRORECEPTIVE AMPULLARY ORGANS — MORPHOLOGY

The electrosensory organs of the lateral line system in lungfishes are ampullary organs, which appear as isolated, tube-like epidermal invaginations with a canal that connects to the outer surface. The sensory cells are seen in the ampullary end of the tube (Figs. 11 and 13). Both the ampulla and its canal are filled with a mucoid substance. Some ampullae reach deep in the connective tissue below the

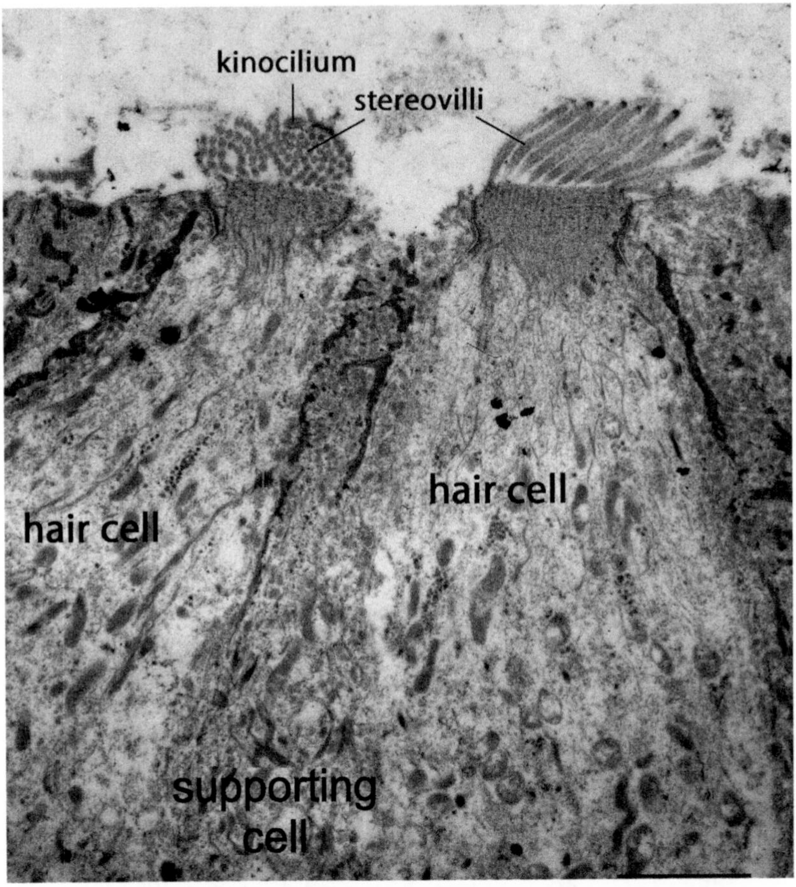

Fig. 10 Transmission electronmicrograph of the top of a neuromast in the epidermis of an aestivating African lungfish (*Protopterus annectens*, 28 cm TL). Note that the hair cells appear intact. Bar scale = 2 μm.

epithelium, while most end at the level of the basal lamina or are situated in the upper and medial parts of the epithelium. Unlike mechanoreceptive neuromasts, the electrosensory ampullary organs in lungfishes appear to lack efferents, as do other electrosensory organs in fishes and amphibians (Fritzsch and Wahnschaffe 1983; Jørgensen 2005).

The fully developed sensory epithelium of the electrosensory ampullae consists of sensory cells and supporting cells. The sensory cells are pear-shaped cells in the luminal half or two-thirds of the epithelium (Fig. 12). The apical part, as seen in transmission and scanning electron microscopy, is restricted to a small area with a central cilium and a few short microvilli. The cilium contains microtubules

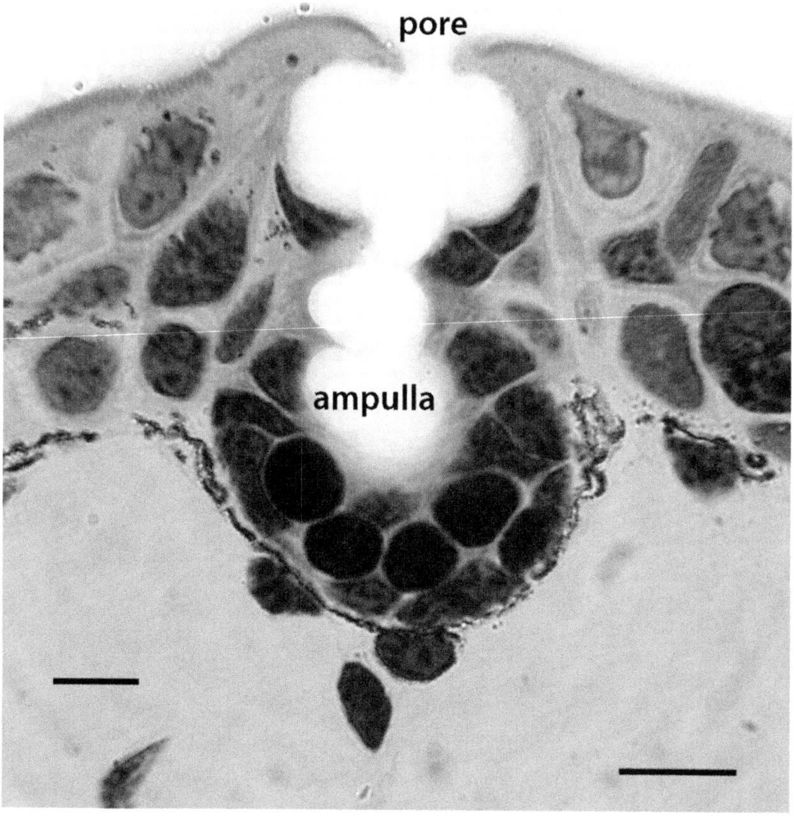

Fig. 11 Photomicrograph of an electrosensory ampullary organ from a stage 52 larva of the Australian lungfish, *Neoceratodus forsteri*. A pore connects the ampullary lumen with the outside water. Bar scale = 20 μm.

in an irregular arrangement. The sensory cilia often have an irregular swelling. One or more synaptic sheets, surrounded by clear vesicles, characterize the afferent synapse. The dark staining sheet often ends in a protrusion, which fills a corresponding invagination in the nerve endings. No efferent nerve endings have been seen.

In spite of the many similarities between lungfishes and amphibians, their electroreceptors are quite different morphologically. Lungfishes, as described above, have only a single cilium and maybe a few microvilli apically, while all amphibians that have been described may or may not have a cilium, but all seem to have numerous long microvilli sitting in a rather wide apical, luminal part of the cell. In amphibians, a large presynaptic sphere is seen opposite the nerve endings,

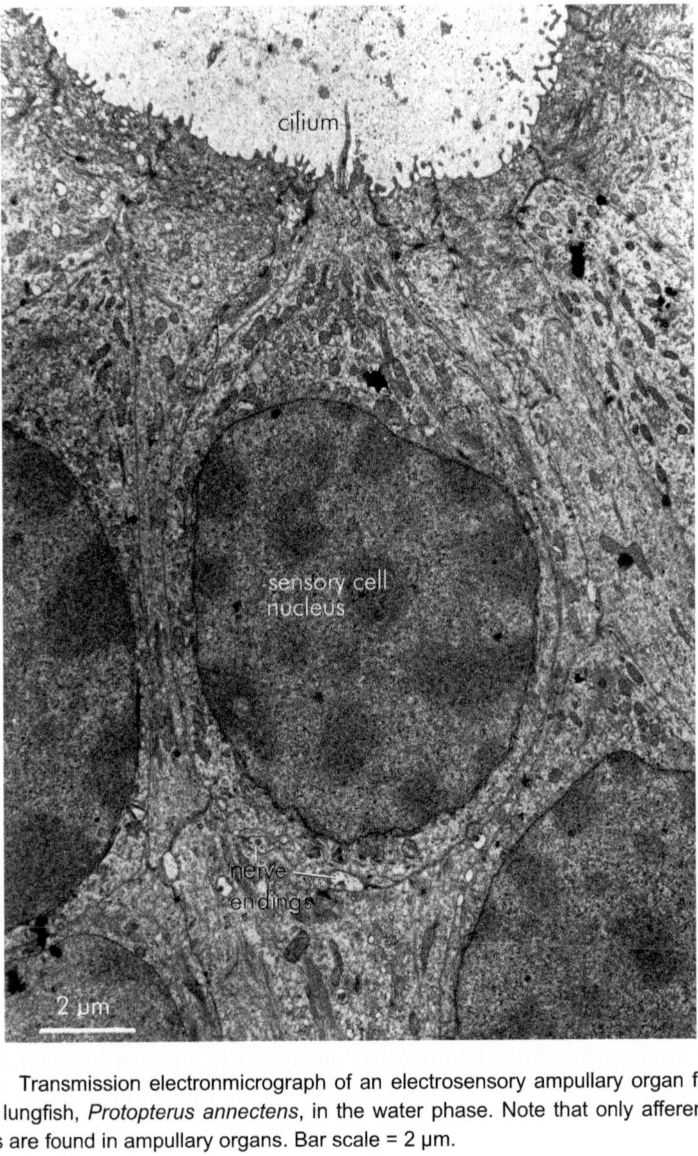

Fig. 12 Transmission electronmicrograph of an electrosensory ampullary organ from the African lungfish, *Protopterus annectens*, in the water phase. Note that only afferent nerve endings are found in ampullary organs. Bar scale = 2 µm.

while cartilaginous and non-teleost bony fishes, including lungfishes, have a platelike synaptic sheet at this type of synapse (Fritsch and Wahnschaffe 1983; Fritsch and Münz 1986).

The sensory cells of the ampullae are separated by supporting cells, which often reach from the basal lamina to the luminary surface. Here numerous irregular

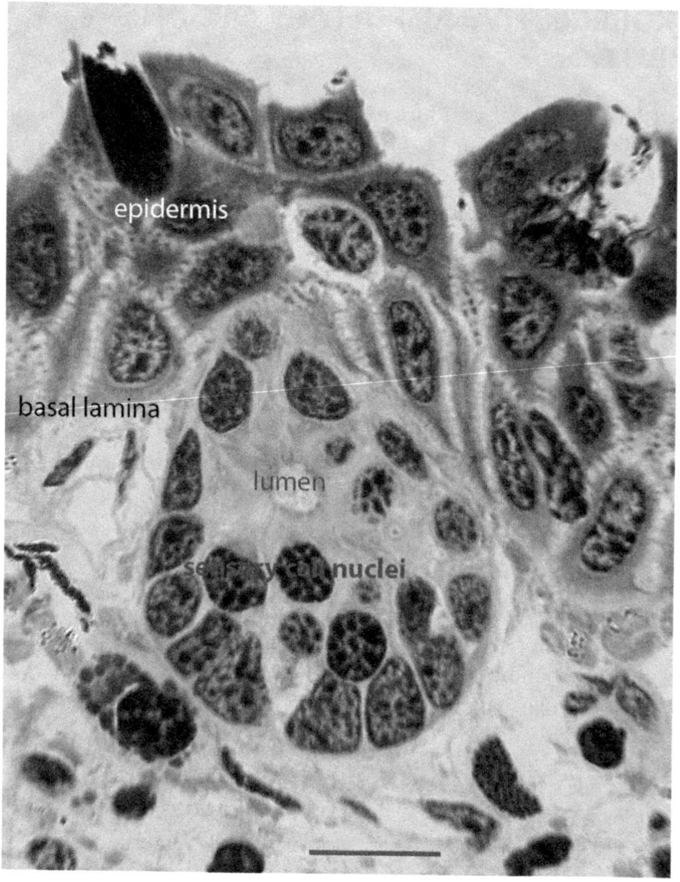

Fig. 13 Photomicrograph of a section through an ampullary organ from an aestivating African lungfish, *Protopterus annectens*. The opening to the outside is closed, and a small lumen is left, with the apical parts of the sensory cells. Bar scale = 20 μm.

microvilli are found. In the apical cytoplasm of the supporting cells, there are numerous vesicles that are believed to secrete the mucoid substance in the ampulla (Jørgensen 1984). The ampullae are easily identified in sections of skin from aestivating *Protopterus annectens*. The canal to the outside is almost absent in most of these ampullae, since they are covered with flattened epidermal cells.

In general, lungfish electrosensory organs are quite similar to the ampullary electroreceptors described in cartilaginous fishes and non-teleost bony fishes.

ELECTRORECEPTIVE AMPULLARY ORGANS — DISTRIBUTION

The electrosensory ampullae in lungfishes are distributed on the head and trunk but tend to be concentrated on the former. In skin preparations from embryos of the Australian lungfish, *Neoceratodus*, ampullae can be seen close to the grooves or canals that constitute the mechanosensory neuromasts of the lateral line system. This is probably not surprising, as it has been shown that both electroreceptive ampullae and mechanoreceptive neuromasts arise from a single placode in axolotls, (Northcutt *et al.* 1995). In the urodele amphibian *Salamandra salamandra*, the ampullary organs are close to the lines of neuromasts (Fritzsch and Wahnschaffe 1983), and in other amphibians, such as the gymnophionan *Ichthyophis*, neuromasts and ampullae are also situated close to one another (Wahnschaffe *et al.* 1985). In older lungfishes of all three genera, the ampullary organs become more widespread but retain their concentration on the snout and around the eyes.

ELECTRORECEPTIVE AMPULLARY ORGANS — DEVELOPMENT

In *Neoceratodus*, single, scattered ampullary organs first appear on the skin of the snout and anterior mandible of embryos, and around the eyes. Later, at stage 44, a line of smaller ampullary organs can be seen above and below the lateral line.

ELECTRORECEPTIVE AMPULLARY ORGANS — PHYSIOLOGY

Very little research has been done on the physiology of the electrosensory ampullae in lungfishes. Roth (1973) demonstrated that afferent fibres from the ampullae of the South American lungfish, *Lepidosiren*, responded with a change in frequency to a current density of 0.03 $\mu A/cm^2$, which is sufficient to detect external bioelectrical fields surrounding freshwater fishes and invertebrates.

The Australian lungfish, *Neoceratodus*, has been shown to detect the electrical field surrounding a crayfish, *Cherax destructor*, hidden in a box that prevented emission of chemical, mechanical and visual cues. If the electrical fields were reduced or abolished, the lungfish had difficulties in detecting the crayfish (Watt *et al.* 1999).

Material and Acknowledgements

Because essential parts of the descriptions above have not been published previously, the material examined will be listed below (TL = total length).

Neoceratodus forsteri :

1 male, 105 cm TL,

1 male, 80 cm TL

26 embryos from stage 45 to juveniles, from 65 mm to 77 mm TL

Parts of 1 subadult, 320 mm TL.

Protopterus annectens:

water phase:

1 female 42 cm TL, 1 male 30 cm TL

aestivation phase:

16 unsexed specimens, 23-31 cm TL

Lepidosiren paradoxa:

1 male, 70 cm TL

1 female, 40 cm TL.

Most of the embryological material from *Neoceratodus* was a gift from Dr. Anne Kemp, and Dr. Tobias Wang put material from the aestivating *Protopterus annectens* at my disposal. I am grateful to both researchers. Many thanks to Mary Sue Northcutt who waded valiantly through my turgid prose and translated it to intelligible English.

References

Dawson, A.B. (1936). Changes in the lateral-line organs during the life of the newt, *Triturus viridescens*. A consideration of the endocrine factors involved in the maintenance of differentiation. Journal of Experimental Zoology 74: 221-237.

Delfino, G., Brizz, R. and Calloni, C. (1984). Lateral line organs in *Salamandrina terdigitata* (Lacépède, 1788) (Amphibia: Urodela). Zeitschrift für Mikroskopisch-anatomishes Forschung Leipzig 98: 161-183.

Escher, K. (1925). Das verhalten der Seitenorgan der Wirbeltiere und ihrer Nerven beim übergang zum Landleben. Acta Zoologica 6: 307-414.

Flock, Å. (1965). Electron microscopic and electrophysiological studies on the lateral line canal organ. Acta Oto-Laryngologica, suppl. 199: 1-90.

Flock, Å. and Jørgensen, J.M. (1974). The ultrastructure of lateral line sense organs in the juvenile salamander *Ambystoma mexicanum*. Cell and Tissue Research 152: 283-292.

Fritzsch, B. and Wahnschaffe, U. (1983). The electroreceptive ampullary organs of urodeles. Cell and Tissue Research 229: 483-503.

Fritzsch, B. and Münz, H. (1986). Electroreception in amphibians. In: T.H. Bullock and W. Heiligenberg (eds). Electroreception. John Wiley & Sons, New York, USA. pp. 483-496.

Hetherington, T.E. and Wake, M.H. (!979). The lateral line system in larval *Ichthyophis* (Amphibia: Gymnophiona). Zoomorphologie 93: 209-225.

Jørgensen, J.M. (1984). On the morphology of the electroreceptors of the two lungfish: *Neoceratodus forsteri* Krefft and *Protopterus annectens*. Videnskabelige Meddelelser fra dansk Naturhistorisk Forening 145: 77-85.

Jørgensen, J.M. (2005). Morphology of electroreceptive sensory organs. In: T.H. Bullock, C.D. Hopkins, A.N. Popper and R.R. Fay (eds). Electroreception. Springer Verlag. New York, USA. pp. 47-67.

Jørgensen, J.M. and Flock, Å. (1973). The ultrastructure of lateral line sense organs in the adult salamander *Ambystoma mexicanum*. Journal of Neurocytology 2: 133-142.

Jørgensen, J.M. and Pickles, J.O. (2002). The lateral line canal sensory organs of the Epaulette shark (*Hemiscyllium ocellatum*). Acta Zoologica (Stockholm) 83: 337-343.

Kemp, A. (1999). Ontogeny of the skull of the Australian lungfish *Neoceratodus forsteri* (Osteichthyes: Dipnoi). Journal of Zoology London 248: 97-137.

Kingsbury, B.F. (1895). The lateral line system of sense organs in some American amphibians, and comparison with the dipnoans. Transactions of the American Microscopical Society 17: 115-154.

Northcutt, R.G. (1986). Electroreception in nonteleost bony fishes. In: T.H. Bullock and W. Heiligenberg (eds). Electroreception. John Wiley & Sons, New York, USA. pp. 257-285.

Northcutt, R.G. (2005). Ontogeny of electroreceptors and their neural circuitry. In: T.H. Bullock , C.D. Hopkins, A.N. Popper and R.R. Fay (eds). Electroreception. Springer. New York, USA. pp. 112-131.

Northcutt, R.G., Catania, K.C. and Criley, B.B. (1994). Development of lateral line organs in the axolotl. Journal of Comparative Neurology 340: 480-514.

Northcutt, R.G., Brandle, K. and Fritzsch, B. (1995). Electroreceptors and mechanoreceptor lateral line organs arise from single placodes in axolotls. Developmental Biology 168: 358-373.

Pehrson, T. (1949). The ontogeny of the lateral line system in the head of Dipnoans. Acta Zoologica 30: 153-182.

Reno, H.W. and Middleton, H.H. III. (1973). Lateral line system of *Siren intermedia* Le Conte (Amphibia: Sirenidae), during aquatic activity and aestivation. Acta Zoologica 54: 21-29.

Roth, A. (1973). Electroreceptors in Brachiopterygii and Dipnoi. Naturwissenschaften 60: 106.

Wahnschaffe, U., Fritsch, B. and Himstedt, W. (1985). The fine structure of the lateral-line organs of larval *Ichthyophis* (Amphibia: Gymnophiona). Journal of Morphology 186: 369-377.

Watt, M., Evans, C.S. and Joss, M.P. (1999). Use of electroreception during foraging by the Australian lungfish. Animal Behaviour 58: 1039-1045.

Webb, J.F. and Northcutt, R.G. (1997). Morphology and distribution of pit organs and canal neuromasts in non-teleost bony fishes. Brain, Behavior and Evolution 50: 139-151.

CHAPTER 19

The Inner Ear of Lungfishes

Jørgen Mørup Jørgensen[1] and Arthur N. Popper[2,*]
[1]Zoophysiology, Department of Biology, Building 1131, University of Aarhus, DK-8000 Aarhus, Denmark
[2]Department of Biology, University of Maryland, College Park, MD 20742, USA

ABSTRACT

The inner ear of lungfishes is unusual among fish inner ears because of a very large utricular macula situated in a closed recessus which is of similar size as the single recessus containing the saccular and lagenar maculae. All three maculae are also unusual among vertebrates because they appear to consist of narrow bands with large hair cells corresponding to the striola in other vertebrates and a wide surrounding area with dwarf hair cells around this band. The otoconia almost fill the utricular and sacculolagenar recessus. An endolymphatic sac covers parts of the brain. Overall, the lungfish ear has many similarities with the ears of other nonteleost bony fishes and the ears of cartilaginous fishes.

Keywords: inner ear, crista, macula, hair cells, statoconia

INTRODUCTION

The lungfish inner ear (Fig. 1) has the same components as other vertebrate inner ears. It is traditionally divided in a *pars superior* and a *pars inferior*. The *pars superior* consists of three semicircular canals (Fig. 2) and the utriculus. A sensory crista is found in all three ampullae which are located at one end of each semicircular canals and a crista neglecta (syn. macula neglecta) which is found in the utricle just below the common opening of the two vertical semicircular canals.

*Corresponding author: E-mail: apopper@umd.edu

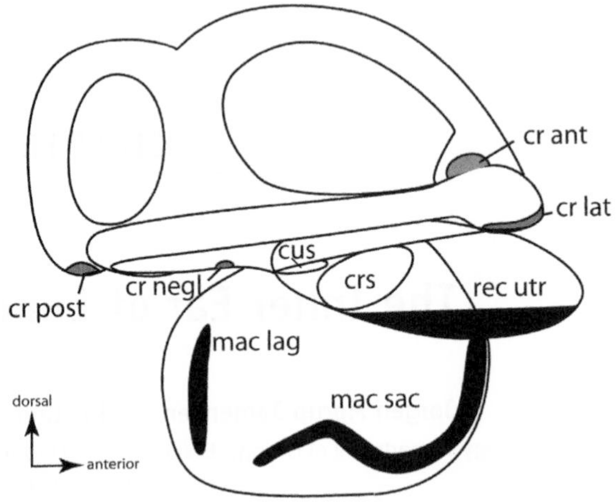

Fig. 1 Drawing of the right inner ear of the Australian lungfish *Neoceratodus*, seen laterally. The endolymphatic duct and sac are not shown, since the duct departs from the utricle medially, not visible from the lateral side. cr ant: crista anterior, cr lat: crista lateralis, cr post: crista posterior, cr negl: crista neglecta, rec utr: recessus utriculi, crs: canalis recessus utriculi-sacculus-lagenae, cus: canalis utriculi-saccularis, mac sac: macula sacculi, mac lag: macula lagenae.

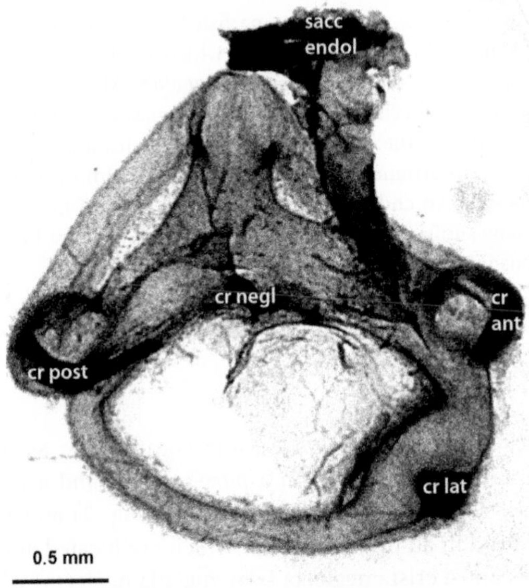

Fig. 2 Light micrograph of the tubelike utriculus with crista neglecta (cr negl) and semicircular canals with their cristae (abbreviations as in Fig. 1) of a 70 mm larva of the Australian lungfish *Neoceratodus*.

The utricle has two major components, a tubelike utriculus proper (Fig. 2) and a pouch, the recessus utriculus. The utricular macula is situated in the utricular recessus.

The *pars inferior* part is formed by a pouch containing the macula sacculi and macula lagenae. Both the utricular recess and the pouch containing the saccular and lagenar maculae are filled with otoconia, which are numerous small crystals held together as a united otolith.

In many chapters in this volume, it is emphasized how many lungfishes characteristics are similar to those of amphibians. Indeed, this similarity is very clear in most of the different organs. The ear of lungfishes, however, is more like that of chondrichthyans. As in most sharks and rays, the utricular recess found in all three lungfish genera is very large and forms a closed pouch, with comparatively small openings to the other compartments of the inner ear (Retzius 1881a). A similar utricular recess was described in the sturgeon *Acipenser sturio* by Retzius (1881a).

Another interesting feature of the lungfish inner ear is that there is also a common sac that contains both the saccular and lagenar maculae. This is similar to that described from the ratfish *Chimaera monstrosa* (Retzius 1881a; personal observations) and a number of other cartilaginous fishes (Werner 1930; Tester *et al.* 1972). Among bony fishes, a single sacculolagenar recess has been described in a number of non-teleosts, for example, the shovel-nose sturgeon *Scaphirhynchus platorhynchus*, the bichirs *Polypterus bichir* and *Polypterus senegalus*, the gar *Lepisosteus osseus* and the bowfin *Amia calva* (Retzius 1881b; Greve 1964; Popper 1978; Mathiesen and Popper 1987; Thomot and Bauchot 1987).

Sensory epithelia. The sensory epithelia are composed of a layer of sensory cells, each of which is separated from one another by supporting cells, and of nerve fibers. In the striolae, the sensory cells extend from the luminal surface to one-half to two-thirds of the way to the basal lamina, while most of the supporting cells reach from the luminal surface to the basal lamina. All sensory cells are hair cells, each of which is equipped with an eccentric kinocilium arising from a basal body like ordinary cilia. The central part of the hair cell surface is occupied by a group of stereovilli (syn. stereocilia), which stand like the pipes in an organ, with the tallest stereovilli close to the kinocilium (Fig. 3). This gives the hair cell a polarity, which reflects its physiology (Flock 1965).

The lengths of the sensory hairs differ considerably. Crista hair cells have the tallest kinocilia (≥ 30 μm), and macular hair cells have shorter sensory cilia. The macular striolae have hair bundles similar to those in other vertebrates, and previously designated as F1 in Popper (1977) or K5s4 in the useful system of Platt (1983).

The basal part of the pear-shaped or roundish hair cell body receives synaptic connection with afferent nerve endings. Some, but not all, also have an efferent innervation.

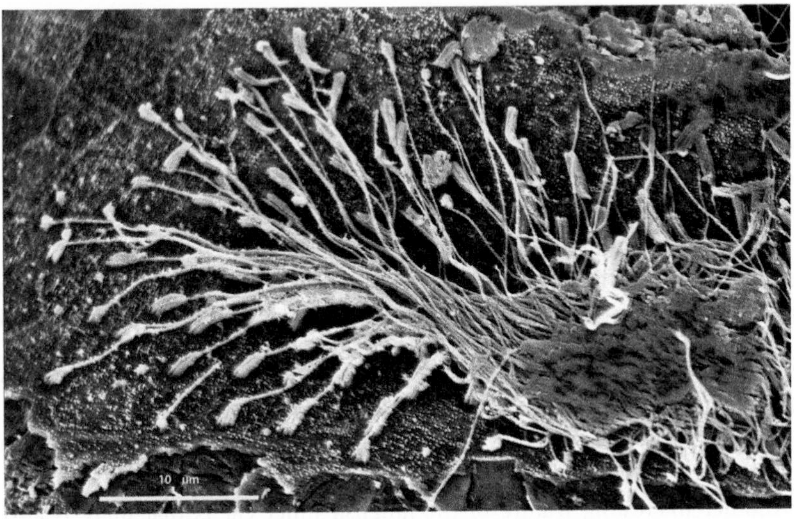

Fig. 3 Scanning electron micrograph of the crista neglecta in the inner ear of an adult Australian lungfish.

PARS SUPERIOR

The three semicircular canals are of normal length and width for the general vertebrate ear. They are positioned almost perpendicular to one another. Each is connected to an enlargement, an ampulla, which contains a sensory organ, the crista. They all connect to the tubelike utriculus. The ampullary end of the anterior and lateral canals and the posterior canal are in direct connection with the utriculus. The other ends of the anterior and posterior canals unite in a sinus superior, which connects to the utriculus from a dorsal position (Fig. 2).

Connected to the tubular utriculus is a ductus endolymphaticus which can be followed to a saccus endolymphaticus. This ductus is filled with otoconia. The saccus is an irregular sac which can be seen lateral to or partly covering the medulla oblongata. Such large endolymphatic sacs, which cover parts of the brain, have previously been described for the bony fish *Polypterus senegalus* by Thomot and Bauchot (1987) and is characteristic for the inner ears of many amphibians (Retzius 1881a) and some reptiles (Retzius 1884).

A little sensory spot, the crista (syn. macula) neglecta is situated in the utriculus just below the sinus superior. Retzius (1881a) did not find the neglecta in the Australian lungfish but presumed its presence in this species as in other fishes. And he was right, since we can demonstrate its presence in the *Neoceratodus* (Figs. 2, 3). The crista neglecta contains relatively few hair cells compared to the other cristae. The kinocilia are quite long and look very much like the sensory cells of the other cristae.

The utricular pouch. Lateral to the utricle and just below the lateral crista is a large pouch, the recessus utriculus. It has an opening to the sacculus-lagena, the canalis recessusaccularis. In the utricular recess is found the utricular macula. The recessus is almost completely filled with an otoconial mass, as in many amphibians. But unlike teleosts, which typically have a solid stone, termed otolith (Retzius 1881a; Gauldie 1986).

The utricular macula lies in the horizontal plane when the head is held in normal position. It is composed of a rampa which follows the anterior and lateral walls of the recess. In this marginal band can be found a few small hair cells situated distantly from each other. Centrally to this is a *striolar part* with few rows of large hair cells situated rather close to one another. It closely follows the anterolateral wall of the recessus (Fig. 4).

The hair bundle of these cells has robust stereovilli of around 0.15 μm and a relatively short kinocilium, only slightly longer than the longest stereovilli (Fig. 5). The number of stereovilli is 60-80. Medially to the striola lies another extrastriolar part, the *cotillus* of Werner (1933) with small, spread hair cells (Platt *et al.* 2004).

The striola is a narrow strip, only 6-10 cells wide. It has two oppositely oriented groups of hair cells, an outer rim that has its hair cells oriented with the kinocilia on the medial side of their stereovilli (Fig. 6).

The hair cells on the inner rim have their kinocilia positioned anteriorly and laterally, and they are thus oppositely oriented to the cells of the outer rim. As shown in Fig. 7, a few striolar hair cells have very short, low and thin stereovilli; we presume that these are young hair cells. They can be found in all parts of the striola.

The main part of the floor in the recessus is the extrastriolar utricular macula. Werner (1933) called this the cotillus. It lies medially to the striolar hair cells. The dwarf hair cells in the cotillus are situated 10-15 μm from one another (Fig. 8) and are very different from the striolar cells. In the extrastriolar areas, the epithelium is low, in sections found to be between 100 and 150 μm. The dwarf hair cells also here extend between one-half and two-thirds of the luminal part of the epithelium. Their hair bundles are composed of a long kinocilium of 5-12 μm next to a tiny bundle of thin stereovilli, which only measure 2.5 μm or less in length (Fig. 8). Such an area with spread dwarf hair cells was also found in the gar *Lepisosteus osseus* by Mathiesen and Popper (1987).

PARS INFERIOR

This part forms in all lungfish a closed sac or recessus, connected with the utriculus by a canalis utriculosaccularis and with the recessus utriculi by a canalis recessusaccularis. The flattened recessus sacculo-lagenae have an almost vertical

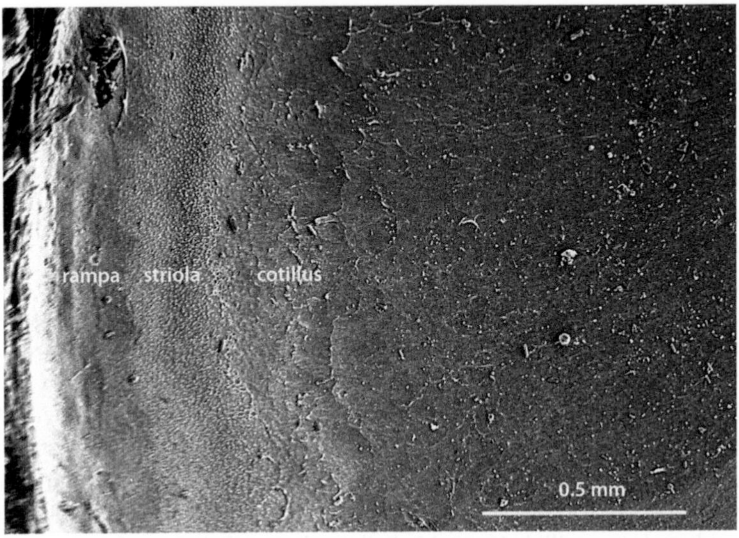

Fig. 4 Scanning electron micrograph of a part of the floor of recessus utriculi, showing the rampa, striola and cotillus.

Fig. 5 Scanning electron micrograph of a hair cell bundle from the striola of the utricular macula.

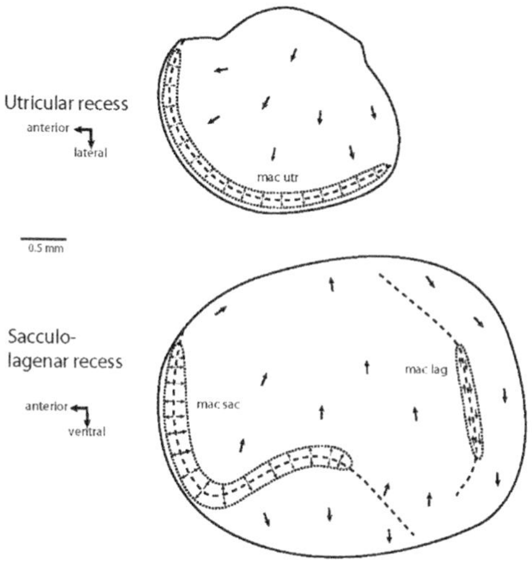

Fig. 6 Drawing showing the orientation of hair cells in the inner ear of an African lungfish *Protopterus* (from Platt *et al.* 2004).

position. Its medial wall contains two sensory maculae, the saccular and lagenar maculae (Fig. 6). The striola of the saccular macula has the shape of an S. Posterior to this is positioned the striola of the lagenar macula. The hair cells look very much like the hair cells in the utricle.

The saccular striola is composed of two narrow bands of hair cells of almost equal size, with the hair hundles oriented away from each other (Fig. 6). The lagenar striola is also formed by two narrow bands of hair cells, an interior with a dorsal orientation and a posterior with a ventral orientation. The spread dwarf hair cells outside the striolae have hair bundles similar to those described for the extramacular dwarf hair cells in the utricular recess. They are oriented dorsally in most of the recessus except for a relatively little ventral population with ventrally oriented dwarf hair cells. These surround the saccular and lagenar striolae and this way one may say that the sacculolagenar recess forms a single large sensory area containing two striolae.

Material and Acknowledgements

Dr. Christopher Platt assisted with valuable information.

Since essential parts of the description above have not been published before, it may be of interest to know the material. At our disposal was the following material:

Fig. 7 Scanning electron micrograph of two hair cell bundles from the striola of the utricular macula. The bundle to the right has a short kinocilium and short and thin stereovilli and belongs supposedly to a new hair cell (NHC).

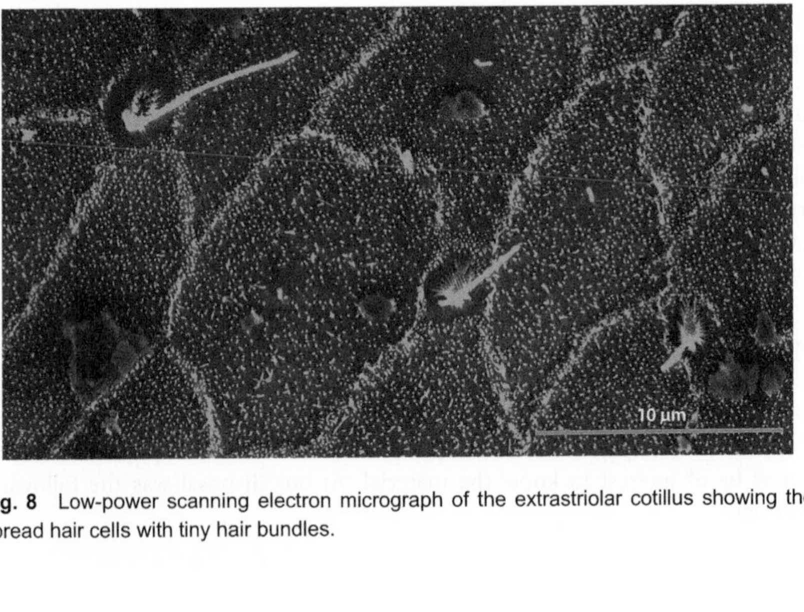

Fig. 8 Low-power scanning electron micrograph of the extrastriolar cotillus showing the spread hair cells with tiny hair bundles.

Neoceratodus forsteri male TL 105 cm, male TL 80 cm and 26 embryos from stage 45 to juveniles with TL 65, 70 and 77 mm. Also parts of a subadult TL 320 mm.

Protopterus annectens: water phase: a female TL 42 cm, a male TL 30 cm, 13 specimen with total total lengths of 25-35 cm. In aestivating phase 16 specimens with total lengths of 23-31 cm. *Lepidosiren*: a female TL 40 cm.

Most the embryological material from *Neoceratodus* was a gift from Dr. A. Kemp, and the aestivating *Protopterus* were given by Dr. T. Wang Nielsen.

References

Bemis, W.E., Burggren, W.W. and Kemp, N.E. (eds). (1987). The biology & evolution of lungfishes. Alan R. Liss, New York.

Flock, Å. (1965). Electron microscopic and electrophysiological studies on the lateral line canal organ. Acta Oto-Laryngologica, Suppl. 199: 1-90.

Gauldie, R.W., Dunlop, D. and Tse, J.C. (1986). The remarkable lungfish otolith. New Zealand Journal of Marine and Freshwater Research 20: 81-92.

Greve, M. (1964). The membranous labyrinth of *Calamoichthys calabaricus* Smith, 1965 (Polypteridae, Pisces). Videnskabelige Meddelelser fra Dansk naturhistorisk Forening 127: 275-281.

Mathiesen, C. and Popper, A.N. (1987). The utrastructure and innervations of the ear of the gar, *Lepisosteus osseus*. Journal of Morphology 194: 129-142.

Platt, C. (1983). The peripheral vestibular system in fishes. In: R.G. Northcutt and R.E. Davis (eds). Fish Neurobiology. University of Michigan Press, Ann Arbor, MI, USA. pp. 89-124.

Platt, C., Jørgensen, J.M. and Popper, A.N. (2004). The inner ear of the lungfish *Protopterus*. Journal of Comparative Neurology 471: 277-288.

Popper, A.N. (1977). A scanning electron microscopic study of the sacculus and lagena in the ears of fiftten species of teleost fishes. Journal of Morphology 153: 397-418.

Popper, A.N. (1978). Scanning electron microscopic study of the otolithic organs in the bichir (*Polypterus bichir*) and shovel-nose sturgeon (*Scaphirhynchus platorynchus*). Journal of Comparative Neurology 181: 117-128.

Popper, A.N. and Northcutt, R.G. (1983). Structure and innervation of the inner ear of the bowfin, *Amia calva*. Journal of Comparative Neurology 213: 279-286.

Retzius, G. (1881a). Das Gehörorgan der Wirbelthiere, Volume I. Samson and Wallin. Stockholm, Sweden.

Retzius, G. (1881b). Das membranöse Gehörorgan von *Polypterus bichir* und *Calamoichthys calabaricus*. Biologische Untersuchungen 4: 61-66.

Retzius, G. (1884). Das Gehörorgan der Wirbelthiere, Volume II. Samson and Wallin. Stockholm, Sweden.

Tester, L., Kendall, J.I. and Milisen, W.B. (1972). Morphology of the ear of the shark genus *Carcharhinus*, with particular reference to the macula neglecta. Pacific Science 26: 264- 274.

Thomot, A. and Bauchot, R. (1987). L'organogenèse du labyrinth membraneux chez, *Polypterus senegalus* Cuvier, 1829 (Pisces, Holostei, Polypteridae). Anatomischer Anzeiger, Jena 164: 189-211.

Werner, C.F. (1930). Das Ohrlabyrinth der Elasmobranchier. Zeitschrift für Wissenschaftliche Zoologie 136: 485-579.

Werner, C.F. (1933). Die Differenzierung der Maculae im Labyrinth, insbesondere bei Säugetieren. Zeitschrift für Anatomie und Entwicklungsgeschichte 99: 696-709.

Werner, C.F. (1960). Das Gehörorgan der Wirbeltiere und des Menschen. G. Thieme. Leipzig, Germany.

CHAPTER 20

Hearing in the African Lungfish, *Protopterus annectens*

Jakob Christensen-Dalsgaard[1,*], Christian Brandt[1], Maria Wilson[2], Magnus Wahlberg[1,3] and Peter Teglberg Madsen[2,4]

[1]Institute of Biology, University of Southern Denmark, Campusvej 55, DK-5230 Odense M, Denmark

[2]Department of Zoophysiology, Building 1131, Biological Institute, University of Aarhus, DK-8000 Århus C, Denmark

[3]Fjord&Belt, Margrethes Plads 1, DK-5300 Kerteminde, Denmark

[4]Woods Hole Oceanographic Institution, Woods Hole, MA 02543, USA

ABSTRACT

The lungfishes are believed to be the closest living relatives to the tetrapods and important for the understanding of evolution of hearing in the tetrapods.

We review results from a non-invasive study of hearing and vibration sensitivity in the West African lungfish (*Protopterus annectens*) using auditory brainstem responses (ABR) to sound and vibration stimulations. The experiments show that the sensitivity to particle motion generated by sound stimulation in water is similar to the sensitivity to direct vibration of the head. The hearing range is limited to frequencies below 300 Hz. Therefore, the West African lungfish is an auditory generalist that detects the particle motion component of sound and not sound pressure. The air volumes in lungfish, most notably the lungs, are therefore not functionally connected to the inner ear as is the swimbladder in auditory specialists among ray-finned fishes.

If the hearing of the African lungfishes resembles that of the tetrapod ancestors, tetrapods may have been very insensitive to airborne sound before the emergence of the tympanic ear, and hearing mediated chiefly through bone conduction (sound-induced skull vibrations). However, it is also possible that the African lungfishes are secondarily reduced with concomitant changes to the auditory system.

Keywords: hearing, vibration sensitivity, sound, frequencies, inner ear

*Corresponding author: E-mail: jcd@biology.sdu.dk

INTRODUCTION

The recent sarcopterygians or lobefin fish are the closest living relatives of tetrapods and are therefore interesting subjects in studies of the evolutionary history of soft-body characters such as physiological traits in the tetrapods. One such trait is the evolution of hearing, where the evolution of the auditory organs in the tetrapods reflects the major change from aquatic to terrestrial lifestyles. Recent palaeontological evidence shows that the tympanic ear of most recent tetrapods developed in the Triassic, almost 100 mya after the emergence of the tetrapods (Clack 1997), and the physiology of hearing in lungfish may thus provide important information on hearing in the early tetrapods. It should be borne in mind, however, that any extant species cannot be expected to embody the ancestral or primitive characters, since it is also product of many years of independent evolution and therefore will be a mixture of ancestral and derived characters. Furthermore, the recent sarcopterygians are fairly distant relatives of the tetrapodomorph sarcopterygians, with an estimated time of divergence dating back 430 million years. The tetrapodomorph tetrapod ancestors (e.g *Tiktaalik*, Daeschler *et al.* 2006) did not bear great resemblances to the extant lungfish morphologically, but were a group of specialized animals with flattened skulls and robust appendages. Finally, in the extant lungfish some of the advanced characters may be secondary reductions generated by paedomorphosis (Joss 2006). If the recent lungfish are paedomorphic, some of the traits that occur latest in development, for example ear structures, could also be reduced.

In water, hearing is governed by the fact that the sensory hair cells in the inner ear respond to displacement. Hair cells in combination with accessory structures such as a cupular membrane and otoliths make the unaided auditory systems of fish act as accelerometers that respond to the particle motion component of the sound field (Sand and Karlsen 1986). Beyond the resonant frequency of the otolith-cupula-hair cell complex, fish will have very poor hearing sensitivity, but that can be alleviated by mechanical coupling to an air volume that may act as a pressure to particle motion converter. Because a pressure sensitive ear will be much more sensitive than a particle motion sensitive ear at higher frequencies, auditory specializations among aquatic vertebrates usually entail the coupling of the inner ear to pressure sensitive (air-filled) structures (Popper *et al.* 2003).

In arctinopterygians, the swimbladder often serves the dual purpose of maintaining neutral buoyancy of the fish and acting as a pressure to particle motion transducer for hearing. Similarly, the lungs of lungfish may also be hypothesized to act as a pressure to particle motion transducer.

ANATOMICAL STUDIES OF THE LUNGFISH EAR

The ear of lungfish has not been extensively studied. The classical anatomical work of Retzius (1881, Fig. 1) shows that the ear has the same general organization as other vertebrate ears, with three dorsal semicircular canals and ventral otolithic organs. In contrast to actinopterygian fish, lungfish only have two otolithic organs, an utricle and a fused sacculo-lagenar organ. Interestingly, a fused sacculo-lagenar organ has been found also in fossil sarcopterygians such as *Eusthenopteron*, that are more closely related to tetrapods than recent lungfish (Clack and Allin 2004). Also, unlike the otoliths of actionpterygian fish the otoliths are not solid, but instead an otoconial mass like in the tetrapods. More recently, the ultrastructure of the ear of *Protopterus* has been investigated (Platt *et al.* 2004), showing that the ear contains two types of hair cells, both with long kinocilia and similar to the types found in other teleosts. The orientation of cells in the sacculo-lagenar macula is predominantly dorso-ventral, with two opposing hair cell groups. Finally, Platt *et al.* note that the structure and ultrastructure of the inner ear is generally similar across the recent lungfish species. The lungfish ear thus seem to differ from that of most bony fishes in some respects, but the presence of air-filled lungs opens the

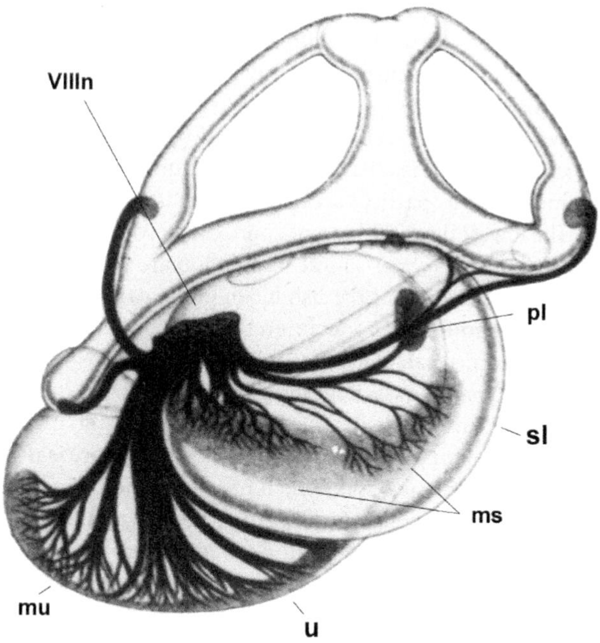

Fig. 1 Drawing of the African lungfish inner ear in medial view from Retzius (1881). Note the large, fused sacculus and lagena. Abbreviations: u, utriculus; mu, macula utriculi; sl, fused sacculus-lagena; ms, macula sacculi; pl, papilla ac. lagenae; VIIIn, 8th nerve

possibility that lungfish evolved a capability to detect sound pressure, as did some of the arctinopterygian hearing specialists.

PHYSIOLOGICAL INVESTIGATION OF LUNGFISH AUDITORY SENSITIVITY

To investigate the sensitivity and modality (pressure versus particle motion) of hearing in the African lungfish we measured the auditory brainstem response (Corwin *et al.* 1982; Kenyon *et al.* 1998) to sound in air and water, and vibrations only. The ABR is an evoked potential response measured by two differential electrodes inserted in the skin, one above the ear and one above the brain stem. A third ground electrode was placed dorsally behind the head. The fish were lightly anesthetized by brief immersion prior to the experiments in MS-222 (tricaine methane sulfonate). To get a robust ABR, it is necessary to get synchronized response from a large number of neurons, so we used a sound or vibration impulse. Such an impulse is broad-band; however, detailed frequency response can be obtained by masking the impulse by a low-frequency tone or narrow-band noise and compare the masked to the unmasked ABR response (see Brandt *et al.* 2008 for method).

For the airborne sound field we were limited by the equipment to frequencies above 100 Hz. Here, even at sound pressures of approximately 110 dB dB re 20 μPa (rms) no response could be measured for any of the frequencies of up to 1 kHz. However, robust ABR responses were measured to head vibrations and underwater sound, where we got reliable ABR-derived audiograms from five individuals (Figure 2). The audiogram in Fig. 2B is almost flat at frequencies below 100 Hz (when thresholds are measured as particle accelerations) as seen for teleosts (Sand and Karlsen 1986). The lowest thresholds are at approximately 2 cm/s^2. The sensitivities to head vibration and underwater sound stimulation are very close, showing that the ear responds to the motion of the fish head, in the case of underwater sound driven by the particle motion component. In the audiogram shown in Fig. 2B the vibration thresholds are recalculated as equivalent far field underwater sound pressure levels. The resulting audiogram is V-shaped with best frequency at approximately 50 Hz. Thresholds increase above 100 Hz, and are approximately 10 dB higher at 200 Hz than at frequency of best hearing. There is no evidence for special sensitivity to the pressure component of sound, and no increased pressure sensitivity at the resonance frequency of the lungs (approximately 300 Hz, judged by their volume). Thus, the apparent ABR derived sensitivity in particle velocity would translate to 150 to 190 dB re 1 μPa of pressure in an acoustic free field. The audiogram in hearing generalists, such as salmon (Hawkins and Johnstone 1978), as measured by heart-rate conditioning have generally the same shape as found here for the lungfish. The thresholds are considerably higher in the lungfish than in salmon. The main explanation for this discrepancy is likely that ABR

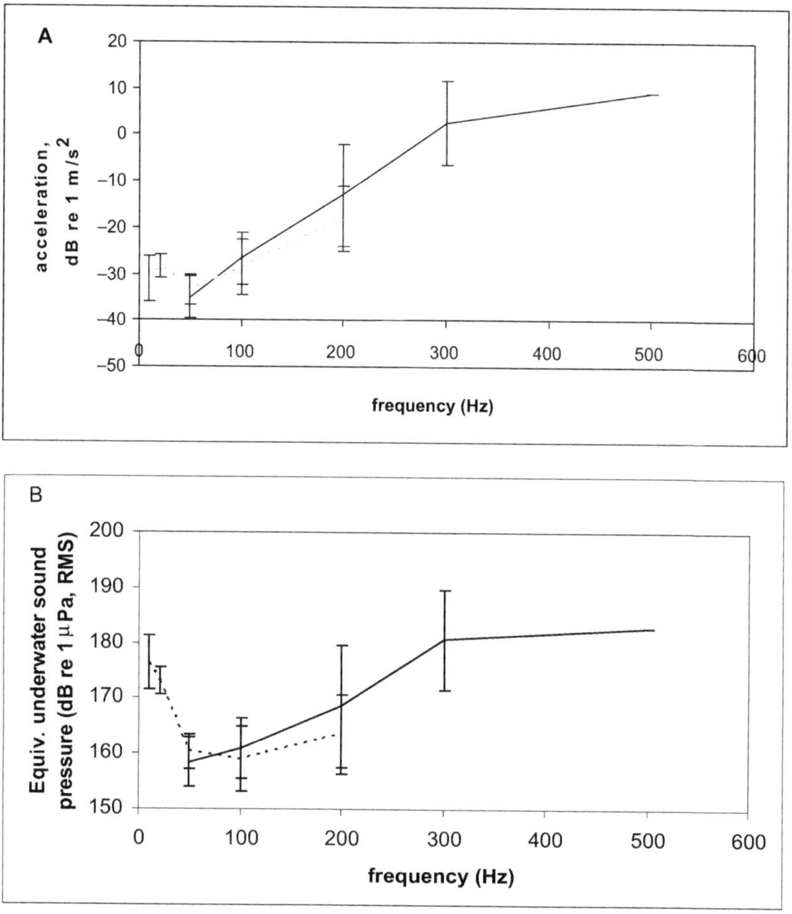

Fig. 2 ABR-derived audiogram of the African lungfish (N=5). The two curves are measurements of auditory brainstem responses to head vibration (dashed line) and to underwater sound (continuous line). A: Thresholds measured as particle acceleration (the parameter that the lungfish otolith most likely responds to). B: The thresholds plotted are equivalent free field sound pressures in dB RMS re 1 µPa, i.e. the free field sound pressure corresponding to the vibration thresholds as derived with ABR. Note that these thresholds are likely at least 25 dB above psychophysical thresholds.

measurements depends on synchroneous responses in large groups of nerve cells and probably are at least 25 dB above the thresholds of the most sensitive neurons (Ngan and May 2001). In this case, we conclude that lungfish hear like the hearing generalists among the actinopterygians. In the hearing generalists, the ear operates as an accelerometer, and there is little or no coupling to gas-filled structures that can act as a pressure to particle motion transducer.

EVOLUTIONARY PERSPECTIVES

The vibration sensitivity is similar to the sensitivity of frogs as shown by pilot ABR measurements, suggesting that the pronounced vibration sensitivity of the frog ear (Christensen-Dalsgaard and Jørgensen 1988) may be an ancestral trait. The finding of a lungfish ear that is unresponsive to higher frequencies in the closest relative to tetrapods leads to two scenarios for the evolution of the tetrapod inner ear. Both are hypothetical, since the structure of the early tetrapod inner ear is presently unknown (Clack and Allin 2004); see also Christensen-Dalsgaard and Carr (2008):

(1) The tetrapodomorph lungfish had a more diversified inner ear with patches of hair cells uncovered by otoliths and the structure of the ear of recent lungfish reflects independent evolutionary history (reduction, paedomorphosis).

(2) The ear of early tetrapods resembled the ear of recent lungfish with essentially no biologically relevant hearing sensitivity above 100 Hz, and a reasonably well-developed vibration sensitivity resembling the sensitivity of modern-day tetrapods and hearing generalists among the arctinopterygian bony fishes.

It has been suggested that the coelacanth ear has some similarities with the tetrapod ear. For example, Fritzsch (1999) has proposed that the basilar papilla in frogs should be homologous to an area in the inner ear of the coelacanth. This hypothesis would agree with scenario 1.

In connection with scenario 2 we propose that the evolution of high-frequency sensitivity could have proceeded by initial diversification of the sacculo-lagenar organ, creating a patch of sensory cells responding to higher frequencies. In this context, inner-ear sensory regions like the extrastriolar region reported by Platt *et al.* (2004) could be important, and it is interesting that the extrastriolar region between the saccule and lagena receives separate innervation (Platt *et al.* 2004). Once high-frequency sensitivity appeared, the ear would respond to airborne sound received via skull vibrations. The next step could be increased sensitivity of the ear to skull vibrations, for example by using the middle ear bone as an inertial element (as suggested by Lombard and Bolt (1979)), which would require a movable link between middle ear bone and otic capsule. The final step would be the tympanic ear. The current consensus that the tympanic ear has originated independently in the different tetrapod groups (Clack 1997; Grothe *et al.* 2004) suggests that there was a relatively long stretch of time (roughly 100 mya, from the Carboniferous to the Triassic) when tetrapods were terrestrial, but atympanate. It is likely that the tetrapods would have improved their sensitivity to airborne sound during this period (Christensen-Dalsgaard and Carr 2008).

CONCLUSION

In summary, we have shown that lungfish detect the particle motion component of a sound field, and that they have an audiogram comparable to generalist actinopterygian fish, with no specialized sensitivity to sound pressure. Also, we have found no evidence of sensitivity to air-borne sound (although also in air, sound pressure-induced vibrations of the substrate or the skull may stimulate the ear). We propose that this lack of pressure sensitivity is due to 1) poor coupling between the lungs and the auditory system and 2) lack of specialized sensory regions in the inner ear responding to frequencies above 100 Hz.

Acknowledgements

We are very grateful to Tobias Wang for providing the lungfish for these experiments and to Jørgen Mørup Jørgensen for the print of the Retzius figure. Supported by WIDEX A/S (CB).

References

Brandt, C., Andersen, T. and Christensen-Dalsgaard, J. (2008). Demonstration of a portable system for auditory brainstem recordings based on pure tone masking differences. In: T. Dau, J.M. Buchholz, J.M. Harte and T.U. Christiansen (eds.). Auditory Signal Processing in Hearing-impaired Listeners. 1st International Symposium on Auditory and Audiological Research. Centertryk. Copenhagen, Denmark. pp. 241-247.

Christensen-Dalsgaard, J. and Jørgensen, M.B. (1988). The response characteristics of vibration-sensitive saccular fibers in the grassfrog, *Rana temporaria*. Journal of Comparative Physiology A 162: 633-638.

Christensen-Dalsgaard, J. and Carr, C.E. (2008). Evolution of a sensory novelty: Tympanic ears and the associated neural processing. Brain Research Bulletin 75: 365-370.

Clack, J.A. (1997). The evolution of tetrapod ears and the fossil record. Brain, Behavior and Evolution 50: 198-212.

Clack, J.A. and Allin, E. (2004). The evolution of single- and multiple-ossicle ears in fishes and tetrapods. In: G.A. Manley, A.N. Popper and R.R. Fay (eds.). Evolution of the Vertebrate Auditory System. Springer, New York, USA. pp. 128-163.

Corwin, J.T., Bullock, T.H. and Schweitzer, J. (1982). The auditory brainstem response in five vertebrate classes. Electroencephalography and Clinical Neurophysiology 54: 629-641.

Daeschler, E.B., Shubin, N.H. and Jenkins, F.A. (2006). A Devonian tetrapod-like fish and the evolution of the tetrapod body plan. Nature 440: 757-763.

Fritzsch, B. (1999). Hearing in two worlds. Theoretical and actual adaptive changes of the aquatic and terrestrial ear. In: R.R. Fay and A.N. Popper (eds.). Comparative Hearing: Fish and Amphibians. Springer, New York, USA. pp. 15-42.

Grothe, B., Carr, C.E., Casseday, J.H., Fritzsch, B. and Köppl, C. (2004). The evolution of central pathways and their neural processing patterns. In: G.A. Manley, A.N. Popper and R.R. Fay (eds.). Evolution of theVertebrate Auditory System. Springer, New York, USA. pp. 289-359.

Hawkins, A.D. and Johnstone, A.D.F. (1978). Hearing of the Atlantic Salmon, *Salmo Salar*. Journal of Fish Biology 13: 655-660.

Joss, J.M.P. (2006). Lungfish evolution and development. General and Comparative Endocrinology 148: 285-289.

Kenyon, T.N., Ladich, F. and Yan, H.Y. (1998). A comparative study of hearing ability in fishes; the auditory brainstem response approach. Journal of Comparative Physiology A 182: 307-318.

Lombard, R.E. and Bolt, J. (1979). Evolution of the tetrapod ear: an analysis and reinterpretation. Biological Journal of the Linnean Society 11: 19-76.

Ngan, E.M. and May, B.J. (2001). Relationship between the auditory brainstem response and auditory nerve thresholds in cats with hearing loss. Hearing Research 156: 44-52.

Platt, C., Jørgensen, J.M. and Popper, A.N. (2004). The inner ear of the lungfish *Protopterus*. Journal of Comparative Neurology 471: 277-288.

Popper, A.N., Fay, R.R., Platt, C. and Sand, O. (2003). Sound detection mechanisms and capabilities of teleost fishes. In: T. Bullock, S.P. Collin, N.J. Marshall and J. Atema (eds.). Sensory Processing in Aquatic Environments. Springer-Verlag, New York, USA. pp. 3-38.

Retzius, G. (1881). Das Gehörorgan der Wirbelthiere, Volume I. Samson and Wallin, Stockholm, Sweden.

Sand, O. and Karlsen, H. (1986). Detection of infrasound by the Atlantic cod. Journal of Fish Biology 125: 197-204.

Index

G

H

I

K

L

M

N

O

P

Color Plate Section

Chapter 3

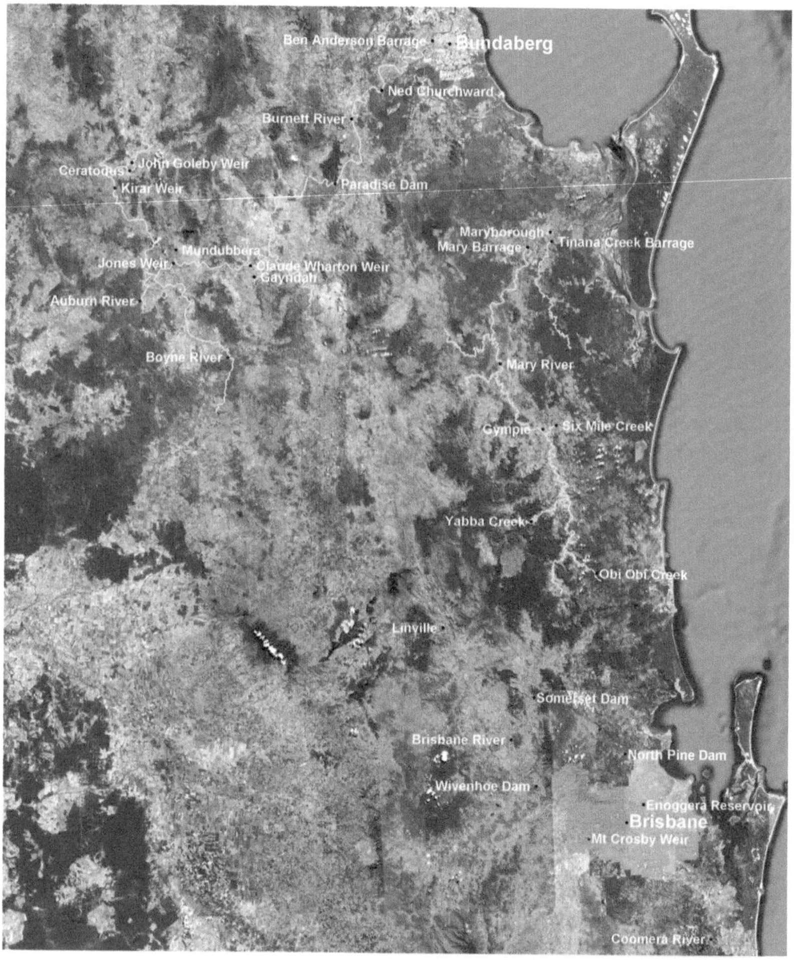

Fig. 3 Map of south east Queensland illustrating the presumed natural distribution of *N. forsteri* (yellow) and other catchments inhabited by the species (orange).

Fig. 8 Aerial breathing is used to supplement oxygen supply during periods of increased activity (Photo by Gordon Hides).

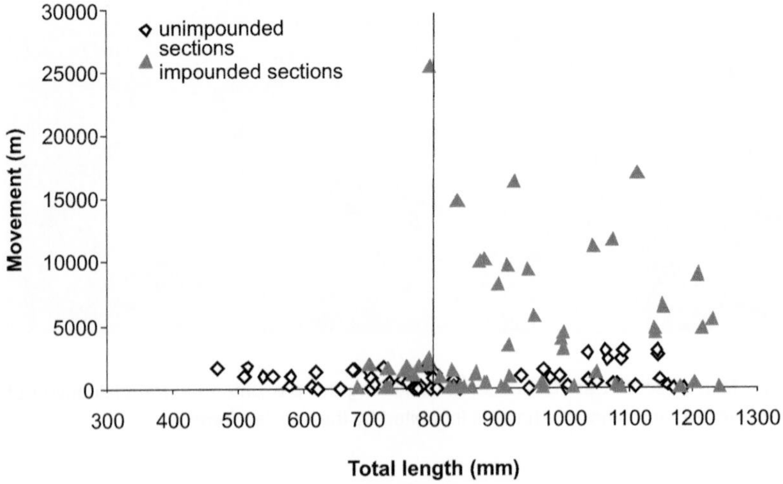

Fig. 9 Movements of recaptured lungfish from impounded and flowing sections of the Burnett River (n = 124). Vertical line at 800 mm represents the approximate threshold for sexual maturity (data from Brooks and Kind (2002)).

Fig. 12 Lungfish stranded downstream of a weir on the Burnett River after water releases were shut down.

Chapter 5

Fig. 3 *Lepidosiren paradoxa* "awaken" in the aquarium in the laboratory (top) and aestivating in the field (bottom). Photos by Lenise Mesquita-Saad during descending water period, 1997.

Chapter 6

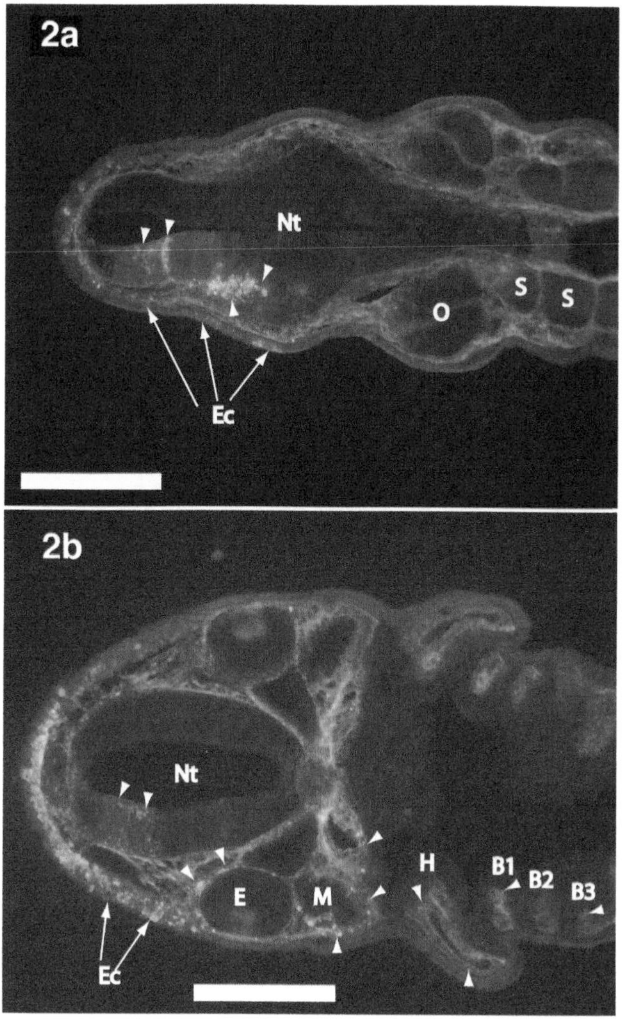

Fig. 2a Dorsal frontal vibratome section of a stage 40 *Neoceratodus forsteri* embryo. Red is DiI, green is fibronectin. Anterior is to the left. Arrowheads indicate DiI in the neural tube and in the mesenchyme. DiI is also seen in the ectoderm covering the neural tube. Ec, Ectoderm; Nt, Neural tube; O, otic vesicle; S, somite. Scale bar = 0.5 mm. (From Ericsson *et al.* 2008.) **2b:** Frontal vibratome section of the same embryo as in Figure 2a, slightly more ventral. Red is DiI, green is fibronectin. Arrowheads indicate DiI in the neural tube, in the mesenchyme surrounding the eye and in the pharyngeal arches. DiI is also found in the epidermis covering the neural tube. B1–B3, Branchial arches; E, Eye; Ec, Ectoderm; H, Hyoid arch; M, Mandibular arch; Nt, neural tube. Scale bar = 0.5 mm. (From Ericsson *et al.* 2008.)

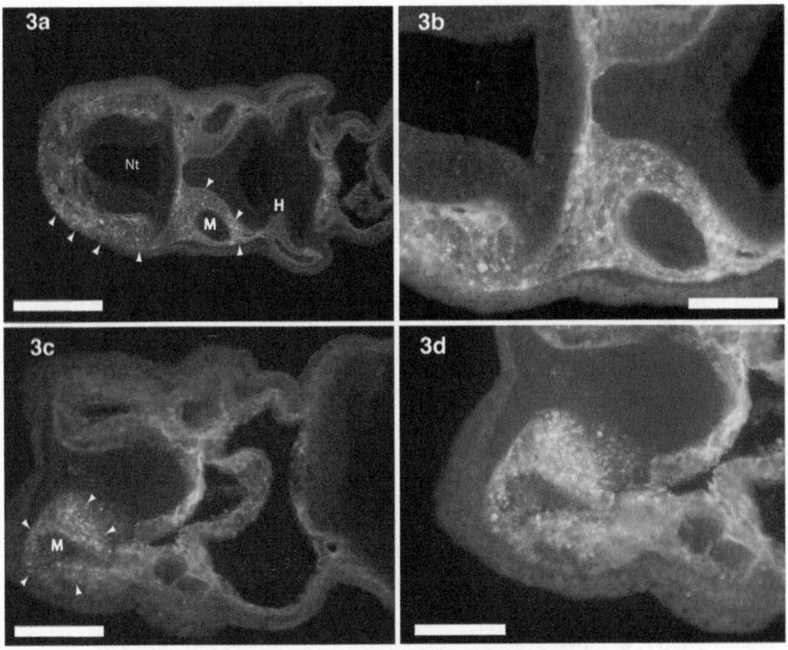

Fig. 3a Frontal vibratome section of the same embryo as in Figure 2a, slightly more ventral than Figure 2b. Red is DiI, green is fibronectin. Arrowheads indicate DiI in the epidermis covering the neural tube and in the mesenchyme of the mandibular arch. H, Hyoid arch; M, Mandibular arch; Nt, Neural tube. Scale bar = 0.5 mm. (From Ericsson *et al.* 2008.) **3b** Magnification of the central area in Figure 4a. Extensive DiI-staining can be observed in mandibular arch tissues surrounding the mesodermal core, especially in the medial part. Scale bar = 0.25 mm. (From Ericsson *et al.* 2008.) **3c** Frontal vibratome section of the same embryo as in Figure 2a, slightly more ventral than Figure 3a. Red is DiI, green is fibronectin. Arrowheads indicate DiI in the mesenchyme surrounding the ventral portion of the mandibular arch. Scale bar = 0.375 mm. (From Ericsson *et al.* 2008.) **3d** Magnification of the DiI-marked area in Figure 3c. DiI labels the medial part very strongly, as it does in Figures 2a and b. Scale bar = 0.5 mm. (From Ericsson *et al.* 2008.)

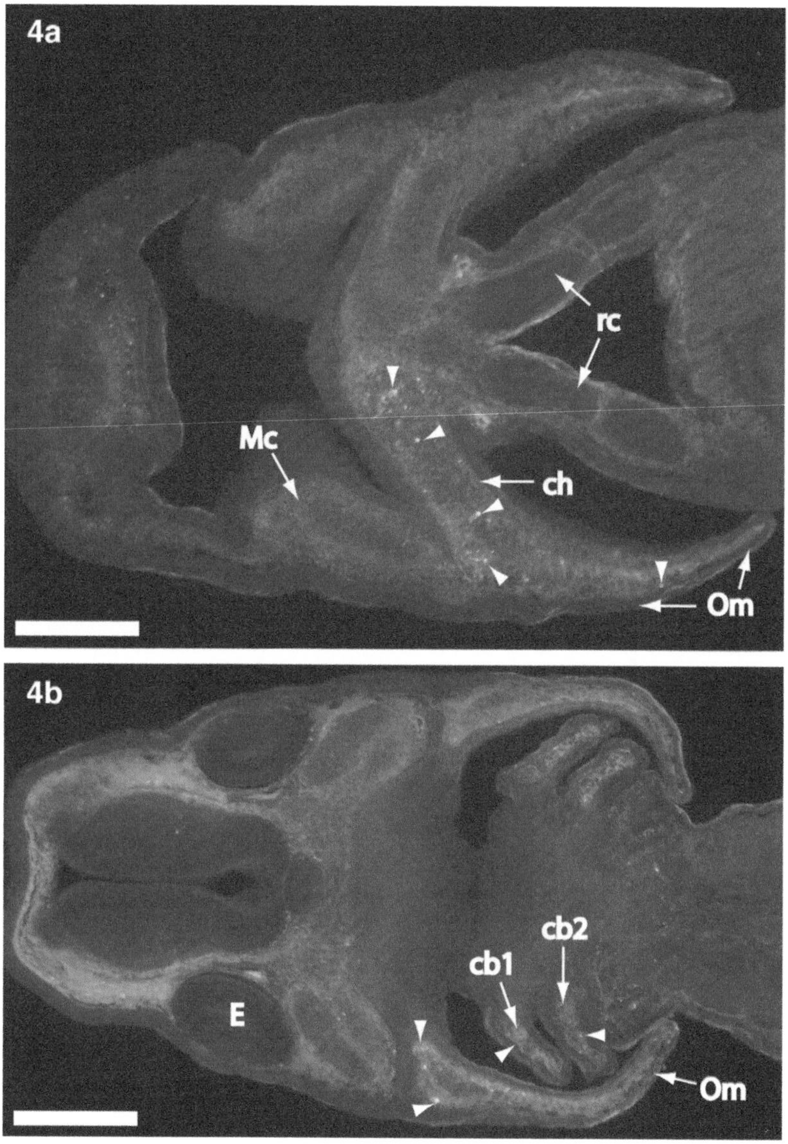

Fig. 4a Ventral frontal vibratome section of a stage 43 embryo. Red is DiI, green is fibronectin. Arrowheads indicate DiI in the cartilage of the ceratohyal and in the operculum. Cells in Meckel´s cartilage are also labelled. ch, ceratohyal; Mc, Meckel's cartilage; Om, Operculum; rc, rectus cervicis. Scale bar = 0.5 mm. (From Ericsson *et al.* 2008.) **4b** Medial frontal vibratome section of a stage 43 embryo. Red is DiI, green is fibronectin. Arrowheads indicate DiI in the cartilage of ceratobranchials 1 and 2, and in the operculum. cb1-2, Ceratobranchials 1 and 2; E, Eye; Om, Operculum. Scale bar = 0.5 mm. (From Ericsson *et al.* 2008.)

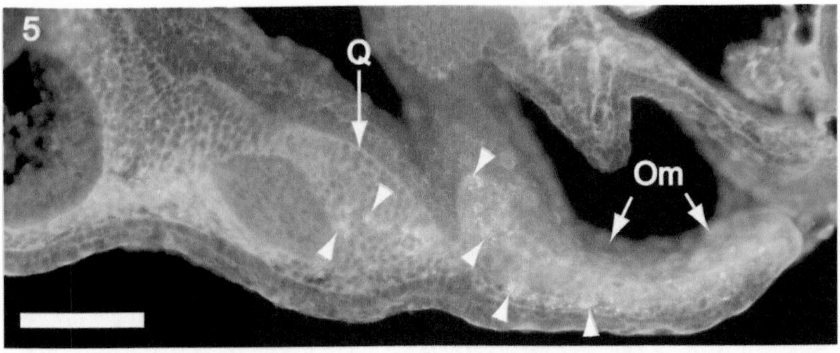

Fig. 5 Lateral frontal vibratome section of a stage 42 embryo, ventral to the eye. Red is DiI, green is fibronectin. Arrowheads indicate DiI in the pre-cartilage condensation of the quadrate and the connective tissue of the operculum. Om, Operculum; Q, Quadrate. Scale bar = 0.25 mm. (From Ericsson *et al.* 2008.)

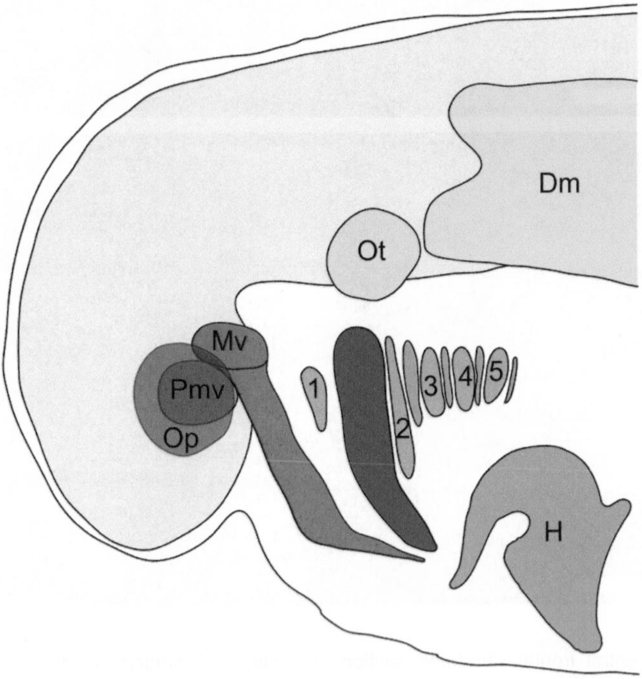

Fig. 6 Schematic view of the head of a stage 39 *Neoceratodus forsteri* embryo. Anterior is to the left. Brown is premandibular mesoderm, red is premandibular/mandibular arch mesoderm, blue is hyoid arch mesoderm, green is branchial arch mesoderm, and yellow is somitic mesoderm. At this stage, the premandibular and mandibular cavities have fused with the mandibular arch mesoderm. Dm, dorsal myotome; H, heart; Mv, mandibular vesicle; O, optic vesicle; Ot, otic vesicle; Pmv, premandibular vesicle; 1–5, pharyngeal pouches 1–5.

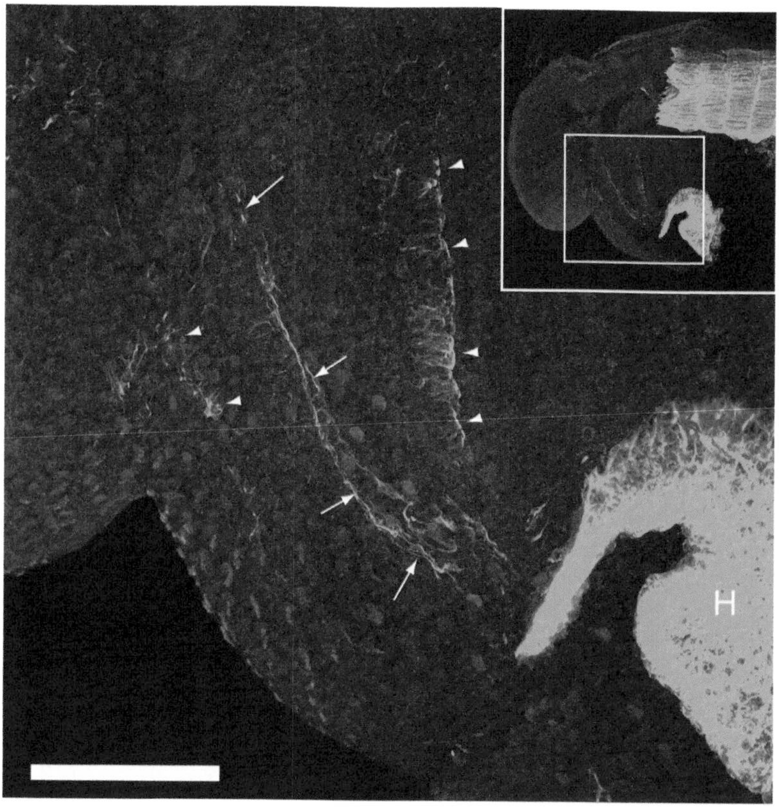

Fig. 7 Lateral view of stage 39 embryo of *Neoceratodus forsteri*, immunostained for desmin. Inset shows overview. Anterior is to the left. Arrowheads show desmin in mandibular and hyoid arch mesoderm. Arrows indicate afferent branchial artery I. H, heart. Scale bar is 250 μm.

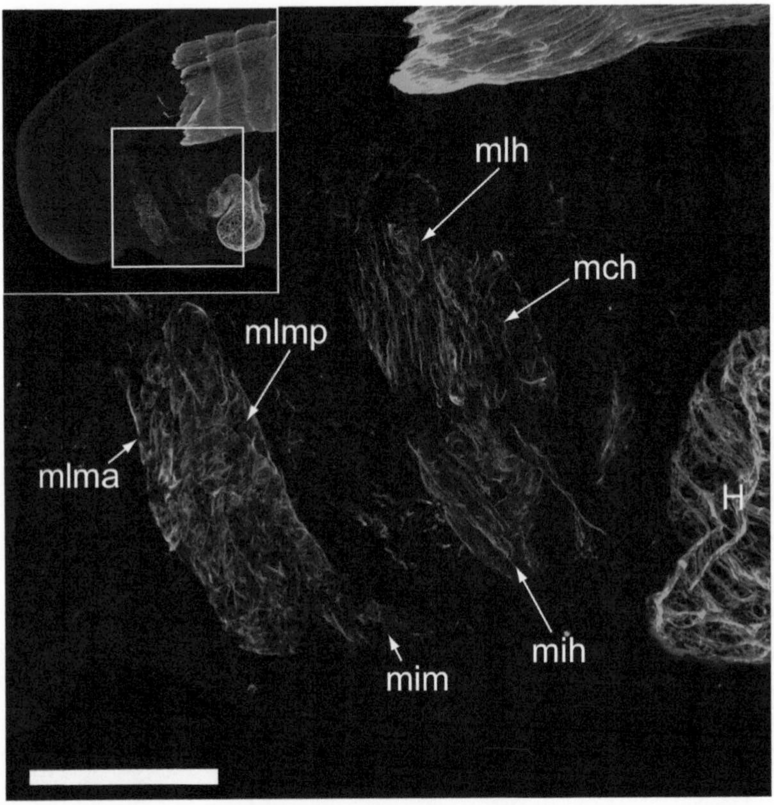

Fig. 8 Lateral view of stage 41 embryo of *Neoceratodus forsteri*, immunostained for desmin. Inset shows overview. Anterior is to the left. Desmin can be seen in the anlagen of all the mandibular and hyoid arch muscles. In mlh, muscle fibres have started forming. H, heart; mch, musculus constrictor hyoideus; mih, musculus interhyoideus; mim, musculus intermandibularis; mlh, musculus levator hyoideus; mlma, musculus levator mandibulae anterior; mlmp, musculus levator mandibulae posterior. Scale bar is 250 μm.

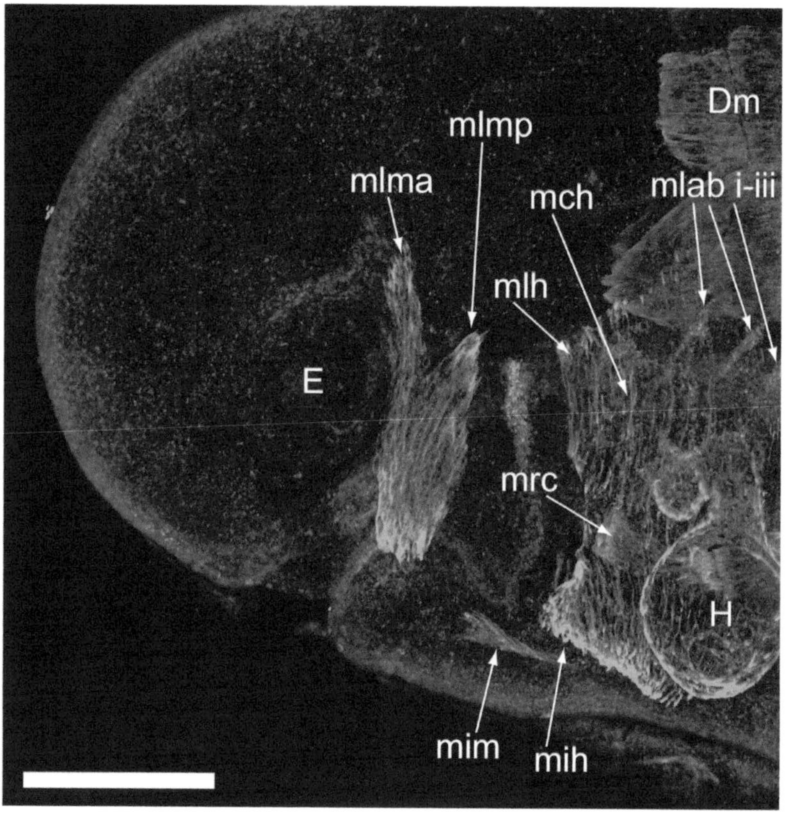

Fig. 9 Lateral view of a stage 43 larva of *Neoceratodus forsteri*, immunostained for desmin. Anterior is to the left. At this stage muscle fibres can be detected in all the mandibular and hyoid arch muscles, and in the hypobranchial rectus cervicis. Desmin is also detected in the anlagen of the levator arcus branchialis i–iii muscles. Dm, dorsal myotome; E, eye; H, heart; mch, musculus constrictor hyoideus; mih, musculus interhyoideus; mim, musculus intermandibularis; mlab i–iii, anlagen of musculus levator arcus branchialis 1–3; mlh, musculus levator hyoideus; mlma, musculus levator mandibulae anterior; mlmp, musculus levator mandibulae posterior; mrc, musculus rectus cervicis. Scale bar is 500 μm.

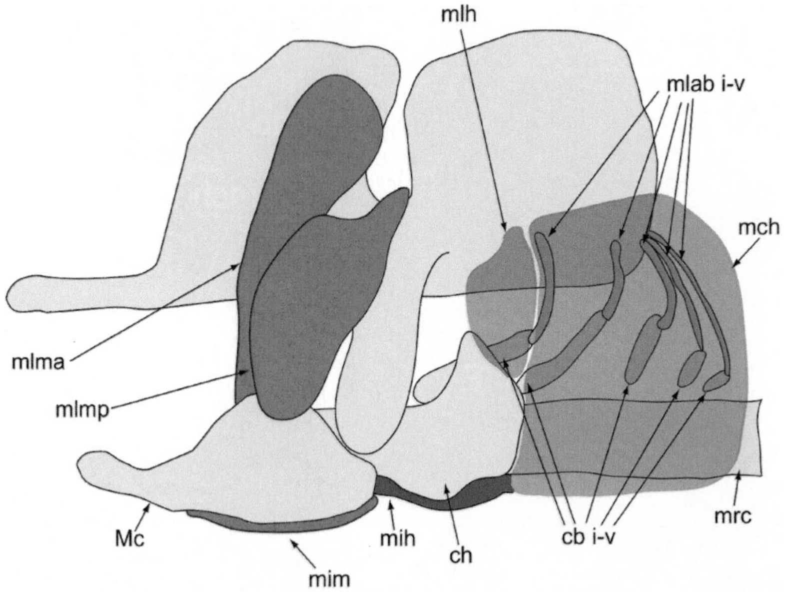

Fig. 10 Schematic, lateral view of the skull and cranial muscles of a stage 45+ *Neoceratodus forsteri* embryo. Anterior is to the left. The coracobranchiales and constrictor branchiales muscles have not yet developed at this stage. The constrictor hyoideus muscle is semitransparent to show the underlying structures. Red is mandibular arch muscles, blue is hyoid arch muscles, green is branchial arch muscles and yellow is hypobranchial muscles. cb i–v, ceratobranchial cartilage 1–5; ch, ceratohyal cartilage; Mc, Meckel's cartilage; mch, musculus constrictor hyoideus; mih, musculus interhyoideus; mim, musculus intermandibularis; mlab i–v, musculus levator arcus branchialis 1–5; mlh, musculus levator hyoideus; mlma, musculus levator mandibulae anterior; mlmp, musculus levator mandibulae posterior; mrc, musculus rectus cervicis. Modified from Fox (1965).

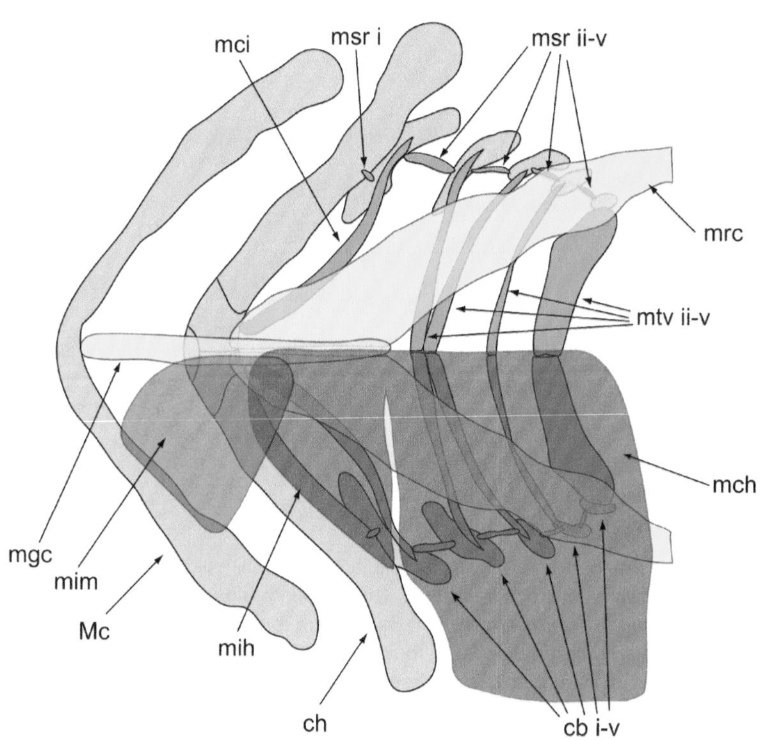

Fig. 11 Schematic ventral view of the lower jaw and associated muscles of a stage 45+ *Neoceratodus forsteri* embryo. Anterior is to the left. The coracobranchiales and constrictor branchiales muscles have not yet developed at this stage. The constrictor hyoideus, intermandibularis, geniocoracoideus and rectus cervicus muscles are semitransparent to show the underlying structures. Red is mandibular arch muscles, blue is hyoid arch muscles, green is branchial arch muscles and yellow is hypobranchial muscles. The mandibular and hyoid muscles are shown only on one side of the jaw. cb i–v, ceratobranchial cartilage 1–5; ch, ceratohyal cartilage; Mc, Meckel's cartilage; mci, musculus ceratohyoideus internus; mgc, musculus geniocoracoideus; mch, musculus constrictor hyoideus; mih, musculus interhyoideus; mim, musculus intermandibularis; mrc, musculus rectus cervicis; msr i–v, musculus subarcualis rectus 1–5; mtv ii–v, musculus transversus ventralis 2–5. Modified from Fox (1965).

Chapter 9

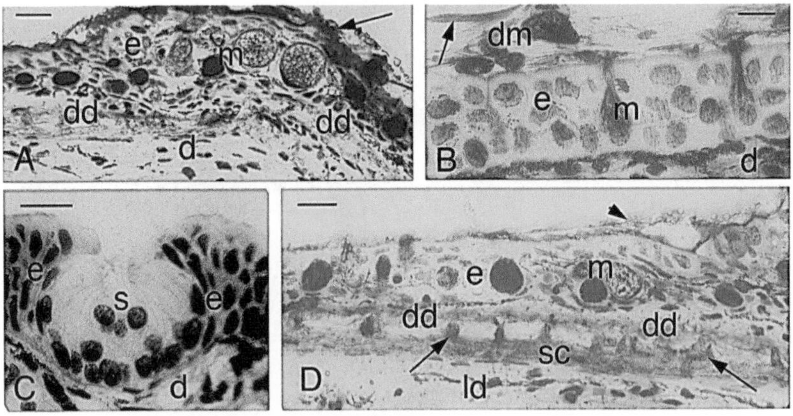

Fig. 2 Histology of the skin of a 5-month old juvenile. A, tail epidermis (e) with numerous mucous cells (m) with dense or pale content. The arrow indicates mucus covering the surface (arrow). d, dermis; dd, dense dermal layer; Bar, 20 μm. B, mucous cells (m) in tail epidermis (e) and degenerated mucus cells (dm) in the cuticular space (arrow). Bar, 10 μm. C, detail of an ampullar-organ (s) in trunk epidermis of sensory or glandular type, which slightly penetrate into the dermis (d). Bar, 20 μm. D, trunk epidermis (e) with numerous mucus cells (m) with underlaying bone tissue of scale (sc). The spinulae of the dermal bone (arrows) contact the overlying dense dermis (dd) and are confined within a loose dermis beneath. Bar, 10 μm.

Chapter 11

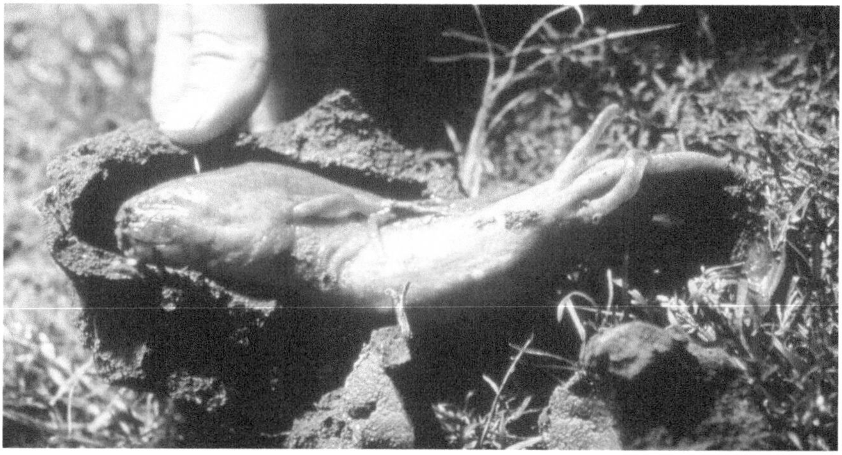

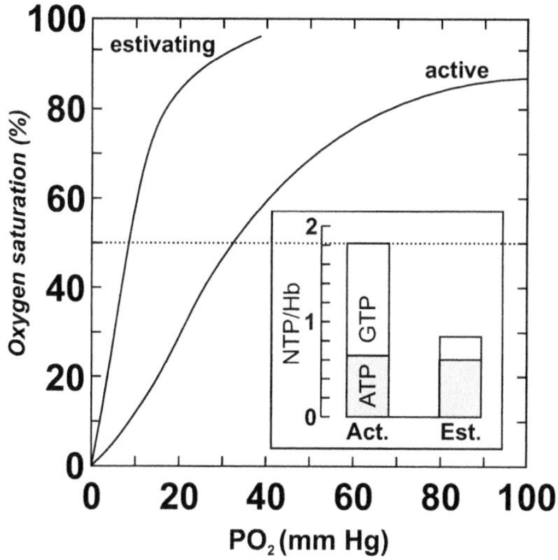

Fig. 1 *Protopterus amphibius* in its natural estivating cocoon (opened, with Kjell Johansen's thumb), O_2 equilibrium curves of whole blood from estivating and active (non-estivating, aquatic) specimens at 26°C, and (inset) blood NTP/Hb ratios, showing that the increased O_2 affinity under estivation correlates with a sharp decrease in the GTP/Hb (tetramer) concentration ratio. Modified after (Johansen *et al.* 1976).

Chapter 14

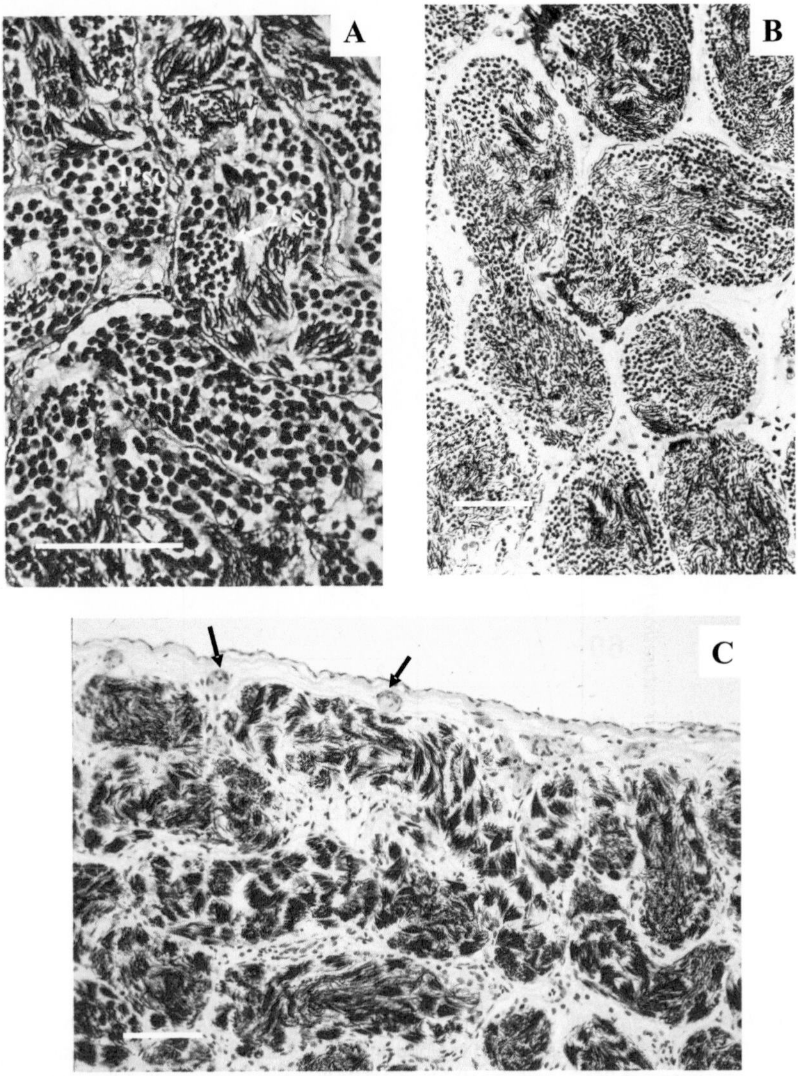

Fig. 2 Histological sections of testis of *Neoceratodus* stained with haematoxylin and eosin. **A.** Section of testis in March, showing most lobules containing primary spermatocytes (1° sc) and some nests within the lobules of secondary spermatocytes (2° sc and arrow). Bar = 20 µm. **B.** Section of testis in late April, showing secondary spermatocytes, spermatids and some tailed sperm. Bar = 100 µm. **C.** Section at the periphery of testis in October, showing all lobules containing mature sperm. Blood vessels can be seen at the periphery (arrows) and Leydig cells between the lobules containing sperm. Bar = 100 µm.

Chapter 17

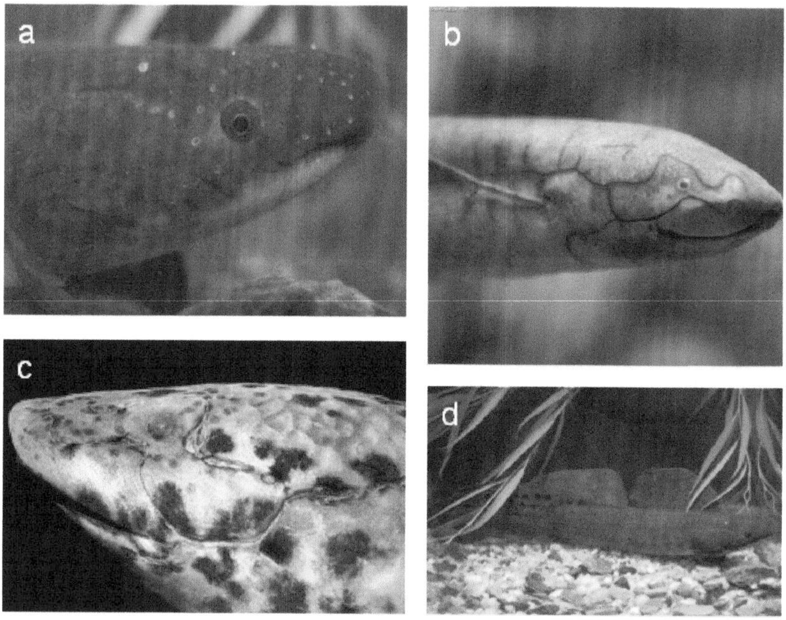

Fig. 1 The heads and eyes of lungfish. (a) Juvenile Australian lungfish, *Neoceratodus forsteri*, at around two years of age (25 cm total length). The body proportions are more similar to the adult at this stage. Note relatively large eye and lateral line pores. (Photograph Justin Marshall and Helena Bales). (b) South American lungfish, *Lepidosiren paradoxa*. Note relatively small eye and pigmented lateral line canals on head. (Photographic source – www.opencage.info/ Image 800_8532). Size unknown. (c) African lungfish *Protopterus annectens*. Note relatively small eye in this species also. (Photographic source – www.Photovault.com). Size unknown. (d) Same fish as (a) demonstrating eye/body proportions.

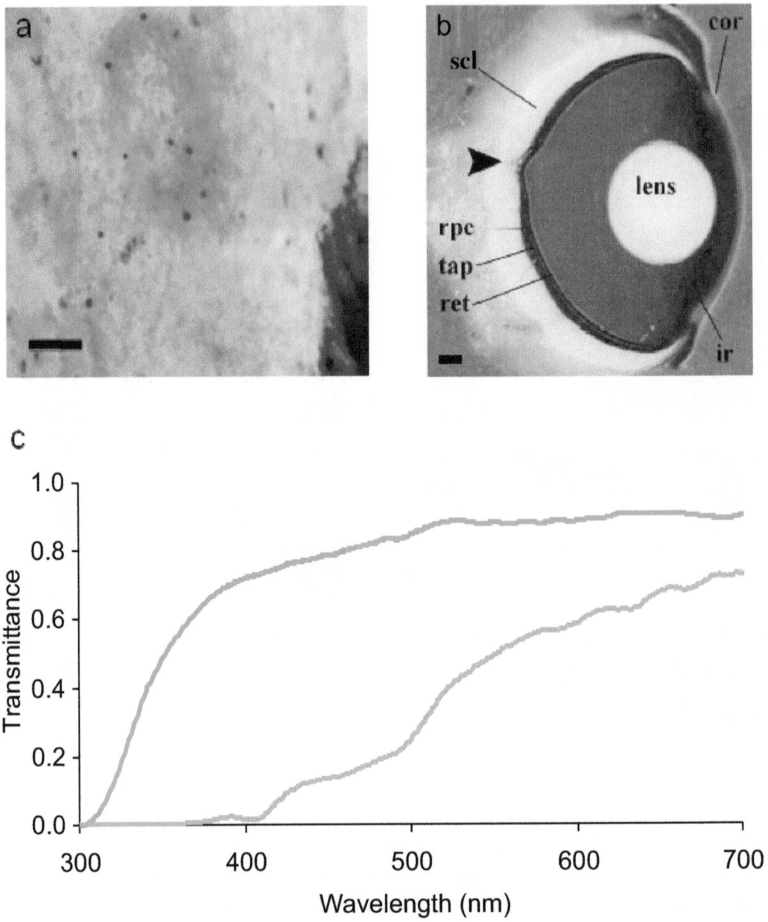

Fig. 2 Some optical features of the eye of *Neoceratodus forsteri*. (a) Freshly dissected cornea of adult lungfish with transmitted light to show intense yellow pigment. Scale 250 µm. (b) Frozen hemisected left eye of adult (127 cm total length), close to the geometrical centre. Note slight flattening of the lens. scl – sclera, ir – iris, cor – cornea, tap – tapetum, rpe – retinal pigment epithelium, ret – retina. Retinal indentation (arrow) close to optic nerve head. Scale - 1 mm.

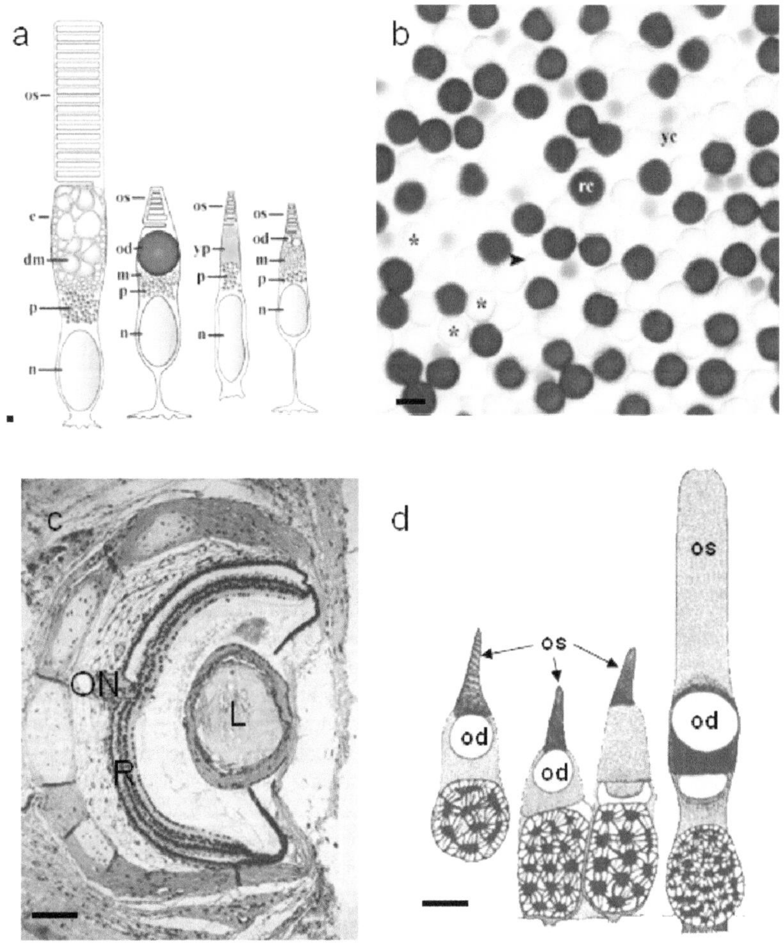

Fig. 3 Photoreceptors and retina of lungfish. (a) Schematic diagram of four morphologically distinct photoreceptors in *Neoceratodus forsteri* from left to right, rod, red cone, yellow cone, clear cone (after Bailes 2006). dm - distended mitochondria, e – ellipsosome, os – outer segment, p – parabaloid, n – nucleus, od – oil droplet, m – myoid, n – nucleus (for explanation of all terms see Walls 1942; Bailes 2006). Scale 10 µm. (b) Retinal whole-mount of fresh *Neoceratodus forsteri* retina showing all four morphological photoreceptor types. Large clear photoreceptors are rods (asterisks), red and yellow photoreceptor inclusions clearly distinguish these cone types. Arrow marks one clear cone type, being notably smaller than the rods. Scale 10 µm. (c) Section through the eye of *Lepidosiren paradoxa*. (From Ali and Anctil 1973, with kind permission). Note thin retinal layers. R – retina, L – lens, ON – optic nerve. scale 250 µm. (d) Schematic diagram of the photoreceptors of *Protopterus aethiopicus*. Note large oil droplet in the rod (far left). Apparent double cone (centre) and single cone (left) (after Walls 1942). Abbreviations as (a). Scale 10 µm.

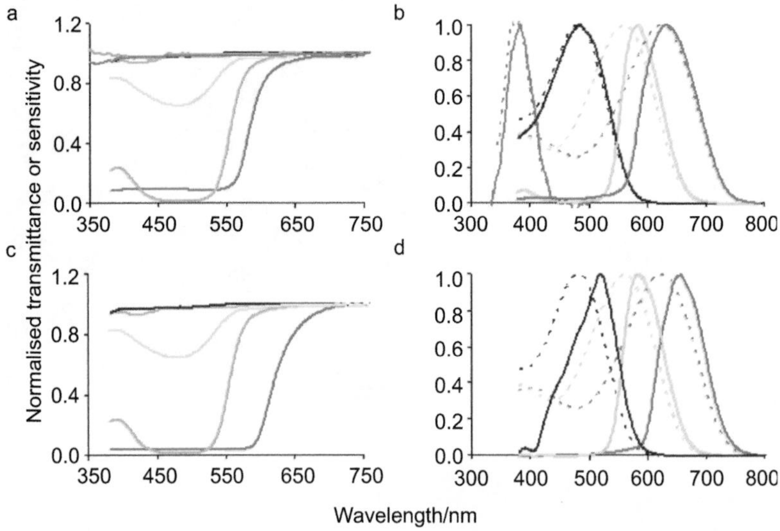

Fig. 5 Oil droplet and visual pigment spectral characteristics (a) Transmittance of oil droplets and yellow pigment in juvenile *Neoceratodus forsteri* photoreceptors relative to maximum transmission at 750 nm. The red oil droplet absorbs strongly below its λ-mid at 585 nm (red line), the yellow ellipsoid pigment λ-mid is at 556 nm (orange line) and the clear oil droplets of both UV (purple line) and short wavelength sensitive clear cones (green line) have negligible absorbance from 350-750 nm. (b) Visual pigment sensitivities of four cones (dotted curves) and the actual photoreceptor spectral sensitivities resulting from filtering by oil droplets and ocular media (Fig. 2) in juvenile *Neoceratodus forsteri*. Visual pigment curves are a nomogram based on peak absorbance (Table 1) and a vitamin A2 template, except for the UV sensitive photoreceptors which do not fit a template well and are just averaged direct measurements (Hart *et al.* 2008). Note loss of spectral sensitivity overlap and peak sensitivities pushed to longer wavelengths. (c) As (a) but for adult *Neoceratodus forsteri*. The red oil droplet is larger and absorbs more light, its λ-mid pushed longer at 623 nm. Yellow and blue pigments are the same as the juvenile. (d) Visual pigment sensitivities of three cones (dotted curves) and the actual photoreceptor spectral sensitivities resulting from filtering by oil droplets and ocular media (Fig. 2) in adult *Neoceratodus forsteri*. Visual pigment curves are a nomogram based on peak absorbance (Table 1) and a vitamin A2 template. Note loss of spectral sensitivity overlap and peak sensitivities pushed to longer wavelengths. Unlike the juvenile (b), the denser yellow adult ocular media (Fig. 2) absorbs strongly below 400 nm.

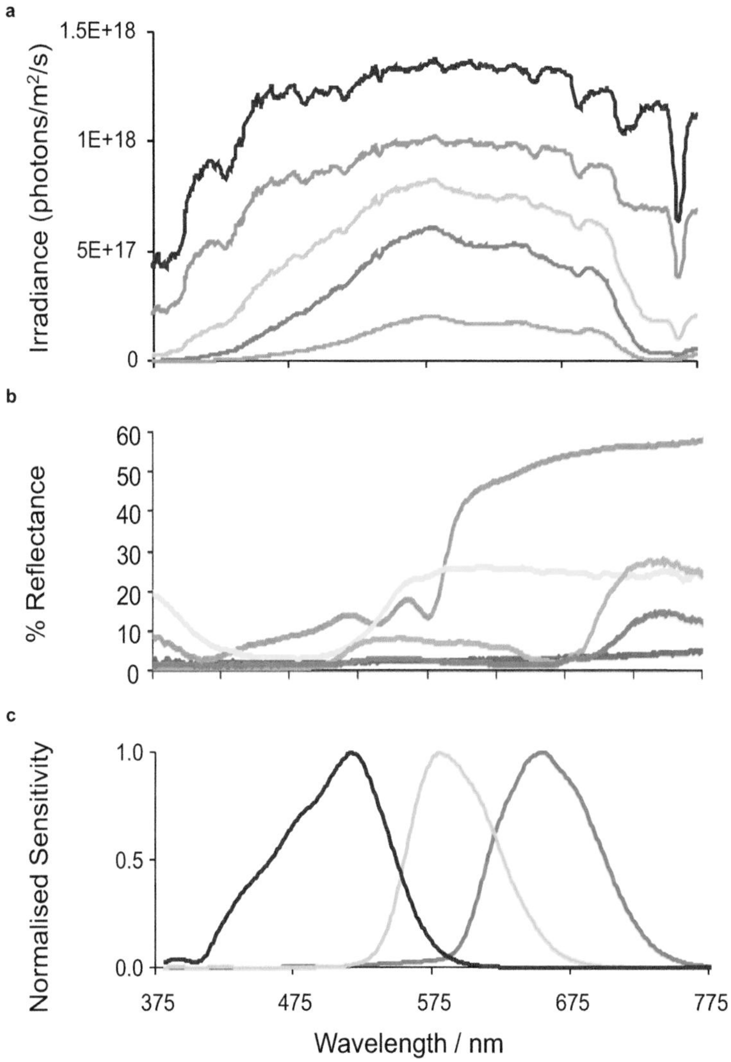

Fig. 6 Aspects of visual ecology in *Neoceratodus forsteri*. (a) Irradiance of light in the Mary River (Queensland, Australia), known habitat of *Neoceratodus forsteri* at: the surface (blue line), 0.05 m (bright green line), 0.5 m (mid-green line), 1 m dark green line, 1.25 m (kaki line). (b) Reflectance of objects in the environment of *Neoceratodus forsteri*: gravid female belly (orange line), non gravid female belly (yellow line), macrophytes from Mary River (green lines – see Hart *et al.* 2008 for details), submerged log from Mary River (brown line). These and other objects are plotted in the colour space of adult and juvenile in Fig. 7. (c) Adult spectral sensitivities (as Fig. 5b) included here for direct comparison with illumination in habitat and to complete complement of components needed for visual ecology characterisation; spectral sensitivities, illuminant on objects of interest and reflectance from objects of interest. Using these 3 factors is the first step towards the colour vision system models of Fig. 7 (Kelber *et al.* 2003).

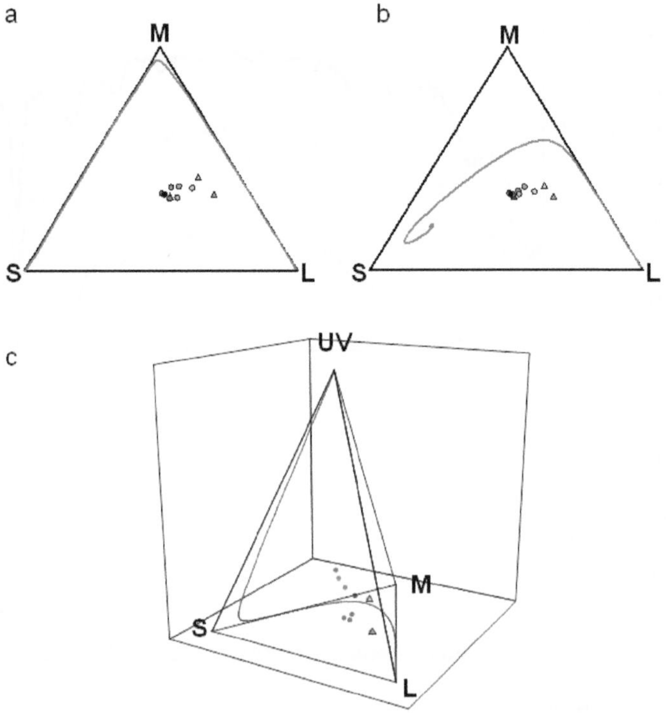

Fig. 7 Modelled colour vision performance in *Neoceratodus forsteri* examining coloured objects from Mary River environment (see Kelber *et al.* 2003 for review and references therein for methods). (a) Colour space based on Maxwell triangle for three cone sensitivities in adult *N. forsteri*. Although behavioural proof is required, this suggests trichromacy. The centre of the triangle is the achromatic point (occupied by flat spectra of e.g. white, grey and black) and areas towards the edges of the triangle are occupied by increasingly saturated spectral chromaticity loci up to the blue line – the monochromatic locus. The shape of the monochromatic locus is set by the spectral sensitivity characteristics and a comparison of (a) and (b) here demonstrates the way filtering by coloured ocular inclusions (oil droplets, ellipsoid pigment and ocular media) expands this space, thus improving the potential colour vision. Each symbol in the triangle (and tetrahedron in (c)) is the position occupied by one reflectance spectrum (Fig. 6b). The corners of the triangle; S, M and L plot the spectral loci of maximal stimulation of the short wavelength, medium wavelength and long wavelength photoreceptors respectively. Thus, for example, red spectra will plot near the L corner. Colour coding same as spectra in Fig. 6b. Female lungfish belly, gravid – orange triangle, female lungfish belly, non-gravid yellow triangle, macrophytes – green circles, log – brown circle, rocks and sand – grey circles (not plotted in Fig. 6b). (b) As (a) except for the hypothetical adult lungfish spectral sensitivities in the absence of filtering by coloured ocular inclusions (Govardovski 1983). Note the contraction of colour space delimited by the monochromatic locus (blue line) and the lesser separation of the loci of the coloured objects from lungfish habitat. (c) Three dimensional tetrahedral colour space of juvenile, potentially tetrachromatic, *N. forsteri*. Conventions and colours the same as triangles, UV, denotes the chromatic locus of the UV sensitive cone (Fig. 5b).

Printed and bound by CPI Group (UK) Ltd, Croydon, CR0 4YY
18/10/2024
01776208-0020